Practices of Irrigation & On-farm Water Management: Volume 2

M.H. Ali

Practices of Irrigation & On-farm Water Management: Volume 2

Foreword by M.A. Salam

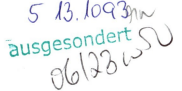

Dr. M.H. Ali
Agricultural Engineering Division
Bangladesh Institute of Nuclear
  Agriculture (BINA)
Mymensingh 2202, BAU Campus
Bangladesh
hossain.ali.bina@gmail.com
(URL: www.mhali.com)

ISBN 978-1-4419-7636-9   e-ISBN 978-1-4419-7637-6
DOI 10.1007/978-1-4419-7637-6
Springer New York Dordrecht Heidelberg London

© Springer Science+Business Media, LLC 2011
All rights reserved. This work may not be translated or copied in whole or in part without the written permission of the publisher (Springer Science+Business Media, LLC, 233 Spring Street, New York, NY 10013, USA), except for brief excerpts in connection with reviews or scholarly analysis. Use in connection with any form of information storage and retrieval, electronic adaptation, computer soft-ware, or by similar or dissimilar methodology now known or hereafter developed is forbidden.
The use in this publication of trade names, trademarks, service marks, and similar terms, even if they are not identified as such, is not to be taken as an expression of opinion as to whether or not they are subject to proprietary rights.

Printed on acid-free paper

Springer is part of Springer Science+Business Media (www.springer.com)

Foreword

Agricultural technologies are very important to feed the growing world population. Scientific principles of agricultural engineering have been applied for the optimal use of natural resources in agricultural production for the benefit of humankind. The role of agricultural engineering is increasing in the coming days at the forthcoming challenges of producing more food with less water, coupled with pollution hazard in the environment and climate uncertainty.

Irrigation is continually straining our limited natural resources. Whether it is through salinity, waterlogging, sedimentation, nutrient transport, or excessive water consumption, irrigation has an impact on our natural ecosystems. It is therefore important that the irrigation system is properly designed, monitored, and executed not only for the benefit of the irrigator but also for the wider community.

I am happy to know that a book (2nd volume in series) entitled *"Practices of Irrigation and On-farm Water Management,"* written by Engr. Dr. M. H. Ali, is going to be published by Springer. This book is designed to cover the major fields of applied agricultural engineering such as designing water conveyance systems, selecting and designing irrigation systems, land and watershed management, performance evaluation of irrigation systems, drainage system, water resources management, management of salt-affected soils, pumps, renewable energy for irrigation, models and crop production functions in irrigation management, and GIS in irrigation management.

This book will be quite useful for the students of agricultural engineering. Students of other related branches of engineering sciences, and engineers working in the field and at research institutes, will also be benefited. The book may serve as a textbook for the students and as a practical handbook for the practitioners and researchers in the field of irrigation and on-farm water management. Utilization of the recent literature in the area and citation of relevant journals/reports have added a special value to this book.

I hope this textbook will be used worldwide to promote agricultural production and conservation of the most important natural resource, water.

(Dr. M.A. Salam)
Director (Research)
Bangladesh Institute of Nuclear Agriculture

Mymensingh, Bangladesh
May, 2010

Preface

Crop production depends on the successful implementation of the agricultural and water management technologies. This is vital to feed the growing world population. The implementation of technologies is also important to minimize environmental degradation resulting from agricultural activities. Agricultural and natural resources engineers are applying scientific principles for the optimal use of natural resources in agricultural production.

Water is the scarcest resource. The importance of the judicious use of water in agricultural sector for sustaining agricultural growth and the retardation of environmental degradation needs no elaboration. Judicious use of water for crop production requires knowledge of water conveyance and application methods, their designing, strategic management of water resources, land and watershed management, etc. Increasing efficiency in conveyance and pumping systems are also of great concern. Irrigation management strategy practiced in normal soils may not be appropriate in problematic soils such as saline soils. This book covers all of the above aspects. In addition, the book covers some recent dimensions such as pollution from agricultural fields, modeling in irrigation and water management, application of the geographical information system (GIS) in irrigation and water management, and renewable energy resources for irrigation. Sample workout problems are provided to explain the design and application methodologies in practice.

The comprehensive and compact presentation of this book will serve as a textbook for undergraduate students in Agricultural Engineering, Biological Systems Engineering, Bio-Science Engineering, Water Resource Engineering, and, Civil and Environmental Engineering. It will also be helpful for the students of relevant fields such as Agronomy, Biological Sciences, and Hydrology. Although the target audience of this book is undergraduate students, postgraduate students will also be benefited from the book. It will also serve as a reference manual for field engineers, researchers, and extension workers in several fields such as agricultural engineering, agronomy, ecology, hydrology, civil, water resource, and environmental engineering.

Effort was made to keep the language as simple as possible, keeping in mind the readers of different language origins. Throughout the book, the emphasis has been on general descriptions and principles of each topic, technical details, and modeling aspects. However, the comprehensive journal references in each area should enable the reader to pursue further studies of special interest. In fact, the book covers broad interdisciplinary subjects.

Mymensingh, Bangladesh   Dr. M.H. Ali

Acknowledgment

I acknowledge the cooperation, suggestions, and encouragement of the faculty members of the Department of Irrigation and Water Management, Bangladesh Agricultural University. I would like to thank Engr. Dr. M. A. Ghani, former Director General of Bangladesh Agricultural Research Institute, and World Bank Country Representative, Bangladesh, who critically reviewed the content and structure of several chapters of the Book. I would also like to thank the scientists and staffs of Agricultural Engineering Division, Bangladesh Institute of Nuclear Agriculture, for their cooperation in various ways.

Thanks are due to Dr. M. A. Salam, Director (Research), Bangladesh Institute of Nuclear Agriculture, for going through the book and writing few words about the same in the form of a "Foreword."

I am grateful to the authority of Soil Moisture Co. for supplying the pictures of their products and giving me permission to use the same in the book.

My sincere thanks are also due to my affectionate wife Anjumanara Begham, daughter, Sanjida Afiate, and son, Irfan Sajid, for their support, understanding, and patience during the preparation of the manuscript.

Mymensingh, Bangladesh Dr. M.H. Ali
May, 2010

Contents

1 Water Conveyance Loss and Designing Conveyance System 1
   1.1 Water Conveyance Loss 2
      1.1.1 Definition of Seepage 2
      1.1.2 Factors Affecting Seepage 2
      1.1.3 Expression of Seepage 3
      1.1.4 Measurement of Seepage 4
      1.1.5 Estimation of Average Conveyance Loss in a Command Area 7
      1.1.6 Reduction of Seepage 8
      1.1.7 Lining for Reducing Seepage Loss 8
   1.2 Designing Open Irrigation Channel 10
      1.2.1 Irrigation Channel and Open Channel Flow ... 10
      1.2.2 Definition Sketch of an Open Channel Section ... 10
      1.2.3 Considerations in Channel Design 11
      1.2.4 Calculation of Velocity of Flow in Open Channel .. 12
      1.2.5 Hydraulic Design of Open Irrigation Channel 14
      1.2.6 Sample Examples on Irrigation Channel Design ... 18
   1.3 Designing Pipe for Irrigation Water Flow 21
      1.3.1 Fundamental Theories of Water Flow Through Pipe 21
      1.3.2 Water Pressure – Static and Dynamic Head 23
      1.3.3 Hydraulic and Energy Grade Line for Pipe Flow .. 25
      1.3.4 Types of Flow in Pipe – Reynolds Number 25
      1.3.5 Velocity Profile of Pipe Flow 26
      1.3.6 Head Loss in Pipe Flow and Its Calculation 26
      1.3.7 Designing Pipe Size for Irrigation Water Flow 31
      1.3.8 Sample Workout Problems 32
   Relevant Journals 32
   Questions 33
   References 34

2 Water Application Methods 35
   2.1 General Perspectives of Water Application 36
   2.2 Classification of Water Application Methods 36

xi

2.3		Description of Common Methods of Irrigation	38
	2.3.1	Border Irrigation	38
	2.3.2	Basin Irrigation	40
	2.3.3	Furrow Irrigation	43
	2.3.4	Sprinkler Irrigation Systems	46
	2.3.5	Drip Irrigation	52
	2.3.6	Other Forms of Irrigation	54
2.4		Selection of Irrigation Method	56
	2.4.1	Factors Affecting Selection of an Irrigation Method	56
	2.4.2	Selection Procedure	63
Relevant Journals			63
Questions			63

3 Irrigation System Designing ... 65
3.1 Some Common Issues in Surface Irrigation System Designing ... 66
3.1.1 Design Principle of Surface Irrigation System ... 66
3.1.2 Variables in Surface Irrigation System ... 67
3.1.3 Hydraulics in Surface Irrigation System ... 67
3.2 Border Irrigation System Design ... 68
3.2.1 Definition of Relevant Terminologies ... 68
3.2.2 General Overview and Considerations ... 69
3.2.3 Factors Affecting Border Performance and Design ... 70
3.2.4 Design Parameters ... 70
3.2.5 Design Approaches and Procedures for Border ... 71
3.2.6 Sample Workout Problems ... 73
3.2.7 Simulation Modeling for Border Design ... 76
3.2.8 Existing Software Tools/Models for Border Irrigation Design and Analysis ... 77
3.2.9 General Guidelines for Border ... 79
3.3 Basin Irrigation Design ... 79
3.3.1 Factors Affecting Basin Performance and Design ... 79
3.3.2 Hydraulics in Basin Irrigation System ... 81
3.3.3 Simulation Modeling for Basin Design ... 82
3.3.4 Existing Models for Basin Irrigation Design ... 84
3.4 Furrow Irrigation System Design ... 84
3.4.1 Hydraulics of Furrow Irrigation System ... 85
3.4.2 Mathematical Description of Water Flow in Furrow Irrigation System ... 86
3.4.3 Some Relevant Terminologies ... 87
3.4.4 Factors Affecting Performance of Furrow Irrigation System ... 90
3.4.5 Management Controllable Variables and Design Variables ... 91

	3.4.6	Furrow Design Considerations	92
	3.4.7	Modeling of Furrow Irrigation System	92
	3.4.8	General Guideline/Thumb Rule for Furrow Design	94
	3.4.9	Estimation of Average Depth of Flow from Volume Balance	95
	3.4.10	Suggestions for Improving Furrow Irrigations	96
	3.4.11	Furrow Irrigation Models	96
	3.4.12	Sample Worked Out Problems	97
3.5	Design of Sprinkler System		98
	3.5.1	Design Aspects	98
	3.5.2	Theoretical Aspects in Sprinkler System	99
	3.5.3	Sprinkler Design	101
Relevant Journals			106
Relevant FAO Papers/Reports			107
Questions			107
References			109

4 Performance Evaluation of Irrigation Projects ... 111

4.1	Irrigation Efficiencies		112
	4.1.1	Application Efficiency	112
	4.1.2	Storage Efficiency/Water Requirement Efficiency	114
	4.1.3	Irrigation Uniformity	114
	4.1.4	Low-Quarter Distribution Uniformity (or Distribution Uniformity)	115
4.2	Performance Evaluation		116
	4.2.1	Concept, Objective, and Purpose of Performance Evaluation	116
	4.2.2	Factors Affecting Irrigation Performance	117
	4.2.3	Performance Indices *or* Indicators	118
	4.2.4	Description of Different Indicators	120
	4.2.5	Performance Evaluation Procedure	126
	4.2.6	Performance Evaluation Under Specific Irrigation System	128
	4.2.7	Improving Performance of Irrigation System	134
Relevant Journals			137
Relevant FAO Papers/Reports			137
Questions			137
References			137

5 Water Resources Management ... 139

5.1	Concept, Perspective, and Objective of Water Resources Management		140
	5.1.1	Concept of Management	140
	5.1.2	Water and the Environment	141
	5.1.3	Increasing Competition in Water Resource	141
	5.1.4	Water As an Economic Good	142

	5.1.5	Purposes and Goals of Water Resources Management	143
	5.1.6	Fundamental Aspects of Water Resources Management	144
5.2	Estimation of Demand and Supply of Water	144
	5.2.1	Demand Estimation	144
	5.2.2	Estimation of Potential Supply of Water	146
	5.2.3	Issues of Groundwater Development in Saline/Coastal Areas	148
	5.2.4	Environmental Flow Assessment	148
5.3	Strategies for Water Resources Management	150
	5.3.1	Demand Side Management	150
	5.3.2	Supply Side Management	161
	5.3.3	Integrated Water Resources Management	170
5.4	Sustainability Issues in Water Resource Management	173
	5.4.1	Concept of Sustainability	173
	5.4.2	Scales of Sustainability	175
	5.4.3	Achieving Sustainability	175
	5.4.4	Strategies to Achieve Sustainability	177
5.5	Conflicts in Water Resources Management	178
	5.5.1	Meaning of Conflict	178
	5.5.2	Water Conflicts in the Integrated Water Resources Management Process	179
	5.5.3	Scales of Conflicts in Water Management	180
	5.5.4	Analysis of Causes of Conflicts in Water Management	184
5.6	Impact of Climate Change on Water Resource	185
	5.6.1	Issues on Water Resources in Connection to Climate Change	185
	5.6.2	Adaptation Alternatives to the Climate Change	186
5.7	Challenges in Water Resources Management	188
	5.7.1	Risk and Uncertainties	188
	5.7.2	International/Intra-national (Upstream–Downstream) Issues	188
	5.7.3	Quality Degradation Due to Continuous Pumping of Groundwater	188
	5.7.4	Lowering of WT and Increase in Cost of Pumping	189
Relevant Journals		189
Questions		190
References		190

6 Land and Watershed Management ... 193
6.1	Concepts and Scale Consideration	194
6.2	Background and Issues Related to Watershed Management	195

		6.2.1	Water Scarcity	196
		6.2.2	Floods, Landslides, and Torrents	196
		6.2.3	Water Pollution	196
		6.2.4	Population Pressure and Land Shrinkage	196
	6.3	Fundamental Aspects of Watershed Management		197
		6.3.1	Elements of Watershed	197
		6.3.2	How the Watershed Functions	198
		6.3.3	Factors Affecting Watershed Functions	198
		6.3.4	Importance of Watershed Management	198
		6.3.5	Addressing/Naming a Watershed	198
	6.4	Land Grading in Watershed		199
		6.4.1	Concept, Purpose, and Applicability	199
		6.4.2	Precision Grading	200
		6.4.3	Factors Affecting Land Grading and Development	201
		6.4.4	Activities and Design Considerations in Land Grading	203
		6.4.5	Methods of Land Grading and Estimating Earthwork Volume	205
	6.5	Runoff and Sediment Yield from Watershed		212
		6.5.1	Runoff and Erosion Processes	212
		6.5.2	Factors Affecting Runoff	213
		6.5.3	Runoff Volume Estimation	214
		6.5.4	Factors Affecting Soil Erosion	219
		6.5.5	Sediment Yield and Its Estimation	221
		6.5.6	Sample Workout Problems on Sediment Yield Estimation	223
		6.5.7	Erosion and Sedimentation Control	225
		6.5.8	Modeling Runoff and Sediment Yield	226
	6.6	Watershed Management		227
		6.6.1	Problem Identification	227
		6.6.2	Components of Watershed Management	228
		6.6.3	Watershed Planning and Management	229
		6.6.4	Tools for Watershed Protection	230
		6.6.5	Land Use Planning	230
		6.6.6	Structural Management	230
		6.6.7	Pond Management	231
		6.6.8	Regulatory Authority	231
		6.6.9	Community-Based Approach to Watershed Management	231
		6.6.10	Land Use Planning and Practices	234
		6.6.11	Strategies for Sustainable Watershed Management	235
	6.7	Watershed Restoration and Wetland Management		236
		6.7.1	Watershed Restoration	236
		6.7.2	Drinking Water Systems Using Surface Water	236

		6.7.3	Wetland Management in a Watershed	237
	6.8	Addressing the Climate Change in Watershed Management		238
		6.8.1	Groundwater Focus	238
	Relevant Journals			238
	Relevant FAO Papers/Reports			238
	Questions			239
	References			239
7	**Pollution of Water Resources from Agricultural Fields and Its Control**			**241**
	7.1	Pollution Sources		242
		7.1.1	Point Sources	242
		7.1.2	Nonpoint Sources	242
	7.2	Types of Pollutants/Solutes		243
		7.2.1	Reactive Solute	243
		7.2.2	Nonreactive Solute	243
	7.3	Extent of Agricultural Pollution		243
		7.3.1	Major Pollutant Ions	243
		7.3.2	Some Relevant Terminologies	244
		7.3.3	Factors Affecting Solute Contamination	244
		7.3.4	Mode of Pollution by Nitrate and Pesticides	247
		7.3.5	Hazard of Nitrate (NO_3–N) Pollution	248
		7.3.6	Impact of Agricultural Pollutants on Surface Water Body and Ecosystem	248
	7.4	Solute Transport Processes in Soil		250
		7.4.1	Transport of Solute Through Soil	250
		7.4.2	Basic Solute Transport Processes	251
		7.4.3	Convection-Dispersion Equation	254
		7.4.4	Governing Equation for Solute Transport Through Homogeneous Media	254
		7.4.5	One-Dimensional Solute Transport with Nitrification Chain	256
		7.4.6	Water Flow and Solute Transport in Heterogeneous Media	257
	7.5	Measurement of Solute Transport Parameters		258
		7.5.1	Different Parameters	258
		7.5.2	Breakthrough Curve and Breakthrough Experiment	259
	7.6	Estimation of Solute Load (Pollution) from Agricultural Field		261
		7.6.1	Sampling from Controlled Lysimeter Box	261
		7.6.2	Sampling from Crop Field	261
		7.6.3	Determination of Solute Concentration	262
	7.7	Control of Solute Leaching from Agricultural and Other Sources		265
		7.7.1	Irrigation Management	265

		7.7.2	Nitrogen Management	265
		7.7.3	Cultural Management/Other Forms of Management	266
	7.8	Models in Estimating Solute Transport from Agricultural and Other Sources		266
	Relevant Journals			267
	Questions			267
	References			269
8	**Management of Salt-Affected Soils**			271
	8.1	Extent of Salinity and Sodicity Problem		272
	8.2	Development of Soil Salinity and Sodicity		273
		8.2.1	Causes of Salinity Development	273
		8.2.2	Factors Affecting Salinity	277
		8.2.3	Mechanism of Salinity Hazard	278
		8.2.4	Salt Balance at Farm Level	278
	8.3	Diagnosis and Characteristics of Saline and Sodic Soils		279
		8.3.1	Classification and Characteristics of Salt-Affected Soils	279
		8.3.2	Some Relevant Terminologies and Conversion Factors	282
		8.3.3	Diagnosis of Salinity and Sodicity	285
		8.3.4	Salinity Mapping and Classification	290
	8.4	Impact of Salinity and Sodicity		293
		8.4.1	Impact of Salinity on Soil and Crop Production	293
		8.4.2	Impact of Sodicity on Soil and Plant Growth	294
	8.5	Crop Tolerance to Soil Salinity and Effect of Salinity on Yield		295
		8.5.1	Factors Influencing Tolerance to Crop	295
		8.5.2	Relative Salt Tolerance of Crops	297
		8.5.3	Use of Saline Water for Crop Production	298
		8.5.4	Yield Reduction Due to Salinity	299
		8.5.5	Sample Examples	300
	8.6	Management/Amelioration of Saline Soil		301
		8.6.1	Principles and Approaches of Salinity Management	301
		8.6.2	Description of Salinity Management Options	302
	8.7	Management of Sodic and Saline-Sodic Soils		317
		8.7.1	Management of Sodic Soil	317
		8.7.2	Management of Saline-Sodic Soil	319
	8.8	Models/Tools in Salinity Management		320
	8.9	Challenges and Needs		323
	Relevant Journals			323
	Relevant FAO Papers/Reports			323
	FAO Soils Bulletins			324

	Questions	324
	References	325
9	**Drainage of Agricultural Lands**	**327**
9.1	Concepts and Benefits of Drainage	329
	9.1.1 Concepts	329
	9.1.2 Goal and Purpose of Drainage	329
	9.1.3 Effects of Poor Drainage on Soils and Plants	329
	9.1.4 Benefits from Drainage	330
	9.1.5 Types of Drainage	330
	9.1.6 Merits and Demerits of Deep Open and Buried Pipe Drains	332
	9.1.7 Difference Between Irrigation Channel and Drainage Channel	334
9.2	Physics of Land Drainage	334
	9.2.1 Soil Pore Space and Soil-Water Retention Behavior	334
	9.2.2 Some Relevant Terminologies	335
	9.2.3 Water Balance in a Drained Soil	338
	9.2.4 Sample Workout Problem	340
9.3	Theory of Water Movement Through Soil and Toward Drain	341
	9.3.1 Velocity of Flow in Porous Media	341
	9.3.2 Some Related Terminologies	341
	9.3.3 Resultant or Equivalent Hydraulic Conductivity of Layered Soil	342
	9.3.4 Laplace's Equation for Groundwater Flow	345
	9.3.5 Functional Form of Water-Table Position During Flow into Drain	346
	9.3.6 Theory of Groundwater Flow Toward Drain	346
	9.3.7 Sample Workout Problems	347
9.4	Design of Surface Drainage System	349
	9.4.1 Estimation of Design Surface Runoff	349
	9.4.2 Design Considerations and Layout of Surface Drainage System	349
	9.4.3 Hydraulic Design of Surface Drain	349
	9.4.4 Sample Work Out Problem	350
9.5	Equations/Models for Subsurface Drainage Design	351
	9.5.1 Steady-State Formula for Parallel Drain Spacing	351
	9.5.2 Formula for Irregular Drain System	355
	9.5.3 Determination of Drain Pipe Size	356
9.6	Design of Subsurface Drainage System	356
	9.6.1 Factors Affecting Spacing and Depth of Subsurface Drain	356
	9.6.2 Data Requirement for Subsurface Drainage Design	357
	9.6.3 Layout of Subsurface Drainage	357

		9.6.4	Principles, Steps, and Considerations in Subsurface Drainage Design	358
		9.6.5	Controlled Drainage System and Interceptor Drain	361
		9.6.6	Sample Workout Problems	362
	9.7	Envelope Materials		365
		9.7.1	Need of Using Envelop Material Around Subsurface Drain	365
		9.7.2	Need of Proper Designing of Envelop Material	365
		9.7.3	Materials for Envelope	365
		9.7.4	Design of Drain Envelope	366
		9.7.5	Use of Particle Size Distribution Curve in Designing Envelop Material	367
		9.7.6	Drain Excavation and Envelope Placement	368
	9.8	Models in Drainage Design and Management		368
		9.8.1	DRAINMOD	368
		9.8.2	CSUID Model	369
		9.8.3	EnDrain	369
	9.9	Drainage Discharge Management: Disposal and Treatment		369
		9.9.1	Disposal Options	369
		9.9.2	Treatment of Drainage Water	370
	9.10	Economic Considerations in Drainage Selection and Installation		371
	9.11	Performance Evaluation of Subsurface Drainage		371
		9.11.1	Importance of Evaluation	371
		9.11.2	Evaluation System	372
	9.12	Challenges and Needs in Drainage Design and Management		373
	Relevant Journals			373
	FAO/World Bank Papers			374
	Questions			374
	References			376
10	**Models in Irrigation and Water Management**			**379**
	10.1	Background/Need of a Model		380
	10.2	Basics of Model: General Concepts, Types, Formulation and Evaluation System		380
		10.2.1	General Concepts	380
		10.2.2	Different Types of Model	381
		10.2.3	Some related terminologies	386
		10.2.4	Basic Considerations in Model Development and Formulation of Model Structure	389
		10.2.5	Model Calibration, Validation and Evaluation	390
		10.2.6	Statistical Indicators for Model Performance Evaluation	391
	10.3	Overview of Some of the Commonly Used Models		393

		10.3.1	Model for Reference Evapotranspiration (ET_0 Models)	393

- 10.3.1 Model for Reference Evapotranspiration (ET_0 Models) 393
- 10.3.2 Model for Upward Flux Estimation 397
- 10.3.3 Model for Flow Estimation in Cracking Clay Soil . . 397
- 10.3.4 Model for Irrigation Planning and Decision Support System 402
- 10.3.5 Decision Support Model 405
- 10.4 Crop Production Function/Yield Model 406
  - 10.4.1 Definition of Production Function 406
  - 10.4.2 Importance of Production Function 406
  - 10.4.3 Basic Considerations in Crop Production Function . 407
  - 10.4.4 Pattern of Crop Production Function 407
  - 10.4.5 Development of Crop Production Function 408
  - 10.4.6 Some Existing Yield Functions/Models 408
  - 10.4.7 Limitations/Drawbacks of Crop Production Function 411
- 10.5 Regression-Based Empirical Models for Predicting Crop Yield from Weather Variables 411
  - 10.5.1 Need of Weather-Based Prediction Model 411
  - 10.5.2 Existing Models/Past Efforts 412
  - 10.5.3 Methods of Formulation of Weather-Based Prediction Model 413
  - 10.5.4 Discussion 415
  - 10.5.5 Sample Example of Formulating Weather-Based Yield-Prediction Model 415
- Relevant Journals 419
- Questions 419
- References 420

11 GIS in Irrigation and Water Management 423
- 11.1 Introduction 424
- 11.2 Definition of GIS 424
- 11.3 Benefits of GIS Over Other Information Systems 424
- 11.4 Major Tasks in GIS 425
- 11.5 Applications of GIS 425
- 11.6 Techniques Used in GIS 427
- 11.7 Implementation of GIS 427
- 11.8 Data and Databases for GIS 428
- 11.9 Sources of Spatial Data 428
- 11.10 Data Input 429
- 11.11 GIS-Based Modeling or Spatial Modeling 429
- 11.12 Remote Sensing Techniques 430
- Relevant Journals 431
- Questions 431
- References 431

Contents

12 Water-Lifting Devices – Pumps 433
- 12.1 Classification of Water-Lifting Devices 435
  - 12.1.1 Human-Powered Devices 435
  - 12.1.2 Animal-Powered Devices 436
  - 12.1.3 Kinetic Energy Powered Device 436
  - 12.1.4 Mechanically Powered Water-Lifting Devices 437
- 12.2 Definition, Purpose, and Classification of Pumps 437
  - 12.2.1 Definition of Pump 437
  - 12.2.2 Pumping Purpose 437
  - 12.2.3 Principles in Water Pumping 438
  - 12.2.4 Classification of Pumps 438
- 12.3 Factors Affecting the Practical Suction Lift of Suction-Mode Pump 442
- 12.4 Centrifugal Pumps 442
  - 12.4.1 Features and Principles of Centrifugal Pumps 442
  - 12.4.2 Some Relevant Terminologies to Centrifugal Pump 443
  - 12.4.3 Pump Efficiency 445
  - 12.4.4 Specific Speed 446
  - 12.4.5 Affinity Laws 446
  - 12.4.6 Priming of Centrifugal Pumps 448
  - 12.4.7 Cavitation 449
- 12.5 Description of Different Types of Centrifugal Pumps 449
  - 12.5.1 Turbine Pump 449
  - 12.5.2 Submersible Pump 451
  - 12.5.3 Mono-Block Pump 454
  - 12.5.4 Radial-Flow Pump 455
  - 12.5.5 Volute Pump 456
  - 12.5.6 Axial-Flow Pump 456
  - 12.5.7 Mixed-Flow Pump 456
  - 12.5.8 Advantage and Disadvantage of Different Centrifugal Pumps 457
  - 12.5.9 Some Common Problems of Centrifugal Pumps, Their Probable Causes, and Remedial Measures 457
- 12.6 Other Types of Pumps 458
  - 12.6.1 Air-Lift Pump 458
  - 12.6.2 Jet Pump 459
  - 12.6.3 Reciprocating Pump/Bucket Pump 461
  - 12.6.4 Displacement Pump 462
  - 12.6.5 Hydraulic Ram Pump 462
  - 12.6.6 Booster Pump 462
  - 12.6.7 Variable Speed Pump 462
- 12.7 Cavitation in Pump 463
  - 12.7.1 Cavitation in Radial Flow and Mixed Flow Pumps 463

		12.7.2	Cavitation in Axial-Flow Pumps	463
	12.8	Power Requirement in Pumping		464
	12.9	Pump Installation, Operation, and Control		465
		12.9.1	Pump Installation	465
		12.9.2	Pump Operation	466
		12.9.3	Pump Control	467
	12.10	Hydraulics in Pumping System		468
		12.10.1	Pressure Vs Flow Rate	468
		12.10.2	Pressure and Head	468
		12.10.3	Elevation Difference	469
	12.11	Pumps Connected in Series and Parallel		469
	12.12	Pump Performance and Pump Selection		469
		12.12.1	Pump Performance	469
		12.12.2	Factors Affecting Pump Performance	469
		12.12.3	Selecting a Pump	470
		12.12.4	Procedure for Selecting a Pump	470
	12.13	Sample Workout Problems on Pump		473
	Questions			476
13	**Renewable Energy Resources for Irrigation**			**479**
	13.1	Concepts and Status of Renewable Energy Resources		480
		13.1.1	General Overview	480
		13.1.2	Concept and Definition of Renewable Energy	481
		13.1.3	Present Status of Uses of Renewable Energy	482
	13.2	Need of Renewable Energy		482
	13.3	Mode of Use of Renewable Energy		483
	13.4	Application of Solar Energy for Pumping Irrigation Water		483
		13.4.1	General Overview	483
		13.4.2	Assessment of Potential Solar Energy Resource	484
		13.4.3	Solar or Photovoltaic Cells – Theoretical Perspectives	485
		13.4.4	Solar Photovoltaic Pump	485
		13.4.5	Uses of Solar System Other than Irrigation Pumping	489
		13.4.6	Solar Photovoltaic Systems to Generate Electricity Around the Globe	490
	13.5	Wind Energy		491
		13.5.1	Wind as a Renewable and Environmentally Friendly Source of Energy	491
		13.5.2	Historical Overview of Wind Energy	491
		13.5.3	Causes of Wind Flow	492
		13.5.4	Energy from Wind	493
		13.5.5	Advantages of Wind Energy	493
		13.5.6	Assessing Wind Energy Potential	494
		13.5.7	Types of Wind Machines	495
		13.5.8	Suitable Site for Windmill	495

	13.5.9	Application of Wind Energy	496
	13.5.10	Working Principle of Wind Machines	497
	13.5.11	Wind Power Plants or Wind Farms	498
	13.5.12	Calculation of Wind Power	498
	13.5.13	Intermittency Problem with Wind Energy	500
	13.5.14	Wind and the Environment	501
	13.5.15	Sample Work Out Problems	501
13.6	Water Energy		502
	13.6.1	Forms of Water Energy	503
	13.6.2	Wave Energy	503
	13.6.3	Watermill	504
	13.6.4	Tide Mill	505
	13.6.5	Exploring the Potentials of Water Power	505
13.7	Bio-energy		506
	13.7.1	Liquid Biofuel	507
	13.7.2	Biogas	508
13.8	Geothermal Energy		508
13.9	Modeling the Energy Requirement		509
13.10	Factors Affecting Potential Use of Renewable Energy in Irrigation		509
	13.10.1	Groundwater Requirement and Its Availability . . .	510
	13.10.2	Affordability of the User	510
	13.10.3	Willingness of the User to Invest in a Renewable Energy Based Pump	510
	13.10.4	Availability of Alternate Energy for Irrigation and Its Cost	511
	13.10.5	Alternate Use of Renewable Energy	511
13.11	Renewable Energy Commercialization: Problems and Prospects		511
	13.11.1	Problems	512
	13.11.2	Prospects/Future Potentials	514
	13.11.3	Challenges and Needs	516
Relevant Journals			516
Questions			517
References			518

Subject Index 519

Chapter 1
Water Conveyance Loss and Designing Conveyance System

Contents

1.1	Water Conveyance Loss	2
	1.1.1 Definition of Seepage	2
	1.1.2 Factors Affecting Seepage	2
	1.1.3 Expression of Seepage	3
	1.1.4 Measurement of Seepage	4
	1.1.5 Estimation of Average Conveyance Loss in a Command Area	7
	1.1.6 Reduction of Seepage	8
	1.1.7 Lining for Reducing Seepage Loss	8
1.2	Designing Open Irrigation Channel	10
	1.2.1 Irrigation Channel and Open Channel Flow	10
	1.2.2 Definition Sketch of an Open Channel Section	10
	1.2.3 Considerations in Channel Design	11
	1.2.4 Calculation of Velocity of Flow in Open Channel	12
	1.2.5 Hydraulic Design of Open Irrigation Channel	14
	1.2.6 Sample Examples on Irrigation Channel Design	18
1.3	Designing Pipe for Irrigation Water Flow	21
	1.3.1 Fundamental Theories of Water Flow Through Pipe	21
	1.3.2 Water Pressure – Static and Dynamic Head	23
	1.3.3 Hydraulic and Energy Grade Line for Pipe Flow	25
	1.3.4 Types of Flow in Pipe – Reynolds Number	25
	1.3.5 Velocity Profile of Pipe Flow	26
	1.3.6 Head Loss in Pipe Flow and Its Calculation	26
	1.3.7 Designing Pipe Size for Irrigation Water Flow	31
	1.3.8 Sample Workout Problems	32
Relevant Journals		32
Questions		33
References		34

The conveyance efficiency in irrigation projects is poor due to seepage, percolation, cracking, and damaging of the earth channel. Seepage loss in irrigation water conveyance system is very significant, as it forms the major portion of the water loss in the irrigation system. Irrigation conveyance losses controlled through lining may reduce the drainage requirement and also enhance irrigation efficiency. As such, reliable estimates of quantities and extent of seepage losses from canals under pre- and post-lining conditions become important. Various methods are used to estimate the canal seepage rate. The loss in conveyance is unavoidable unless the canal is lined. Lining may be done with a large variety of materials. Selection of a suitable one depends mainly on cost, performance, durability, and availability of lining materials.

Irrigation efficiency is greatly dependent on the type and design of water conveyance and distribution systems. Designing of economic cross-sections of various types of irrigation channels is important to minimize cost, water loss, and land requirement. This chapter illustrates these issues with sample design problems.

1.1 Water Conveyance Loss

1.1.1 Definition of Seepage

Seepage may be defined as the infiltration downward and lateral movements of water into soil or substrata from a source of supply such as reservoir or irrigation channel. Such water may reappear at the surface as wet spots or seeps or may percolate to join the groundwater, or may join the subsurface flow to springs or streams.

1.1.2 Factors Affecting Seepage

Many factors are known to have a definite effect on seepage rate. The major factors are

(i) the characteristics of the soil or strata through which the channels are laid (e.g., texture, bulk density, porosity, permeability)
(ii) bulk density, porosity, and permeability of the side soil
(iii) top width and wetted perimeter of the channel
(iv) depth of water in the channel
(v) amount of sediment in the water
(vi) viscosity or salinity of canal water
(vii) aquatic plants
(viii) velocity of water in the channel
(ix) pump discharge
(x) length of time the channel has been in operation (canal age)
(xi) nature of channel like dug or raised (topography)
(xii) channel geometry
(xiii) presence of cracks or holes or piping through the subgrades of the section

1.1 Water Conveyance Loss

(xiv) flow characteristics
(xv) gradient of channel
(xvi) wetness of the surrounding soil or season
(xvii) depth to groundwater table
(xviii) constraints on groundwater flow, e.g., presence of wells, drains, rivers, and/or impermeable boundaries.

Permeability of soil is influenced by both pore size and percentage of pore space (porosity). Soils consisting of a mixture of gravel and clay are almost completely impervious, while coarse gravel may transmit water many times faster; thus a wide range of seepage losses is possible. Seepage loss increases with the increase in water depth in the canal. The distribution of seepage losses across the bed and sides of the canal depends upon the position of the water table or impervious layer. Seepage increases with the increase of the difference in water level in the canal and water table. If the flowing water contains considerable amounts of suspended material, the seepage rate may be reduced in a relatively short time. Even small amounts of sediment may have sealing effects over a period of time. If the velocity is reduced, the sediment-carrying capacity of the water decreases, resulting in the settlement of part of the suspended materials. This forms a thin slowly permeable layer along the wetted perimeter of the canal which decreases the seepage. In seasonally used unlined canals, the seepage rate will be high at the beginning of the season and gradually decrease toward its end. On most lined canals, seepage increases with lapse of time (long period) for a variety of reasons and depending on the material.

1.1.3 Expression of Seepage

The following terms are mostly used to express the amount of seepage:

(i) volume per unit area of wetted perimeter per 24 h or day (m^3/m^2/day)
(ii) volume per unit length of canal per day (m^3/m/day)
(iii) percentage of total flow per km of canal (%/km)

Conveyance losses are sometimes expressed as a percentage of total flow for the scheme or project basis.
Equivalents of the units (i) are

$$1 \text{ m}^3/\text{m}^2/\text{d} = 3.2816 \text{ ft}^3/\text{ft}^2/\text{day}$$
$$1 \text{ ft}^3/\text{ft}^2/\text{day} = 0.3047 \text{ m}^3/\text{m}^2/\text{d}$$

When comparing figures on seepage losses in lined canals with those in unlined, attention should be paid to the following: For equal unit loss, the total volume lost per unit length of canal is greater for an unlined than for a hard surface-lined canal, since the wetted perimeter of a concrete-lined canal is about 30% less than that of an unlined canal.

1.1.4 Measurement of Seepage

Irrigation conveyance losses controlled through lining may reduce the drainage requirement and also enhance irrigation efficiency. As such, reliable estimates of quantities and extent of seepage losses from canals under pre- and post-lining conditions become important. Various methods are used to estimate the canal seepage rate such as empirical formulae, analytical or analogue studies, and the direct seepage measurement techniques. Direct seepage measurement includes seepage meters, ponding tests, and inflow–outflow tests. Each of these methods has merits, demerits, and limitations.

1.1.4.1 Ponding Method

Ponding tests can be carried out during the canal closure period starting immediately after the cessation of normal flow while the canal banks are still almost saturated. A reach of several hundred meters (often 300 m) for the main or distribution canal and 30–100 m for the field channel is isolated by building temporary dykes across the canal, sealing them with a plastic sheet (Fig. 1.1). The water level in the ponded section is recorded at regular interval, usually for several days (6–12 h for small channel) and observing the rate of fall of water level from the initial filling. Rainfall and evaporation are measured in proximity of test site and compensate for the surface area of water in test section. The evaporation loss may be neglected for small time intervals between two successive recordings of water levels from scales. Keeping in view the level of the canal, it is more common to allow the level to drop only a short way and then refill the pond and start again. A series of independent tests are to be conducted and then the value should be averaged. A considerable number of replications reduce the uncertainty in the mean result.

Seepage rate for the ponding method can be computed using the following formula (Ali, 2001):

$$S = \frac{L(d_1 - d_2)W}{L \times P \times t} \times 24 \qquad (1.1)$$

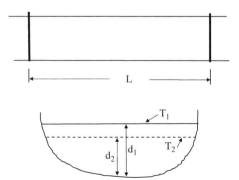

Fig. 1.1 Schematic presentation of measuring seepage loss using ponding method [longitudinal section (*upper*) and cross-section (*lower one*)]

1.1 Water Conveyance Loss

where

S is the seepage rate, m³/m²/day
L is the length of the canal reach (test section), m
W is the average top width of the canal cross-section, m
P is the average wetted perimeter of the canal section, m [Average of the initial and final perimeter = $(P_i + P_f)/2$]
d_1 is the initial water depth, m
d_2 is the final water depth, m
t is the duration of ponding, h

The percentage of seepage losses in small canals and farm ditches is normally greater than in large conveyance canals.

Limitations

Major limitations of this method are as follows:

(i) it cannot be used while canals are operating
(ii) it does not reflect the velocities and sediment loads of operating conditions

Merits

(i) the method is simple to understand
(ii) no special equipment is needed to perform the measurement
(iii) does not need too long a channel section as that of inflow–outflow method
(iv) more accurate result can be obtained than the inflow–outflow method, especially where the seepage rates are fairly small

1.1.4.2 Inflow–Outflow Method

In this method, seepage is determined through measuring the inflow and outflow of a canal test reach. Flow rate can be measured by current meter or by other flow measuring structures such as flumes, weirs.

The water balance for the reach of the canal used in an inflow–outflow test, in the general case where there are off-taking channels that are flowing, is

$$S = Q_1 - Q_2 - Q_f - R - F - U - E \tag{1.2}$$

Each term of the above equation is a discharge, e.g., m³/s,

where

S = rate of water loss due to canal seepage
Q_1 = inflow at upstream end of reach
Q_2 = outflow at downstream end of reach
Q_f = flow in off-takes which are noted and gauged at their measuring points

R = rainfall

F = water losses at off-takes between the parent canal and off-take measuring points

U = the water losses through unmeasured orifices in the canal side (e.g., animal burrows, unauthorized outlets, other sorts of water abstraction)

E = the evaporation from the reach

Steady flow condition is necessary during the conduct of the test. In a small irrigation channel, where the terms Q_f, F, and U are nil, the above equation takes the simplified form as

$$S = Q_1 - Q_2 - R - E \tag{1.3}$$

Merits

This method can reflect actual operating (dynamic) conditions.

Limitations

(i) Sufficiently long test reaches are not available in some cases, which may prevent accurate measurement over short stretches of special interest.
(ii) Steady flow condition is necessary.
(iii) Accurate result cannot be obtained where the seepage rate is fairly small.

1.1.4.3 Seepage Meter Method

Various types of seepage meters have been developed. Here, a seepage meter with submerged flexible water bag is discussed. It is the simplest device in construction as well as in operation. It consists of a water-tight seepage cup connected by a hose to a flexible (plastic) water bag floating on the water surface (Fig. 1.2).

During measurement, the seepage meter is set under water. Water flows from the bag into the cup, where it seeps through the canal subgrade area isolated by the cup. By keeping the water bag submerged, it will adapt itself to the shrinking volume so

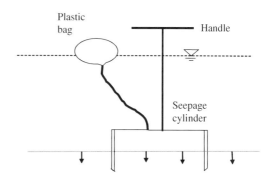

Fig. 1.2 Schematic view of measuring seepage by seepage meter

1.1 Water Conveyance Loss

that the heads on the areas within and outside the cup are equal. The seepage rate is computed from the weight (and then converted to volume) of water lost in a known period of time and the area covered by the meter, i.e.,

$$S = V/(t \times A) \tag{1.4}$$

where

S = seepage rate (m³/m²/h)
V = volume of water lost (m³)
t = time period (h)
A = area covered by the meter (m²)

1.1.5 Estimation of Average Conveyance Loss in a Command Area

Conveyance loss here means the percent of the water lost reaching the field plot on the basis of water diverted and can be calculated as

$$\begin{aligned} C_{AL} &= (Q_d - Q_p) \times 100/Q_d \\ &= (C_L \times L_{av})/Q_d \end{aligned} \tag{1.5}$$

where

C_{AL} is the average conveyance loss in percent
Q_d is the pump discharge or inflow in m³/s
Q_p is the measured discharge at field plot in m³/s
C_L is the average steady state conveyance loss (m³/s) per 100 m
L_{av} is the average channel length of the field plots (m)

To obtain the average channel length, the command area may be divided into n unit areas considering the distance from the pump. A representative diversion point for each unit area may be identified and the length of the channel section from the pump to the diversion point be measured. The average channel length can then be calculated as

$$L_{av} = \left(\sum L_i\right)/n \tag{1.6}$$

where n is the number of the section.

Discharge measurement may be done by a cutthroat flume or other available technique.

1.1.6 Reduction of Seepage

Lining is the straightforward way to reduce seepage from the channel. Besides the channel, sometimes the earthen reservoirs are faced with the problem of seepage. A variety of techniques are available to control seepage from the earthen reservoir or ponds. These include physical, chemical, and biological methods.

1.1.6.1 Physical Method

In this method, the bottom and sides of the ponds are soaked with water until their moisture contents are close to field capacity. Then the soil is physically compacted. Compaction can be done with either manual or tractor-mounted compactors. Walking cattle or buffaloes over the area will help. The amount of compaction achieved depends on the load applied and the wetness of the soil. The soil's physical and chemical properties are also important. The level of compaction can be assessed by measuring the soil bulk density or by the force exhibited by the Penetrometer to enter the soil.

1.1.6.2 Chemical Method

Certain sodium salts such as sodium chloride, tetrasodium pyrophosphate, sodium hexametaphosphate, and sodium carbonate can reduce seepage in earthen ponds. Among them, sodium carbonate performed better (Reginato et al., 1973). Sodium ions cause clay to swell and clay particles to disperse and thereby reduce or plug water-conducting pores in the soil. Seepage losses can be reduced by mixing sodium carbonate with locally available soil and applying the mixture by sedimentation. The recommended rate is 2.5 t/ha, into the top 10 cm soil. The sodic soil, which is naturally high in sodium salts, also do the job.

1.1.6.3 Biological Method

"Bio-plastic," a sandwich made up of successive layers of soil, manure (from pigs, cattle, or others), vegetable materials, and soil can reduce percolation loss. This creates an underground barrier to seepage. Kale et al. (1986) obtained a seepage reduction of approximately 9% by using a mixture of cow dung, paddy husk, and soil.

1.1.7 Lining for Reducing Seepage Loss

1.1.7.1 Benefits of Lining

  (i) savings of water
 (ii) reduced canal dimensions and right of way – cost
(iii) reduced water logging in some cases

Table 1.1 Seepage rates in some typical soils

Soil type	Seepage rate (m³/m²/day)		Reduction by compaction (%)
	Uncompacted	Compacted	
Sandy	4.0	3.8	5
Loam	1.4	1.0	27
Clay	0.35	0.10	71

Water losses occur in earthen channels under both compacted and uncompacted conditions due to seepage. Though in some soils the extent of seepage loss is very low (Table 1.1), canal lining is nevertheless proposed to overcome losses.

Lining a canal will not completely eliminate losses; therefore, it is necessary to measure systematically present losses or estimate the losses that might reasonably be saved by lining before a proper decision can be made. Roughly 60–80% of the water lost in unlined canals can be saved by hard surface lining. Seepage data for different soils and cost of lining materials can serve as a guide in cases where no other data are available and where investigations are extremely difficult.

In canals lined with exposed hard surface materials, such as cement concrete, brick masonry, and other types of lining, greater velocities are permissible than are normally possible in earthen canals. The friction loss is less in such cases. For that reason, to supply a given discharge, the surface area of the concrete lining can be reduced. In addition, steep side-banks can be allowed. As a result, the canal requires lesser cross-sectional area and thus lesser total land wastage.

1.1.7.2 The Decision on Canal Lining

The decision whether or not to line a canal essentially depends upon the permeability of the soil in which the canal is to be excavated, seepage rate of water, cost of lining, durability of the lining, cost of water, opportunity cost of water, and environmental costs (e.g., damage due to waterlogging, salinity). In many practical cases, this decision can be reached from the visual observations of the soil, provided that it is of a type which obviously is very pervious or impervious. When permeability is in doubt, the decision may be reached either by applying comparative seepage data or by measuring seepage (may be in conjunction with the determination of hydraulic conductivity "K", by field tests). Economic analysis may be performed to judge the lining need and to select from alternative options. Details of economic analysis have been discussed in Chapter 12 (*Economics in Irrigation Management*), Volume 1.

1.1.7.3 Lining Materials

Lining may be done with a large variety of materials. Selection of a suitable one depends mainly on cost, performance, durability, and availability of the material. Normally, the brick lining and precast section (both semicircular and rectangular) are durable for about 15 and 10 years, respectively. The soil–cement, asphalt mat, clay lining, and compaction are durable for 3, 2, 1, and 1 year, respectively. Nowadays, irrigation conveyance is being done by low-cost rubber pipes, hose pipes, and underground pipe systems in many developing countries.

1.2 Designing Open Irrigation Channel

1.2.1 Irrigation Channel and Open Channel Flow

An irrigation channel is constructed to convey irrigation water from the source of supply to one or more irrigated areas. A channel or lateral is needed as an integral part of an irrigation water conveyance system.

In an open channel, water flows at atmospheric pressure, under the force of gravity. In most cases, a gentle slope is provided in the open channel to facilitate the flow.

The words "*Canal*" and "*Channel*" are interchangeably used in the literature and also in this book. Basically, "canal" is artificially constructed (man-made), and "channel" is a natural water passage.

1.2.2 Definition Sketch of an Open Channel Section

The cross-sectional view of a trapezoidal channel with definition sketch is shown in Fig. 1.3

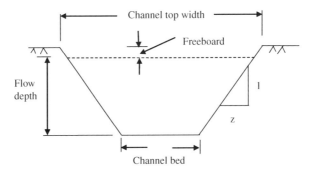

Fig. 1.3 Definition sketch of a trapezoidal channel section

Top width (T): It is the width of the channel at the surface.
Freeboard: It is the additional/extra height of the channel above the design flow depth. It is provided as a safety factor.
Channel bed: It is the bottom width of the channel.
Side slope: Channel side slope is generally expressed as horizontal:vertical (i.e., *H*:*V*). For convenience in computation, the vertical value is reduced to 1, and the corresponding horizontal value is expressed as Z.

Wetted perimeter (P): It is the wetted length of channel across the cross-section of the channel. It is the sum of the channel bed width plus two sloping sides.

Hydraulic radius (R): It is the ratio of wetted area (A) to the wetted perimeter (P) of the channel cross-section, that is, $R = \dfrac{A}{P}$.

1.2.3 Considerations in Channel Design

1.2.3.1 Channel Geometry

The channel geometry design depends on site conditions and conveyance needs. The channel cross-section may be trapezoidal, rectangular, parabolic, V-shaped, or a combination of the geometric shapes.

1.2.3.2 Capacity Requirements

The capacity of canals or laterals should be as follows:

- Sufficient to meet demands of all the irrigation systems served and the amount of water needed to cover the estimated conveyance losses in the canal or lateral
- sized to convey the available water supply in water-short areas, where irrigation water is in demand.
- Capable of conveying surface runoff that is allowed to enter the channel, and
- Such that flow or runoff velocity must be non-erosive.

1.2.3.3 Permissible Velocity/Velocity Limitations

The design of an open channel should be consistent with the velocity limitations for the selected channel lining to satisfy the condition of non-erosive velocity in the channel. The velocity should not be too low to cause siltation in case of surface drainage.

Permissible non-erosive velocity of a channel is dependent upon the stability of lining materials and channel vegetation, as follows:

Material	Maximum velocity (m/s)
Sandy soil	0.6
Loam to silt	1.0
Silty clay	1.2
Stiff clay	1.5
Graded loam to silt	1.5
Hard pan/coarse gravel	1.5
Vegetative channel (grass cover of alfalfa, weeping lovegrass)	1.2

1.2.3.4 Freeboard

The required freeboard above the maximum design water level shall be at least one-fourth of the design flow depth ($0.25d$) and shall not be less than 0.3 m.

1.2.3.5 Water Surface Elevations

Water surface elevations should be designed to provide enough hydraulic head for successful operation of all ditches or other water conveyance structures diverting from the canal or lateral.

1.2.3.6 Side Slopes

Canals, laterals, and field channels should be designed to have stable side slopes. Local information on side-slope limits for specific soils and/or geologic materials should be used if available. If such information is not available, the design of side slopes for the banks of canals or laterals shall not be steeper than those shown below:

Materials	Side slope (horizontal to vertical)
Sandy to loam soil	2:1
Silty clay	1.5:1
Heavy clay	1:1
Loose rock to solid rock	$\frac{1}{4}$:1

1.2.4 Calculation of Velocity of Flow in Open Channel

The irrigation or drainage channel design should be such that it provide adequate capacity for the design discharge or flow resulting from the design storm. The velocity of flow in open channels can be determined by using Chezy's equation or Manning's equation.

1.2.4.1 Chezy's Equation

The earliest formula for open channel was proposed by Chezy (in 1775). The Chezy's equation can be expressed as

$$V = C\sqrt{RS} \tag{1.7}$$

where

V = velocity of flow (m/s)
R = hydraulic radius of the flowing section (m)
S = slope of water surface (taken as equal to the slope of channel bed (m/m))
C = Chezy's constant, which varies with surface roughness and flow rates ($\sim$45–55).

1.2 Designing Open Irrigation Channel

Later on, different scientists and engineers worked on this formula. After conducting a series of experiments, Kutter, Basin, and Manning proposed a method for determining "C" in Chezy's formula. But due to simplicity, Manning's formulation is widely used.

1.2.4.2 Manning's Equation

Manning suggested $C = \frac{1}{N} \times R^{1/6}$ in Chezy's formula. Manning's equation is commonly expressed as

$$V = \frac{1}{N} R^{2/3} S^{1/2} \tag{1.8}$$

where

  V = average flow velocity, m/s
  N = Manning's roughness coefficient
  S = Channel slope, in m per m
  R = Hydraulic radius, m, calculated as $R = A/P$
  A = Flow cross-sectional area, in square meter (m^2)
  P = Wetted perimeter, m

The Manning's equation is best used for uniform steady-state flows. Though these assumptions are rarely achieved in reality, Manning's equation is still used to model most open channel flows. Manning's equation is a semiempirical equation. Thus, the units are inconsistent and handled through the conversion factor.

1.2.4.3 Manning's "N" Values

Manning's "N" value is an important variable describing material roughness in open channel flow computations. Manning's "N" values depend on many physical characteristics of channel surface. Changes in this variable can significantly affect flow discharge, depth, and velocity estimates. So care and good engineering judgment must be exercised in the selection process of "N". The composite "N" value should be calculated where the lining material, and subsequently Manning's "N" value, changes within a channel section (Table 1.2)

Table 1.2 Manning's roughness coefficient for different artificial channels

Channel type/lining type	N value
Concrete	0.011–0.013
Stone masonry	0.03–0.042
Soil cement	0.02–0.025
Bare soil	0.02–0.023
Vegetative waterway[a]	0.15–0.35

[a] For medium, dense, and very dense grass, N should be 0.15, 0.25, and 0.35, respectively.

1.2.5 Hydraulic Design of Open Irrigation Channel

We always search for an efficient and economic channel section. The most economic section is one which can carry maximum discharge for a given cross-sectional area, or, in other words, the channel which requires minimum cross-sectional area (or excavation) for a given discharge. For practical purposes, discharge is fixed, and the minimum cross-sectional area is of interest.

1.2.5.1 Condition for Maximum Discharge Through a Channel of Rectangular Section

The earthen channel of rectangular section is not used except in heavy clay or rocky soils, where the faces of rocks can stand vertically. A concrete channel of rectangular configuration is generally used. Thus, the hydraulically efficient section of the concrete channel is important.

Let us consider a channel of rectangular section as shown in Fig. 1.4. Let

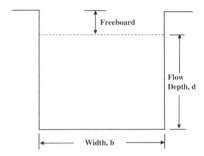

Fig. 1.4 Sketch of a rectangular channel section

b = width of the channel and
d = depth of flow
Then, area of flow, $A = b \times d$
Discharge, $Q = A \times V = AC\sqrt{RS}$ (since $V = C\sqrt{RS}$)
where R is the hydraulic radius, and S is the slope of the channel.
We know, hydraulic radius, $R = \dfrac{A}{P}$.
Thus, $Q = AC\sqrt{\dfrac{A}{P}S}$

Taking S, A, and C constant in the above equation, the discharge will be maximum when A/P is maximum, or the perimeter P is minimum.
We get wetted perimeter $P = b + d + d = b + 2d$ (from Fig. 1.4), and $A = b \times d$, or $b = A/d$
Thus, $P = \dfrac{A}{d} + 2d$

To find the minimum value of P, we have to differentiate the function, set it equal to zero, and solve for variable d. That is,

1.2 Designing Open Irrigation Channel

$$\frac{dP}{dd} = -\frac{A}{d^2} + 2 = 0$$

$$\text{Or,} -\frac{A}{d^2} + 2 = 0$$

$$\text{Or, } A = 2d^2$$

$$\text{Or, } bd = 2d^2 \text{(since } A = b \times d\text{)}$$

$$\text{Or, } b = 2d \tag{1.9}$$

i.e., the width is double the flow depth.

To ensure that P is minimum value rather than maximum value, we have to compute second derivative $\left(\frac{dP^2}{dd^2}\right)$. It will be minimum if the $\frac{dP^2}{dd^2}$ is positive and maximum if $\frac{dP^2}{dd^2}$ is negative.

Here, $\frac{dP^2}{dd^2} = 2\frac{A}{d^3}$ which is positive. Thus, the value of "b" obtained from the first derivative of P is for the minimum value of P.

In this case, the hydraulic mean depth, or hydraulic radius,

$$R = \frac{A}{P} = \frac{bd}{b+2d} = \frac{2d \times d}{2d+2d} \text{ (since } b = 2d\text{)}$$

$$= \frac{2d^2}{4d} = \frac{d}{2}$$

$$\text{i.e., } R = \frac{d}{2} \tag{1.10}$$

Hence, for the maximum discharge conditions, the criteria of channel configurations are as follows:

(a) $b = 2d$ (i.e., width is twice flow depth) and
(b) $R = \frac{d}{2}$ (i.e., hydraulic radius is half of flow depth)

1.2.5.2 Condition for Maximum Discharge Through a Channel of Trapezoidal Section

Many natural and man-made channels are approximately trapezoidal. In practice, the trapezoidal section is normally used in earthen channels. Generally, the side slope in a particular soil is decided based on the soil type. In a loose or soft soil, flatter side slopes are provided, whereas in a harder one, steeper side slopes are allowed.

Fig. 1.5 Schematic presentation of trapezoidal channel section

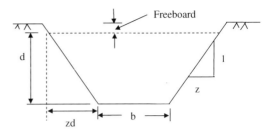

Let us consider a channel of trapezoidal cross-section as shown in Fig. 1.5 Assume that

b = width of the channel at the bottom,
d = depth of the flow, and
$\frac{1}{z}$ = side slope (i.e., 1 vertical to z horizontal)

Thus, Area of flow, $A = (d \times b) + 2\left(\frac{1}{2} \times d \times zd\right) = bd + zd^2 = d(b + zd)$

or $\dfrac{A}{d} = b + zd$

$$b = \frac{A}{d} - zd \tag{1.11}$$

and discharge, $Q = A \times V = AC\sqrt{RS}$

$$= AC\sqrt{\frac{A}{P}S} \quad [\text{since } R = A/P]$$

Keeping A, C, and S constant in the above equation, the discharge will be maximum, when A/P is maximum, or the perimeter P is minimum. From the theory of maxima and minima, P will be minimum when,

$$\frac{dP}{dd} = 0$$

We know that $P = b + 2\sqrt{z^2 d^2 + d^2} = b + 2d\sqrt{z^2 + 1}$.
Substituting the value of b from Eq. (1.11),

$$P = \frac{A}{d} - zd + 2d\sqrt{z^2 + 1}$$

Differentiating the above equation with respect to d and equating the same to zero,

1.2 Designing Open Irrigation Channel

$$\frac{dP}{dd} = -\frac{A}{d^2} - z + 2\sqrt{z^2+1}$$

Or, $-\dfrac{A}{d^2} - z + 2\sqrt{z^2+1} = 0$

Or, $-\dfrac{A}{d^2} + z + 2\sqrt{z^2+1} = 0$

Or, $-\dfrac{d(b+zd)}{d^2} + z = 2\sqrt{z^2+1}$ [since $A = (b+zd)$]

Or, $\dfrac{b+2zd}{d} = 2\sqrt{z^2+1}$

Or, $\dfrac{b+2zd}{d} = 2\sqrt{z^2+1}$

$$\frac{b+2zd}{2} = d\sqrt{z^2+1} \tag{1.12}$$

Here, $(b+2zd)$ is the top width of water; and $(d\sqrt{z^2+1})$ is the single sloping side. That is, the condition is that the sloping side is equal to half of the top width.

In this case, the hydraulic mean depth or hydraulic radius,

$$R = \frac{A}{P} = \frac{d(b+zd)}{b+2d\sqrt{z^2+1}}$$

$$= \frac{d(b+zd)}{b+(b+2zd)} \qquad \text{[from Eq. (1.12)]}$$

$$= \frac{d(b+zd)}{2(b+zd)} = \frac{d}{2}$$

$$\text{That is, } R = \frac{d}{2} \tag{1.13}$$

Hence, for maximum discharge or maximum velocity, the two conditions are as follows:

* $\dfrac{b+2zd}{2} = d\sqrt{z^2+1}$ (the sloping side is equal to half of the top width)

* $R = \dfrac{d}{2}$ (hydraulic radius is half of flow depth)

1.2.6 Sample Examples on Irrigation Channel Design

Example 1.1

Determine the velocity of flow and discharge capacity of an unlined canal branch on a grade of 1 m in 800 m having depth of flow 1.5 m, bottom width 0.80 m, and side slope 1:1.

Solution

Required:

(a) Velocity of flow, $V = ?$
(b) Discharge, $Q = ?$

Given,

> Depth of flow, $d = 1.5$ m
> Bed width, $b = 0.8$ m
> Canal bottom slope, $S = \frac{1}{800} = 0.00125$
> Side slope, $Z : 1 = 1 : 1$
> As the channel is unlined, assume roughness coefficient of Manning's formula, $N = 0.023$
> Area, $A = bd + zd\ 2 = (0.8 \times 1.5) + 1 \times (1.5)^2 = 3.45$ m^2
> Wetted perimeter, $P = b + 2d\sqrt{1+z^2} = 5.042$ m
> We know, hydraulic radius, $R = \frac{A}{P} = 0.684$
> We know, velocity, $V = \frac{1}{n}R^{2/3}S^{1/22} = \frac{1}{0.023} \times (0.684)^{2/3} \times \sqrt{0.00125} = 1.193$ m/s (Ans.)
> Discharge, $Q = AV = 3.45 \times 1.193 = 4.116$ m^3/s (Ans.)

Example 1.2

Design an earthen channel of trapezoidal section for the following conditions:

> Discharge = 2 cumec
> Channel bottom slope: 1 in 1,200
> Side slope: 1.3:1
> Value of N in Manning's equation = 0.02

Solution

Given,

> Discharge, $Q = 2$ m^3/s
> Slope of the channel bed, $S = \frac{1}{1200}$
> Side slope, $z:1 = 1.3:1$
> Manning's $N = 0.02$

1.2 Designing Open Irrigation Channel

For design purposes, always the most economical section is designed.
We know, for most economical trapezoidal section,

Half of the top width = Sloping side
That is, $\frac{b+2zd}{2} = d\sqrt{z^2 + 1}$
Or, $b + 2.6d = 2d\sqrt{(1.3)^2 + 1}$ (∵ $z = 1.3$)
Or, $b = 0.68d$ (A)

We know, for most economical trapezoidal section another condition is

hydraulic radius = half of depth of flow
i.e., $R = \frac{d}{2}$ (B)
Now, area, $A = d(b + zd) = d(0.68d + 1.3d)$ (putting b = 0.68d)
or, $A = 1.98d^2$
The discharge, $Q = AV = A \times \frac{1}{N}R^{2/3}S^{1/2}$ with usual notations.

Putting the values, $2 = 1.98d^2 \times \frac{1}{0.02} \times \left(\frac{d}{2}\right)^{\frac{2}{3}} \times \left(\frac{1}{1,200}\right)^{\frac{1}{2}}$

or, $d^{8/3} = 1.11$
or, $d = 1.04$ m
Then, $b = 0.68d = 0.7072$ m
Assuming freeboard as 20% of flow depth, depth of channel would be

$$d_c = 1.04 \times 1.20 = 1.248 \text{ m} \approx 1.25 \text{ m}$$

Top width $= b + 2zd_c + (2 \times 1.3 \times 1.25) = 3.957$ m

Results

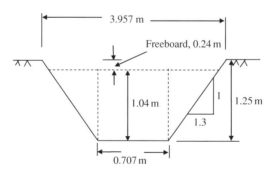

Channel bed width, $d = 1.04$ m
Flow depth, $d = 1.04$ m
Depth of channel = 1.25 m
Top width = 3.957
Side slope = 1.3:1(H:V)

Example 1.3

Design a concrete lined canal from the following data:

Discharge = 400 cumec
Slope = 1 in 4,000
Side slope = 1.5:1 (H:V)
Limiting velocity = 2.5 m/s
Value of N in Manning's equation = 0.013

Solution

Applying Manning's equation, the velocity, V is given by

$$V = \frac{1}{N} R^{2/3} S^{1/2} \quad (A)$$

Here, limiting velocity,

$$V = 2.5 \text{ m/s}$$

$$N = 0.013$$

$$S = \frac{1}{4000}$$

Substituting the above values in Eq. (A), we get

$$2.5 = \frac{1}{0.013} \times R^{2/3} \times \left(\frac{1}{4000}\right)^{1/2}$$

or $R^{2/3} = 2.5 \times 0.013 \times \sqrt{4000} = 2.554$

$$R = (2.489)^{3/2} = 4.08 \text{ m}$$

Cross-sectional area, $A = \frac{Q}{V}$ 400/2.5 = 160 m^2
We know, for most economical section in trapezoidal canal, $R = d/2$
Then, $d = 2R = 2 \times 4.08 = 8.16$ m
We get, $A = bd + zd^2$
That is, $160 = b \times 8.16 + 1.5 (8.16)^2$
Or, $b = 7.36$ m
Considering freeboard as 15% of flow depth, depth of canal, $d_c = 8.16 \times 1.15 = 9.384$ m
Top width, $T = b + 2zd_c = 7.36 + 2 \times 1.5 \times 9.384 = 35.51$ m

Results

Width of channel bed, $b = 7.36$ m
Depth of flow, $d = 8.16$ m
Depth of canal, $d_c = 9.384$ m
Top width, $T = 35.51$ m

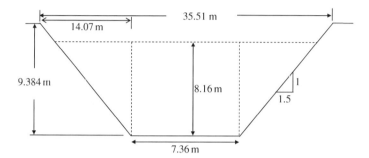

1.3 Designing Pipe for Irrigation Water Flow

1.3.1 Fundamental Theories of Water Flow Through Pipe

1.3.1.1 Theories of Physics

Equations describing fluid flow are based on three fundamental laws of physics. These are as follows:

- Conservation of matter (or mass)
- Conservation of energy
- Conservation of momentum

Conservation of Matter

This law states that matter cannot be created or destroyed, but it can be converted (e.g., by a chemical process). In case of hydraulics, chemical activity is not considered; thus the law reduces to conservation of mass.

Conservation of Energy

This says that energy cannot be created or destroyed, but can be converted from one type to another. For example, potential energy may be converted to kinetic energy or pressure energy.

Conservation of Momentum

This says that a moving body cannot gain or loss momentum unless an external force acts upon it.

1.3.1.2 Theories of Hydraulics and Fluid Flow

Pascal's Law

Pascal's law can be stated as follows: "The intensity of pressure at any point of a closed conduit is equal in all directions."

Consider a closed vessel as of Fig. 1.6. Let the intensity of external pressure on the vessel liquid be "p". According to Pascal's law, the intensity of pressure at every direction of the vessel will be equal and "p". Since water is a noncompressible liquid (not compressed/reduced in volume due to pressure), it exhibits the unique trait of transforming pressure to all directions when in a confined space.

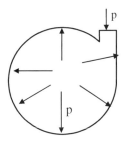

Fig. 1.6 Schematic of pressure distribution in a closed vessel

Continuity Equation/Conservation of Mass

If the flow is steady, and the fluid is incompressible, the mass entering a pipe (Fig. 1.7) is equal to the mass leaving.

That is, mass flow entering = mass flow leaving
Or, $\rho Q_e = \rho Q_l$

Since the volume flow rate is the product of area and mean velocity, consider that

velocity of flow at entering section = V_1
cross-sectional area at entering section = A_1
velocity of flow at leaving section = V_2
cross-sectional area at leaving section = A_2

Then, applying the law of conservation of mass, $\rho \times V_1 \times A_1 = \rho \times V_2 \times A_2$

$$\text{or,} \quad A_1 V_1 = A_2 V_2 \tag{1.14}$$

1.3 Designing Pipe for Irrigation Water Flow

Fig. 1.7 Schematic of fluid flowing in a pipe

This is the Continuity equation.

Bernoulli's Equation/Energy Equation (Conservation of Energy)

In reference to Fig. 1.7 consider that the fluid moves from the inlet to the outlet (over the length L) in time δt.

Bernoulli's equation says that total energy per unit weight of flowing fluid is constant over the flow section. That is,

$$\frac{p_1}{\rho g} + \frac{V_1^2}{2g} + z_1 = \frac{p_2}{\rho g} + \frac{V_2^2}{2g} + z_2 = H = \text{constant} \qquad (1.15)$$

The dimension of each term of the equation is "meter" (m). Each term is called as follows:

$\frac{p}{\rho g}$ = pressure head
$\frac{V^2}{2g}$ = velocity head
z = potential head or elevation head (with respect to a reference datum)

1.3.2 Water Pressure – Static and Dynamic Head

Water pressure at a certain depth of a water column is equal to the weight of water above that point. Water pressure is normally expressed as "feet head" or "lbs/in^2" (PSI). Pressure,

$$P = \rho h \qquad (1.16),$$

where

h = height of water, ft
ρ = density of water, lbs/ft^3
P = pressure, lbs/ft^2 (PSI)

Conversion

Head (ft) × 0.433 = PSI
PSI × 2.304 = head (ft) of water

1.3.2.1 Static and Dynamic Head

In reference to Fig. 1.8 the pressure of water at point "A" in the pipeline, when the pipe is closed (water is static, not flowing), is

$$P = \rho h$$

It does not matter what the distance is of the point from the top of the source, that is, static pressure is just for elevation difference. If we measure the pressure with the water flowing, the pressure would be termed as "dynamic pressure."

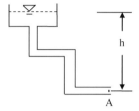

Fig. 1.8 Water tank showing static pressure

1.3.2.2 Pressure Distribution in a Water Column/Tank

Consider a tank of water as shown in Fig. 1.9. At the top surface of water, the pressure is zero.

The pressure increases with the water depth. At the bottom of the tank (having water depth of "h"), the intensity of pressure would be "ρh."

Pressure Vs Flow Rate Relationship

There is an inverse relationship between pressure and flow. For a pipe of particular size, higher pressure means lower flow. Lower pressure results in higher flows. This is because as the total energy is constant, higher pressure head results in lower velocity head (and hence lower velocity), thus lower flow rate.

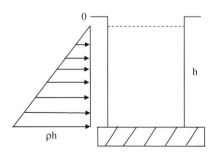

Fig. 1.9 Schematic of pressure distribution in a vertical water tank

1.3.3 Hydraulic and Energy Grade Line for Pipe Flow

Hydraulic calculations are required to design irrigation pipes. A hydraulic grade line analysis is required for all designs to ensure that water flows through the pipes in the manner intended.

We learned that the total energy of flow in a pipe section (with respect to a reference datum) is the sum of the elevation of the pipe center (elevation head), the pressure exerted by the water in the pipe expressed or shown by the height of a column of water (pressure head, or piezometric head, if a piezometer is provided in the pipe) and the velocity head. The total energy of flowing water when represented in figure is termed as *energy grade line* or *energy gradient*. The pressure of water in the pipe represented by elevation when drawn in line is termed as *hydraulic grade line* or *hydraulic gradient* (Fig. 1.10).

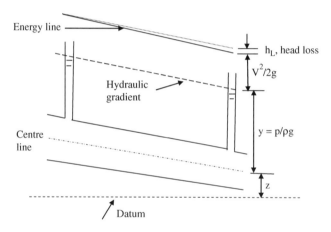

Fig. 1.10 Schematic showing of hydraulic and energy line in pipe flow

1.3.4 Types of Flow in Pipe – Reynolds Number

Flow of water in pipe is of two types: laminar and turbulent. In laminar flow, the fluid moves in layers called laminus. In turbulent flow, secondary random motions are superimposed on the principal flow, and mixing occurs between adjacent sectors. In 1883, Reynolds introduced a dimensionless parameter (which has since been known as Reynolds number) that gives a quantitative indication of the laminar to turbulent transition. Reynolds number R_N is

$$R_N = \frac{\rho V d}{\mu} \qquad (1.17)$$

where

ρ = density of fluid (kg/m³)
V = mean fluid velocity (m/s)
d = diameter of the pipe (m)
μ = coefficient of viscosity of the fluid (kg/m/s)

Generally a flow is laminar if $R_N \leq 2{,}100$. A transition between laminar and turbulent flow occurs for R_N between 2,100 and 4,000 (transition flow). Above 4,000, the flow is turbulent. At turbulence range, the flow becomes unstable, and there is increased mixing that results in viscous losses which are generally much higher than those of laminar flow.

The Reynolds number can be considered in another way, as

$$R_N = \frac{\text{Inertia forces}}{\text{Viscous forces}} \qquad (1.18)$$

The inertia forces represent the fluid's natural resistance to acceleration. The viscous forces arise because of the internal friction of the fluid. In a low Reynolds number flow, the inertia forces are small and negligible compared to the viscous forces. Whereas, in a high Reynolds number flow, the viscous forces are small compared to the inertia forces.

1.3.5 Velocity Profile of Pipe Flow

Typical velocity profile of a pipe flow is shown in Fig. 1.11. The velocity is zero at the surface, increases thereafter and reaches its maximum at the center of the pipe.

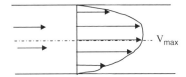

Fig. 1.11 Diagram showing velocity distribution in pipe flow

1.3.6 Head Loss in Pipe Flow and Its Calculation

1.3.6.1 Causes and Components of Head Loss

When fluid flows through the pipe, the internal roughness of the pipe wall can create local eddy currents vicinity to the surface, thus adding a resistance to the flow of the pipe. Pipes with smooth walls have only a small resistance to flow (frictional resistance). Smooth glass, copper, and polyethylene have small frictional resistance, whereas cast iron, concrete, and steel pipe, etc. create larger eddy currents and effect on frictional resistance.

1.3 Designing Pipe for Irrigation Water Flow

Determination of head loss (meaning loss of energy) in the pipe is necessary because pump and motor (power) combination should be matched to flow and pressure requirement. Oversizing makes for inefficiencies that waste energy and cost money.

Components of Head Loss

Head loss in pipe may be divided into the following:

- Major head loss
- Minor head loss

Major head loss consists of loss due to friction in the pipe. Minor loss consists of loss due to change in diameter, change of velocity in bends, joints, valves, and similar items.

1.3.6.2 Factors Affecting Head Loss

Frictional head loss (h_L) in the pipe can be functionally expressed as follows:

$$h_L = f(L, V, D, n, \rho, v) \tag{1.19}$$

where

L = length of pipe
V = velocity of flow
D = diameter of pipe
n = roughness of the pipe surface (internal surface, over which flow occurs)
ρ = density of flowing fluid
v = viscosity of the flowing fluid

The mode of action of the factors affecting head loss is as follows:

- head loss varies directly as the length of the pipe
- it varies almost as the square of the velocity
- it varies almost inversely as the diameter
- it depends on the surface roughness of the pipe wall
- it is independent of pressure

1.3.6.3 Different Head Loss Equations

Darcy-Weisbach Formula for Head Loss

Darcy-Weisbach formula for head loss in a pipe due to friction in turbulent flow can be expressed as

28   1 Water Conveyance Loss and Designing Conveyance System

$$h_f = f \frac{L}{D} \frac{V^2}{2g}, \quad (1.20)$$

where

h_f = head loss due to friction (m)
f = friction factor (or Darcy's friction coefficient)
L = length of pipe (m)
V = velocity of flow (m/s)
g = acceleration due to gravity (m/s^2) = 9.81 m/s^2
D = inner diameter of the pipe (m)

Darcy introduced the concept of relative roughness, where the ratio of the internal roughness of a pipe to the internal diameter of the pipe affect friction factor for turbulent flow.

Colebrook-White Equation

The Colebrook-White equation for calculating friction factor is

$$\frac{1}{\sqrt{f}} = 1.14 - 2 \log_{10}\left(\frac{e}{D} + \frac{9.35}{Re\sqrt{f}}\right) \quad (1.21)$$

where

f = friction factor
e = internal roughness of the pipe
D = inner diameter of pipe work

To find out "f", iteration (trial and error) is required. A value of 0.02 can be assumed as a first step.

The Moody Chart

In 1944, L. F. Moody plotted the data obtained from the Colebrook equation, which is designated as "The Moody Chart" (Fig. 1.12). From this chart, the user can find the friction factor for turbulent flow condition with reasonable accuracy.

Fanning's Friction Factor

Fanning, after many experiments, provided data for friction factors, but with hydraulic radius concept. For full pipe flow, hydraulic radius, R = 1/4th of the diameter of the pipe (i.e., $R = D/4$). Thus, the head loss equation becomes

$$h_f = \frac{4 f_f L V^2}{2gD} \quad (1.22)$$

1.3 Designing Pipe for Irrigation Water Flow

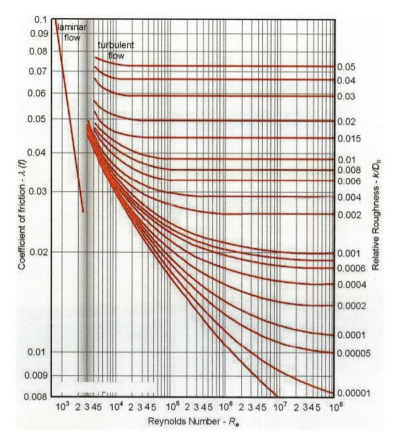

Fig. 1.12 Moody diagram

where f_f is the Fanning friction factor.

From the above equation, it is observed that Darcy's friction coefficient is four times greater than Fanning's friction factor. So, care should be taken in selecting friction coefficient, whether it is for Darcy-Weisbach formula *or* for others.

Head Loss Under Laminar Flow – Hagen-Poiseuille Equation

$$h_f = \frac{32\mu VL}{wD^2} \tag{1.23}$$

where

V = velocity of flow (m/s)
L = length of pipe (m)

D = inner diameter of the pipe (m)
w = specific wt. of the fluid (kg/m³)
μ = viscosity of the flowing fluid (kg/s/m²)

1.3.6.4 Calculation of Minor Loss

Minor loss can be expressed as

$$h_{minor} = c \frac{V^2}{2g} \tag{1.24}$$

where c is the minor loss coefficient. Thus, the total minor loss can be calculated by summing the minor loss coefficients and multiplying the sum with the dynamic pressure head. Minor loss coefficients of different components/fittings are given in Table 1.3.

Table 1.3 Minor loss coefficient for different fittings

Fittings	Minor loss coefficient (c)
Fully open ball valve	0.05
Threaded union	0.08
Fully open gate valve	0.15
½ closed gate valve	2.1
Fully open angle valve	2
Threaded long radius 90° elbows	0.2
Flanged 180° return bends	0.2
Flanged Tees, line flow	0.2
Threaded Tees, line flow	0.9
Threaded Tees, branch flow	2.0
Fully opened globe valve	10

1.3.6.5 Minimizing Head Loss in Pipe

One of the main aims of pipe design is to minimize the head losses associated with pipe length (frictional loss), bends, diameter change, and transitions. Minimization of head loss will keep the diameter of the pipeline to the minimum (necessary to achieve the design flow capacity), and therefore its cost will be reduced.

Head losses in pipe can be minimized by

(1) Using large diameter pipe in the mainline
(2) Minimizing bends or turns
(3) Making/selecting internal surface of the pipe smoother

1.3 Designing Pipe for Irrigation Water Flow

1.3.6.6 Sample Workout Problem

Example 1.4

Determine the head loss due to friction in an irrigation pipe having 150 m length and 25 cm diameter. The velocity of flowing water in the pipe is 2.0 m/s

Solution

Head loss according to Darcy-Weisbach formula:

$$h_f = f \times \frac{L}{D} \times \frac{V^2}{2g}$$

Given,
 V = 2.0 m/s
 Assuming $f = 0.025$ (as of cast iron)
 $h_f = 0.025 \times \frac{150}{0.25} \times \frac{(2)^2}{2 \times 9.81}$
 = 3.058 m of water (Ans.)

1.3.7 Designing Pipe Size for Irrigation Water Flow

Selection of pipe size should be based on the following:

- hydraulic capacity (discharge) requirement
- head loss, and
- economy

In the short run, a small diameter pipe may require lower initial cost, but due to excessive head loss, it may require higher cost in the long run.

Pipe size based on hydraulic capacity can be found as

$$A = \frac{Q}{V}$$

where

 A = cross-sectional area of the pipe (m^2)
 Q = required discharge (m^3/s)
 V = permissible velocity of flow (m/s)

Diameter of the pipe can be found from the relation, $A = \pi D^2/4$, that is, $D = \sqrt{\frac{4A}{\pi}}$.

The pipe must have the capacity to supply peak demand (Q). After calculating the maximum size required, the second step is to calculate the head loss per unit length (say 100 m) and also for the whole irrigation farm. Extra power and cost requirement for the head loss should be calculated for the entire effective life of the pipe. Similar calculations should be performed for the next available pipe sizes (2 or 3 nos).

Now, compare the prices of the pipes (present or first installment) and extra cost for head loss for the entire life (transferred into present value). Choose the least-cost one.

1.3.8 Sample Workout Problems

Example 1.5

Compute the size of best quality cast-iron pipe that will carry 0.03 m³/s discharge with a head loss of 2 m per 1,000 m.

Solution

We get, $h_f = f \times \frac{L}{D} \times \frac{V^2}{2g}$
Given,

$Q = 0.03$ m³/s
for $L = 1{,}000$ m, $h_f = 2$ m
$V = Q/A = Q/(\pi D^2/4)$

thus,

$$2 = 0.023 \times \frac{1000}{D} \times \frac{(0.03)^2}{(\pi D^2/4)^2} = 0.023 \times \frac{1000 \times 16 \times (0.03)^2}{\pi^2 \times D^5}$$

Or, $D^5 = 0.016779$
Or, $D = 0.4415$ m $= 44.15$ cm
That is, pipe dia $= 44.15$ cm (Ans.)

Relevant Journals

– Journal of Soil and Water Conservation
– Journal of Hydraulic Engineering
– Journal of Hydrology
– Agricultural Water Management

- Water Resources Management
- Water Resources Research
- Irrigation Science
- Trans. ASAE
- Irrigation and Drainage System

Questions

(1) What is seepage? What are the factors affecting seepage in earthen conveyance system?
(2) Describe in brief the different methods of seepage measurement.
(3) How can the average seepage loss in an irrigation command area be estimated?
(4) Discuss the various means of reducing seepage in an earthen channel.
(5) How will you decide on lining a canal?
(6) What are the different configurations/geometries of irrigation channel?
(7) Define the following: permissible velocity, wetted perimeter, hydraulic radius, freeboard, and side slope.
(8) Describe Chezy's and Manning's equation for calculating velocity in open channel.
(9) What is the most economical cross-section of a channel?
(10) Derive the conditions for the most economical cross-section for (i) trapezoidal and (ii) rectangular channel section.
(11) Determine the velocity of flow and discharge capacity of an unlined canal section on a grade of 1 m on 900 m having depth of flow 1.2 m, bottom width 0.9 m, and side slope 1.2:1.
(12) Design an earthen channel of trapezoidal section for the following conditions:
    Discharge = 2.5 cumec
    Channel bottom slope: 1 in 1,400
    Side slope: 1.25:1
    Value of N in Manning's equation = 0.022
(13) Design a concrete canal from the following information:
    Discharge = 50 cumecs
    Slope = 1 in 3,600
    Side slope = 1:1(H:V)
    Value of N in Manning's equation = 0.011
(14) Draw the velocity diagram of pipe flow.
(15) Draw the hydraulic and energy grade line of pipe flow.
(16) What do you mean by "head loss" in pipe flow? What are the factors affecting head loss? How can the head loss be minimized?
(17) Write down the principles of selecting pipe size for irrigation purposes.
(18) Determine the head loss due to friction in an irrigation pipe having 300 m length and 30 cm diameter. The velocity of flowing water in the pipe is 2.5 m/s.
(19) Compute the size of best quality concrete pipe that will carry 0.05 m^3/s discharge with a head loss of 1 m per 1,000 m.

References

Ali MH (2001) Technical performance evaluation of Boyra deep tube-well – a case study. J Inst Eng Bangladesh 28(AE No.1):33–37

Kale SR, Ramteke JR, Kadrekar SB, Charp PS (1986) Effects of various sealent material on seepage losses in tanks in lateritic soil. Indian J Soil Conserv 14(2):58–59

Reginato RJ, Nakayama FS, Miller JB (1973) Reducing seepage from stock tanks with uncompacted sodium-treated soils. J Soil Water Conserv 28(5):214–215

Chapter 2
Water Application Methods

Contents

2.1	General Perspectives of Water Application	36
2.2	Classification of Water Application Methods	36
2.3	Description of Common Methods of Irrigation	38
	2.3.1 Border Irrigation	38
	2.3.2 Basin Irrigation	40
	2.3.3 Furrow Irrigation	43
	2.3.4 Sprinkler Irrigation Systems	46
	2.3.5 Drip Irrigation	52
	2.3.6 Other Forms of Irrigation	54
2.4	Selection of Irrigation Method	56
	2.4.1 Factors Affecting Selection of an Irrigation Method	56
	2.4.2 Selection Procedure	63
Relevant Journals		63
Questions		63

The application of water to soils for crop use is referred to as irrigation. Irrigation systems differ greatly depending on what they are going to be used for. They range from the simple hand watering method used in most home gardens and many nurseries to the huge flood and furrow irrigation systems found in large-scale production. Surface (gravity-driven surface irrigation), sprinkler, drip/micro, and subsurface are types of irrigation methods that are used by growers to irrigate various crops. Each system has its advantages and disadvantages. But with good design, they can be very successful for appropriate cases. Water losses from irrigation vary with the type of irrigation method. The water management decisions strongly influence how uniform water can be applied through different irrigation methods to provide optimal soil water conditions for crop growth and marketable yields. The most appropriate irrigation method for an area depends upon physical site conditions, the crops being grown, amount of water available, and management skill. This chapter gives some very broad guidance and indicates several important criteria in the selection of a suitable irrigation method.

2.1 General Perspectives of Water Application

Several decisions must be made before an irrigation system is installed in a field. Some determinations are technical in nature, some economic, and others involve a close scrutiny of the operation and crop to be irrigated.

Location, quantity, and quality of water should be determined before any type of irrigation system is selected. No assumptions should be made with the water supply. Make sure that the water source is large enough to meet the irrigation system's demand by test pumping groundwater sources or measuring flow rate of streams. Securing of water rights on groundwater wells should be taken beforehand.

Numerous irrigation systems are on the market. Each system has advantages, disadvantages, and specific uses. A discussion of different systems, their roles, and capabilities in irrigated crop production are given in the next sections.

2.2 Classification of Water Application Methods

Water application methods can be classified based on different themes:

A Based on energy/pressure required
B Based on placement of irrigation water
C Based on wetted area by irrigation

Classification system – A

Based on energy/pressure requirement, irrigation methods can be grouped as

- gravity irrigation and
- pressurized irrigation

Again, gravity irrigation may be subdivided based on mode of application as

- border irrigation
- basin irrigation
- furrow irrigation

Pressure irrigation system may be subdivided based on mode of application as

- drip irrigation
- sprinkler irrigation

Classification system – B

Based on the placement of irrigation water (whether on, above, or below the soil surface), irrigation methods may be grouped as

2.2 Classification of Water Application Methods

- surface irrigation
- subsurface irrigation
- overhead irrigation

Surface irrigation system may be subdivided based on mode of water application as

- border irrigation
- basin irrigation
- furrow irrigation
- drip irrigation

Overhead irrigation includes sprinkler irrigation and hand watering.

Classification system – C

Based on wetted area of crop root zone by irrigation, irrigation methods can be grouped as

- flood irrigation
- drip (or trickle or localized) irrigation
- sprinkler irrigation

Flood irrigation can be further grouped as

- basin irrigation
- border irrigation
- furrow irrigation

Basin irrigation may be either "check basin" or "contour basin".

The definition of different types of irrigation systems are outlined below:

> *Gravity irrigation*: Irrigation in which the water is not pumped but flows and is distributed to the crop field by gravity.
> *Pressurized irrigation*: Irrigation system in which water is pumped and flows to the crop field by pressure.
> *Surface irrigation*: A form of irrigation where the soil surface is used as a conduit.
> *Subsurface irrigation or subirrigation*: Applying irrigation water below the ground surface either by raising the water table within or near the root zone.
> *Border irrigation*: Border irrigation is defined as the application of water to an area typically downslope and surrounded by two border ridges or dikes to the ends of the strip.
> *Basin irrigation*: Basin irrigation is defined as the application of water to an area typically leveled to zero slope and surrounded by dikes or check banks to prevent runoff.

Furrow irrigation: A partial surface flooding method of irrigation in which water is applied in furrows (narrow channels dug between the rows of crops) or "rows of sufficient capacity" to contain the designed irrigation system, instead of distributing water throughout the whole field.

Sprinkler irrigation: A system in which water is applied by means of nozzle or perforated pipe that operates under pressure in the form of a spray pattern.

Drip irrigation: An irrigation system in which water is applied directly to the root zone of plants.

Flood irrigation: A system in which the entire soil surface of the field is covered by ponded water.

2.3 Description of Common Methods of Irrigation

2.3.1 Border Irrigation

2.3.1.1 Concept and Features

Border irrigation is a modern method of surface irrigation. Border irrigation uses land formed into strips, bounded by ridges or borders (Fig. 2.1). Borders are generally prepared with zero side slope and a small but uniform longitudinal slope not exceeding 1%. The borders are divided by levees running down the slope at uniform spacing. The lower end of the border is opened to a drainage ditch or closed with a levee to create ponding on the end of the border. Levees are pulled across the end on steeper borders.

In this method, water is applied at the upper end of the border strip, and advances down the strip. Irrigation takes place by allowing the flow to advance and infiltrate along the border. After a time, the water is turned off, and a recession front, where standing water has soaked into the soil, moves down the strip. Smaller inflow discharges and longer time duration of application are utilized in graded fields to reduce

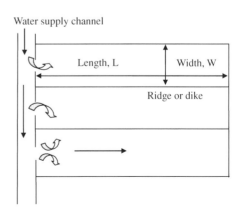

Fig. 2.1 Schematic of border irrigation system

2.3 Description of Common Methods of Irrigation

downstream losses, like for furrow irrigation. Larger inflow rates are utilized when the field slope is very small with management becoming similar to that of basins. Automated water control is often applied.

If it is possible, irrigate each block of border individually when irrigating. To maintain control with border irrigation, discharge the well into the top block until the desired application is achieved. The total well discharge then is moved to the second border and the first levee cut. Erosion control measures may be required if large stream sizes are used.

2.3.1.2 Suitability, Capabilities, and Limitations of Border Method

Crop Suitability

Border irrigation is best adapted to grain and forage crops where there are large areas of flat topography and water supplies are large. Border irrigation could be used on precision leveled rice fields, where beans or other grain crops are grown in rotation with rice.

Soil and Land Suitability

Border irrigation system performs better when soils are uniform, and the slope is mild. Undulating topography and shallow soils do not respond well to grading to a plane. Steep slopes and irregular topography increase the cost of land leveling and reduce border size. Deep cuts may expose areas of nonproductive soils, requiring special fertility management.

Economy and Financial Involvement

The major investment in border irrigation is that of land grading or leveling. The cost is directly related to the volume of earth that must be moved, the area to be finished, and the length and size of farm canals. Border irrigation is relatively inexpensive to operate after installation.

Attainable Irrigation Efficiency

Reasonable irrigation efficiency is possible with border irrigation method. Typical efficiencies for border strip irrigation ranges from 70 to 85%. With the border method, runoff return flow systems may be needed to achieve high water use efficiency.

The system designer and operator can control many of the factors affecting irrigation efficiency, but the potential uniformity of water application with surface irrigation is limited by the variability of soil properties (primarily infiltration rate) throughout the field. Results of field studies indicate that even for relatively uniform soils, there may be a distribution uniformity of infiltration rates of only about 80%.

Labor Requirement

Border systems may be automated to some degree to reduce labor requirements. It requires skilled irrigators to obtain high efficiencies. The labor skill needed for setting border flows can be decreased with equipment of higher cost. The setting of siphons or slide openings to obtain the desired flow rate requires skillness, but that one can learn.

Advantages and Disadvantages

Advantages

 (i) Easy to construct and maintain
 (ii) Operational system is simple and easy
 (iii) High irrigation efficiencies are possible if properly designed, but rarely obtained in practice due to difficulty of balancing the advance and recession phases of water application
 (iv) Natural drainage is facilitated through downward slope
 (v) Comparatively less labor is required

Limitations

 (i) Requires flat and smooth topography
 (ii) Not suitable for sandy soils
 (iii) Not suitable for crops which requires ponding water
 (iv) Higher amount of water is required compared to sprinkler or drip irrigation.

2.3.2 Basin Irrigation

2.3.2.1 Concept and Characteristics

In this method, water is applied to leveled surface units (basins) which have complete perimeter dikes to prevent runoff and to allow infiltration after cutoff (Fig. 2.2). Basin irrigation is the simplest of the surface irrigation methods. Especially if the basins are small, they can be constructed by hand or animal traction. Their operation and maintenance are simple.

The best performance is obtained when advance time is minimized by using large non-erosive discharges, and the basin surface is precision leveled. This method is the most commonly practiced worldwide, both for rice and other field crops, including orchard tree crops. In general, basins are small and uneven and water application is manually controlled. Level basin irrigation using large laser-leveled units with automated or semiautomatic control is practiced in few areas in developed countries.

In this approach, water is applied to a completely level area enclosed by dikes or borders (called basins). This method of irrigation is used successfully for both field and row crops. The floor of the basin may be flat, ridged, or shaped into beds,

2.3 Description of Common Methods of Irrigation

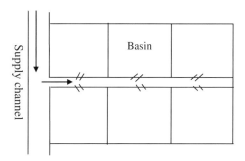

Fig. 2.2 Schematic layout of a basin irrigation system

depending on crop and cultural practices. Basins need not be rectangular or straight sided, and the border dikes may or may not be permanent. This irrigation technique is also called by a variety of other names such as check-basin irrigation, level-basin irrigation, level borders, check flooding, check irrigation, dead-level irrigation. Basins may be of different types: rectangular, ring, and contour.

Basin size is limited by available water stream size, topography, soil factors, and degree of leveling required. Basin may be quite small or as large as 15 ha or so. Level basins simplify water management, since the irrigator need only supply a specified volume of water to the field. With adequate stream size, the water will spread quickly over the field, minimizing nonuniformities in inundation time. Basin irrigation is most effective on uniform soils, precisely leveled, when large stream sizes (relative to basin area) are available. High efficiencies are possible with low labor requirements.

If it is possible, irrigate each basin individually when irrigating. To maintain control with basin irrigation, divert the well discharge into the top basin until the desired application is achieved. The total well discharge is then moved to the second basin and the first basin cut so it drains into the second basin. This process requires a certain amount of labor as the water "steps down" the field. This gives better water control for application amounts and increases the amount of water available for irrigation.

2.3.2.2 Suitabilities and Limitations of Basin Irrigation Method

Crop

Basin irrigation is suited to irrigate close growing crops (e.g., paddy). Paddy (rice) is always grown in basins. Many other crops can also be grown in basins: e.g., maize, sorghum, trees. Those crops that cannot stand a very wet soil for more than 12–24 h should not be grown in basins.

Basin and border strip irrigations flood the soil surface, and will cause some soils to form a crust, which may inhibit the sprouting of seeds.

Soil and Topography

For basin irrigation, basin size should be appropriate for soil texture and infiltration rate. Basin lengths should be limited to 100 meter (m) on very coarse textured soils, but may reach 400 m on other soils. Flat lands, with a slope of 0.1% or less, are best suited for basin irrigation: little land leveling will be required. If the slope is more than 1%, terraces can be constructed. However, the amount of land leveling can be considerable.

In areas of high intensity rainfall and low intake rate soils, surface drainage should be considered with basin irrigation, to reduce damage due to untimely inundation.

Water Quantity

It is important that irrigation stream size be properly matched to basin or border size for uniform irrigation. Basin systems are suitable for leaching of salts for soil reclamation, since the water can be held on the soil for any length of time. Under normal operating conditions, leaching fractions adequate for salinity control can be maintained with basin.

Required Depth of Irrigation Application

When the irrigation schedule is determined, it is known how much water (in mm depth) has to be given per irrigation application. It must be checked that this amount can indeed be given. Field experience has shown that the highest water can be applied per irrigation application when using basin irrigation. In practice, in small-scale irrigation projects, usually 40–70 mm of water is applied in basin irrigation.

Attainable Efficiencies

The system designer and operator can control many of the factors affecting irrigation efficiency. Properly designed and maintained basin systems are capable of obtaining moderately high efficiencies. Some basins are typically designed to pond the water on their surfaces and prevent tail water; they are usually the most efficient surface irrigation method. Design efficiencies should be on the order of 70–85%. With reasonable care and maintenance, field efficiencies in the range of 80–85% may be expected.

Labor and Energy Requirement

Basin irrigation requires accurate land leveling. Some labor and energy will be necessary for land grading and preparation. Basin irrigation involves the least labor of the surface methods, particularly if the system is automated. With surface irrigation, little or no energy is required to distribute the water throughout the field, but some energy may be needed to bring the water to the field, especially when water is

2.3 Description of Common Methods of Irrigation

pumped from groundwater. In some instances, these energy costs can be substantial, particularly with low water use efficiencies.

Cost and Economic Factor

Basin irrigation is generally the most expensive surface irrigation configuration to develop and maintain but often the least expensive to operate and manage. Basin system's costs can vary greatly, depending on crop and soil. A major cost in basin irrigation is that of land grading or leveling, if required. The cost is directly related to the volume of earth that must be moved. Typical operation and maintenance costs for basin irrigation systems vary greatly, depending on local circumstances and irrigation efficiencies achieved.

Advantage and Disadvantage

Advantages

 (i) One of the major advantages of the basin method is its utility in irrigating fields with irregular shapes and small fields
 (ii) Best suited for lands/crops where leaching is required to wash out salts from the root zone
 (iii) Water application and distribution efficiencies are generally high

Limitations

 (i) It requires accurate land leveling to achieve high application efficiency
 (ii) Comparatively high labor intensive
 (iii) Impedes surface drainage
 (iv) Not suitable for crops which are sensitive to waterlogging
 (v) Border ridges interfere with the free movement of farm machineries
 (vi) Higher amount of water is required compared to sprinkler or drip irrigation.

2.3.3 Furrow Irrigation

2.3.3.1 Concepts and Features

Furrow irrigation is one of the oldest controlled irrigation methods. A furrow is a small, evenly spaced, shallow channel installed down or across the slope of the field to be irrigated parallel to row direction (Fig. 2.3). In this method, water is applied to furrows using small discharges to favor water infiltration while advancing down the field. Furrow irrigation can thus be defined as a partial surface flooding method of irrigation (normally used with clean-tilled crops), where water is applied in furrows or rows of sufficient capacity to obtain the designed irrigation system.

Fig. 2.3 Schematic of furrow irrigation system

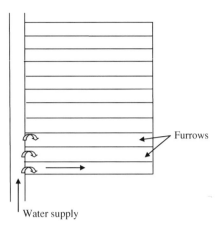

The furrow method is an efficient system if properly managed, but a most inefficient one if improperly managed. For this method, fields must have a mild slope and inflow discharge must be such that advance is not too fast and produce excessive runoff losses, nor too slow to induce excessive infiltration in the upper part of the field. Short blocked furrows with manually controlled water applications are practiced by traditional irrigators. Nowadays, long and precisely leveled furrows with automated or semi-automated control have become increasingly popular.

2.3.3.2 Suitability and Limitations

Crop Suitability

Furrow irrigation is best used for irrigating widely spaced row crops such as potato, maize, vegetables, and trees.

Soil and Topography

Loam soil is best suited for furrow irrigation. Sandy soils can cause excess infiltration at the upper end of the furrow, clay soils may need extra standing water to infiltrate.

Steeper land compared to basin or border (mild slopy topography, 0.5–2%) is needed to establish furrow irrigation. Undulated/zigzag topography is not suited for this method.

Water Quantity

Depending on surface conditions, stream size used should be as large as possible to move water through the field quickly without causing erosion.

2.3 Description of Common Methods of Irrigation

Attainable Efficiencies

When using properly designed row slopes, row lengths, set times, stream sizes, and a reuse system, furrow irrigation efficiency can be high as 90%.

Labor and Energy Requirement

To establish the furrow, some labors are required. After that, the least labor required among the surface methods. If the system is automated, the labor requirement is reduces to minimum.

Required Depth of Irrigation Application

If only little water is to be applied per application, e.g., on sandy soils and a shallow rooting crop, furrow irrigation would be most appropriate (However, none of the surface irrigation methods can be used if the sand is very coarse, i.e., if the infiltration rate is more than 30 mm/h).

Cost and Economic Factor

Major initial cost in furrow system is the construction of furrows. The cost is directly related to the number of furrows (i.e. furrow spacing), volume of soil to be removed, and the unit labor/instrument charge.

Level of Technology

Furrow irrigation, with the possible exception of short, level furrows, requires accurate field grading. This is often done by machines. The maintenance – plowing and furrowing – is also often done by machines. This requires skill, organization, and frequently the use of foreign currency for fuel, equipment, and spare parts.

Advantages and Disadvantages

Advantages

(i) developed gradually as labor or economics allows
(ii) developed at a relatively low cost after necessary land-forming activities are accomplished
(iii) erosion is minimal
(iv) adaptable to a wide range of land slopes

Limitations

(i) Not suitable for high permeable soil where vertical infiltration is much higher than the lateral entry
(ii) Higher amount of water is required, compared to sprinkler or drip irrigation
(iii) Furrows should be closely arranged

2.3.4 Sprinkler Irrigation Systems

2.3.4.1 Concept and Features

In sprinkler irrigation, water is delivered through a pressurized pipe network to sprinklers, nozzles, or jets which spray the water into the air, to fall to the soil as an artificial "rain" (Fig. 2.4). Sprinkler irrigation can be defined as a pressurized system, where water is distributed through a network of pipe lines to and in the field and applied through selected sprinkler heads or water applicators.

The basic components of any sprinkler system are

- a water source
- a pump (to pressurize the water)
- a pipe network (to distribute the water throughout the field)
- sprinklers (to spray the water over the ground) and
- valves (to control the flow of water)

In addition, flow meters and pressure gauges are sometimes added to monitor system performance.

The sprinklers, when properly spaced, give a relatively uniform application of water over the irrigated area. Sprinkler systems are usually (there are some

(a)

(b)

Fig. 2.4 Sprinkler system (**a**) view of a sprinkler and (**b**) sprinklers irrigating a field

2.3 Description of Common Methods of Irrigation

exceptions) designed to apply water at a lower rate than the soil infiltration rate so that the amount of water infiltrated at any point depends upon the application rate and time of application but not the soil infiltration rate.

Sprinkler irrigation systems are normally used under more favorable operational conditions than surface systems because farmers may control the discharge rates, duration, and frequency. Many sprinkler systems have independent water supply or are connected to networks which may be operated on demand. However, the pressure from the hydrants or farm pumps is often not appropriate resulting in lower (or higher) discharges than those envisaged during the design phase. Pressure head (and discharge) variations at the hydrant should be identified by the user when appropriate equipment is available.

Sprinklers can be moved manually to ensure an even distribution of water over the ground, but a series of small fixed sprinklers are commonly used in an irrigation system. The "throw" of a sprinkler is the area of land which receives water from it, and sprinklers are placed "head to head," meaning that they are placed sufficiently close together so that there are no gaps of dry land between them.

2.3.4.2 Types of Sprinkler Systems

Many types of sprinkler devices and sprinkler systems are available. Sprinkler irrigation systems exist in various shapes, sizes, costs, and capabilities. Descriptions of the more common types are given below.

Portable (or Hand-Move) Sprinkler System

These systems employ a lateral pipeline with sprinklers installed at regular intervals. The lateral pipe is often made of aluminum, with 20-, 30-, or 40-feet sections, and special quick-coupling connections at each pipe joint. The sprinkler is installed on a pipe riser so that it may operate above the crop being grown (in orchards, the riser may be short so that these types of sprinklers operate under the tree canopy). The risers are connected to the lateral at the pipe coupling, with the length of pipe section chosen to correspond to the desired sprinkler spacing. The sprinkler lateral is placed in one location and operated until the desired water application has been made. Then the lateral line is disassembled and moved to the next position to be irrigated. This type of sprinkler system has a low initial cost but a high labor requirement. It can be used on most crops, though with some, such as corn, the laterals become difficult to move as the crop reaches maturity.

Solid Set and Permanent Systems

Sprinklers irrigate at a fixed position. Solid set systems are similar in concept to the hand-move lateral sprinkler system, except that enough laterals are placed in the field so that it is not necessary to move the pipe during the season. The laterals are controlled by valves, which direct the water into the laterals irrigating at any particular moment. The pipe laterals for the solid set system are moved into the

Fig. 2.5 Permanent system sprinkler

field at the beginning of the season and are not removed until the end of the irrigation season. The solid set system utilizes labor at the beginning and ends of the irrigation season but minimizes labor needs during the irrigation season.

A permanent system is a solid set system where the main supply lines and the sprinkler laterals are buried (Fig. 2.5) and left in place permanently (this is usually done with PVC plastic pipe).

Side Roll System

The side roll sprinkler system is best suited for rectangular fields. The lateral line is mounted on wheels, with the pipe forming the axle (Fig. 2.6). The wheel height

Fig. 2.6 Side roll sprinkler system

2.3 Description of Common Methods of Irrigation

is selected so that the axle clears the crop as it is moved. A drive unit (usually an air-cooled gasoline-powered engine located near the center of the lateral) is used to move the system from one irrigation position to another by rolling the wheels.

Traveling Gun System

This system utilizes a high volume, high pressure sprinkler (called "gun") mounted on a trailer, with water being supplied through a flexible hose or from an open ditch along which the trailer passes (Fig. 2.7). The gun may be operated in a stationary position for the desired time and then moved to the next location. However, the most common use is as a continuous move system, where the gun sprinkles as it moves. The gun used is usually a part-circle sprinkler, operating through 80–90% of the circle for best uniformity and allowing the trailer to move ahead on dry ground. These systems can be used on most crops, though due to the large droplets and high application rates produced, they are best suited to coarse soils having high intake rates and to crops providing good ground cover.

Fig. 2.7 Traveling gun type sprinkler

Center Pivot and Linear Move Systems

The center pivot system consists of a single sprinkler lateral supported by a series of towers. The towers are self-propelled so that the lateral rotates around a pivot point in the center of the irrigated area (Fig. 2.8). The time for the system to revolve through one complete circle can range from a half a day to many days. The longer the lateral, the faster the end of the lateral travels and the larger the area irrigated by the end section. Thus, the water application rate must increase with distance from the pivot to deliver an even application amount. A variety of sprinkler products

Fig. 2.8 Center pivot sprinkler system

have been developed specifically for use on these machines to better match water requirements, water application rates, and soil characteristics. Since the center pivot irrigates a circle, it leaves the corners of the field unirrigated (unless additions of special equipment are made to the system). Center pivots are capable of irrigating most field crops but have on occasion been used on tree and vine crops. Linear move systems are similar to center pivot systems in construction, except that neither end of the lateral pipeline is fixed. The whole line moves down the field in a direction perpendicular to the lateral.

Continuous Move Laterals

These systems are well adapted to apply to small and frequent irrigations.

LEPA Systems

Low Energy Precision Application (LEPA) systems are similar to linear move irrigation systems but are different enough to deserve separate mention of their own. The lateral line is equipped with drop tubes and very low pressure orifice emission devices discharging water just above the ground surface into furrows. This distribution system is often combined with micro-basin land preparation for improved runoff control (and to retain rainfall which might fall during the season). High-efficiency irrigation is possible but requires either very high soil intake rates or adequate surface storage in the furrow micro-basins to prevent runoff or nonuniformity along a furrow.

2.3.4.3 Capabilities and Limitations of Sprinkler System

Soil Type

Sprinklers adapt to a range of soil and topographic conditions. Light sandy soils are well suited to sprinkler irrigation systems. Most soils can be irrigated with the

2.3   Description of Common Methods of Irrigation   51

sprinkler method, although soils with an intake rate below 0.2 in./h may require special measures. Sprinklers are applicable to soils that are too shallow to permit surface shaping or too variable for efficient surface irrigation.

Field Shape and Topography

In general, sprinklers can be used on any topography that can be farmed or cropped. Land leveling is not normally required. Odd-shaped fields cannot be easily irrigated with certain types of sprinkler systems such as center pivots.

Crops

Nearly all crops can be irrigated with some type of sprinkler system, though the characteristics of the crop, especially the height, must be considered in system selection. Sprinklers are sometimes used to germinate seed and establish ground cover for crops like lettuce, alfalfa, and sod. The light, frequent applications that are desirable for this purpose are easily achieved with some sprinkler systems.

Water Quantity and Quality

Leaching salts from the soil for reclamation can be done with sprinklers using much less water than is required by flooding methods (although a longer time is required to accomplish the reclamation). This is particularly important in areas with a high water table.

Efficiencies

Both the center pivot and the linear move systems are capable of very high efficiency water application.

Financial Involvement and Labor Requirement

Sprinkler irrigation requires high capital investment but has low irrigation labor requirements.

Advantages and Disadvantages

Advantages of sprinkler systems include the following: readily automatable, facilitates to chemigation and fertigation, reduced labor requirements needed for irrigation. LEPA type systems can deliver precise quantities of water in a highly efficient manner and are adaptable to a wide range of soil and topographic conditions.

   A disadvantage of sprinkler irrigation is that many crops (citrus, for example) are sensitive to foliar damage when sprinkled with saline waters. Other disadvantages of sprinkler systems are the initially high installation cost and high maintenance cost thereafter (when needed).

2.3.4.4 Choosing a Sprinkler Type

When choosing a sprinkler type for irrigation, there are several considerations:

- Adaptability to crop, terrain, and field shape
- Labor availability and requirements
- Economics
- Automation facility
- Ability of the system to meet crop needs

2.3.5 Drip Irrigation

2.3.5.1 Concept and Features

Drip irrigation system is traditionally the application of a constant steady flow of water to soil at low pressure. In this system, water is applied directly to the root zone of plants by means of applicators (orifices, emitters, porous tubing, perforated pipe, etc.) operated under low pressure with the applicators being placed either on or below the surface of the ground (Fig. 2.9). Water loss is minimized through these measures, as there is very little splash owing to the low pressure and short distance to the ground.

Drip systems tend to be very efficient and can be totally automated. Of the irrigation systems available, drip is the most ideally suited to high-value crops such as the vegetables and fruits. Properly managed systems enable the production of maximum yields with a minimum quantity of water. These advantages often help justify the high costs and management requirements. A typical drip irrigation system is shown in Fig. 2.9. There are many types of drip products on the market designed to meet the demands for just about any application.

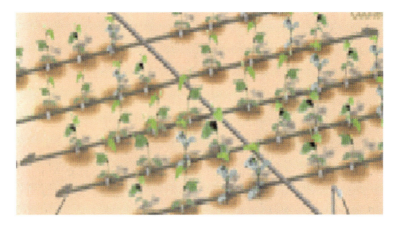

Fig. 2.9 Drip irrigation system

2.3.5.2 Suitabilities, Capabilities, and Limitations

Crop Suitability

Drip irrigation is most suited to high-density orchards, tree crops, and high-value horticultural crops. Drip systems allow accurate amounts of water to be supplied regularly to a small area of the root zone. Such a system can be used to restrict the vegetative growth of the trees, an important part of management in high density planting.

Drip irrigation is more suited to areas where cooler climates and higher rainfall reduce the need for high volumes of water application.

Water Supply

Drip irrigation is not designed for applying water to large root systems. To obtain adequate water distribution and application rates, two to three dripper lines per row of trees are required. As only a small area of the total field is wetted, drip irrigation is especially suited for situations where the water supply is limited. Drip tubing is used frequently to supply water under plastic mulches.

Fertilizer Application

Applying nutrients (fertilizers) through the drip system is very effective and may reduce the total amount of fertilizer needed. Some chemical insecticides can also be efficiently applied (precision of amounts) via drip irrigation system. This can lead to significant savings in money and maintenance time of the garden and field. In this case, care must be taken to ensure that the product is suitable for this type of application and will not damage the irrigation system.

Utilities of Buried Drip System

Burying the drip system reduces water loss even further by preventing runoff across the surface, which can occur at very high rates on dry impervious ground. It also reduces the chance of damaging the system while weeding. The soil surface is also kept dry, which can reduce invasion by weeds.

Attainable Efficiency

It is the most efficient irrigation system as the water is supplied directly to the root system, an important consideration where water supplies are limited.

Advantages and Disadvantages

There are several disadvantages and potential problems with a drip irrigation system. Costs of the product and its installation can be relatively high compared to more simple alternatives, although these may eventually be outweighed by savings in water bills. The systems are vulnerable to blockage by organic matter, either in the water supply or algal growth in the pipes themselves. Chemicals and filtering systems can be used to minimize these problems. In a wide spaced orchard, supplying large trees with sufficient water can pose problems with a drip system, particularly in the 4 weeks prior to harvest.

The advantages of drip irrigation are as follows:

- Highly efficient system
- Saves water
- Limited water sources can be used
- Correct volume of water can be applied in the root zone
- The system can be automated and well adapted to chemigation and fertigation
- Reduces nutrient leaching, labor requirement, and operating cost
- Other field operations such as harvesting and spraying can be done while irrigating
- Each plant of the field receives nearly the same amount of water
- Lower pressures are required to operate systems resulting in a reduction in energy for pumping

The disadvantages/limitations of the drip system are as follows:

- High initial cost
- Technical skill is required to maintain and operate the system
- The closer the spacing, the higher the system cost per hectare
- Damage to drip tape may occur
- Cannot wet the soil volume quickly (to recover from moisture deficit) as other systems
- Facilitates shallow root zone
- Needs clean water

2.3.6 Other Forms of Irrigation

Besides the above-mentioned methods, other categories of water application methods include the following:

- Hand watering
- Capillary irrigation
- Localized irrigation
- Trickle irrigation
- Micro-irrigation
- Subsurface irrigation

Hand Watering

The hand watering method is probably the most basic or earliest type of irrigation method. Water is applied to the plant root zone (close to or directly at the root area)

2.3 Description of Common Methods of Irrigation

by means of a container or bucket. In the present age of automation, people do not consider hand watering a viable alternative. However, many horticultural enterprises, such as nurseries and fruit trees, cannot use the automated fixed irrigation system efficiently due to the random location of the plants and therefore use hand watering.

Capillary Irrigation

Water is applied beneath the root zone in such a manner that it wets the root zone by capillary rise. Buried pipes or deep surface canals are used for this purpose.

Localized Irrigation

Water is applied around each plant or a group of plants so as to wet locally and the root zone only. The application rate is adjusted to meet evapotranspiration needs so that percolation losses are minimized.

Trickle Irrigation

The term trickle irrigation is general and includes several more specific methods. Trickle irrigation is the slow, frequent application of water to the soil through emitters placed along a water delivery line. It includes drip irrigation, subsurface irrigation, and bubbler irrigation.

Trickle irrigation is best suited for tree, vine, and row crops. The main limitation is the cost of the system, which can be quite high for closely spaced crops. Complete cover crops, such as grains or pasture cannot be economically irrigated with trickle systems. Trickle irrigation is suitable for most soils, with only the extremes causing any special concern. With proper design, using pressure compensating emitters and pressure regulators if required, trickle irrigation can be adapted to virtually any topography. In some areas, trickle irrigation is successfully practiced on such steep slopes that cultivation becomes the limiting factor.

Trickle irrigation uses a slower rate of water application over a longer period of time than other irrigation methods. The most economical design would have water flowing into the farm area throughout most of the day and every day during peak use periods. If water is not available on a continuous basis, on-farm water storage may be necessary. Trickle irrigation can be used successfully with waters of some salinity, although some special caution is needed. Salts will tend to concentrate at the perimeter of the wetted soil volume.

Subsurface Irrigation

Applying irrigation water below the ground surface either by raising the water table within or near the root zone or by using a buried perforated or porous pipe system that discharges directly into the root zone is termed subsurface irrigation.

2.4 Selection of Irrigation Method

Decision must be made regarding the type of irrigation method before an irrigation system is installed in a field. To choose an appropriate irrigation method, one must know the advantages and disadvantages of the various methods. He or she must know which method suits the local conditions best. Unfortunately, in many cases there is no single best solution, as all methods have their relative advantages and disadvantages. Trials of the various methods under the prevailing local conditions provide the best basis for a sound choice of irrigation method.

2.4.1 Factors Affecting Selection of an Irrigation Method

Factors determining irrigation method are some in technical nature, some economic, and others involve a close scrutiny of the operation and crop to be irrigated.

In selecting an irrigation method, the following factors should be considered:

- Soil type
- Field shape/geometry and topography
- Climate – evaporation rates, wind, and rainfall
- Water availability and its price
- Water quality
- The particular crop to be grown – physical requirements, crop layout, and water use characteristics
- Required depth and frequency of irrigation application
- Labor requirements and its availability
- Energy requirement
- Economic factor – cost–benefit ratio, initial investment
- Compatibility with existing farm equipments
- Attainable irrigation efficiency of the proposed system
- Relative advantages and disadvantages of the available systems
- Type/level of technology at the locality
- Cultural factor/previous experience with irrigation
- Automation capacity
- Fertigation capability
- Environmental conditions – impact and regulations
- Farm machinery and equipment requirements

Soil Type

Light sandy soils are not well suited to furrow or basin irrigation systems. Sandy soils have a low water storage capacity and a high infiltration rate. They therefore need frequent but small irrigation applications, in particular when the sandy soil is also shallow. Under these circumstances, sprinkler or drip irrigation are more suitable than surface irrigation. On loam or clay soils all three irrigation methods

2.4 Selection of Irrigation Method

can be used, but surface irrigation is more commonly used. Clay soils with low infiltration rates are ideally suited to surface irrigation.

When a variety of soil types exists within one irrigation scheme, sprinkler or drip irrigation is recommended as they will ensure a more even water distribution. Sprinkler or drip irrigation are preferred to surface irrigation on steeper or unevenly sloping lands, as they require little or no land leveling. An exception is rice grown on terraces on sloping lands.

Field Shape/Geometry and Topography

Topography of a field is a decision-making aid in the selection of the type of irrigation system, or in determination of size of the irrigation system to be installed. Sprinklers fit rolling topography, but surface irrigation systems require graded fields. Odd-shaped fields cannot be easily irrigated with certain types of sprinkler systems such as center pivots. Rolling topography prohibits the use of furrow or surface systems because water cannot run uphill. Basins can be adopted in irrigating fields with irregular shapes and small sizes.

Climate

Local climate greatly influences the choice of an irrigation system. In a very hot, dry climate, a significant amount of water is evaporated during irrigation through sprinklers. Strong wind can disturb the uniform distribution of water from sprinklers. Under very windy conditions, drip or surface irrigation methods are preferred. In areas of supplementary irrigation, sprinkler or drip irrigation may be more suitable than surface irrigation because of their flexibility and adaptability to varying irrigation demands on the farm.

Water Availability

An adequate water supply to meet crop demand is important for ease of operation and for management of an irrigation system. With low probability of rainfall, a water supply should be large enough to meet crop demand. Location of water source, quantity, and quality of water should be determined before any type of irrigation system is selected. No assumptions should be made with the water supply. The amount of water available and the cost of the water (due to pumping or direct purchase) will determine the type of system you should use. If the supply is sufficient, assured and low cost, labor and/or energy saving irrigation method may be employed. On the other hand, if the supply is scarce/limited and very expensive, then consider only the most efficient type of irrigation system (e.g., sprinkler, drip).

Water Quality

Surface irrigation is preferred if the irrigation water contains much sediment. The sediments may clog the drip or sprinkler irrigation systems. Water having high salt content may cause foliar damage if sprayed directly on the plants (e.g., sprinkler irrigation). In this case, consider systems that deliver water directly on or below the surface such as drip, surface, or LEPA (low energy precision agriculture) systems. In these methods, less water is applied to the soil (and hence less salt) than with surface methods. Special consideration is also needed in the placement of drip tubing and emitters when irrigating with saline water.

Labor Requirement

The labor requirement and skill required for operation and maintenance varies greatly between systems. Labor availability and cost are prime considerations for a labor-intensive system. For example, studies have shown that about one-man-hour per acre is required for a hand-move sprinkler system. Mechanical move systems require 1/10 to 1/2 as such labor. Automated systems are more expensive but may be more profitable when the labor costs over the life of the system are considered.

Surface irrigation often requires a much higher labor input – for construction, operation, and maintenance – than sprinkler or drip irrigation. Surface irrigation requires accurate land leveling, regular maintenance, and a high level of farmers' organization to operate the system. Sprinkler and drip irrigation require little land leveling; system operation and maintenance are also less labor intensive.

Energy Requirement

With surface irrigation, little or no energy is required to distribute the water throughout the field, but some energy may be needed to bring the water to the field, especially when water is pumped from the ground. In some instances, these energy costs can be substantial, particularly with low water use efficiencies. Some labor and energy will be necessary for land grading and preparation.

Economic Factors

Costs and Benefits

Before choosing an irrigation system, an estimate must be made of the costs and benefits of the available options. On the cost side, not only the construction and installation but also the operation and maintenance (per hectare) should be taken into account. These costs should then be compared with the expected benefits (price of yields). It is obvious that farmers will only be interested in implementing a certain method if they consider this economically attractive (higher benefit–cost ratio).

Initial Investment/Development Cost

Although a method is found to be economical, it cannot be implemented due to limitation of fund for initial development cost. Sprinkler and drip systems require higher initial costs. Among surface irrigation configurations, basin irrigation is generally expensive to develop and maintain.

2.4 Selection of Irrigation Method

Compatibility with Existing Farm Equipments

Choose a system that is compatible with your farming operations, equipment, field conditions, and crops and/or crop rotation plan.

Attainable Irrigation Efficiency of the System

Water application efficiency is generally higher with sprinkler and drip irrigation than surface irrigation, so these methods are preferred when water is in short supply. However, it must be remembered that efficiency is just as much a function of the irrigator as the method used.

Relative Advantage and Disadvantages of the Available Methods

Several irrigation systems are on the market. Each system has advantages and disadvantages. A discussion of suitability/capabilities of different systems has been explained earlier. These points should be taken into consideration.

Crop to Be Irrigated/Type of Crop

The crop type influences the selection of the irrigation method. Surface irrigation can be used for all types of crops. Sprinkler and drip irrigation, because of their high capital investment per hectare, are mostly used for high-value cash crops, such as vegetables and fruit trees. They are seldom used for the lower value staple crops. Widely spaced crops do not require total field soil volume to be wetted, and thus basin or border irrigation in this case is less useful. Instead, a mini-basin can be formed around each tree. Drip irrigation is suited to irrigating individual plants or trees or row crops. It is not suitable for close growing crops (e.g., paddy). Paddy (rice) is always grown in basins. Many other crops can also be grown in basins (e.g., maize, sorghum). If paddy is the major crop, basins will be the logical choice. Those crops that cannot stand a very wet soil for more than 12–24 h should not be grown in basins. Row crops such as maize, vegetables, and trees are best suited to furrow irrigation. Close growing crops such as wheat, mustard, and alfalfa are best suited to border irrigation.

Required Depth and Frequency of Irrigation Application

The depth of water (mm) required per irrigation and seasonal total water requirement influence the irrigation method. Field experience has shown that most water can be applied per irrigation application when using basin irrigation, less with border irrigation, and least with furrow irrigation. Usually 40–70 mm of water is applied in basin irrigation, 30–60 mm in border irrigation, and 20–50 mm in furrow irrigation (in large-scale irrigation projects, the amounts of water applied may be much higher). This means that if only little water is to be applied per application, e.g., on sandy soils and a shallow rooting crop, furrow irrigation would be most appropriate.

On the other hand, if a large amount of irrigation water is to be applied per application, e.g., on a clay soil and with a deep rooting crop, border or basin irrigation would be more appropriate.

Farm Machinery and Equipment Requirement

If an irrigation system requires heavy farm machinery and equipment to install and for maintenance, it will be less preferred by the irrigators having low- and medium-level technology.

Level of Technology at the Locality

The level of technology in the locality affects the choice of irrigation method. In general, drip and sprinkler irrigation are technically more complicated methods. The purchase of equipment requires high capital investment per hectare. To maintain the equipment a high level of "know-how" has to be available. Also, a regular supply of fuel and spare parts must be maintained which, together with the purchase of equipment, may require foreign currency.

Surface irrigation systems, in particular small-scale schemes, usually require less sophisticated equipment for both construction and maintenance (unless pumps are used). The equipment needed is often easier to maintain and less dependent on the availability of foreign currency.

Basin irrigation is the simplest of the surface irrigation methods. Especially if the basins are small, they can be constructed by hand or animal traction. Their operation and maintenance is simple. Furrow irrigation, with the possible exception of short, level furrows, requires accurate field grading. This is often done by machines. The maintenance – plowing and furrowing – is also often done by machines. This requires skill, organization, and frequently the use of foreign currency for fuel, equipment, and spare parts. All these factors affect the selection process of irrigation methods.

Tradition/Previous Experience with Irrigation

The choice of an irrigation method also depends on the irrigation tradition within the region or country. Introducing a previously unknown method may lead to unexpected complications. It is not certain that the farmers will accept the new method. Most irrigators tend to stay with practices that have been used previously in their area rather than take the risk associated with a new technology. The uncertainties with the new method include the following: the servicing of the equipment may be problematic and the costs may be high compared to the benefits. Often it is easier to improve the traditional irrigation method than to introduce a totally new method.

Personal Preference/Cultural Factor

Select a system that you can live with. If you do not like your system, chances are you will not operate or maintain it properly.

2.4 Selection of Irrigation Method

Table 2.1 Comparison of irrigation systems in relation to site and situation factors

Site and situation factors	Suitability/preferred factor under the irrigation system					
	Basin	Border	Furrow	Sprinkler	Drip	Sub-irrigation
Soil	Loam to heavy soil	Loam to heavy soil	Loam to heavy soil	Sandy soil	Sandy soil	Sandy soil
Infiltration rate	Moderate to low	Moderate	Moderate	Moderate to high	Moderate to high	Moderate to high
Topography	Flat/nearly level ground	Flat-to-small slope	Flat-to-small slope	All category (flat to rolling)	Flat	Flat
Crop	Close growing crops, suited to standing water	Close growing crops, not suited to standing water	Widely spaced row crops	Generally short crops	Widely spaced row crops, generally high value crops	Row crops
Water supply/stream size	Large stream	Medium-to-large stream	Medium stream	Small stream	Small stream	Small stream
Water quality	All category	All category	All category	Clean water	Clean water	Clean water
Windy climate	No problem	No problem	No problem	Problem	No problem	No problem
Attainable irrigation efficiency	80–90%	70–85%	65–75%	85–95%	85–95%	85–95%
Capital required/initial investment	Medium cost for establishment of basin	Low cost	Medium cost	High initial cost	High initial cost	High initial cost
Labor requirement	High for establishment, but low for operation	Medium	Medium, low if automated	Medium, low if automated	Low	High for establishment, but low for operation

Table 2.1 (continued)

Site and situation factors	Suitability/preferred factor under the irrigation system					
	Basin	Border	Furrow	Sprinkler	Drip	Sub-irrigation
Energy requirement	No energy required (only if groundwater is to be supplied)	No energy required (only if groundwater is to be supplied)	No energy required (only if groundwater is to be supplied)	Energy required	Energy required	No energy required
Skill required	Skill required to establish basin	No skill required	Moderate skill required	High skill required	High skill required	High skill required
Epidemic diseases	No problem	No problem	No problem	Problem	No problem	No problem
Operation and maintenance	Easy; low operation and maintenance cost	Easy; low operation and maintenance cost	Easy; low operation and maintenance cost	Not easy, require skill; high operation and maintenance cost	Not easy, require skill; high operation and maintenance cost	Not easy, require skill; high operation and maintenance cost

2.4.2 Selection Procedure

To choose an irrigation method, the farmer must know the advantages and disadvantages of the various methods. He or she must know which method suits the local conditions best. Unfortunately, in many cases, there is no single best solution: all methods have their advantages and disadvantages. Testing of the various methods under the prevailing local conditions provides the best basis for a sound choice of irrigation method. Based on the local soil, climate, crop and water availability, and the suitability and limiting criteria of the methods (described in earlier sections, and also summarized in Table 2.1), the irrigation engineer will prescribe the appropriate method for the particular area.

Relevant Journals

– Irrigation Science
– Agricultural Water Management
– Irrigation and Drainage System
– Journal of Irrigation and Drainage Division, ASCE
– Transactions of the American Society of Agricultural Engineers
– ICID Bulletins
– Agronomy Journal

Questions

(1) What are the different methods of applying water to crops?
(2) Describe in brief the characteristic features, suitabilities, and limitations of the following irrigation methods: (a) Border, (b) Basin, (c) Furrow, (d) Sprinkler, (e) Drip, and (f) Trickle.
(3) Describe the factors influencing selection of an irrigation method.
(4) Compare different irrigation systems in relation to site and different situation factors.
(5) As an irrigation engineer, you are asked to advise regarding irrigation method in a new irrigable farming area. What points will you consider and what steps will you follow to materialize your job.

Chapter 3
Irrigation System Designing

Contents

3.1	Some Common Issues in Surface Irrigation System	66
	3.1.1 Design Principle of Surface Irrigation System	66
	3.1.2 Variables in Surface Irrigation System	67
	3.1.3 Hydraulics in Surface Irrigation System	67
3.2	Border Irrigation System Design	68
	3.2.1 Definition of Relevant Terminologies	68
	3.2.2 General Overview and Considerations	69
	3.2.3 Factors Affecting Border Performance and Design	70
	3.2.4 Design Parameters	70
	3.2.5 Design Approaches and Procedures for Border	71
	3.2.6 Sample Workout Problems	73
	3.2.7 Simulation Modeling for Border Design	76
	3.2.8 Existing Software Tools/Models for Border Irrigation Design and Analysis	77
	3.2.9 General Guidelines for Border	79
3.3	Basin Irrigation Design	79
	3.3.1 Factors Affecting Basin Performance and Design	79
	3.3.2 Hydraulics in Basin Irrigation System	81
	3.3.3 Simulation Modeling for Basin Design	82
	3.3.4 Existing Models for Basin Irrigation Design	84
3.4	Furrow Irrigation System Design	84
	3.4.1 Hydraulics of Furrow Irrigation System	85
	3.4.2 Mathematical Description of Water Flow in Furrow Irrigation System	86
	3.4.3 Some Relevant Terminologies	87
	3.4.4 Factors Affecting Performance of Furrow Irrigation System	90
	3.4.5 Management Controllable Variables and Design Variables	91
	3.4.6 Furrow Design Considerations	92
	3.4.7 Modeling of Furrow Irrigation System	92
	3.4.8 General Guideline/Thumb Rule for Furrow Design	94
	3.4.9 Estimation of Average Depth of Flow from Volume Balance	95
	3.4.10 Suggestions for Improving Furrow Irrigations	96

	3.4.11 Furrow Irrigation Models	96
	3.4.12 Sample Worked Out Problems	97
3.5	Design of Sprinkler System	98
	3.5.1 Design Aspects	98
	3.5.2 Theoretical Aspects in Sprinkler System	99
	3.5.3 Sprinkler Design	101
Relevant Journals		106
Relevant FAO Papers/Reports		107
Questions		107
References		109

Irrigation scheduling is the decision process related to "when" to irrigate and "how much" water to apply to a crop. The irrigation method concerns "how" that desired water depth is applied to the field. To achieve high performance in an irrigation system, it must be designed to irrigate uniformly, with the ability to apply the right depth at the right time. Properly designed, installed, maintained and managed irrigation systems greatly reduce the volume of irrigation water and hence save energy and money. Besides, it improves the crop yield and quality. This chapter discusses the detailed design aspects of different types of irrigation system. The design procedures are explained through sample examples.

3.1 Some Common Issues in Surface Irrigation System Designing

The study of surface irrigation can be classified into two basic categories (Alazba, 1997): design and analysis. Determination of the water advance or infiltration advance is an analysis problem, whereas computation of the inflow rate or system layout (e.g., length, width, slope) is a design problem.

The analysis of flow in surface irrigation is complex due to the interactions of several variables such as infiltration characteristics, inflow rate, and hydraulic resistance. The design is more complex due to interactions of input variables and the target output parameters such as irrigation efficiency, uniformity, runoff, and deep percolation.

In most cases, the aim of the surface irrigation system design is to determine the appropriate inflow rates and cutoff times so that maximum or desired performance is obtained for a given field condition.

3.1.1 Design Principle of Surface Irrigation System

The surface irrigation method (border, basin, and furrow) should be able to apply an equal depth of water all over the field without causing any erosion. To minimize the percolation losses, the opportunity time (difference between advance and recession periods) should be uniform throughout the plot and also equal to the time required

3.1 Some Common Issues in Surface Irrigation System Designing

to put the required depth of water into the soil. Runoff from the field can be eliminated through controlling the inflow rate at which inflow decays with time exactly coincides with decay of the average infiltration rate with time for the entire length of the field. Inflow is usually cut back in discrete steps.

3.1.2 Variables in Surface Irrigation System

Important variables in surface irrigation system include the following: (i) infiltration rate, (ii) surface roughness, (iii) size of stream, (iv) slope of land surface, (v) erosion hazard, (vi) rate of advance, (vii) length of run, (viii) depth of flow, (ix) depth of water to be applied, (x) infiltration depth. These are schematically presented in Fig. 3.1.

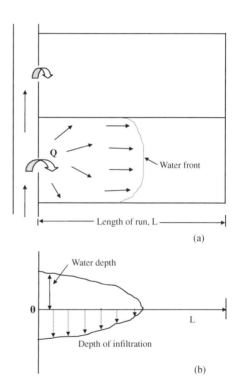

Fig. 3.1 Schematic showing of different variables in a surface irrigation system

3.1.3 Hydraulics in Surface Irrigation System

In general, there are the following three phases of water-front in a surface irrigation system:

- advance
- wetting (or ponding) and
- recession

The advance phase starts when water first enters the field plot and continues up to the time when it has advanced to the end of plot (Fig. 3.2). The period between the time of advance completion and the time when the inflow is cut off or shut off is referred to as wetting or ponding or storage phase. After termination of the inflow, the ponding water or the water-front recedes from the field by draining and/or into the next field by infiltration. This is the recession phase.

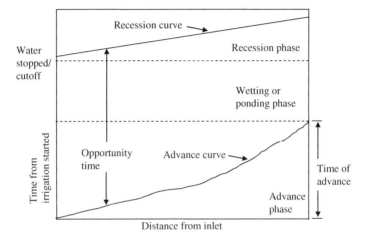

Fig. 3.2 Schematic presentation of phases of water-front in surface irrigation system

Unsteady overland flow analysis is required for the design and management of surface irrigation systems. When sufficient water is released over a porous medium in surface irrigation, part of this water infiltrates into the soil (Fig. 3.1) and the remainder moves over the field as overland flow. Hydraulic analysis of surface flow during all the phases of irrigation from advance to recession is important for successful design and operation of a surface irrigation system.

3.2 Border Irrigation System Design

3.2.1 Definition of Relevant Terminologies

Border strip: The area of land bounded by two border ridges or dikes that guide the irrigation stream from the inlet point of application to the ends of the strip.

Cutback: Reduction of flow rate (partially or fully) is referred to as cutback. The objective of cutback is to reduce runoff from the field.

Cutoff time: Cutoff time is the time at which the supply is turned off, measured from the onset of irrigation. Cutoff time has no impact on advance as long as the former is taken up to equal or is larger than the advance time. Cutoff, however, has an influence on recession. The most important effect of cutoff is reflected on the amount of losses, deep percolation and surface runoff, and hence efficiency as well as adequacy of irrigation.

Advance ratio: Advance ratio quantifies the relative proportion of advance time to that of cutoff time.

Cutback ratio: The cutback ratio represents the ratio of post-advance flow rate to advance flow rate.

Tailwater recovery ratio: Tailwater recovery ratio (R_{TWR}) represents the ratio of the volume of runoff that can be recovered for use in subsequent sets to that of the total volume of surface runoff resulting from a single application, that is,

$$R_{TWR} = \frac{runoff \cdot recovered}{total \cdot runoff}$$

Return flow: Water that reaches a surface water source after release from the point of use, and becomes available for use again.

Irrigation efficiency: It is the ratio of average depth of irrigation water beneficially used to the average depth applied, and normally expressed as percentage.

Application efficiency: It is the ratio of the average depth of irrigation water stored in the crop root zone to the average depth applied, expressed as percentage.

Farm irrigation efficiency: It is the percentage of the water applied at the farm inlet which is stored in the root zone for crop use.

Distribution uniformity: It is a measure of evenness (or unevenness) of application and has a significant effect on application efficiency.

Recession flow: After the inflow stream is cut off, the tail water recedes from the plot downward. This flow of water is termed as recession flow.

3.2.2 General Overview and Considerations

Borders generally are prepared with zero side slope and a small but uniform longitudinal slope not exceeding 1%. The borders are divided by levees running down the slope at uniform spacing. The lower end of the border is opened to a drainage ditch or closed with a levee to create ponding on the end of the border. Levees are pulled across the end on steeper borders. Border irrigation is best adapted to grain and forage crops where there are large areas of flat topography, and water supplies are large. The major investment is land preparation, and border irrigation is relatively inexpensive to operate after installation.

The main design considerations for border irrigation include flow rate, width, length, slope, and outlet conditions. Water requirements range from 15 to 45 gal per min per foot of width depending on length, slope, and soil type of the border. Sometimes crops are planted across borders or planted parallel to the border or parallel to the border with shallow furrows or corrugations to help guide the water down the border, especially if there is a small side slope.

3.2.3 Factors Affecting Border Performance and Design

3.2.3.1 Soil Type and Infiltration Characteristics

Soil infiltration characteristics have the biggest influence on design border inflow, time of inflow for specific field and crop condition, length of border, and application uniformity. Infiltration rate for a soil type and surface texture varies from farm to farm, field to field, and throughout the growing season; typically because of the field preparation, cultural practice (such as weeding, irrigation), and field traffic. To approximate the infiltration amount based upon advance and opportunity time for a border, a correlation is made using cylinder infiltration test data.

3.2.3.2 Border Inflows

For a particular soil and crop condition, the length of border run and width varies with the available flow rate. Insufficient water supply can cause the water advance to prolong, resulting in a reduction of efficiency. On the other hand, time of advance decreases with the increase in flow rate. Thus, an increase in unit inflow rate can reduce deep percolation losses and improve application efficiency.

3.2.3.3 Longitudinal Slope

Optimum longitudinal slope may aid in achieving better uniformity and efficiency.

3.2.3.4 Irrigation Depth

The duration of irrigation is dependent on the depth to be applied.

3.2.4 Design Parameters

The main design parameters for border irrigation system include the following:

- unit flow rate, Q
- length of border, L
- width of border, W
- slope, S
- cutoff time, t_{co}

The slope of a border is controlled by the natural slope of the field and can be modified only if the orientation of the field with respect to the main slope of the land is changed or when land grading is applied. Appropriate slope for border depends on soil (type, profile depth) and crop combination.

3.2.5 Design Approaches and Procedures for Border

3.2.5.1 Approaches

Design parameters of border irrigation system can be determined using either of the following two approaches:

(a) Design discharge (Q) for a predetermined border size and slope (L, W, S)
(b) Design L, W, S for a given Q

In designing surface irrigation system, several difficulties are encountered. This is because the output requirements such as application efficiency, storage efficiency, and distribution uniformity (DU) have interaction with the input parameters. The storage efficiency is expected to be above 95% and DU is above 90%.

3.2.5.2 Empirical Models for Designing Border Irrigation System

SCS Method

To ensure adequate spread of water over the entire border, a minimum allowable inflow rate q_{min} must be used. The following equation was proposed by SCS (USDA, 1974) to estimate q_{min}:

$$q_{min} = (5.95 \times 10^{-6}) \times \frac{L \times S_0^{0.5}}{n} \qquad (3.1)$$

where

q_{min} = discharge per unit width, m³/s/m
L = border length, m
S_0 = border slope, m/m
n = roughness coefficient (0.15–0.25, the higher the rougher, the higher the n value)

When the soil erodibility causes restrictions on q, the maximum allowable inflow rate q_{max} can be obtained using the empirical method proposed by SCS (USDA, 1974), where q_{max} is expressed as a function of field slope S_0 and type of crop, sod, and nonsod, by

$$q_{max} = CS_0^{-0.75} \qquad (3.2)$$

where

q_{max} is in m³/s/m
S_0 = field slope in m/m
C = empirical coefficient equal to 3.5×10^{-4} for sod and 1.7×10^{-4} for nonsod.

When the dike height causes the restrictions on q, the maximum allowable inflow rate can be obtained using Manning's equation:

$$q_{max} = \frac{1}{n} y_{max}^{5/3} S_0^{1/2} \qquad (3.3)$$

where

y_{max} = maximum allowable depth of flow assumed to equal 0.15 m
n = roughness coefficient (~0.4–0.25)
S_0 = field slope (m/m) [0.1–0.5%]

Alazba's Empirical Model

Alazba (1998) derived empirical equation for border inflow rate and application time as follows:

$$q_{apl} = CU_q \frac{L^{1.0562} \times n^{0.1094} \times k^{1.225} \times a^{3.832}}{S^{0.09} \times D_{req}^{0.823}} \qquad (3.4)$$

where

q_{apl} = inflow rate, m³/h/m
CU_q = unit conversion factor, equal to 0.642 for q_{appl} in m³/h/m, L in m, k in m/hª, and D_{req} in m (and 3.39 ×10⁻⁴ for FPS unit)
L = border length in m
k, a = empirical parameter of Kostiakov infiltration equation ($z = kt^a$, z is infiltration depth (m), t is time of infiltration (h))
S_0 = field slope (m/m)
n = roughness coefficient (0.1–0.2)
D_{req} = required depth of infiltration (m)

The application time corresponding to design application rate:

$$T_{apl} = CU_T \times \frac{L^{1.1} \times n^{0.0093} \times S_0^{0.0203} \times k^{0.387} \times D_{req}^{0.952}}{q_{appl}^{1.0885} \times a^{0.75}} \qquad (3.5)$$

where

T_{apl} = application time, h
CU_q = unit conversion factor, equal to 2.5 for T_{appl} in h, L in m, k in m/hª, and D_{req} in m (and 1.47 ×10⁻² for FPS unit)
q_{apl} = inflow in m³/h/m
D_{req} = required depth of infiltration (m)

Other variables are defined earlier.

3.2 Border Irrigation System Design

3.2.6 Sample Workout Problems

Example 3.1

Design a border strip with the following characteristics:

Field length, $L = 180$ m
Field slope, $S_0 = 0.003$
Infiltration family, IF = 0.5
$k = 0.033$ m/h^a
$a = 0.63$
Roughness coefficient, $n = 0.15$
Required depth of infiltration, $D_{req} = 0.08$ m

Solution

We get the unit flow rate to be applied:

$$q_{apl} = CU_q \frac{L^{1.0562} \times n^{0.1094} \times k^{1.225} \times a^{3.832}}{S^{0.09} \times D_{req}^{0.823}}$$

Here, given

$L = 180$ m
$n = 0.15$
$k = 0.033$ m/h^a
$a = 0.62$
$S = 0.003$ m/m
$D_{req} = 0.08$ m
CU_q = conversion factor. To convert into SI unit, $CU_q = 0.642$

(i) Putting the values, we obtain,

$$q_{apl} = 0.642 \times \frac{(180)^{1.0562}(0.15)^{0.1094}(0.033)^{1.225}(0.63)^{3.832}}{(0.003)^{0.09}(0.08)^{0.823}}$$

$$= 4.157 \text{ m}^3/\text{h-m (Ans.)}$$

(ii) We get application time for designed flow rate:

$$T_{apl} = CU_T \times \frac{L^{1.1} \times n^{0.0093} \times S_0^{0.0203} \times k^{0.387} \times D_{req}^{0.952}}{q_{appl}^{1.0885} \times a^{0.75}}$$

Given,

$q_{apl} = 4.157$ m^3/h-m
$CU_T = 2.5$ (coefficient to convert into SI unit)

Putting the values,

$$T_{apl} = 2.5 \times \frac{(180)^{1.1}(0.1)^{0.0093}(0.003)^{0.0203}(0.033)^{0.387}(0.08)^{0.952}}{(4.157)^{1.0885}(0.62)^{0.75}}$$

$$= 4.76 \text{ h (Ans.)}$$

Example 3.2

In an irrigation command, it is decided to implement border irrigation system. From the field observation, the following field characteristics are gathered:

- Field length, $L = 250$ m
- Field slope, $S_0 = 0.005$
- Infiltration family, IF $= 0.5$
  $k = 0.04$ m/h^a
  $a = 0.6$
- Roughness coefficient, $n = 0.12$
- Required depth of infiltration, $D_{req} = 0.11$ m

Design a border strip with the above information.

Solution

We get unit flow rate to be applied:

$$q_{apl} = CU_q \frac{L^{1.0562} \times n^{0.1094} \times k^{1.225} \times a^{3.832}}{S^{0.09} \times D_{req}^{0.823}}$$

Here, given

$L = 250$ m
$n = 0.12$
$k = 0.04$ m/h^a
$a = 0.6$
$S = 0.005$ m/m
$D_{req} = 0.09$ m
$CU_q =$ conversion factor. To convert into SI unit, $CU_q = 0.642$

(a) Putting the values, we obtain

$$q_{apl} = 0.642 \times \frac{(250)^{1.0562}(0.12)^{0.1094}(0.04)^{1.225}(0.60)^{3.832}}{(0.005)^{0.09}(0.09)^{0.823}}$$

$$= 5.5546 \text{ m}^3/\text{h - m (Ans.)}$$

$$= 1.543 \text{ l/s - m}$$

3.2 Border Irrigation System Design

(b) We get application time for designed flow rate:

$$T_{apl} = CU_T \times \frac{L^{1.1} \times n^{0.0093} \times S_0^{0.0203} \times k^{0.387} \times D_{req}^{0.952}}{q_{appl}^{1.0885} \times a^{0.75}}$$

Given,

$q_{apl} = 5.5546$ m³/h-m
$CU_T = 2.5$ (coefficient to convert into SI unit)
Putting the values,

$$T_{apl} = 2.5 \times \frac{(250)^{1.1}(0.12)^{0.0093}(0.005)^{0.0203}(0.04)^{0.387}(0.09)^{0.952}}{(5.5546)^{1.0885}(0.6)^{0.75}}$$

$= 6.22\text{h (Ans.)}$

Example 3.3

In a wheat field, a farmer has made border strip of 150 m length. The roughness of the field is estimated as $n = 0.10$, and the average field slope along the border is 0.08%. Determine the required flow rate per unit width of the border.

Solution

Minimum flow rate based on length of run, field slope, and roughness is (SCS):

$$q_{min} = (5.95 \times 10^{-6}) \times \frac{L \times S_0^{0.5}}{n}$$

Given,

$L = 150$ m
$S_0 = 0.08\% = 0.008$
$n = 0.10$

Putting the values in the above equation, we get

$$q_{min} = 0.000798 \text{ m}^3/\text{s - m (Ans.)}$$

Example 3.4

A soybean field is to be designed for irrigation through border method. The field configuration allows 300 m long border strip. The grade of the field toward the field drainage channel is 0.3%. Calculate the unit flow rate based on minimum criteria.

Solution

We get

$$q_{min} = (5.95 \times 10^{-6}) \times \frac{L \times S_0^{0.5}}{n}$$

Here

$L = 300$ m
$S_0 = 0.3\% = 0.003$
Assuming $n = 0.15$
We get $q_{min} = 0.000978$ m³/s-m (Ans.)

Check

According to SCS, max. non-erosive flow rate $q_{max} = CS_0^{-0.75}$
Taking $C = 0.00017$, $S_0 = 0.003$; $q_{max} = 0.013262$ m³/s-m
Assuming limiting case for ponding (~0.15 m) and adopting Manning's formulation,

$$q_{max} = \frac{1}{n} y_{max}^{5/3} S_0^{1/2}$$

$n = 0.1$
$y_{max} = 0.15$ m
$S_0 = 0.003$
Putting the values, $q_{max} = 0.0232$

The minimum flow rate (q_{min}) is lower than the q_{max} under different limiting conditions; thus it is safe.

3.2.7 Simulation Modeling for Border Design

Design and management of border layouts (and also for other surface irrigation systems) requires knowledge of the hydraulics of overland flow, infiltration, and drainage behavior. The simulation model is useful in integrating all the relevant and interacting processes.

Border irrigation system can be modeled using one-dimensional or two-dimensional flow analysis. In one-dimensional analysis, the pattern of water flow over and under the soil surface is assumed to be repeated across the width of the field. With respect to field behavior, this assumption cannot be fully justified. But for simplicity, one-dimensional assumption is frequently used in border or other surface irrigation systems.

3.2 Border Irrigation System Design

In actual field condition, infiltration occurs along the border strip. Thus, the flow with constant inflow discharge becomes unsteady and gradually varied. Hence, the steady flow equation does not match the reality. Therefore, two-dimensional unsteady gradually varied surface water flow equation (with appropriate initial and boundary conditions) can be solved for modeling the border irrigation system (and also other surface irrigation systems) analysis and simulation.

3.2.8 Existing Software Tools/Models for Border Irrigation Design and Analysis

3.2.8.1 BORDEV

BORDEV is a Border Design-management and EValuation tool. The BORDEV software package is based on the solution of the volume balance model (Zerihun and Feyen, 1996 BORDEV: BORder Design-management and Evaluation manual & FURDEV: FURow Design-management manual. Unpublished manuals. Institute for Land and Water Management, Katholieke Universiteit Leuven, Leuven, Belgium). The volume balance model consists of a spatially and temporally lumped form of the continuity equation and is applied primarily to the advance phase. The basic output of the model is the information for drawing the distance-elapsed time diagram of the irrigation cycle, showing the advance and recession curve and providing the opportunity time versus distance. The program has advanced developed user interface. The design and management approach which governed the design of the structure of BORDEV is based on the notion that a design and management scenario must be able to "maximize" E_a, given the system parameters, while maintaining the other two performance indices above certain threshold levels.

3.2.8.2 SIRMOD

The SIRMOD (Surface IRrigation computer simulation MODel) model simulates the hydraulics of surface irrigation (border, basin, and furrow) at the field level (Walker, 1989). The simulation routine used in SIRMOD is based on the numerical solution of the Saint-Venant equations for conservation of mass and momentum.

Inputs required for the model to simulate an irrigation event in furrow include infiltration characteristic, hydraulic resistance (Manning's n), furrow geometry, furrow slope, furrow length, inflow rate, and advance cutoff time. The output from the model includes the advance and recession characteristics, ultimate distribution of infiltrated water, and parameters related to water application efficiency, storage efficiency, and runoff hydrographs.

3.2.8.3 SIRMOD II

It is an updated version of SIRMOD. SIRMOD II is a simulation, evaluation, and design program for surface irrigation systems. It employs user-selectable kinematic-wave, zero-inertia, and hydrodynamic analyses. Design output includes field dimensions, optimal inflow and cutoff time, and field subdivisions.

3.2.8.4 WinSRFR

WinSRFR is an integrated software package for analyzing surface irrigation systems (border, basin, furrow). Founded on an unsteady flow hydraulic model, the software integrates event analysis, design, and operational analysis functionalities, in addition to simulation (USDA-ARS-ALARC, 2006; Bautista et al., 2009). Except for the Event Analysis, WinSRFR's capabilities are based on those provided by the programs SRFR, BORDER, and BASIN. Procedures in the Event Analysis world are used to evaluate the performance of irrigation events from field measured data and to estimate infiltration parameters needed for evaluation, simulation, physical design, and operational analysis.

WinSRFR was designed with two important organizational features. First, WinSRFR has four major defined functionalities (referred to as Worlds in the software):

> Event Analysis World – Irrigation event analysis and parameter estimation functions
>
> Physical Design World – Design functions for optimizing the physical layout of a field
>
> Operations Analysis World – Operations functions for optimizing irrigations
>
> Simulation World – SRFR simulation functions for testing and sensitivity analysis

The second organizational feature is that scenarios run with these functions are stored in separate data folders. This structure organizes the data into logical groups and allows outputs generated in one World to be used as inputs in a different World.

Performance measures analyzed by WinSRFR include distribution uniformity, potential application efficiency, runoff and deep percolation fractions, minimum infiltrated depth, total applied depth, the ratio of advance distance at cutoff time relative to field length, or the ratio of cutoff time to final advance time. The tool allows users to search for combinations of the decision variables that will result in high levels of uniformity and efficiency while taking into account practical and hydraulic constraints.

3.2.8.5 SADREG

SADREG is a Web decision support system for surface irrigation design, an Internet application to assist designers and managers in the process of design and planning improvements in farm surface irrigation systems – furrow, basin, and border irrigation (Muga et al., 2008). It allows creating a large set of alternative solutions, their impact evaluation and multicriteria selection analysis through an integrated framework of user knowledge, database, and simulation models.

3.2.9 General Guidelines for Border

The unit flow rate, border length, width, and cutoff time should be selected such that the application efficiency, water storage efficiency, and distribution uniformity are maximized or optimal (higher than the prescribed threshold lower limit). In addition, the system needs to monitor or evaluate on a regular basis.

Typical border parameters for different soil types and slopes are given in Table 3.1.

Table 3.1 Typical border parameters under different conditions (USDA, 1955)

Soil type	Slope (%)	Depth applied (mm)	Border width (m)	Border length (m)	Flow rate (l/s)
Coarse	0.25	50	15	150	240
		100	15	250	210
	1.0	50	12	100	80
		100	12	150	70
Medium	0.25	50	15	250	210
		100	15	400	180
	1.0	50	12	150	70
		100	12	300	70
Fine	0.25	50	15	400	120
		100	15	400	70
	1.0	50	12	400	70
		100	12	400	35

3.3 Basin Irrigation Design

In the basin, water is applied in such a way that it covers the basin relatively quickly. Check bands/dikes around the field keep the water within the basin until all the water infiltrates. Thus, the water remains in all parts of the basin for about the same duration with only minor differences.

Mainly two types of basin layouts are practiced worldwide: closed single basins (with or without outflow or runoff), and multiple basin layouts which are sequentially connected through inter-basin flow.

3.3.1 Factors Affecting Basin Performance and Design

The shape and size of basins are mainly determined by the land slope, soil type, available stream size (the water flow to the basin), required depth of irrigation application, and farming practices.

3.3.1.1 Flow Rate

Supply channel discharge is an important parameter in the design and management of contour basin layouts as it determines the boundary inflow depth, irrigation uniformity and efficiency, and ultimately it is the key controllable design and management parameter that determines how fast a basin can be irrigated to the target depth.

An increase in unit inflow rate can reduce deep percolation losses and improve application efficiency. The time of advance decreases with the increase in flow rate. Advance time in contour layouts is a very important factor for crops that do not require or tolerate ponding or when the target depth of irrigation is relatively small. The objective in this situation is to irrigate and drain the basin as quickly as possible.

In summary, it can be said that higher inflows lead to higher efficiencies and uniformity but should be used judiciously to avoid an excessive application depth.

3.3.1.2 Soil Type

The efficiency and uniformity of basin irrigation depend on the relative magnitude of the soil infiltration rate and flow rate. A soil with a relatively high infiltration rate will require a substantially higher flow rate to achieve the same uniformity and efficiency as for a heavier soil.

3.3.1.3 Basin Longitudinal Slope

By definition, basins are irrigation units graded to zero slope in both directions. However, it is a common practice among designers to provide some slope in the longitudinal direction to facilitate advance. But the research results (Khanna et al., 2003a) suggests that while some longitudinal slope might aid in achieving better uniformity and efficiency, the selection of the best slope for a basin requires careful analysis for each case.

3.3.1.4 Aspect Ratio

Aspect ratio is the ratio of width to length of the basin. With the increase in aspect ratio, the application efficiency and irrigation uniformity decrease, meaning a reduction in irrigation performance. The reason is that increase in aspect ratio leads to greater deep percolation losses as the time of advance also increases. The indicators are particularly sensitive at low aspect ratio, whereas these two parameters remain largely unchanged for greater aspect ratios.

3.3.1.5 Local Surface Micro-topography

The local undulations are commonly termed as micro-topography. Local undulations on the basin's surface are important factors affecting advance and recession.

They are significant in basin irrigation because they cause local stagnation of water and irregular advance of the waterfront.

3.3.1.6 Number of Check Bank Outlets

The amount of drainage between basins depends on the number of outlets installed in the check bank. Drainage of the basin is very important for good water and crop management. It is a common practice that in a multiple-basin operation, runoff from the upstream basin drains into the downstream basin. Typically, designers use either one or two outlets in their basin designs.

3.3.1.7 Elevation Difference (Vertical Interval) Between Adjacent Basins

The vertical interval between adjacent basins is an important parameter in the design of contour layouts because it affects ponding of water and drainage of excess water from the upstream basin. The vertical difference in elevation between contour basins is primarily dictated by the natural land topography. Through land forming practices, a designer can alter the existing elevation intervals between adjacent basins to better suit other features of the design including the elevation of the water source, supply channel, and reuse pond.

3.3.1.8 Irrigation Depth

The duration of irrigation is dependent on the depth to be applied.

3.3.2 Hydraulics in Basin Irrigation System

Overland flow in surface irrigation systems is commonly described using a one-dimensional analysis, in which the pattern of water flow over and into the soil surface is assumed to be repeated across the width of the field (Clemmens and Strelkoff, 1979). This assumption produces good results in cases when the flow can be considered linear such as in furrow and border irrigation. However, in a basin irrigation configuration, as well as in contour basin layouts, a one-dimensional approach is difficult to justify especially if the field geometry is irregular or if water does not enter the field uniformly along one of its sides. Hence, it is more appropriate to simulate the hydraulic processes in contour layouts using a two-dimensional flow simulation approach.

There are two main processes involved in flow over porous media. One is the surface flow and the other is the vertical movement (infiltration) of water into the soil. A typical advance process in basin irrigation layouts is shown in Fig. 3.3 The figure shows the waterfront lines during the advance phase in a basin.

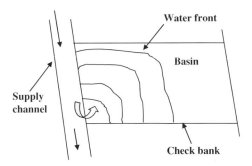

Fig. 3.3 Water flow pattern in basin during inflow and advance

3.3.3 Simulation Modeling for Basin Design

There are many design parameters which influence hydraulic processes during an irrigation event in basin layouts. It is very difficult to predict and compare the performance of alternative design layouts without using a physically based simulation model to describe the process.

Generally, simulation models for basin irrigation design are based on governing equations in the form of full hydrodynamic Saint-Venant equations or the simplified zero-inertia approximation (neglecting inertial terms).

3.3.3.1 Hydrodynamic Model

In this approach, the overland flow is described by the depth-averaged hydrodynamic flow equations. These equations consist of the continuity equation and the momentum equation. The two-dimensional continuity equation for shallow water flow is written as (Chaudhry, 1993):

$$\frac{\partial h}{\partial t} + \frac{\partial (hu)}{\partial x} + \frac{\partial (hv)}{\partial y} + I_s = 0 \qquad (3.6)$$

where u and v are the velocities in x- and y-directions (m/s), h the water depth (m), I_s the volumetric rate of infiltration per unit area (m/s) and t is the time (s). The momentum equations in the x- and y-directions are as follows:

$$\frac{\partial (uh)}{\partial t} + \frac{\partial (u^2 h)}{\partial x} + \frac{\partial (uvh)}{\partial y} + gh\frac{\partial H}{\partial x} + ghS_{fx} = 0 \qquad (3.7)$$

$$\frac{\partial (vh)}{\partial t} + \frac{\partial (uvh)}{\partial x} + \frac{\partial (v^2 h)}{\partial y} + gh\frac{\partial H}{\partial y} + ghS_{fy} = 0 \qquad (3.8)$$

where g is the acceleration due to gravity (m/s^2), $H=h+z_0$ the water surface elevation above the datum, z_0 the bottom elevation above an arbitrary datum (m), and S_{fx} and S_{fy} are the components of friction slope in x- and y-directions. The first term in both Eqs. (3.7) and (3.8) relates to the temporal acceleration term, and the following two terms stem from the advective accelerations in x and y-directions; these account for the inertia effects.

Based on these shallow water flow equations, numerous models have been developed for basin irrigation design. But the solution approaches of the governing equations differ from each other. The two-dimensional hydrodynamic model has been extended to incorporate micro-topography (Playan et al., 1996), irregular boundaries (Singh and Bhallamudi, 1997), non-level basin (Bradford and Katopodes, 2001), and many other particular situations.

3.3.3.2 Zero-Inertia Model

Probably Strelkoff et al. (1996, 2003) were the first to attempt to develop a simulation model for basin irrigation in two-dimensions using the zero-inertia approximation. The developed model simulates two-dimensional flow from a point or line source in an irrigated basin with a non-level soil surface. The governing equations used were obtained by simplifying the full hydrodynamic from of the equations (Eqs.3.6, 3.7 and 3.8) by neglecting the inertial terms. The effect of the inertial terms becomes small compared with those describing the effect of depth gradient gravity and friction in shallow water flow. This is typical of agricultural fields where the flow process is more diffusional in nature due to the low velocities.

The continuity equation can be obtained by expressing $q_x=uh$ and $q_y=vh$, discharge per unit width (m^2/s) in the x- and y-directions, respectively. Using $\partial h/\partial t = \partial H/\partial t$ and substituting q_x and q_y in Eq. (3.6) yields

$$\frac{\partial H}{\partial t} + \frac{\partial q_x}{\partial x} + \frac{\partial q_y}{\partial y} + I_s = 0 \qquad (3.9)$$

By neglecting the inertial terms in Eqs. (3.2) and (3.3) become

$$\frac{\partial H}{\partial x} + S_{fx} = 0 \qquad (3.10)$$

$$\frac{\partial H}{\partial y} + S_{fy} = 0 \qquad (3.11)$$

3.3.3.3 Other Approaches

Clemmens and Strelkoff (1979) developed dimensionless advance curves, using the zero-inertia approach for level basin irrigation design. Khanna et al. (2003b, c) developed a two-dimensional model for the design of rectangular basin, and irregular shape and multiple basins. In both the cases, the model's governing equations are based on a zero-inertia approximation to the two-dimensional shallow water equations. For the first case, the governing equations were solved by using a split-operator approach. For the second case, the governing equations were solved by using a split-operator approach using the method of characteristics coupled with two-dimensional Taylor series expansion for interpolation and calculation of diffusion terms.

3.3.4 Existing Models for Basin Irrigation Design

3.3.4.1 COBASIM

A two-dimensional simulation model, titled "contour basin simulation model" (COBASIM) was developed by Khanna et al. (2003a, b). It simulates the overland flow hydraulics and infiltration processes that occur in contour basin layouts. The model is capable of simulating contour basin layouts of regular and irregular shape and size. The main objective of developing the simulation tool was to enable designers and practitioners to simulate the behavior of multiple design scenarios.

3.3.4.2 BASCAD

It was developed by Boonstra and Jurriens (1988) for level basin design. BASCAD simulates advance with a zero-inertia model in real time, then uses a volume balance to determine a single recession time and the final distribution of infiltrated water. The program allows the user to start by providing the computer with very limited data, have the computer provide "ballpark" estimates for unknown parameters, and proceed with the user providing values for more parameters. The user does not have direct control over performance measures.

3.3.4.3 SIRMOD, WinSRFR, SADREG

These models have been described in border irrigation section.

3.4 Furrow Irrigation System Design

Furrows are sloping channels formed in the soil. The amount of water that can be applied in a single application via furrow (or in other conventional surface irrigation, that is, flood or border irrigation or, to some extent, sprinkler irrigation) depends upon the ability of the soil to absorb water. The irrigation process in a furrow is identical to the irrigation process in a border, with the only difference that the geometry of the cross-section, and as such the infiltration process, is different. Among surface irrigation systems, furrow irrigation with cutback is commonly used because of its potential higher irrigation efficiency, lower cost and relative simplicity.

3.4.1 Hydraulics of Furrow Irrigation System

Furrow irrigation involves the application of irrigation water at the top end of a field into furrows (Fig. 3.4). The water then flows along these furrows to the bottom of the field, infiltrating into the soil along the length of the furrow (Fig. 3.5). The length of time the soil is exposed to this water is known as the infiltration opportunity time. Infiltration occurs laterally and vertically through the wetted perimeter of the furrow (Fig. 3.6). Wetting patterns in furrows may vary considerably depending on soil type. In an ideal situation, adjacent wetting patterns overlap each other, and there is an upward movement of water (capillary rise) that wets the entire ridge. To achieve the ultimate in furrow irrigation performance, the infiltration opportunity time should equal the amount of time necessary to apply the required depth of water (to fulfill the moisture deficit).

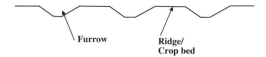

Fig. 3.4 Schematic view of furrow

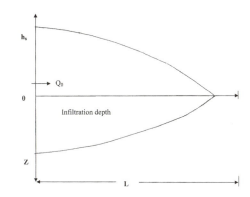

Fig. 3.5 Schematic of infiltration advance throughout the furrow length

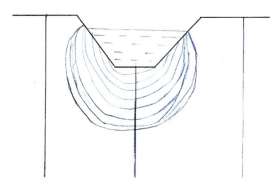

Fig. 3.6 Schematic of infiltration into the cross-section of the furrow

3.4.2 Mathematical Description of Water Flow in Furrow Irrigation System

Different approaches to simulate surface flow and infiltration in furrows have been developed by the researchers. Hall (1956) and Davis (1961) based their models on the solution of the mass conservation equation under the hypothesis of normal depth of flow. Other authors, such as Katapodes and Strelkoff (1977), Walker and Humphreys (1983), Wallender and Rayej (1990), and Schmitz and Seus (1992) based their models on the numerical solution of the partial differential equations of mass and momentum in open-channel flow applied to borders and furrows. Levien and de Souza (1987) presented algebraic model simulating furrow irrigation.

In furrow system, as infiltration occurs along the furrow, flow with constant inflow discharge is unsteady and gradually varied. The velocity of water advancing on the surface of the furrow is normally higher than the infiltration velocity (infiltration rate). Thus, water moving into and inside the soil at each cross-section can be considered two-dimensional and occurring perpendicular to the direction of the flow on the furrow.

The water flow in a furrow is similar to the flow in an open porous channel initially dry. Therefore, the mathematical formulation describing water flow in a furrow should take into account the wave propagation in the furrow during the advance phase, the flow discharge variation during the supply and the recession phases, and the movement of water penetration and redistribution in the soil.

3.4.2.1 Unsteady Gradually Varied Surface Water Flow

Unsteady gradually varied flow can be described by the partial differential equations of Saint Venant (1871):

Continuity equation

$$\frac{\partial Q}{\partial y} + \frac{\partial A}{\partial t} + \frac{\partial I}{\partial t} = 0 \qquad (3.12)$$

Dynamic equation

$$\frac{1}{g}\frac{\partial Q}{\partial t} + \frac{\partial}{\partial y}\left(\frac{Q^2}{Ag}\right) + \frac{\partial P}{\partial y} + D - AS_0 = 0 \qquad (3.13)$$

with $D = AS_f$

$$S_0 = \frac{n^2 Q |Q|}{A^2 R^{4/3}} \qquad (3.14)$$

3.4 Furrow Irrigation System Design

where Q is the flow rate (L^3/T), A is the cross-sectional flow area (L^2), I is the infiltrated volume per unit of length (L^3/L) in the furrow, y is the distance along the furrow in the direction of flow (L), g is the acceleration of gravity (L/T^2), t is the time (T), P is the pressure force per specific weight of the water (L^3), S_f is the friction slope (L/L), S_0 is the slope of the furrow bottom (L/L), R is the hydraulic radius (L), n is the Manning's roughness coefficient, and D is the drag force per specific weight of water per unit length (L^3).

An approximate solution to Eqs. (3.12) and (3.13) can be generated by rewriting these equations in the integral form and numerically integrating in the plane using weighted averages. The integral can be defined in either Lagrangian or Eulerian coordinate systems.

According to the latter approach, the discrimination of the Eqs. (3.12) and (3.13) yields the following pair of non-linear algebraic equations (Tabuada et al., 1995):

$$\begin{aligned} RC = &\left[\Theta(Q_k^{t+1} - Q_{k+1}^{t+1}) + (1-\Theta)(Q_k^t - Q_{k+1}^t)\right]\Delta t \\ &+ \Phi(A_k^t - A_k^{t+1}) + (1-\Phi)(A_{k+1}^t - A_{k+1}^{t+1})]\Delta y \\ &+ \left[\Phi(I_k^t - I_k^{t+1}) + (1-\Phi)(I_{k+1}^t - I_{k+1}^{t+1})\right]\Delta y = 0 \end{aligned} \quad (3.15)$$

and

$$\begin{aligned} RD = &\tfrac{1}{g}[\varphi(Q_k^{t+1}) + (1-\varphi)(Q_k^{t+1} - Q_{k+1}^{t+1})]\Delta y \\ &+ \left\{\Theta\left[\left[\tfrac{Q^2}{Ag}\right]_k^{1+1} - \left[\tfrac{Q^2}{Ag}\right]_{k+1}^{1+0}\right] + (1-\Theta)\left[\left[\tfrac{Q^2}{Ag}\right]_k^t - \left[\tfrac{Q^2}{Ag}\right]_{k+1}^t\right]\right\}\Delta t \\ &+ \Theta\left[\left(P_k^{t+1} - P_{k+1}^{t+1}\right) + (1-\Theta)\left(P_k^t - P_{k+1}^t\right)\right]\Delta t \\ &- \left[\Theta\varphi D_k^{t+1} + \Theta(1-\varphi)D_{k+1}^{t+1} + (1-\Theta)\varphi D_k^t + (1-\Theta)(1-\Phi)D_{k+1}^t\right]\Delta y \Delta t \\ &+ S_0[\Theta\Phi A_k^{t+1} + \Theta(1-\Theta)A_{k+1}^{t+1} + (1-\Theta A_k^t + (1-\Theta)(1-\Theta)A_{k+1}^t]\Delta y \Delta t \\ = &\ 0 \end{aligned} \quad (3.16)$$

which are the residuals of continuity (RC) and dynamic (RD).

Equations (3.15) and (3.16) are expressions of finite differences which allow an approximate solution.

3.4.3 Some Relevant Terminologies

3.4.3.1 Intake Rate

The rate at which water is absorbed by the soil of furrow is termed as intake rate. It varies with time. Initially, water is absorbed by the soil at a higher rate and then decreases over time. The fairly constant intake rate is termed as basic intake rate.

3.4.3.2 Infiltration Opportunity Time

Infiltration opportunity time is defined as the amount of time that water has the opportunity to infiltrate the soil. Or, it is the difference between the times when the water recedes and when it advanced to a specific location.

The infiltration opportunity time is different at different points along the length of the furrow. This is because the length of time that the water is present on the surface of the soil at any location is the difference between the time the water arrives (advance) and the time the water leaves (recession). The rate at which water advances down the field is different to the rate at which it recedes. Making opportunity time more uniform down a furrow is a desirable strategy for improving distribution uniformity for furrow irrigation.

3.4.3.3 Distribution Uniformity

The measure of how evenly the infiltration occurs is called distribution uniformity (DU). In general, depth of infiltration at low quarter is considered as reference for expressing DU (Burt et al., 1997). That is,

DU = (average low quarter infiltration depth) × 100/(average infiltration depth)

During a furrow irrigation event, distribution uniformity is primarily influenced by the soil infiltration characteristic (soil type, variability, and moisture content), rate of water flow into the furrow (inflow) and the length of time this water is flowing (time to cut off). Other factors that influence uniformity include field slope and variability and field length.

Irrigation efficiencies are directly related to the uniformity of water application (distribution uniformity) on the individual fields. Furrow-irrigated field distribution uniformity is directly related to the advance ratio and the average depth of water infiltrated per hour. For a uniform soil with good land grading, the distribution uniformity of water infiltration for furrows is dependent upon the uniformity of opportunity times. Making opportunity times more uniform down a furrow is a desirable strategy for improving DU for furrow irrigation.

A DU value of 80% is considered excellent for all irrigation methods.

3.4.3.4 Time Ratio

Time ratio (R_t) is defined as the ratio of the time required for the infiltration of total net amount of water required for the root zone to the time when the water front reaches the end of the run. It plays a key role in determining optimum furrow length to achieve maximum irrigation efficiency.

$$R_t = \frac{t_i}{t_L}$$

where

t_i = time required for infiltration of net irrigation depth
t_L = time required for waterfront to reach the end of run

R_t value of 1.0 represents the length of run in which the time required to infiltrate the irrigation depth is equal to the advance time.

The advance ration is an important factor for managing a furrow irrigation system. Generally, water should get to the end of a furrow in less than ½ of the set time to achieve good distribution uniformity. Whether that should be as quickly as ¼ of the set time would depend on the soil texture and conditions.

3.4.3.5 Cutoff Ratio *or* Advance Ratio

Advance ratio or cutoff ratio is defined as the ratio of time to reach the waterfront at the end of furrow, to the time set for irrigation. Mathematically, it can be expressed as

$$R_{cut} = \frac{t_e}{t_{irr}}$$

where

t_e = time required for water advance to reach the end of furrow
t_{irr} = time of irrigation

Runoff and the uniformity of water infiltrated along the furrow are related to the cutoff ratio.

A cutoff ratio of 0.5 is desired. For example, for an 8-h irrigation set time, the advance time should be about 4 h. The easiest way to change the advance time is by altering the furrow stream size, i.e., by changing the size of the irrigation set. This will affect the cutoff ratio and hence the uniformity of water application.

For both level and sloping furrow systems, high uniformities (greater than 85%) require a reasonably small advance ratio.

3.4.3.6 Irrigation Set Time

It is the total time for irrigation. Irrigation set time is determined by furrow inflow rate, furrow shape, roughness, and length.

3.4.3.7 Advance Rate

It is the rate at which the waterfront advances through the furrow. Advance rate is influenced by both soil conditions (size, slope, and roughness of the furrow) and inflow rate.

3.4.3.8 Advance Function

Both the infiltration depth and water advance rate on soil surface in furrow irrigation is a function of irrigation time. These relationships, expressed in empirical forms, are known as advance function.

3.4.3.9 Surge Flow/Irrigation

Surge irrigation is the practice of applying water to a set for a while, then switching the water to another set, then switching back and forth between the sets periodically. Surge flow is identified as a strategy to improve the performance of the furrow irrigation system.

3.4.3.10 Gross Water Needed for Furrows

Gross water to apply

$$\mathrm{WR_G} = \frac{\mathrm{WR_N}}{\mathrm{DU}} \times \left(1 - \frac{R}{100}\right) \times F_S$$

where

$\mathrm{WR_G}$ = gross water to be applied (mm)
$\mathrm{WR_N}$ = net water required for root zone soil (determined with any recommended technique) (mm)
R = percent of water which runs off field and is not re-circulated to that field
DU = distribution uniformity expressed as a decimal (DU = 0.80 is often used)
F_S = salinity factor, a factor to account for the increased irrigation requirement due to maintenance leaching. Guidelines are according to the salinity of the irrigation water. If salinity is not involved, $F_S = 1$.

3.4.4 Factors Affecting Performance of Furrow Irrigation System

3.4.4.1 Soil Characteristics

Soil characteristics and field conditions are major factors controlling the efficiency of furrow irrigation systems. Advance rates are influenced by both soil conditions and furrow inflow rates.

3.4.4.2 Stream Size

When selecting the furrow stream size, consider furrow erosion. Use a furrow stream that does not cause serious erosion. In general, the maximum non-erosive stream size decreases as furrow slope increases. The use of high inflow rates (stream size) will result in more runoff but less deep percolation losses or vice versa.

3.4.4.3 Length of Run

Time required for advance increases with furrow length. Irrigation runs which are too long result in water being lost by deep percolation at the head of the furrow by the time the lower end is adequately irrigated.

3.4.4.4 Cutoff Ratio

Deep percolation and runoff depends on cutoff ratio.

3.4.4.5 Tailwater Reuse

In most cases, tailwater reuse systems are essential to properly manage furrow irrigation systems so that the best distribution uniformity and irrigation efficiency may be achieved.

3.4.4.6 Wetted Perimeter

Furrow intake increases with average wetted perimeter and decreases with distance from the water source, because wetted perimeter decreases as the flow depth declines, assuming homogeneous soil and hydraulic conditions. Likewise, for lower inflow rates, steeper slopes, and hydraulically smoother surfaces, wetted perimeter and thus intake will decrease.

3.4.5 Management Controllable Variables and Design Variables

3.4.5.1 Management Controllable Variables

In the furrow system, the water should reach the end of the field in about one-half of the total irrigation time, and the irrigator can manipulate that time by controlling the outflow volume, slope, number and shape of furrows, and field length.

Factors the farmer can readily vary or manage are as follows:

- furrow shape
- roughness
- length of furrow
- irrigation set time
- flow rate (stream size for furrow)
- cutoff time

3.4.5.2 Design Variables

A furrow irrigation system has several design variables. These are as follows:
- the inflow rate
- the length of the run in the direction of the flow
- the time of irrigation cutoff

Because of the many design and management controllable parameters, furrow irrigation systems can be utilized in many situations, within the limits of soil uniformity and topography (<2% slope). With runoff return flow systems, furrow irrigation can be a highly uniform and efficient method of applying water. However, the uniformity and efficiency are highly dependent on proper management, so mismanagement can severely degrade system performance.

3.4.6 Furrow Design Considerations

Furrow systems may be designed with a variety of shapes and spacings. Optimal furrow lengths are primarily controlled by the soil intake rate, furrow slope, set time, and stream size. For most applications the stream size should be as large as possible without causing erosion.

When the intake rate is slow, the maximum application efficiency can be attained providing a relatively longer furrow length. For soils with high intake rate, the length of the furrow should be selected shorter. The maximum slope of 0.1% (0.1/100 m) should be maintained for block ends furrow.

Optimal furrow irrigation performance requires understanding of application efficiency and distribution uniformity and the methods for improving both. Improving the efficiencies involve careful management of flow rates and irrigation duration and appropriate timing (scheduling) of irrigation events:

- Select a stream size appropriate for the slope, intake rate, and length of run. Or alternatively, optimal furrow length and irrigation cutoff can be determined, as related to soil infiltration characteristics, by the time ratio.
- With the proper cutoff ratio and gross application, you can achieve uniform water application and minimize deep percolation and runoff. Try different combinations of furrow stream size and set time. The best combination is the one which moves water to the end of the furrow within the requirements of the cutoff ratio, is less than the maximum erosive stream size, and results in gross applications that are not excessive.

3.4.7 Modeling of Furrow Irrigation System

3.4.7.1 Theoretical Considerations

Surface irrigation processes are governed by general physical laws such as conservation of mass, energy, and momentum, which are expressed as a function of physical quantities. Simulation models of furrow irrigation rely on the knowledge of furrow infiltration and hydraulic characteristics. These models solve equations of mass and motion conservation which describe unsteady, nonuniform surface flow over a permeable bed. Flow rate, slope, hydraulic roughness, and geometry affect flow depth and therefore wetted perimeter in time and space. Wetter perimeter and flow depth, and their temporal and spatial variation affect infiltration. Other factors such as initial soil moisture, cracks and other voids in the soil, soil layering, and heterogeneity, and water quality, including chemistry and temperature also influence infiltration.

Furrow irrigation performance depends on a number of variables, namely, inflow rate, cutoff time, furrow length, spacing and shape, roughness, slope, infiltration characteristics, and irrigation requirement. Variables such as furrow length and slope are constrained by the field shape and size, infiltration characteristics and roughness by soil type, and furrow shape and size by equipment. The irrigation requirements are driven by climate, soil, and crop conditions. All these factors need to be considered in the decision-making process, but they are generally given, not decision variables. Thus, for a given set of field and crop conditions, furrow inflow rate and cutoff time are the decision variables.

3.4.7.2 Simulation of Furrow Design Variables

Simulation Alternatives

Furrow irrigation has been simulated under different considerations. Wu and Liang (1970) optimized furrow run length using the minimum cost criterion, whereas, Reddy and Clyma (1981) optimized furrow irrigation system design without considering the irrigation schedule or the minimum discharge to assure the advance of water to the end of the run. Holzapfel et al. (1986, 1987) used linear and nonlinear optimization models to design surface irrigation systems, considering homogeneous soils and regression-derived relationships between irrigation performance and design variables. Raghuwanshi and Wallender (1998) used kinematic-wave model to optimize furrow irrigation.

Kinematic-Wave Model

A kinematic-wave model can be used to represent unsteady and spatially varied flow in a sloping and free draining furrow. The model consists of the continuity and a simplified form of the hydrodynamic equation in that friction force is balanced by furrow bottom slope. Inertial and water depth gradient terms are negligible:

$$\frac{\partial A}{\partial t} + \frac{\partial Q}{\partial x} + \frac{\partial Z}{\partial \tau} = 0 \qquad (3.17)$$

$$S_\mathrm{r} = S_0 \qquad (3.18)$$

where A = the cross-sectional flow area in m^2, Q = the flow rate in m^3/min, Z = infiltration per unit furrow length in m^3/m, x = distance in the direction of flow in m, t = elapsed time in min, T = intake opportunity time in min, S_r = friction slope, and S_0 = furrow bottom slope. Infiltration per unit furrow length (Z) can be computed using Eqs. (3.19) and (3.20), which empirically accounts for differences in infiltration rate along the wetted perimeter section (Bautista and Wallender, 1992):

$$Z_i = Z_{i-1} + (\zeta_i - \zeta_{i-1})\text{WP}_{i-1} + \zeta(\delta t_i)(\text{WP}_i - \text{WP}_{i-1}) \quad \text{for WP}^{i-1} \quad (3.19)$$

or

$$Z_i = Z_{i-1} + (\zeta_i - \zeta_{i-1})\text{WP}_{i-1} < \text{WP}_{i-1} \quad \text{for WP}^i \quad (3.20)$$

where WP = the wetted perimeter, and ζ is given by the extended Kostiakov equation:

$$\zeta = K\tau^a + C\tau$$

where τ = the intake opportunity time; k, a, C = field-measured coefficients; and subscripts i and $i-1$ denote consecutive time lines at which the solution is computed.

For surface irrigation modeling, Eq. (3.17) can be solved by considering either the fixed time step or the fixed space step.

3.4.8 General Guideline/Thumb Rule for Furrow Design

Furrow irrigation has limitations or field constraints to acknowledge as a guideline when considering this method. The general limits of different parameters are described below.

3.4.8.1 Furrow Length

Row length should not exceed 400 m on heavy soils and, depending on soil texture and slope, can be shorter. Although furrows can be longer when the land slope is steeper, the maximum recommended furrow slope is 0.5% to avoid soil erosion. The coarser the soil or the steeper the slope, the shorter will be the run length.

3.4.8.2 Slope

Row slopes should be between 0.05 and 0.5%. Cross slope should be less than or equal to row slope, except in a permanent-graded furrow design. Furrows can also be level and are thus very similar to long narrow basins. However, a minimum grade of 0.05% is recommended so that effective drainage can occur following irrigation or excessive rainfall. If the land slope is steeper than 0.5%, then furrows can be set at an angle to the main slope or even along the contour to keep furrow slopes within the recommended limits. Furrows can be set in this way when the main land slope does not exceed 3%. Beyond this there is a major risk of soil erosion following a breach in the furrow system. On steep land, terraces can also be constructed, and furrows should be constructed along the terraces.

3.4.8.3 Stream Size/Flow Rate

In general, the larger the stream size available, the larger the furrow must be to contain the flow. When larger stream sizes are available, water will move rapidly down the furrows and so generally furrows can be longer. Normally stream sizes up to 0.5 l/s will provide an adequate irrigation provided the furrows are not too long. The maximum stream size that will not cause erosion will obviously depend on the furrow slope. In any case, it is advisable not to use stream sizes larger than 3.0 l/s.

Michael (1978) suggested the maximum non-erosive flow rate based on furrow slope as

$$q_m = \frac{0.60}{S} \quad (3.21)$$

where

q_m = maximum non-erosive flow rate in individual furrow (l/s)
S = slope of furrow (%)

For normal conditions, flow rate can be estimated from

$$q_m = \frac{0.50}{S} \quad (3.22)$$

The units are the same as that of the earlier.

3.4.8.4 Furrow Shape

In sandy soil, water moves faster vertically than sideways (or lateral). Narrow, parabolic, or deep V-shaped furrows are desirable to reduce the soil area through which water percolates. However, sandy soils are less stable and tend to collapse, which may reduce the irrigation efficiency.

In clay soils, the infiltration rate is much less than for sandy soils, and there is much more lateral movement of water. Thus, a wide, shallow furrow is desirable to obtain a large wetted area to encourage infiltration.

3.4.8.5 Spacing of Furrow

The spacing of furrows is influenced by the soil type and the cultivation practice. Based on the crop and soil, it normally ranges from 50 to 80 cm.

3.4.9 Estimation of Average Depth of Flow from Volume Balance

Average depth of applied (or infiltrated) water can be estimated from volume balance approach as follows:

Water in = Water stored

i.e., flow rate × application time = Furrow cross-sectional area × average depth of water

i.e., $q \times t = A \times d$
$= (W \times L) \times d$

$$\text{Or,} \quad d = \frac{q \times t}{W \times L} \tag{3.23}$$

where
$q =$ flow rate in individual furrow, m^3/s
$t =$ time, sec (s)
$W =$ width of furrow, meter (m)
$L =$ length of furrow, m

3.4.10 Suggestions for Improving Furrow Irrigations

- Always look to measure and improve distribution uniformity (DU) first. Water must infiltrate as evenly as possible across a field. Then, try to improve control over the total infiltration and reduce or reuse surface runoff.
- Utilize an irrigation scheduling system so that you have reasonably accurate estimates of "WHEN" and "HOW MUCH" to irrigate.
- Reduce the length of the furrow – it improves down row uniformity by helping water to get to the end of furrow quicker in relation to the total time of irrigation.
- Install a "tailwater" reuse system – it improves overall irrigation efficiency by saving "tailwater" for reuse.
- Increase the flow per furrow – it improves down row uniformity by helping water to get to the end of the furrow quicker in relation to the total time of irrigation.
- Irrigate in every other furrow – this helps to reduce overapplication when either small irrigations are desired or the soil has a high infiltration rate.
- Use cutback furrow flows – it reduces the amount of surface runoff.
- Utilize surge irrigation – this might occur from three times to as many as 10 depending on the situation, equipment, and experience. Surging water acts to reduce the infiltration rate of the soil quickly so that differences between compacted and uncompacted furrows are minimized. It can also help in very light soils as it will act to reduce overapplications.

3.4.11 Furrow Irrigation Models

3.4.11.1 FURDEV

FURDEV is a software tool for furrow irrigation system design and evaluation. The FURDEV software package is based on the solution of the volume balance model (Zerihun and Feyen, 1996 BORDEV: BORder Design management and Evaluation manual & FURDEV: FURow Design – management manual. Unpublished manuals. Institute for Land and Water Management, Katholieke Universiteit Leuven, Leuven, Belgium). The volume balance model consists of a spatially and temporally lumped form of the continuity equation and is applied primarily to the advance phase. The basic output of the model is the information for drawing the distance-elapsed time diagram of the irrigation cycle, showing the advance and recession curve and providing the opportunity time versus distance. The program has advanced developed user interface. The design and management approach which governed the design of the structure of FURDEV is based on the notion that a design and management scenario must be able to "maximize" application efficiency (E_a), given the system parameters, while maintaining the other two performance indices above certain threshold levels.

3.4.11.2 SIRMOD, SIRMODII, WinSRFR

These are described in the border irrigation section.

3.4.12 Sample Worked Out Problems

Example 3.5

Furrows of 100 m length and 0.80 m width and having a slope of 0.3% are irrigated for 30 min with a stream size of 0.005 m³/s. Determine the average depth of water.

Solution

$$\text{We know, } d = \frac{q \times t}{W \times L}$$

Given,

$L = 100$ m
$W = 0.80$ m
$S = 0.3\%$
$q = 0.005$ m³/s
$t = 30$ min $= 30 \times 60$ s $= 1{,}800$ s
Putting the above values in equation, $d = 0.1125$ m (Ans.)

Example 3.6

Furrows of 120 m length and 0.70 m width and having a slope of 0.3% are initially irrigated for 40 min with a stream size of 0.005 m³/s. The stream size is then reduced to half and continued for 30 min. The furrow end is closed (no outflow from furrow). Determine the average depth of infiltrated water.

Solution

Given,

$L = 120$ m
$W = 0.70$ m
$S = 0.3\%$
$q_1 = 0.005$ m³/s
$t_1 = 40$ min $= 40 \times 60$ s $= 2{,}400$ s

$$\text{We get, } d = \frac{q \times t}{W \times L}$$

Thus, putting the values, depth of infiltration from initial stream, $d_1 = 0.1428$ m
Now, $q_2 = q_1/2 = 0.005/2 = 0.0025$ m³/s
$t_2 = 30$ min $= 30 \times 60$ s $= 1{,}800$ s
Putting the values, $d_2 = 0.0535$ m
Thus, total depth, $d = d_1 + d_2 = 0.1428$ m $+ 0.0535$ m
$= 0.1964$ m (Ans.)

Example 3.7

Water is applied in a furrow using non-erosive maximum stream size. The length, width, and slope of the furrow are 150, 0.75 m, and 0.4%, respectively. The stream is continued for 2 h. Estimate the average depth of irrigation.

Solution

We get non-erosive maximum stream size, $q_{max} = 0.60/S$ l/s
Here, $S = 0.4\%$, thus $q_{max} = 0.60/0.4 = 1.5$ l/s $= 0.0015$ m³/s
We get, $d = \frac{q \times t}{W \times L}$
Given,

$L = 150$ m
$W = 0.75$ m
$T = 2$ h $= 2 \times (60 \times 60)$ s $= 7{,}200$ s
and $q = 0.0015$ m³/s
Putting the values, $d = 0.096$ m (Ans.)

3.5 Design of Sprinkler System

3.5.1 Design Aspects

Design aspects of sprinkler irrigation system are as follows:

- System layout
- Operating pressure, nozzle diameter, sprinklers discharge, and wetted diameter

3.5 Design of Sprinkler System

- Spacings between sprinklers and laterals
- Design of main line and sublines
- Sprinkler line azimuth
- Pivot or ranger length
- System capacity for water supply
- Pump design

The most common design criteria for sprinkler laterals is that sprinkler discharge should not vary by more than 10% between the points of highest and lowest pressure in the system.

3.5.2 Theoretical Aspects in Sprinkler System

3.5.2.1 Water Distribution Pattern

In sprinkler system, the precipitation rate of water decreases from the center of the irrigated circle to its edges (Fig. 3.7). To overcome the problem, sprinklers are spaced in such a way that their application rates overlap each other and cover the nonuniformities.

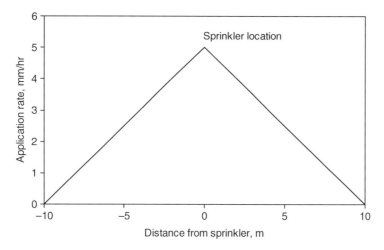

Fig. 3.7 Schematic of application pattern of a sprinkler (after Zazueta and Miller, 2000)

3.5.2.2 Factors to Be Considered in Sprinkler Design

For proper design of sprinkler, factors to be considered (in addition to spacing, nozzle size) are as follows: crop root zone depth, water use rate, wind, air temperature, and humidity. Wind affects the distribution pattern of water by both wind speed and direction.

Equipment and design factors affecting uniformity and efficiency include nozzle type and size, operating pressure, and spacing. The length of the irrigation time can also affect uniformity (Solomon, 1990).

3.5.2.3 Definition of Some Relevant Terminologies

Precipitation Rate

Precipitation rate (P_r) is the rate at which water is delivered from the nozzle, averaged as millimeters per hour, over the area covered by one nozzle (Hill, 2008). It can be calculated by the following formula:

$$P_r \text{ (mm/h)} = \text{(nozzle flow rate, l/h)/(area covered, m}^2\text{)} \quad (3.24)$$

Application Rate

Application rate (A_r) is the average rate at which water is stored in the soil.

$$A_r = \text{Precipitation rate} \times \text{Application efficiency} \quad (3.25a)$$

Application rate should be related to infiltration rate of the soil, should not be greater than the infiltration rate in order to prevent surface runoff and/or ponding. The application rate of an existing system can be calculated as follows:

$$I = \frac{3,600 \times q_s}{S_m \times S_l} \quad (3.25b)$$

where

I = application rate (mm/h)
q_s = discharge per sprinkler (l/s)
S_m = lateral spacing along mainline (m)
S_l = sprinkler spacing along lateral (m)
3,600 = unit conversion factor

Duration of Irrigation

The duration of irrigation (D_{ir}) needed to store the crop irrigation requirement in the root zone is as follows:

$$\text{Irrigation duration (h)} = \text{Crop irrigation requirements (mm)/Application rate (mm/h)} \quad (3.26)$$

3.5 Design of Sprinkler System

Average Application Depth

The average application depth is the average amount applied throughout the field. It can be computed as

$$d_{av} = \frac{V}{A} \tag{3.27}$$

where d_{av} is the average application depth (m), V is the volume pumped (m³), and A is the application area (m²).

3.5.3 Sprinkler Design

3.5.3.1 Considerations in Sprinkler Design

- Main line capacity should be based on the water demand of the target crop considering peak use rate, minimum irrigation interval, and minimum operating hours per day.
- Application rate should not exceed soil intake rate.
- The minimum design application rate should meet the maximum total daily wet soil evaporation rate (for deficit irrigation plan, it will be different).
- When determining capacity requirements, allowance should be made for reasonable water losses during application (i.e., should consider application efficiency).
- Maximum spacing of sprinklers should not be greater than 75%, and no closer than 50% of wetted diameter listed in manufacturer's performance tables.
- Riser pipes used in lateral lines should be high enough to minimize interference with the distribution pattern.
- The velocity and direction of prevailing winds and the timing of occurrence should be considered when planning a sprinkler system.
- Coefficient of uniformity (CU) data or distribution uniformity (DU) should be used in selecting sprinkler spacing, nozzle size, and operating pressure.

3.5.3.2 Design Principles

- Estimate application rate based on planned crop(s)/cropping patterns, atmospheric water demand, and soil intake rate.
- Draw a layout.
- Optimize sprinkler spacing (between sprinklers and laterals), nozzle size, and operating pressure that provide the design application rate and distribution pattern.
- Design sub-mains, main lines, and supply lines such that required water quantities can be conveyed to all operating lateral lines at required pressures.
- Design pump and power units such that they are adequate to efficiently operate the sprinkler system at design capacity and total dynamic head.

3.5.3.3 Design Steps and Procedures

(1) Determine the daily maximum supply requirement for an area (A) for the target crop as

$$V_A = \frac{A \times ET_{max}}{E_a} \qquad (3.28)$$

where

V_A = Required volume of water for the area A (m³)
A = Specific area that is to be irrigated (m²)
ET_{max} = Daily maximum evapotranspiration (m)
E_a = Design application efficiency of the sprinkler

ET_{max} can be calculated as

$$ET_{max} = K_c \times ET_0 \qquad (3.29)$$

where

K_c = crop-coefficient at peak water demand period
ET_0 = reference evapotranspiration at peak water demand period

Note: For specific application, when some soil moisture deficit is allowed, ET can be calculated as follows:

$$ET_{max} = K_c \times K_s \times ET_0 \qquad (3.30)$$

where K_s is the soil moisture stress coefficient.

Detailed procedure for calculating ET_0 and ET has been described in Chapter 7 (*Field Water Balance*), Volume 1.

(2) Determine discharge rate (Q) for the area A based on the minimum operating hour.

$$Q_A(m^3/h) = \frac{V_A(m^3)}{t(h)} \qquad (3.31)$$

(3) Optimize sprinkler and lateral spacing for the individual sprinkler discharge rate and application rate (which is constrained by the soil infiltration rate) from the following relationship:

$$q = S_m \times S_l \times I \qquad (3.32)$$

where

S_l = sprinkler spacing along laterals (m)
S_m = lateral spacing along mainline (m)
I = average application rate (m/h)
q = discharge rate for the individual sprinkler (m³/h) [for the area $(S_m \times S_l)$ m²

3.5 Design of Sprinkler System

$$I = \frac{Q_A}{A}, \text{ if not limited by soil intake rate.}$$

$S_l = D_{ml}\left[1 - \frac{F}{2}\right]$, where D_{ml} is the manufacturer's rated wetting diameter of lateral sprinkler, F is the overlapping factor

$S_m = D_{mm}\left[1 - \frac{F}{2}\right]$, where D_{mm} is the manufacture's rated wetting diameter of mainline sprinkler, F is the overlapping factor

Overlapping factor is normally taken as 0.5–0.75. For windy condition, overlapping factor may be as high as 1.0.

(4) Number of sprinklers, $n = \dfrac{A}{S_l \times S_m}$

(5) Determine system capacity as

$$Q_{ST} = \sum_{i=1}^{n} Q_{Ai} \qquad (3.33)$$

where i is the number of sub-area like "A"

(6) Determine the power requirement to pump the water for a sprinkler system as

$$P = Q_{ST} \times 9.81 \times H_T$$

where

P = power, KW
Q_{ST} = total discharge rate for the system, m³/s
H_T = total pumping head, m

Total head consists of the following: $H_T = H_m + H_f + H_r + H_s + H_{sf}$

where

H_m = pressure head required to operate the sprinklers at minimum required pressure (m)
H_f = total frictional head in the lines (m)
H_r = maximum riser height from the pump level (m)
H_s = suction head (vertical difference between pump level and source water level after drawdown) (m) (if needed)
H_{sf} = friction head loss in suction line (if suction line exists)

Procedure for calculating friction loss has been described in Chapter 1 (*Conveyance Loss and Designing Conveyance System*), this volume.

General Guideline for Minimum Pressure of Sprinkler Irrigation System

The following pressure estimates can be used as a general guide or thumb rule for sprinkler pressure:

Spray type sprinkler head = 40 PSI (93 ft head)
Rotor type sprinkler head = 45 PSI (104 ft head)

3.5.3.4 Sample Workout Problems

Example 3.8

In a sprinkler irrigation system, the lateral spacing along the mainline is 20 m and sprinkler spacing along laterals is 15 m. The application rate for fulfilling the peak demand of the proposed crop should be 8 mm/d. Find the discharge rate per sprinkler.

Solution

We know application rate, $I(\text{mm/hr}) = \dfrac{3600 \times q_s(\text{l/s})}{S_m(m) \times S_l(m)}$

Or, $q_s = \dfrac{I \times S_m \times S_l}{3600}$

Given,

$I = 8$ mm/d
Assuming that the sprinkler will operate 12 h a day
Then, $I = 8$ mm/12 h $= 0.667$ mm/h
$S_m = 20$ m
$S_l = 15$ m
Putting the values, $q_s = 0.055$ l/s
Or, 200 l/h (Ans.)

Another Mode of Operation

If we consider that the daily demand should be provided within a certain practical irrigation period, say in 4 h, to avoid excessive evaporation loss, then the application rate would be $I = 8$ mm/4 h $= 2$ mm/h
Thus, $q_s = 0.1667$ l/s or 600 l/h (Ans.)

Example 3.9

In a sprinkler irrigation system, the required total capacity of the system is 0.5 m³/s. Determine the pump capacity. Assume that head loss in pipe and bends and velocity head required $= 3$ m of water.

Solution

Pump capacity, $P = (Q \times 9.81 \times H)$ [KW]
Here $Q = 0.5$ m³/s
Total head $= 3$ m
Putting the values, $P = 0.5 \times 9.81 \times 3 = 14.7$ KW (Ans.)

3.5 Design of Sprinkler System

Example 3.10

A farm of 25 ha is planned to be brought under sprinkler irrigation. The textural class of the soil is loam-to-silt loam, having moisture content at field capacity (FC) and permanent wilting point (WP) of about 42% (by volume) and 26% (by volume), respectively. An infiltration test data showed that constant (basic) infiltration rate is 2 mm/h. A hardpan (relatively impervious layer) exists at a depth of 2.0 m below the soil surface. Long-term average reference evapotranspiration (ET_0) rate in that area is 4.5 mm/d. Vegetable crops are planned to grow in the farm, and the crop coefficient (K_c) at maximum vegetative period is 1.1. The climate is moderately windy in a part of the season. Design the sprinkler irrigation system (various components) for the farm. Assume standard value of any missing data.

Solution

Given,

    Area, $A = 25$ ha $= 250,000$ m^2
    FC $= 42\%$ (by vol.)
    WP $= 26\%$
    $I_c = 2$ mm/h
    $ET_0 = 4.5$ mm/d
    $K_c = 1.1$
    $D_{imp} = 2$ m below soil surface
    Wind status: moderately windy
    Now, the solution steps:

(1) $ET_{max} = ET_0 \times K_c = 4.5 \times 1.1 = 4.95$ mm/d
(Assuming depletion of soil-moisture up to readily available level, so that ET occurs at its maximum rate, i.e., soil moisture stress factor, $K_s = 1$)
(2) Daily water requirement for the area, A (i.e., for whole farm here) is

$$V_A = \frac{A \times ET_{max}}{E_a}$$

Assuming application efficiency, $E_a = 80\%$, i.e., 0.8

Then, $V_A = \dfrac{250,000 \times (4.95/1000)}{0.8} = 1,546.875$ m^3

(3) Discharge rate, $Q_t = \dfrac{V_A}{t}$
Here, $t =$ irrigation period $= 4$ h (assuming for the prevailing windy condition)
Thus, $Q_t = 1,546.875/(4 \times 3,600) = 0.1074$ m^3/s
(4) Discharge rate of individual sprinkler, $q = S_m \times S_l \times I$
    $S_m = D_{mm}(1 - F/2)$
    $S_l = D_{ml}(1 - F/2)$

Assuming overlapping factor, $F = 0.7$ (higher for windy condition)
Taking a manufacturer rated wetting diameter for mainline and lateral sprinkler as 12 m and 10 m, respectively, we get
$S_m = 12 (1 - 0.7/2) = 7.8$
$S_l = 10 (1 - 0.7/2) = 6.5$
Application rate, $I = Q_A/A = (0.1074 \times 3600)/250,000 = 4.296 \times 10^{-7}$ m/h = 1.5468 mm/h, which is less than the soil infiltration rate.
Here, assuming $I = 2$ mm/h (to minimizing evaporation loss in windy climate)
Putting the values, $q = [7.8 \times 6.5 \times (2/1,000)] \times (1,000/60) = 1.69$ l/min

(5) Number of sprinklers, $n = \dfrac{A}{S_m \times S_l} = \dfrac{250,000}{7.8 \times 6.5} = 4{,}930.9 \approx 4{,}931$ nos

Note: The above calculation is for fixed lateral. If moving lateral is used, no. of laterals should be based on the maximum working /pump operating period. Note that each setting requires 4 h for the above calculation, so $16/4 = 4$ settings can be operated if 16 h is the working period.
Besides, number of laterals should be based on the dimension of the land, lateral size available in the market, etc.

(6) Power required (motor capacity required), P (KW) $= Q$ (m³/s) $\times 9.81 \times H_T$ (m)
Here, $Q = 0.1074$ m³/s
$H_T = H_m + H_f + H_r + H_s + H_{sf}$
H_m = pressure head required to operate the sprinklers at minimum required pressure (m) = 28.37 m (= 40 psi) (assuming)
H_f = total frictional head in the lines (m) $\approx$ 5% of H_m = 1.71 m (assuming/estimating)
H_r = maximum riser height from the pump level (m) = 1.5 m (assuming)
H_s = suction head (vertical difference between pump level and source water level after drawdown) (m) (if needed) = 0 (assuming that water is pumped from the supply canal)
H_{sf} = friction head loss in suction line (if suction line exists) = 0
Thus, $H_T = 31.26$ m
Thus, $P = 0.1074 \times 9.81 \times 31.26 = 32.94$ KW

(7) Summary design parameters are as follows:
Taking for fixed laterals:
Pump capacity: $Q_t = 0.1074$ m³/s
Irrigation period = 4 h
Motor capacity: $P = 32.94$ KW
Lateral spacing along mainline: $S_m = 7.8$ m
Sprinkler spacing along lateral: $S_l = 6.5$ m
Number of total sprinklers: $n = 4{,}931$ nos (Ans.)

Relevant Journals

– Agricultural Water Management
– Irrigation Science

- Irrigation and Drainage System
- ICID Bulletins
- Agricultural Systems
- Journal of Irrigation and Drainage Division, ASCE
- Transactions of the American Society of Agricultural & Bio-system Engineers (former ASAE)
- Applied Engineering in Agriculture

Relevant FAO Papers/Reports

- FAO Irrigation and Drainage Paper 45 (Guidelines for designing and evaluating surface irrigation systems, 1989)

Questions

Common in Surface Irrigation

(1) What are the principles of designing a surface irrigation system?
(2) What are the variables in designing a surface irrigation system?
(3) Discuss the hydraulic factors associated with surface irrigation system?

Border Irrigation

(4) Briefly describe the factors affecting border performance and design.
(5) What are the points to be considered in border design?
(6) Define the following terms: cutoff time, advance ratio, return flow, recession flow, distribution uniformity, and application efficiency.
(7) What are the design parameters in the case of border system? What are the approaches of the modern border design?
(8) Describe the following empirical model for border design: (a) SCS method and (b) Alazba's method.
(9) Why is the simulation model useful? Write down the names of some simulation model/software tools for the border irrigation system design and highlight their principal features/working principles.
(10) Design a border strip with the following characteristics: field length, $L = 120$ m, field slope, $S_o = 0.005$, infiltration family, IF $= 0.5$: $k = 0.031$ m/h^a, $a = 0.58$; roughness coefficient, $n = 0.12$, required depth of infiltration, $D_{req} = 0.06$ m.
(11) In a cotton field, a farmer has made border strip of 200 m long. The roughness of the field is estimated as $n = 0.11$, and the average field slope along the border is 0.09%. Determine the required flow rate per unit width of the border.

Basin Irrigation

(12) Briefly describe the factors affecting basin performance and design.
(13) Define the following terms: aspect ratio, check bank.
(14) What are the points to be considered in a basin design?
(15) Describe the hydraulics of the basin system.
(16) Briefly discuss different simulation approaches of the basin design.
(17) Describe the governing equations of Zero-inertia and Hydrodynamic model for basin simulation.
(18) Write down the name of some simulation model/software tools for basin irrigation system design and highlight their principal features/working principles.

Furrow Irrigation

(19) Describe the hydraulics of the furrow irrigation system.
(20) Describe different approaches of simulating surface flow and infiltration process in the furrow irrigation system.
(21) Define the following terms: infiltration opportunity time, time ratio, cutoff ratio, advance rate, surge flow.
(22) What are management controllable variables and design variables in the furrow system?
(23) What are the considerations in the furrow design?
(24) Describe the governing equations of kinematic-wave model with respect to furrow irrigation.
(25) Discuss some general guidelines for the furrow design. Have you any suggestions for improving furrow irrigation system?
(26) Name some software tools for furrow design, and highlight their principal features/working principles.
(27) Furrows of 120 m long and 0.75 m wide and having a slope of 0.3% are irrigated for 35 min with a stream size of 0.005 m^3/s. Determine the average depth of water.
(28) Furrows of 110 m long and 0.70 m wide and having a slope of 0.25% are initially irrigated for 40 min with a stream size of 0.0045 m^3/s. The stream size is then reduced to half and continued for 20 min. The furrow end is closed (no outflow from furrow). Determine the average depth of infiltrated water.

Sprinkler Irrigation

(29) What are the design parameters in sprinkler irrigation system?
(30) Define precipitation rate and application rate.
(31) Draw a sketch of water distribution pattern in the sprinkler system.
(32) What are the factors to be considered in the sprinkler design?
(33) Write down the general consideration in the sprinkler design.

(34) What are the design principles in the sprinkler system?
(35) Briefly describe the design steps and procedures of the sprinkler irrigation system.

References

Alazba AA (1997) Design procedure for border irrigation. Irri Sci 18:33–43
Alazba AA (1998) Quantitative management variable equations for irrigation borders. Appl Eng Agric 14(4):583–589
Bautista E, Clemmens AJ, Strelkoff T, Niblack M (2009) Modern analysis of surface irrigation systems with WinSRFR. Agric Water Manage 96:1146–1154
Bautista E, Wallender WW (1992) Hydrodynamic furrow irrigation model with specified space steps. J Irrig Drain Engg ASCE 118(3):450–465
Boonstra J, Jurriens M (1988) BASCAD. A mathematical model for level basin irrigation. ILRI Publication 43, Int. Inst. for Land Reclamation & Improvement, Wegeningen, The Netherlands, 30 pp
Bradford SF, Katopodes ND (2001) Finite volume model for no level basin irrigation. J Irrig Drain Eng 127(4):216–223
Burt CM, Clemens AJ, Strelkoff TS, Solomon KH, Bliesner RD, Hordy LA, Howell TA, Eisenhauer DE (1997) Irrigation performance measures: efficiency and uniformity. ASCE J Irrig Drain Div 123(6):423–442
Chaudhry MH (1993) Open-channel flow. Prentice-Hall, Englewood Cliffs, NJ
Clemmens AJ, Strelkoff T (1979) Dimensionless advance for level-basin irrigation. J Irrig Drain ASCE 105(IR3):259–293
Davis JR (1961) Estimating rate of advance for irrigation furrows. Trans ASAE 4:52–57
Hall WA (1956) Estimating irrigation border flow. Agric Engg 37:263–265
Hill RW (2008) Management of sprinkler irrigation systems. Utah State University, Logan, Utah
Holzapfel EA, Marino MA, Morales JC (1986) Surface irrigation optimization models. J Irri Drain Div ASCE 112(1):1–19
Holzapfel EA, Marino MA, Morales JC (1987) Surface irrigation non-linear optimization models. J Irri Drain Div ASCE 113(3):379–392
Katapodes ND, Strelkoff T (1977) Hydrodynamics of border irrigation – complete model. J Irrig Drain Div ASCE 113(IR3):309–324
Khanna M, Malano HM, Fenton JD, Turral H (2003a) Design and management guidelines for contour basin irrigation layouts in southern Australia. Agric Water Manage 62:19–35
Khanna M, Malano HM, Fenton JD, Turral H (2003b) Two-dimensional simulation model for contour basin layouts in southeast Australia. I. Rectangular basins. J Irrig Drain ASCE 129(5):305–316
Khanna M, Malano HM, Fenton JD, Turral H (2003c) Two-dimensional simulation model for contour basin layouts in southeast Australia. II. Irregular shape and multiple basins. J Irrig Drain ASCE 129(5):317–325
Levien SLA, de Souza F (1987) Algebraic computation of flow in furrow irrigation. J Irri Drain Eng ASCE 113(3):367–377
Michael MA (1978) Irrigation theory and practices. Vikas Publishing, India
Muga A, Goncalves JM, Pereira LS (2008) Web decision support system for surface irrigation design. CIGR – International Conference of Agricultural Engineering, XXXVII Congresso Brasileiro de Engenharia Agricola, Brazil, Aug 31–Sept 4
Playan E, Faci JM, Serreta A (1996) Modeling micro-topography in basin-irrigation. J Irrig Drain ASCE 122(6):339–347
Raghuwanshi NS, Wallender WW (1998) Optimization of furrow irrigation schedules, designs and net return to water. Agric Water Manage 35:209–226

Reddy M, Clyma W (1981) Optimal design of furrow irrigation system. Trans ASAE 24(3):617–623

Saint Venant AJCB (1871) Theorie du movement non permanent des eaux avec application aux crues des rivieres et a l'introduction des marees dans leur lits. Comtes Rendues des Seances de l'Academie des Sci 73:147–154

Schmitz G, Seus GJ (1992) Mathematical zero-inertia modeling of surface irrigation: advance in furrows. J Irrig Drainage Eng ASCE 118(1):1–18

Singh V, Bhallamudi SM (1997) Hydrodynamic modeling of basin irrigation. J Irrig Drain ASCE 123(6):407–414

Solomon KH (1990) Sprinkler irrigation uniformity. Irrigation Notes, California State University, Fresno, California 93740-0018

Strelkoff TS, Al-Tamaini AH, Clemmens AJ (2003) Two-dimensional basin flow with irregular bottom configuration. J Irrig Drain Eng ASCE 129(6):391–401

Strelkoff TS, Al-Tamaini AH, Clemmens AJ, Fangmeier DD (1996) Simulation of two-dimensional flow in basins. International Meeting, Phoenix Givic Plaza, Arizona, July 14–18

Tabuada MA, Rego ZJC, Vachaud G, Pereira LS (1995) Modelling of furrow irrigation. Advance with two-dimensional infiltration. Agric Water Manage 28:201–221

USDA (1955) U.S. Department of Agriculture Yearbook, 1955, 'Water'

USDA (1974) Border irrigation. National Engineering Handbook, Chapter 4, Section 15. US Soil Conservation Service (SCS), Washington, DC

USDA-ARS-ALARC (2006) WinSRFR 2.1 User manual – Draft. Surface Irrigation Analysis, Design & Simulation. U.S. Department of Agriculture, Agricultural Research Service, Arid-Land Agricultural research Centre, Maricopa, AZ 85238

Walker WR, Humphreys AS (1983) Kinematic-wave furrow irrigation model. J Irrig Drainage Eng ASCE 109(IR4):377–392

Walker W (1989). The surface irrigation simulation model: user's guide (SIRMOD). Irrigation Software Engineering Division, Department of Biological and Irrigation Engineering, Utah State University, Logan, Utah

Wallender WW, Rayej M (1990) Shooting method for Saint Venant equations of furrow irrigation. J Irrig Drainage Eng ASCE 116(1):114–122

WU IP, Liang T (1970) Optimal design of furrow length of surface irrigation. J Irri Drain Div ASCE 96(3):319–332

Zazueta FS, Miller G (2000) Turf irrigation with a hose and sprinkler. University of Florida & IFAS Extension, AE265

Chapter 4
Performance Evaluation of Irrigation Projects

Contents

4.1	Irrigation Efficiencies	112
	4.1.1 Application Efficiency	112
	4.1.2 Storage Efficiency/Water Requirement Efficiency	114
	4.1.3 Irrigation Uniformity	114
	4.1.4 Low-Quarter Distribution Uniformity (or Distribution Uniformity)	115
4.2	Performance Evaluation	116
	4.2.1 Concept, Objective, and Purpose of Performance Evaluation	116
	4.2.2 Factors Affecting Irrigation Performance	117
	4.2.3 Performance Indices *or* Indicators	118
	4.2.4 Description of Different Indicators	120
	4.2.5 Performance Evaluation Procedure	126
	4.2.6 Performance Evaluation Under Specific Irrigation System	128
	4.2.7 Improving Performance of Irrigation System	134
Relevant Journals		137
Relevant FAO Papers/Reports		137
Questions		137
References		137

Any water applied above that needed to grow a crop is inefficient use of water. In order to determine how much irrigation water to apply, it is needed to estimate irrigation efficiency. There are many definitions of irrigation efficiency. Which one to use depends on which aspect one is interested in. Efficiency can be measured at the scale of a whole catchment, at the individual plant scale, and at almost any level in between. The scale of measurement depends on the focus of the person doing the measurement. A range of issues affect irrigation efficiency.

For successful design, implementation, and execution of command area development programme, systematic evaluation of various components of the existing system is necessary. Adequate monitoring and evaluation of performance are needed to improve water management practices in order to achieve an increase in overall efficiency. In general, evaluation helps to identify problems and the measures required to correct them. No single indicator is satisfactory for all descriptive purposes. In general, a set of indices are used for evaluating the performance of an irrigation scheme. Most commonly used indices are described in this chapter. In addition, evaluation procedures for specific irrigation systems (e.g., furrow, border, sprinkler) are described in details.

4.1 Irrigation Efficiencies

In order to determine how much irrigation water to apply, you need to estimate the efficiency of the irrigation system. There are many ways of thinking about, determining, and describing concepts relating to irrigation efficiency. Simply speaking, the "efficiency" implies a ratio of something "in" to something "out". Many efficiency terms related to irrigation efficiency are in use or have been proposed.

Efficiency can be measured at the scale of a whole catchment, at the individual plant scale, and at almost any level in between. The scale of measurement is of critical importance in tackling the issue of improving efficiency and must be matched with the specific objective. For example, when measuring on-farm efficiency, too broad a scale makes it difficult to determine what the causes of low efficiency are and what can be done to improve the situation. Going to a smaller scale excludes the consideration of wider issues, such as delivery system losses and inefficiencies but is necessary to clearly identify real opportunities for improvement at the individual property/manager scale. Commonly used irrigation efficiencies are described below.

4.1.1 Application Efficiency

Water application efficiency expresses the percentage of irrigation water contributing to root zone requirement. It indicates how well the irrigation system can deliver and apply water to the crop root zone. Hence, the application efficiency takes into account losses such as runoff, evaporation, spray drift, deep drainage, and application of water outside the target crop areas. Of these factors, deep drainage and runoff are probably the major causes of inefficiency and are generally due to overwatering. Whenever more water is applied than can be beneficially used by the crop, water is wasted and efficiency is low.

Application efficiency defined by different researchers varies slightly in the expression (Bos and Nugteren, 1974; ASCE, 1978; Jensen et al., 1983; Walker and Skogerboe, 1987; Bos et al., 1993; Solomon, 1988; Burt et al., 1997; Heermann et al., 1990). In broad term, application efficiency is the percentage of water delivered to the field that is ready for crop use. As the application efficiency is a measure of how efficiently water has been applied to the root zone of the crop, this parameter relates the total volume of water applied by the irrigation system to the volume of water that has been added to the root zone and is available for use by the crop. Thus, the application efficiency (E_a) is calculated as (Wingginton and Raine, 2001):

4.1 Irrigation Efficiencies

$$E_a = \text{(irrigation water available to the crop)} \times 100/\text{(water delivered to the field)} \quad (4.1)$$

where

Irrigation water available to the crop = root zone soil moisture after irrigation − root zone soil moisture prior to irrigation

Water delivered to the field = flow meter reading or nozzle flow rate

Kruse (1978) defined application efficiency as

$$E_a = \text{(average depth of water stored in the root zone)} \times 100/\text{(average depth applied)} \quad (4.2)$$

For in-field evaluations where the depth of water applied is less than the root zone moisture deficit prior to irrigation and runoff is not evident, the irrigation water available to the crop can be assumed to be equal to the average depth of water applied as measured at the soil surface (e.g., with catch cans in sprinkler system). In these cases,

$$E_a = [\text{average depth applied (mm)} \times \text{area (ha)}/10] \times 100/[\text{water delivered to the field}(m^3)] \quad (4.3)$$

Note: $(1 \text{ mm} \times 1 \text{ ha})/10 = 1 \text{ m}^3$

Application efficiency is primarily affected by the management of the irrigation and may vary significantly between irrigation events.

Attainable water application efficiencies of different irrigation systems under normal condition are given in Table 4.1.

Table 4.1 Attainable application efficiencies under different irrigation systems (adapted from Solomon, 1988)

Type of irrigation system	Attainable efficiency range
Surface irrigation	
Border	75–85%
Basin	80–90%
Furrow	65–80
Sprinkler	
Solid set or permanent	75–85%
Hand move or portable	75–85%
Center pivot & linear move	75–90%
Traveling gun	65–75%
Trickle irrigation	
Point source emitters	80–90%
Line source	75–85%

4.1.2 Storage Efficiency/Water Requirement Efficiency

Storage efficiency indicates how well the irrigation satisfies the requirement to completely fill the target root zone soil moisture. Thus, storage Efficiency (E_S) is represented as

$$E_S = \text{(change in root zone soil moisture)} \times 100/\text{(target change in root zone soil moisture)} \quad (4.4)$$

where the change in the root zone soil moisture content is not measured directly, the storage efficiency can be approximated by relating the average depth of water applied over the field to the target root zone deficit. The root zone deficit is calculated using soil type, crop root zone, and soil moisture content data. In this case, the storage efficiency is calculated as

$$E_S = \text{(average depth applied)} \times 100/\text{(root zone deficit)} \quad (4.5)$$

The maximum storage efficiency is 100%. Calculations with a result above 100% indicate losses due to runoff or deep drainage.

4.1.3 Irrigation Uniformity

Irrigation uniformity is a measure of how uniform the application of water is to the surface of the field. That is, it is an expression that describes the evenness of water application to a crop over a specified area, usually a field, a block, or an irrigation district. The value of this parameter decreases as the variation increases. It applies to all irrigation methods, as all irrigation systems incur some nonuniformity. An irrigation uniformity of 100% would mean that every point within the irrigated area received the same amount of water as every other point.

An important component of the evaluation of in-field irrigation performance is the assessment of irrigation uniformity. If the volume of water applied to a field is known, then the average applied depth over the whole field can be calculated. In most cases, one half of the field receives less than the average depth and one half more than the average depth applied. Hence, if the average volume applied is the target application required to meet the crop requirements, one half of the field has been over-irrigated (reducing the efficiency of application) while the other half of the field has been under-irrigated (potentially reducing yield). Thus, a major aim of irrigation management is to apply water with a high degree of uniformity while keeping wastage to a minimum.

The uniformity of application is primarily a function of the irrigation system design and maintenance. Low levels of uniformity limit the maximum efficiency achievable. Numerous irrigation uniformity coefficients are used in performance evaluation. Commonly used irrigation uniformities are as follows: Christiansen's

4.1 Irrigation Efficiencies

uniformity coefficient and low-quarter distribution uniformity (or simply distribution uniformity).

4.1.3.1 Uniformity Coefficient

Uniformity coefficient, introduced by Christiansen (1942), is defined as the ratio of the difference between the average infiltrated amount and the average deviation from the infiltrated amount, to the average infiltrated amount. That is,

$$\text{UCC} = \left[1 - \frac{\sum_{i=1}^{i=N} |Z_i - Z_{av}|}{Z_{av} \times N} \right] \times 100 \quad (4.6)$$

where

UCC = Christiansen uniformity coefficient (or simply uniformity coefficient)
Z_i = infiltrated amount at point i
Z_{av} = average infiltrated amount
N = number of points used in the computation of UCC

Christiansen developed uniformity coefficient to measure the uniformity of sprinkler systems, and it is most often applied in sprinkler irrigation situation. It is seldom used in other types of irrigation. Values of UCC typically range from 0.6 to 0.9.

4.1.4 Low-Quarter Distribution Uniformity (or Distribution Uniformity)

Low-quarter distribution uniformity (DU_{lq}) is defined as the percentage of the average low-quarter infiltrated depth to the average infiltrated depth. Mathematically,

$$DU_{lq} = 100 \times \frac{LQ}{M} \quad (4.7)$$

where

DU_{lq} = distribution uniformity at low quarter (or simply distribution uniformity, DU)
LQ = average low-quarter depth infiltrated (mm)
M = average depth infiltrated (mm)

The "average low-quarter depth infiltrated" is the average of the lowest one-quarter of the measured values where each value represents an equal area.

For calculation of DU of low one-half, substitute "low quarter" by "average low-half depth received or infiltrated."

The DU_{lq} has been applied to all types of irrigation systems. In trickle irrigation, it is also known as "Emission Uniformity." In sprinkler situation, it is termed "Pattern Efficiency."

The relationship between DU and UCC can be approximated by (USDA, 1997)

$$UCC = 100 - 0.63\,(100 - DU)$$
$$DU = 100 - 1.59\,(100 - UCC)$$

Distribution uniformity is primarily influenced by the system design criteria. Poor uniformity of application is often easily identified by differences in crop response and/or evidence of surface waterlogging or dryness. The part of the field receiving more than the average depth may suffer from inefficiencies due to waterlogging and/or runoff, while the other part receiving less than the average may suffer from undue water stress. Thus, uniform irrigation is important to ensure maximum production and minimum cost.

4.2 Performance Evaluation

4.2.1 Concept, Objective, and Purpose of Performance Evaluation

4.2.1.1 Concept

Performance terms measure how close an irrigation event is to an ideal one. An ideal or a reference irrigation is one that can apply the right amount of water over the entire area of interest without loss.

Evaluation is a process of establishing a worth of something. The "worth" means the value, merit, or excellence of the thing.

Performance evaluation is the systematic analysis of an irrigation system and/or management based on measurements taken under field conditions and practices normally used and comparing the same with an ideal one.

Traditionally, irrigation audits are conducted to evaluate the performance of existing irrigation systems. A full irrigation audit involves an assessment of the water source characteristics, pumping, distribution system, storage, and in-field application systems. However, audits are also conducted on several components of on-farm irrigation system.

4.2.1.2 Objectives

The modernization of an irrigated area must start with a diagnosis of its current situation. Following this procedure, the specific problem affecting water use can be addressed and that may lead to a feasible solution. The specific objectives of performance evaluation are as follows:

4.2 Performance Evaluation

- To identify the causes of irrigation inefficiencies
- To identify the problem/weak point of irrigation management
- To diagnose the water management standard of the irrigation project
- To determine the main principles leading to an improvement of irrigation performance

4.2.1.3 Purposes

The purpose of performance assessment is to measure, through consistently applied standards, various factors that indicate either by comparison across systems whether a system is performing 'well' or 'badly' in a relative sense or by a system-specific analysis to see how the system is operating in relation to its own objectives. The specific purposes are as follows:

- to improve irrigation performance
- to improve management process
- to improve sustainability of irrigated agriculture

4.2.1.4 Benefits of Evaluation

Evaluation leads to the following benefits:

- Improved quality of activities
- Improved ability of the managers to manage the system
- Savings of water and energy
- Ensure maximum production/benefit and minimum cost

4.2.2 Factors Affecting Irrigation Performance

The performance of an irrigation system at field scale depends on several design variables, management variables, and system variables *or* factors. These factors characterize an irrigation event. Mathematically, it can be expressed as

$$P_{ir} = f(q_{in}, A, L, W, N, S_0, I_n, t_{cutoff}, S_w, D_{ru}, P, R_d, ET, W.....), \quad (4.8)$$

where

P_{ir} = performance of an irrigation event
f = function
q_{in} = inflow rate or application rate (to the furrow, or per unit width of border or basin, or per emitter or sprinkler)
A = sectional form of the unit plot to be irrigated (specially for furrow)
L = length of run of the flow

W = width of the section *or* unit plot
N = roughness coefficient of flow for the plot (Manning's N)
S_0 = longitudinal slope of the plot
I_n = infiltration characteristics of the soil
t_{cutoff} = time of cutoff
S_w = soil water status at the time of irrigation (i.e., condition of deficit)
D_{ru} = reuse of drainage runoff (if applicable)
P = pressure of the flow system (specially for sprinkler)
R_d = root zone depth of the crop during the irrigation event
ET = atmospheric water demand or evapo-transpirative demand (specially for sprinkler)
W = wind factor or windy condition (specially for sprinkler)

Irrigation performance may vary from irrigation event to event, based on the dynamics of some factors such as infiltration characteristics, roughness coefficient, root zone depth, soil water deficit.

4.2.3 Performance Indices or Indicators

Activities of irrigation systems start at the point of water supply head-work or pump. Impacts of irrigation are not limited to the field but also extend to the socioeconomic conditions of the target audience. In general, a set of indices or indicators are used for evaluating the performance of an irrigation scheme. The indicators are termed as performance indicators. No single indicator is satisfactory for all descriptive purposes. Moreover, there are uncertainties about the exact values of some indicators. Several indicators can give an overall picture of the irrigation project.

For convenience in understanding and application, the indicators can be grouped as

- Engineering
- Field water use
- Crop and water productivity and acreage
- Socioeconomic

4.2.3.1 Engineering Indicators

Engineering indicators are those which are related to pump, water headwork, water supply, water conveyance system, and energy use. Indices under this category include (Sarma and Rao, 1997; Ali, 2001) the following:

  (i) Pumping plant efficiency
 (ii) Headworks efficiency
(iii) Water conveyance efficiency

4.2 Performance Evaluation

(iv) Water delivery performance
(v) Irrigation system efficiency (or overall efficiency)
(vi) Equity of water delivery
(vii) Channel density
(viii) Water supply – requirement ratio
(ix) Water availability and shortage
(x) Energy use efficiency

4.2.3.2 Field Water Use Indicators

These indicators concern the efficiency of on-farm water application and the uniformity of water distribution along the irrigated field. Indicators under this category are as follows:

- On-farm water loss
- Deep percolation fraction/deep percolation ratio
- Runoff fraction/tailwater ratio
- Water application efficiency
- Storage efficiency/water requirement efficiency
- Application efficiency of low quarter
- Distribution efficiency or uniformity
- Low-quarter distribution uniformity

4.2.3.3 Crop and Water Productivity and Acreage

Indicators under this category are as follows:

- Area irrigated
- Irrigation intensity
- Duty of discharge/supply water
- Crop productivity (Yield rate)
- Water productivity
- Irrigation water productivity

4.2.3.4 Socioeconomic Indicators

In some cases, cost–benefit or social uplift and social acceptance aspects are measured. These are called socioeconomic indicators. Indicators under this category include the following:

- Irrigation benefit–cost ratio
- Cost per unit production
- Irrigation cost per unit area
- Farmers income ratio

4.2.4 Description of Different Indicators

4.2.4.1 Engineering Indicators

Pumping Plant Efficiency

Pumping plant efficiency (E_{pp}) is calculated as

$$\begin{aligned} E_{pp} &= \text{(Output horsepower)} \times 100/\text{(Input horsepower)} \\ &= \text{(water horsepower)} \times 100/\text{(Input horsepower)} \\ &= [\{(Q \times \omega) \times H\}/550] \times 100/\text{(Input horsepower)} \end{aligned} \quad (4.9)$$

where

Q is the discharge rate (cfs)
ω is the density of water (1b/ft^3)
H is the head of water (ft) [here, head indicates the velocity head]
"550" is the factor to convert "ft-lb/s" to horse power

In SI unit, the above formula can be expressed as

$$\begin{aligned} E_{pp} &= \text{(Water power)} \times 100/\text{(Input power)} \\ &= [(Q \times 9.81 \times H) \times 100]/\text{(Input power)} \end{aligned} \quad (4.10)$$

where

"Input power" in Kilowatt
Q = discharge rate (m^3/s)
H = head of water (m)

Headworks Efficiency

It expresses how much energy (pressure) is lost through the system's headwork.

$$E_{\text{headworks}} = \text{(energy of water after passing the headworks)} \\ \times 100/\text{(energy of water before entering the headworks)} \quad (4.11)$$

Conveyance Efficiency

Conveyance efficiency (E_c) means the percent of the water reaching the field plot on the basis of water diverted and is calculated as

$$E_c = \text{(water reached to the plot)} \times 100/\text{(water diverted from the source)} \quad (4.12)$$

For the whole command area of a watercourse, average E_c can be computed as

4.2 Performance Evaluation

$$E_c = \frac{Q_d - \frac{C_{SL} \times L_{av}}{100}}{Q_d} \times 100 \quad (4.13)$$

where

E_c is the conveyance efficiency in percent,
Q_d is the pump discharge in cumec (m³/s).
C_{SL} is the average steady state conveyance loss (m³/s)/100 m
L_{av} is the average channel length of the field plots (m)

To obtain the average channel length, the command area should be divided into n unit areas considering the distance from the pump. A representative diversion point for each unit area is identified, and the length of the channel section from the pump (or field inlet channel) to the diversion point is measured. The average channel length is then calculated as

$$L_{av} = \frac{\sum L_i}{n}$$

where n is the number of sections.

The steady state conveyance loss may be determined by using inflow-outflow method (described in Chapter 1, this volume). In this method, discharge measurement may be done by a cut-throat flume (water flow measurement has been described in Chapter 10, Volume 1).

Water Supply – Requirement (SR) Ratio

SR ratio = (water supplied to the scheme/water required to the scheme)

Channel Density

Channel density means the channel length per unit of cultivated irrigated area.

Channel density = (Total channel length, m)/ (Total irrigated area, ha) (4.14)

Water Delivery Performance (WDP)

$$\text{WDP} = \sum_{t=1}^{N} \frac{K(t) * V(t)}{V^*(t)} \quad (4.15)$$

where

$K(t)$ = weightage factor indicating the relative importance of irrigation at the tth period (say week) of the crop
$V(t)$ = volume of water delivered to the farm during the tth period
$V^*(t)$ = target volume of water to be delivered to the farm during the tth period
N = number of time periods for the crop growth

Equity of Water Delivery (EWD)

Equity of water delivery is expressed as

$$\text{EWD} = \frac{\text{WDP}_t}{\text{WDP}_h} \tag{4.16}$$

where

WDP_t = WDP (Water delivery performance) value at the tail of the outlet
WDP_h = WDP value at the head of the outlet

Overall Efficiency (OE)

OE = (volume of water needed to maintain the soil moisture above a minimum level required for crop + nonirrigation deliveries from the distribution and conveyance system)/(Volume of water diverted into the system + inflows from other sources)

(4.17)

Project Application Efficiency

Project application efficiency (E_{pa}) is defined (USDA, 1997) as the ratio of the average depth of irrigation water infiltrated and stored in the plant root zone to the average depth of irrigation water diverted or pumped. That is

E_{pa} = (average depth infiltrated) × 100/(average irri. water diverted or pumped)

(4.18)

It includes the combined efficiencies from conveyance and application. It can be the overall efficiency of only on-farm facilities.

Energy Use Efficiency

It relates to how much energy we need to produce one ton of crop.

$E_{\text{energy use}}$ = (crop yield in one hectare land)
  × 100/(energy needed for successful cultivation of one hectare land)

(4.19)

4.2 Performance Evaluation

4.2.4.2 Field Water Use Indicators

Although field irrigation water use performance can be influenced by a large array of factors, several measures together can provide a picture of how well irrigation water is being used. Three main indices are commonly used to measure the application system performance:

- Application efficiency
- Distribution uniformity
- Storage efficiency

The above three irrigation performance measures should be used together to provide an adequate representation of the irrigation system. These are described in an earlier section (Irrigation efficiencies).

It is generally difficult to design and manage irrigation systems in a manner that maximizes all three of these indices simultaneously due to the conflicting nature of each index. For example, high application efficiencies may be obtained through significant under-irrigation producing a low storage efficiency, irrespective of the uniformity of this application.

Other indicators are as follows:

Application Efficiency of Low Quarter

Application efficiency of low quarter ($E_{a,lq}$) is defined as (Merriam and Keller, 1978)

$$E_{a,lq} = 100 \times \frac{d_{a,lq}}{D} \tag{4.20}$$

where

$d_{a,lq}$ = average low quarter depth of water added to root zone storage (mm) [$d_{a,lq} \leq SWD$]
D = average depth of water applied (mm)
$E_{a,lq}$ allows to take into consideration the nonuniformity of water application when under-irrigation is practiced.
Both E_a and $E_{a,lq}$ are applicable to surface, sprinkler, and trickle irrigation.

On-Farm Water Loss

It is the amount (depth) of water lost from the crop field per day. Specially, this indicator is used for ponding water applications (e.g., in rice).

The amount of water loss from the plot (from ponding depth) due to seepage and percolation per day or a certain period can be measured by installing vertical or inclined gauges in the plot. In case of inclination, gauge should be inclined by a fixed angle, and then the reading should be transformed to vertical depth by trigonometric formula. Where the change of depth is small, inclined gauge gives accurate reading. In such case, cumulative value for several days may be taken, and then it should be averaged. In addition, magnifying glass may be used for easy and accurate reading.

Deep Percolation Fraction or Deep Percolation Ratio

Deep percolation fraction (D_F) or deep percolation ratio is defined (USDA, 1997) as the ratio of the volume of water percolated below the bottom boundary of the crop root zone to the total volume entered into the soil.

$$D_F = \text{(volume of water percolated below the bottom boundary of the crop root zone)/(total volume entered into the soil)} \quad (4.21)$$

Runoff Fraction or Tailwater Ratio

Runoff fraction (R_F) can be defined as the ratio of the volume of runoff to the volume of water diverted to the plot.

$$R_F = \text{(volume of runoff)/ (volume of water diverted to the plot)} \quad (4.22)$$

4.2.4.3 Crop and Water Productivity and Acreage

Irrigated Area

Irrigated area refers to the area irrigated in a season under the facility provided by the irrigation system. It is expressed in hectares (ha).

Irrigation Intensity

Irrigation intensity is defined as the percentage of irrigable command area irrigated for a season. That is

$$\text{Irrigation intensity} = \text{(Actual area irrigated from a source)} \times 100/\text{(total irrigable area under the scheme)} \quad (4.23)$$

Duty of Discharge

The term "Duty" relates to the command area coverage by the water source and the stream size. Duty (D) refers to the irrigation capacity of the unit discharge or flow.

$$D = \frac{A}{Q} \quad (4.24)$$

where

A = area irrigated with the available supplied water source (ha)
Q = supply capacity (or the discharge of the pump) (m^3/s)

4.2 Performance Evaluation

If a 2.0 cusec supply source irrigates 50 ha land, the duty of the flow is 50/2.0 = 25 ha/cusec.

Crop Productivity/Yield Rate

Crop productivity (CP) refers to the yield per unit area. That is,

$$CP = \frac{Y}{A} \tag{4.25}$$

where

CP = crop productivity, (t/ha)
Y = yield from specific area, A (t)
A = area irrigated with the available supplied water source (ha)

Yield per hectare is the traditional way of representing the performance of agricultural enterprise. It is of interest to irrigators too. Sometimes it can give a false impression of efficiency, when other inputs of production are not being used, or not provided at right amount and/or right time.

Water Productivity (WP) or Water Use Efficiency (WUE)

For a single crop,

$$WP = (\text{Total dry matter or seed yield})/(\text{Water used by the crop}) \tag{4.26}$$

For yearly basis,

$$WP = (\text{Total production of all the crops})/(\text{Total water used by the crops})$$

Here, "total water" represents "rainfall + irrigation + soil moisture depletion from the root zone." Different crops should be converted to equivalent one specific crop.

Irrigation Water Productivity (IWP)

Irrigation water productivity is defined as

$$IWP = \frac{Y}{IW} \tag{4.27}$$

where

Y = Total dry matter or seed yield, t/ha
IW = Irrigation water applied to the crop, mm
IWP = Irrigation water productivity, t/ha-mm

It can also be expressed in yearly basis.

4.2.4.4 Socioeconomic Indicators

Irrigation Benefit–Cost Ratio

It is the ratio of benefit obtained (in monetary form) from unit land to the cost of production for that land. That is

$$r_{B-C} = \frac{I}{C} \quad (4.28)$$

where

r_{B-C} = benefit–cost ratio
I = total monetary benefit *or* income obtained from one hectare land, US$
C = cost of production for that land, US$

For multiple crops or yearly calculation, different crops should be converted to the equivalent of a particular crop.

Cost per Unit Production

It is the cost of production for unit yield. Yield may be total harvestable yield (e.g., forage, grass), or grain, *or* seed yield (for cereals). Its unit is $/ton.

Irrigation Cost per Unit Area

It is the cost of irrigation for unit area for successful crop production. That is

$$\text{Cost per unit area, \$/ha} = \frac{\text{(cost of irrigation for certain area, in\$)}}{\text{(quantity of the irrigated area, ha)}} \quad (4.29)$$

Farmer's Income Ratio

It is the ratio of income of a farmer before irrigation scheme initiation to the income of the same farmer after the irrigation scheme has been established. For this, benchmark income is necessary.

4.2.5 Performance Evaluation Procedure

4.2.5.1 Steps and Techniques

The performance evaluation procedure consists of planning, field work, analysis, and recommendation. The following steps may be followed to carry out a performance evaluation of an irrigation system:

(1) As a first step, identify all factors affecting water use in the irrigation scheme.
(2) In a second step, select appropriate performance indicators for the prevailing conditions of irrigation method, irrigation system, and socioeconomic condition.
(3) Thirdly, measure the components of indicators and compute the indicator values.
(4) As a fourth step, assess the irrigation performance by analyzing/comparing the indicator values with the ideal one.
(5) Finally, analyze the irrigation modernization alternatives using technical and economic criteria, make comparison of alternatives, and suggest/adopt the most appropriate one.

A range of factors affect water use at farm level. These include soil type, farmers irrigation system, climatic condition, crop type, crop rotation, type of farmer (owner or leaser), farming practice, farmers' water management practice, farmers' economic condition, amount of land, education level of the farmers.

Additional information may be recorded through farmer's interview about the size of the test plot, name and code of the owner, the name of the irrigator, etc.

Multiple regression analysis may be performed to assess the relationship between water use (WU) and the factors, first including all the factors and then removing insignificant factors individually and interactively.

4.2.5.2 Queries That Should Be Answered

Performance evaluation of an irrigation project should enable us to answer the following questions:

- Does the supply of water meet the demand (especially at peak demand period) of the irrigators?
- Is the quality of water acceptable for the intended use?
- What is the pumping plant efficiency?
- How much water is lost in supply canal (conveyance) and in the field (deep percolation and runoff)?
- How is the demand of water estimated?
- How frequently is the water applied?
- How much water (in depth) is applied per application?
- Are other crop management events done at the right time and in the right way?
- Are there any pollution problems from the project?
- Is the quality of drainage water reasonable?
- What are the values of water application efficiency, storage efficiency, and distribution uniformity?
- What is the overall irrigation efficiency?

4.2.5.3 Ideal Condition for Evaluation of Irrigation System

The field (soil) and crop condition should represent the ideal/normal field condition during the evaluation of an irrigation system. The conditions can be summarized as follows:

(i) The field soil should be stable, not new, refilled or a developed one
(ii) The crop condition should be representative, not just after emergence or at ripening stage but in between (good coverage)
(iii) The soil should be dry enough – appropriate time for irrigation
(iv) Water supply/water pressure should be sufficient enough to apply inflow in the designed rate

4.2.6 Performance Evaluation Under Specific Irrigation System

4.2.6.1 Pumping Plant Evaluation

Pumping system efficiency can contribute substantially to energy saving. Pumping plant evaluation requires a pump test, which checks the flow rate capacity, lift, discharge pressure and/or velocity, rated discharge capacity, and input horsepower.

Pump discharge can be measured by flow meter, flume (at the vicinity of the pump outlet) or by the coordinate method. Pump lift can be estimated by measuring the depth to water table during non-pumping period using water-level indicator, or at least by inserting a rope with load up to the water table and measuring the distance with a tap. Rated discharge capacity of the pump can be read from the manufacturer's manual *or* the pump rating written on the pump body. If a mechanical engine is used to power the pump, its capacity can be read from its rating seal or manual. If an electrical motor is used to operate the pump, power consumption by the motor can be measured by "Clip-On meter" or "Multi-meter" or from the change in power reading in the "electric meter" for a certain period. Rated capacity of the motor can be read from its body.

Knowing the above information, overall pumping plant efficiency and efficiency of each component (such as motor or engine efficiency, pump efficiency) can be calculated.

4.2.6.2 Border Irrigation Evaluation

Field observations and measurements required for conducting a border irrigation system evaluation include the following:

- Border dimension
- Slope of the border
- Inflow rate
- Runoff rate and volume (if any)
- Irrigation time (duration)

4.2 Performance Evaluation

- Advance phases and time
- Recession time
- Topography of the field
- Crop type and stage of the crop

The measurement steps and procedures are as follows:

- The border dimensions can be measured using a "measuring tape."
- Soil surface elevations (at different points, 10–30 m interval along the borders) can be determined using a "total topographic station" or "Level instrument." Slope and standard deviation of soil surface elevations can be determined from the measured data.
- The "inflow" or "irrigation discharge" can be measured using suitable flow measuring device such as mini-propeller meter and flume.
- The advance phase can be determined from recording of the advance time to reference points located along the border (every 10–30 m).
- A number of flow depth measurements are to be performed across the border, every 5–10 m. The average of all measurements is used to represent the flow depth at this point and time.
- The flow depth at the upstream end of the border is to be measured shortly before cutoff.
- In open border, surface runoff (if conditions permit) is to be monitored. The runoff can be measured using the mini-propeller meter or a flume.

A hydrograph is to be established from discharge measurements, and its time integration will yield the runoff volume.

- Infiltration in "ring infiltrometer" and border infiltration can be correlated, and a relationship can be established. Then the infiltration parameters can be determined.
- To estimate the infiltrated depths of water (required for computing uniformity and efficiency indices), field data from evaluation can be utilized to derive the infiltration parameters of a Kostiakov type infiltration equation. The infiltration parameters (K, a) and roughness coefficient (N) can be determined through the solution of inverse surface irrigation problem (Katapodes et al., 1990).

    For that, a hydrodynamic one-dimensional surface irrigation model (e.g., SIRMOD) can be used. Such a model is to be executed using tentative values of the coefficient "K" and the exponential "a" from the Kostiakov infiltration equation, and the Manning's N. The parameters should be adjusted until the model satisfactorily reproduces the experimental values of flow depth and irrigation advance for each evaluation.
- Performance indices – application efficiency and the low-quarter distribution uniformity – should be determined using the formula described in an earlier section.

4.2.6.3 Basin Irrigation Evaluation

Basins have no global slope, but the undulations of the soil surface can have an important effect on the advance and recession process of an irrigation event.

For evaluation of basin irrigation, measurements should be made during representative irrigation events. The required measurements are as follows:

- advance, water depths at selected locations
- surface drainage or recession

Commonly measured performance indices for basin irrigation are as follows:

- application efficiency
- distribution uniformity
- deep percolation ratios
- requirement efficiency or storage efficiency

For basin irrigation, tailwater ratio is zero

Clemmens and Dedrick (1982) defined distribution uniformity (DU) for basin irrigation as

$$DU = (\text{minimum depth infiltrated})/(\text{average depth infiltrated}) \qquad (4.30)$$

Application efficiency is defined in Eq. (4.2).

4.2.6.4 Furrow Irrigation Evaluation

Generally, the evaluation of furrow irrigation system is restricted to a single or small number of adjacent furrows due to intensive measurement process. Complete inflow, advance, and runoff measurements are used to accurately determine soil infiltration rate for a small number of furrows.

The working steps and procedures for the evaluation of furrow irrigation system are as follows:

- Measure the length and spacing of furrow
- Measure the soil moisture (before irrigation)
- Install the equipments (e.g., flume, scale, moisture measuring equipment)
- Start irrigation
- Record the flow rate (at 5–10 min intervals, until the constant flow rate is achieved)
- Record the advance data after 6, 12, and 24 h from the starting of irrigation
- Record the water depth at different points (10, 20, 50 m) at several time intervals
- Record the cutoff time
- Record the recession data (water depth) at several distances (10, 20, 50 m) from the starting point at several time intervals (1/2, 1, 2, 5 h)
- Record the depth of ponding at lower 1/4th part of furrow
- Record the runoff volume (if the process permits)

4.2 Performance Evaluation

- Measure the soil moisture up to the desired depth (root zone) at different points throughout the furrow after reaching field capacity
- Determine the wetted cross-section of the furrow at several sections and average them

Data Processing

– determine infiltration function
– determine different performance indices (such as distribution uniformity, application efficiency, deep percolation ratio, deficit ratio), as defined earlier

Advance and recession curves can be drawn from the recorded data (distance vs advance/recession). The difference between advance time and recession time (calculating from the starting of test) at each point represents the infiltration opportunity time (Fig. 4.1).

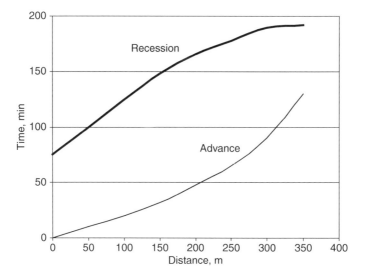

Fig. 4.1 Advance and recession curve

Volume balance approach can be applied to find out different components of water balance (e.g., infiltration, deep percolation). Volume balance approach is based on the principle of mass conservation. At any time, the total volume of water that has entered the furrow must be equal to the sum of the surface storage, subsurface storage (infiltrated), deep percolation (if any), and runoff (if any).

4.2.6.5 Sprinkler Irrigation Evaluation

A sprinkler water distribution pattern depends on system design parameters and on environmental variables (such as wind speed and direction). Wind speed affects not only uniformity, but also evaporation and wind drift losses.

Most widely used and useful performance indicator for sprinkler irrigation system is "Distribution Uniformity." The distribution uniformity (DU) can be functionally expressed as (Pereira, 1999)

$$\text{DU} = f(P, \ \Delta P, \ S, \ d_n, \text{WDP}, \text{WS}) \tag{4.31}$$

where

$P =$ the pressure available at the sprinkler
$\Delta P =$ variation of the pressure in the operating set or along the moving lateral
$S =$ spacings of the sprinklers along the lateral (and between laterals) or the spacings between travelers
$d_n =$ nozzle diameter, which influences the discharge q_s and wetted diameter D_w (and the coarseness of water drops) for a given P
$\text{WDP} =$ water distribution pattern of the sprinkler and
$\text{WS} =$ wind speed and direction.

All the above variables are set at the design stage (including a forecasted WS).

Recordable Information During Evaluation

Numerous information should be collected to enable effective comparisons between systems and more detailed analysis. The information required includes the following:

- Traveler location
- Description of traveler
- Description of gun
- Angle of gun rotation
- Nozzle type
- Nozzle diameter
- Length of run
- Lane spacing
- Run duration (start and finish times)
- Pump details

Measurements That Should Be Taken

The measurements that should be made in the performance test include the following:

- Sprinkler spacing and pattern
- Number of sprinklers
- Nozzle and tail jet diameters
- Application rate
- Irrigation duration
- Sprinkler height

4.2 Performance Evaluation

Data Analysis

Distribution uniformity is often the primary measure of system performance, as an irrigation system requires a high distribution uniformity in order to maintain good crop yields, even though the water can be uniformly overapplied. Another performance measure is application efficiency. Application efficiency is required with relation to irrigation scheduling.

Solid set sprinkler

Determination of Distribution Uniformity (DU)

The DU is usually determined by measuring the depth of water falling into a grid of catch cans during an irrigation event (Fig. 4.2) and analyzing the variation of water depths in the catch cans. Sorting the catch can reading in ascending order, the lowest quarter of values is determined. Then average the lowest quarter readings. This average is then divided by the average of all the readings to give the distribution uniformity as a percentage.

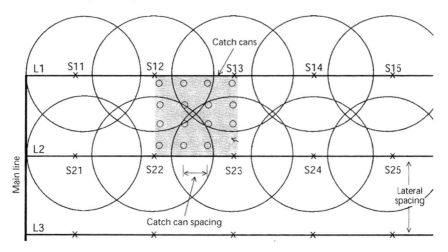

Fig. 4.2 Schematic of can setting for sprinkler evaluation

Application Efficiency

Application efficiency is calculated using the average depth of water applied, as calculated above, and the volume of water applied. In order to convert the depth of water applied to a volume, the depth must be multiplied by the irrigation target area. Flow is measured as a nozzle discharge, and the average flow from the four nozzles bounding the catch can grid is used as the measure of flow. The formula representing application efficiency becomes

E_a = (catch can depth × grid area)/(average nozzle flow rate × irrigation duration)

4.2.6.6 Drip/Micro-irrigation Evaluation

The most useful performance indicators for micro-irrigation field evaluations are the statistical uniformity coefficient and the distribution uniformity.

Statistical Uniformity Coefficient

The statistical uniformity coefficient, U_s (%), is defined as (Pereira, 1999)

$$U_s = 100(1 - V_q) = 100\left(1 - \frac{S_q}{q_a}\right) \quad (4.32)$$

where

V_q = coefficient of variation (CV) of emitter flow
S_q = standard deviation (SD) of emitter flow (l/h)
q_a = average emitter flow rate (l/h).

Distribution Uniformity

The distribution uniformity, DU (%), is defined as:

$$\text{DU} = 100 \frac{Z_{lq}}{Z_{av}} \quad (4.33)$$

where

Z_{lq} = average observed applied depths in the low quarter of the field (mm) and
Z_{av} = average observed applied depths in the entire field (mm).

Distribution uniformity is often named emission uniformity in trickle irrigation.

4.2.7 Improving Performance of Irrigation System

The most obvious way to improve the performance of an irrigation system is to take remedial measures for correcting the fault/deficiency, which has been identified during evaluation/diagnosis process. Besides, a number of techniques can be used in the design of a system to increase its uniformity and efficiency. For surface irrigation systems, the inflow rate can be matched with the soil intake rate, slope, and length of run; and the cutoff time can also be matched thereby. For pressurized systems, the technique includes using larger pipe sizes to minimize pressure differences due to friction losses, using pressure regulators to minimize pressure differences due to elevation differentials, using appropriate closer spacings or trickle emitters with low manufacturing variations. Another technique is that water use is more efficient with afternoon irrigations, as the evaporative loss is minimal.

Some common problems/faults and suggestive measures for improving the performances are summarized below:

4.2 Performance Evaluation

Sl no.	Problem identified by diagnosis/ evaluation	Suggestion(s) for improving performance
1	Pumping plant efficiency is low	• Renovate the moving parts • In case of Deep Tubewell, wash out the well screen
2	Water conveyance efficiency is low	• Renovate/perform lining the conveyance channel • Reduce the field channel density • Perform efficient/economic channel design
3	Water delivery performance is not satisfactory	• Recast/ensure delivery system
4	Channel density is high	• Reduce the channel length by straightening through the command area
5	On-farm water loss is high	• Compact the borders of each plot • Improve the water-holding capacity of the soil by adding organic manures • Reduce relative percentage of sand by adding silt or clay soil
6	Water supply – requirement ratio is not good	• Recast the supply amount, or • Change the cropping pattern (if possible), altering high water-demanding crops; or • Search for new source of supply
7	Deep percolation fraction is high	• Line the channels • Improve physical condition/water-holding capacity of the soil
8	Runoff fraction is high	• Maintain correct slope of land • Apply correct flow rate and time for flow (cutoff time) • Take care of the borders; construct high levees
9	Water application efficiency is low	• Minimize on-farm water loss • Estimate correct amount of water demand • Apply correct flow rate based on infiltration characteristics • Level the land with appropriate slope • Maintain correct slope toward the water run considering infiltration rate and flow rate • Improve water-holding capacity of the soil
10	Water storage efficiency is not satisfactory	• Correctly estimate the crop root zone depth before irrigation • Estimate correct amount of water demand • Apply correct flow rate for correct duration • Improve water-holding capacity of the soil
11	Distribution uniformity is low (poor distribution of infiltrated water over the field)	• Apply correct flow rate based on infiltration rate and slope of the run • Design the length of run based on infiltration rate, slope, and flow rate • Cut off the flow at proper time

(continued)

Sl no.	Problem identified by diagnosis/evaluation	Suggestion(s) for improving performance
12	Irrigation efficiency of sprinkler is low	• Adjust the "set time" and "interval between irrigations" such that irrigation amount matches the soil moisture deficit • Change operating conditions to increase water droplet size, or operate the system under conditions of low climatic demand (e.g., at night, morning, and evening)
13	Low-quarter distribution uniformity is low	• Apply correct flow rate based on infiltration rate • Design the length of run based on infiltration rate, slope, and flow rate • Cut off the flow at proper time (after reaching the water front at tail end)
14	Area irrigated per unit flow (Duty) is not satisfactory	• Reduce conveyance, seepage, and percolation loss • Schedule irrigation properly (apply correct amount of water based on need) • Improve water-holding capacity of the soil
15	Intensity of irrigation is low	• Reduce all possible losses • Increase irrigation efficiency • Schedule crops and crop rotations
16	Crop productivity is low	• Ensure proper irrigation • Ensure proper management of other inputs (like balance fertilizer) • Ensure other cultural management (proper population, weeding, pesticide, and insecticide application, if needed)
17	Water productivity is below the normal range	• Schedule irrigation properly • Reduce tailwater runoff • Minimize on-farm water loss • Maximize utilization of stored soil moisture • Ensure other crop management aspects
18	Irrigation water productivity is below the desired limit	• Schedule irrigation properly • Reduce tailwater runoff • Minimize on-farm water loss • Maximize utilization of stored soil moisture • Ensure other crop management aspects
19	Irrigation benefit–cost ratio (B-C ratio) is low	• Minimize irrigation cost by proper scheduling and reducing all sorts of water loss • Maximize production by proper management of other inputs and selecting appropriate crop type and variety • Maximize utilization of stored soil water and rainwater, if available
20	Cost per unit production is high	Similar to that of B-C ratio
21	Irrigation cost per unit area is high	Similar to that of B-C ratio
22	Farmers income ratio is not satisfactory	Similar to that of B-C ratio

Relevant Journals

- Agricultural Water Management
- Journal of Irrigation and Drainage Division, ASCE
- Transactions of the American Society of Agricultural Engineers
- Irrigation and Drainage System
- Irrigation Science
- ICID Bulletins

Relevant FAO Papers/Reports

- FAO Irrigation and Drainage Paper 45 (Guidelines for designing and evaluating surface irrigation systems, 1989)
- FAO Irrigation and Drainage Paper 59 (Performance analysis of on-demand pressurized irrigation systems, 2000)

Questions

(1) What is meant by "irrigation efficiency"? Explain different types of irrigation efficiencies.
(2) What do you mean by "performance evaluation of an irrigation project?" Describe the objectives, purposes, and benefits of performance evaluation of irrigation systems.
(3) What is "performance indicators?" Name the indicators under Engineering, Field water use, Crop and water productivity, and Socioeconomic category.
(4) Define and write down the equations for different types of performance indices.
(5) Write down the principles and procedures of performance evaluation.
(6) Briefly discuss the principles of evaluating a pumping plant.
(7) Discuss specific considerations and procedures for the following irrigation systems: (a) border, (b) basin, (c) furrow, (d) sprinkler and (e) drip.
(8) How can the performance of an irrigation system be improved?

References

Ali MH (2001) Technical performance evaluation of Boyra Deep Tube-Well – a case study. J Inst Eng Bangladesh 28/AE(1):33–37

ASCE (1978) Describing irrigation efficiency and uniformity. ASCE J Irrig Drainage 104(IR1): 35–42

Bos MG et al (1993) Methodologies for assessing performance of irrigation and drainage management. Paper presented to workshop of the working group on irrigation and drainage performance, 15th congress of the international commission on irrigation and drainage, The Hague, The Netherlands

Bos MG, Nugteren J (1974) On irrigation efficiencies. International Institute for Land Reclamation and Improvement. Wageningen, The Netherlands

Burt CM, Clemmens AJ, Strelkoff TS, Solomon KH, Hardy L, Howell T, Eisenhauer D, Bleisner R (1997) Irrigation performance measures – efficiency and uniformity. J Irrig Drainage Eng 123(6)

Christiansen JE (1942) Irrigation by sprinkling. California Agric Expt Station Bulletin No 570

Clemmens AJ, Dedrick AR (1982) Limits for practical level-basin design. J Irrig Drain Div ASCE 108(IR2):127–141

Heermann DF, Wallender WW, Bos GM (1990) Irrigation efficiency and uniformity. In: Hoffman GJ, Howell TA, Solomon KH (eds) Management of farm irrigation systems. ASAE, St. Joseph, pp 125–149

Jensen ME, Harrison DS, Korven HC, Robinson FE (1983) The role of irrigation in food and fibre production. In: Jensen ME (ed) Design and operation of farm irrigation systems. ASAE, St. Joseph, MI

Katapodes ND, Tang JH, Clemmens AJ (1990) Estimation of surface irrigation parameters. J Irrig Drain Eng ASCE 116(5):676–696

Kruse EG (1978) Describing irrigation efficiency and uniformity. J Irrig Drain Div ASCE 104(IR1):35–41

Merriam JL, Keller J (1978) Farm irrigation system evaluation: a guide for management. Department of Agricultural Irrigation Engineering Utah State University, Logan

Pereira LS (1999) Higher performance through combined improvements in irrigation methods and scheduling: a discussion. Agric Water Manage 40:153–169

Sarma PBS, Rao VV (1997) Evaluation of an irrigation water management scheme – a case study. Agric Water Manage 32:181–195

Solomon KH (1988) Irrigation systems and water application efficiencies. Center for irrigation technology research notes, CAIT Pud # 880104. California State University, California

USDA (1997) Irrigation systems evaluation procedures. National Engineering Handbook, Chapter 9, Part 652. Natural Resources Conservation Service USDA, Washington, DC

Walker WR, Skogerboe GV (1987) Surface irrigation: theory and practice. Prentice-Hall, Inc., Englewood Cliffs, NJ, 386p

Wigginton DW, Raine SR (2001) Measuring irrigation system performance in the Queensland Dairy Industry. National Centre for Engineering in Agriculture Publication 179729/5, Toowoomba, Australia

Chapter 5
Water Resources Management

Contents

5.1	Concept, Perspective, and Objective of Water Resources Management	140
	5.1.1 Concept of Management	140
	5.1.2 Water and the Environment	141
	5.1.3 Increasing Competition in Water Resource	141
	5.1.4 Water As an Economic Good	142
	5.1.5 Purposes and Goals of Water Resources Management	143
	5.1.6 Fundamental Aspects of Water Resources Management	144
5.2	Estimation of Demand and Supply of Water	144
	5.2.1 Demand Estimation	144
	5.2.2 Estimation of Potential Supply of Water	146
	5.2.3 Issues of Groundwater Development in Saline/Coastal Areas	148
	5.2.4 Environmental Flow Assessment	148
5.3	Strategies for Water Resources Management	150
	5.3.1 Demand Side Management	150
	5.3.2 Supply Side Management	161
	5.3.3 Integrated Water Resources Management	170
5.4	Sustainability Issues in Water Resource Management	173
	5.4.1 Concept of Sustainability	173
	5.4.2 Scales of Sustainability	175
	5.4.3 Achieving Sustainability	175
	5.4.4 Strategies to Achieve Sustainability	177
5.5	Conflicts in Water Resources Management	178
	5.5.1 Meaning of Conflict	178
	5.5.2 Water Conflicts in the Integrated Water Resources Management Process	179
	5.5.3 Scales of Conflicts in Water Management	180
	5.5.4 Analysis of Causes of Conflicts in Water Management	184
5.6	Impact of Climate Change on Water Resource	185
	5.6.1 Issues on Water Resources in Connection to Climate Change	185

	5.6.2	Adaptation Alternatives to the Climate Change	186
5.7		Challenges in Water Resources Management	188
	5.7.1	Risk and Uncertainties	188
	5.7.2	International/Intra-national (Upstream–Downstream) Issues	188
	5.7.3	Quality Degradation Due to Continuous Pumping of Groundwater	188
	5.7.4	Lowering of WT and Increase in Cost of Pumping	189
Relevant Journals			189
Questions			190
References			190

Together with energy, water is one of the major fuels of economic development. A development plan, especially in a water-short area, such as an arid zone belt, or a flat area underlain by hard rocks without major surface streams, cannot be drawn unless a clear idea of water availability and costs has been reached.

Water is distributed in the earth unevenly in time and space. Although the natural water is adequate in absolute terms, it is not available at the right time, place, quantity, and quality. With an increasing population and its legitimate demand for an improved standard of living, requiring increased economic development and agricultural production, most of the regions of the world are facing an enormous challenge in how to allocate, use, and protect this limited resource. Improved and efficient water management practices can help to maintain farm profitability in an era of increasingly limited and more costly water supplies. Improved water management may also reduce the impact of irrigated production on offsite water quantity and quality, and conserve water for growing nonagricultural demands.

5.1 Concept, Perspective, and Objective of Water Resources Management

5.1.1 Concept of Management

According to some sources, the word "management" originated from the Italian word "*Maneggiare,*" which means "to train up the horses." Other sources said "management" came from the French word "*Menager*" and "*Menage.*" Here, "*Menager*" means "to direct a household," and "*Menage*" means an act of guiding or leading. Thus, it means to plan, organize, command, coordinate and control. According to Griffin (1997), "Management is a set of activities (including planning and decision making, organizing, leading, and control) directed at an organization's resources (e.g., human, financial, physical, information) with the aim of achieving organizational goals in an efficient and effective manner."

Water management is concerned with improving access to and the efficient use of water. Such measures will increase the volume of goods produced by increasing the availability of water and will also improve its productivity by increasing the returns to water by, for example, applying water more efficiently to irrigated crops.

In essence, water resources management are concerned with how the available supply of water can be better allocated and utilized to fulfill the demand. It includes the intervention of humans in the manner in which surface and/or ground water is captured, conveyed, utilized, and drained in a certain area. Moreover, it is a process of social interaction between different stakeholders, each employing different methods, resources, and strategies, around the issue of water control.

5.1.2 Water and the Environment

Increased agricultural production to feed a growing world population along with many other forces driving modern societies has placed many resources at risk. Water availability is a major problem in many areas, and even in areas of abundant water, quality being jeopardized by salinization, pesticide and nitrate contamination, arsenic contamination, etc. and altering of biological activity. In the case of environmental unsustainability, change is slow and the process is varied.

In many areas of the world, conditions of overexploitation of groundwater aquifers have resulted in serious ecological consequences. The lowering of groundwater together with less rainfall has intensified the phenomenon of drought affecting the main crops and has started the process of desertification. The needs of water for the environment (or the nature) in considerations of water allocation is often neglected. The impacts of water use by different sectors on environment are as follows:

(a) *Agriculture*: The agriculture sector is most important as a user of water and impacts most heavily on ecosystems' water share. Abstraction of water for agriculture is leading to dried-up rivers, falling groundwater table, salinated soil, and polluted waterways.
(b) *Urban water use*: Urban water use pollutes water by waste water effluents, pollutes downstream ecosystems (if not properly treated). The treatment of the effluent is often costly, but when due consideration is given to the ecosystems, effluent recycling and reuse may be cost-effective conservation measures.
(c) *Industry*: Very often Industry has substantial impacts on ecosystems downstream through water use and pollution.

5.1.3 Increasing Competition in Water Resource

Competition in limited water resources increasingly occurs among agriculture, rural, urban, industrial, and environmental uses. At the same time, disparity in the economic conditions between the urban and rural areas in a country and among countries continues to increase.

Water can be used for a great number of purposes: domestic needs (drinking, washing, bathing, toilet flushing, etc.), agriculture, industry, power generation, fishing, forestry, recreation, transport, and so on. These are alternative uses, which become competing uses if there is not enough water to satisfy them all.

5.1.4 Water As an Economic Good

Water should be regarded as "an economic good." The water at source may be free in the sense that nobody owns it; but if it is not in abundant supply, then it is scarce and that condition has important consequences that must be taken account. More specifically, there are competing demands for water and there must, therefore, be some mechanism for allocating it. Pricing is one, but not the only, means of effecting allocation.

We know from the hydrological cycle that water is a renewable resource. Using water therefore does not typically mean "using it up," but making it unusable for other purposes or at least unusable without incurring additional costs, e.g., by polluting or diverting it. Investment in a complete and expensive drinking water supply would typically involve three components: transmission, treatment, and distribution. Water occurs naturally. At its source, before it is transmitted, treated, and distributed, no expenditure is incurred. At that point, therefore, it has not involved costs for anyone. In contrast, if one is talking about a drinking water supply, which provides a specific quantity and quality of water to a specific geographic area, that is a different matter. Each of the above three components requires investment, and the money (for both the initial investment and routine maintenance/regular operation are equally important) will have to come from elsewhere. Thus all concerned would surely agree that careful consideration must be given to where the money is to be acquired from, including the issue of whether those benefiting will pay some or all of the costs. However, it is crucial to point out that cost recovery has nothing to do with the general water resources question of whether or not water should be treated as an economic good.

If water is used for one particular purpose, e.g., agriculture, it cannot be used for another, e.g., drinking. More specifically, the opportunity cost of water is its value in alternative use. This implies that there is some sort of hierarchy of demand. For example, water is most valuable for drinking, followed by industry and then agriculture. Moving from drinking water supply to other uses, the situation changes substantially as the broader water resources issue emerges. Suppose an estimate was made, in the manner described above, of the value of water not only for agriculture but also for other purposes, say, industry and fishing. The result may be that water for industry or, less likely, fishing is more viable than for agriculture. This implies that priority in use, in this particular instance, should be given to industry rather than agriculture. Mechanisms other than pricing may be used to enforce this priority. But if it is not followed, economic losses will ensue. Those losses will not be reflected in direct financial terms; rather the net value of output will be less than it may have been.

The question now arises, how would an economist estimate the value of water for agriculture? This can be done by estimating the value of agricultural output from, say, a hectare of irrigated land and then estimating the contribution by water to that value. This is a rough and ready calculation, with a rather wide margin of error. But it is important to note that the value of water per cubic meter thus calculated will be minimal by comparison with the cost of transmission, treatment, and distribution.

5.1 Concept, Perspective, and Objective of Water Resources Management

It is relevant here to interpose a point about wastage of water. For example in irrigation, the potential for reducing water losses is typically quite considerable. If water is to be regarded as a resource, then this should imply a need to discourage wastage. If losses can be reduced, then there may be sufficient water for all requirements.

In conclusion, to say that water should be regarded as an economic good does not necessarily imply that a "market price" must be paid for it or even must be paid for at all. It means simply that water is a scarce resource, a valuable resource that should not be wasted. From the economic point of view, alternatives that should be properly considered include the following:

- reducing losses/wastage
- reducing agricultural production in water-scarce area (bringing in food, instead of water, from outside the area)
- changing the pattern of agriculture to less water-intensive crops
- reducing industrial production (demanding water) in the area (bringing in goods instead of water, from outside the area)

5.1.5 Purposes and Goals of Water Resources Management

Water resources management is required to make water of right quantity and right quality available at the right time and at the right places. Management is used in its broadest sense. It emphasizes that we must not only focus on the development of water resources but that we must consciously manage water development in a way that ensures long-term sustainable use for future generations. The main purpose of water resources management is to meet needs of humans and nature.

Broader goals and objectives of water resources management are as follows:

- general welfare of human being and improvement of the quality of life
- regional economic development
- health and safety
- income distribution
- cultural and educational opportunities

Two broad classes of purpose and function of water resources management are as follows:

Water use	Water control
– Agriculture, irrigation	– Water quality management
– Water supply for municipal, rural and industrial uses	– drainage, sedimentation control, erosion control
– Hydro-electric power	– Flood control
– Fishing	– Watershed control
– Navigation	
– Recreation	

5.1.6 Fundamental Aspects of Water Resources Management

Water resources should be managed in the context of a national water strategy that reflects the nation's social, economic, and environmental objectives and is based on an assessment of the country's water resources. The assessment would include a realistic forecast of the demand for water, based on projected population growth and economic development, and a consideration of options for managing demand and supply, taking into account existing investment and likely to occur in the private sector. The strategy would spell out priorities for providing water services; establish policies on water rights, water pricing and cost recovery, public investment, and the role of the private sector in water development; and institutional measures for environmental protection and restoration.

In water resources management, two fundamental aspects are to be considered:

i. Planning
ii. The implementation process

The basic approach to planning is to determine:

- the needs,
- the resources, and
- ways to develop the resources to meet needs, on the basis of the technological, financial and human resources, which are available.

In the past, a water resources planning exercise was applied mainly to a river basin, or to political and geographical boundaries such as a region or country. However, this concept is shifting to economic consideration, for example, to metropolitan areas or large industrial areas.

5.2 Estimation of Demand and Supply of Water

For proper planning and management of any resource, knowledge of demand and supply is prerequisite.

5.2.1 Demand Estimation

The evaluation of future effective water consumption is more difficult as it implies knowledge of interventions by man. For example, a country may change its fundamental economic options: agriculture versus industrialization following a change in government. New technological development may promote the conservation of water, such as inexpensive methods for surface water treatment in tropical countries or by changing water needs through the use of drip irrigation.

An evaluation of future demand should be based on the results of a recent comprehensive survey of water users. Forecasts of future uses of water resources on the basis of various "scenarios" for exploitation should also be undertaken. These can be achieved by means of mathematical models.

Demands for water in a region can be classified into the following sectors:

(i) demand for domestic, industrial, and commercial uses
(ii) irrigation demand
(iii) evaporative demand for fisheries, forestry, and environment, and
(iv) in-stream demand

5.2.1.1 Demand for Domestic, Industrial, and Commercial Uses

Domestic water per capita varies substantially between urban and rural areas, and depends on income, standard of living, mode of water supply, availability and quality of water, etc. An approach of demand estimation may be based on standard of living. Average per capita consumption for metropolitan, town, and rural areas should be estimated separately. In urban area, typically the per capita consumption varies between 100 and 150 l/day; and in rural areas it varies between 50 and 100 l/day. In long-term planning purposes, change in income and a cultural shift in water uses of the rural population should be considered. A system loss of about 10–15% should also be taken into account.

Commercial and industrial demands are typically 8–11% and 10–15%, respectively, of the total water supply for the piped distribution system (World Bank, 1997). A return of 30–50% may be considered, but this is not useful in most cases.

5.2.1.2 Irrigation Demand

The usefulness of an estimate of future irrigation demand largely depends on how closely prediction can be made of areas under different crops, cropping sequences and intensity, and crop calendar (the time and length of growing period). The probable climatic scenario is also required to compute crop evapo-transpirative demand. Agricultural crops also require water for seedbed preparation, land preparation, salt leaching etc. These water requirements are fulfilled either from rainfall or from irrigation supply. Not all the rains received over the year or growing period become useful for these purposes. Effective rainfall should be calculated or estimated using appropriate techniques. In meeting evapo-transpirative demand, some seepage and percolation loss will occur from the field, specially in case of rice crop. The measurement and/or estimation techniques of these components have been described in detail in an earlier chapter (in Volume 1). There is a difference between the quantity of water diverted from a source for irrigation and that reached at the field plot or used in actual evapotranspiration. This difference is a loss from the viewpoint of irrigation water supplier. The irrigation efficiency, which is a ratio of usage to supply, varied depending on soil, climate, crop and method of application, and usually between 60 and 70%. The International Water Management Institute (IWMI) used irrigation efficiency up to 70% for country level demand assessment (Seckler, 1996). Considering all these factors, the gross irrigation demand should be estimated.

5.2.1.3 Nonirrigated Evaporative Demand

Nonagricultural evaporative demand arises from areas under fisheries, forestry, and environmental uses. Areas under fisheries include major and regional rivers, perennial and seasonal standing water bodies (haor, baor, and beel), and fishponds.

Forest abstracts water from high and perched groundwater tables. The quantity of water required to meet evapotranspiration demands of such trees is a net demand and needs to be accounted for. Environmental demand arises from rivers as well as urban, rural, and other areas (excluding areas under fisheries and forestry). Urban areas under environmental use include parks, gardens, and playgrounds. Rural areas include household trees, garden, graveyards, and playgrounds.

5.2.1.4 In-Stream Demand

Some flows are to be maintained in the major and regional rivers for navigation, fisheries, salinity control, chemical and biological dilution, and sustenance of aquatic flora and fauna. These constitute in-stream demand, but they are not additive. The in-stream flow is required to push the salinity front toward the salinity source (the sea, which contains saline water) to arrest the environmental degradation. To salinity control, a minimum flow should be maintained.

5.2.2 Estimation of Potential Supply of Water

At the same time that water needs are being assessed, water availability has to be evaluated as to quantity, quality, and costs for various sources of water. While surface water surveys are relatively easy and inexpensive, groundwater surveys are more complex and costly as they involve test drilling, pumping, and geophysical surveys, in addition to continuous data collection through yield and water level measurements on wells.

5.2.2.1 Surface Water Resource

It is very important for surface water estimation that spatial as well as temporal distribution of water is properly accounted for. It is not appropriate to calculate the average for the purpose of planning.

In estimating effective volume of water in different available water bodies (such as ponds, lakes, rivers), volume of water should be determined first during periods (dry and wet) of the year. Then the assessment of environmental flow requirement (such as fish culture, navigation, and other purposes) should be done. Effective storage/volume of water for a particular season is the difference between actual volume and environmental flow (EF) requirement, i.e.,

5.2 Estimation of Demand and Supply of Water

Effective volume of water = actual volume − EF requirement

For estimation of actual volume of water in water bodies, different available techniques (such as Simpson's rule, trapezoidal rule) may be employed. For flowing rivers and streams, stage–discharge relationship should be determined.

The effective available water may be used for different purposes, such as irrigation (supplemental irrigation during monsoon period, and full irrigation during dry season crops) and hydroelectric power.

5.2.2.2 Groundwater Resource

Assessment of future groundwater development potential is a pre-requisite for the proper and sustainable use of the resource. Groundwater assessment requires conceptualization of the physical process and methodology for assessment. The following two approaches are generally employed to estimate the groundwater potentials for long-term sustainability:

Groundwater Availability for Pumping in Terms of Potential Recharge

Groundwater availability for pumping in terms of potential recharge may be estimated using a simplified hydrological balance:

$$P = PET + U + R_e$$

Or,

$$R_e = P - PET - U \tag{5.1}$$

where

P = Rainfall
PET = Potential evapotranspiration
U = Runoff
R_e = Potential recharge

That is, potential recharge is the excess of rainfall over runoff and potential evapotranspiration. This estimate may be carried out monthly basis, which is more judicious than the seasonal estimates.

In the absence of data on surface runoff, runoff may be estimated as a percentage of rainfall (20–50%), depending on the rainfall amount, intensity of rainfall, topography, physical and hydraulic characteristics of the topsoil, and geo-hydrologic condition. It is to be mentioned here that runoff rate will not be constant but will depend upon antecedent soil moisture, land use and topographic condition. A complex model, which is capable of incorporating interacting factors on the amount of runoff, may assess the runoff more closely.

Groundwater Availability in Terms of Safe Yield

Groundwater availability in a region may be estimated based on safe yield concept. According to this concept,

$$\text{Safe yield} = \text{maximum allowable fluctuation of water table (MAWT)} \\ \times \text{specific yield of the aquifer}$$

Here, MAWT refers the difference between maximum allowable depletion or lowering of water table (WT) during pumping, and the present depth to water table, i.e.,

$$\text{MAWT} = \text{Maximum allowable lowering of WT} - \text{present depth to WT}$$

Here, it is to be mentioned that the use of historically observed maximum fluctuation in annual water table in safe-yield or recharge estimation may result in an underestimated value of recharge. This is because the annual fluctuation can be increased by withdrawal of groundwater for irrigation during dry season, thus creating further scope for increased recharge. That is, the potential recharge, which may be defined as the mean annual/seasonal volume of surface water that could reach the aquifer, is the sum of actual recharge to the aquifer and the rejected recharge. Rejected recharge is that fraction of water available at the surface which can not infiltrate because the water table is near/at the surface.

Specific yield should be determined by free drainage from the collected samples of the aquifer (through bore logs). Specific yield is an important parameter in safe yield or groundwater potential estimation using this approach. The value of specific yield varies widely within the physiographic zones and even in a zone. So, the specific yield should be determined for each zone and the mean of representative number of samples should be used, that the data represent the actual aquifer characteristics.

For each of the approaches mentioned above, the total area of a country or a region should be divided into suitable physiographic units on the basis of some criteria such as geological condition, topography, physical and hydraulic characteristics of the topsoil, and/or depth of flooding. Available or potential recharge for each unit should be calculated.

5.2.3 Issues of Groundwater Development in Saline/Coastal Areas

Freshwater in the saline and coastal area may occur in single or two zones (shallow and deep freshwater) (Saleh and Nishat, 1989; Rashid, 2008). The development of freshwater in such area requires careful planning. Localized pressure reduction caused by tube-well abstraction may cause upconing of saline water interface. To mitigate salinization of the aquifer, it is essential to control groundwater abstraction, adopt suitable well design procedures and have a thorough knowledge of the hydrogeology of the aquifer.

5.2.4 Environmental Flow Assessment

5.2.4.1 Concept of Environmental Flow

Management of river flows should attempt to ensure minimum natural flows in order to maintain the conditions that supply goods and services and ensure biodiversity.

5.2 Estimation of Demand and Supply of Water

Environmental flow is the water regime provided within a river, wetland or coastal zone to maintain ecosystems and their benefits where there are competing water uses and where flows are regulated.

Key concepts in environmental flow assessment are:

- river systems can be maintained at different levels of health
- different flows play different roles in maintaining river systems
- ecological and social consequences of flow manipulation can be predicted
- complexity and variability are vital to ecosystem health

5.2.4.2 Methods of Environmental Flow Assessment

Generally the following two methods are used to assess environmental flow:

(i) Building block method
(ii) Drift method

Drift Method

It is a scenario based method. Scenario is used to identify tradeoffs that can be used in negotiations between water users to balance needs of the ecosystem with other needs. A set of scenarios (produced as output from different water levels) or options for a river is used in decision and policy making.

Building Block Method

It is an objective based method. In this method, a specific ecological status of a river is maintained. The future desired condition of the river is identified. Environmental flow regime is then constructed based on premise that riverine species rely on basic elements of flow regime (building block). The main phases of building block method (BBM) are

a. Comprehensive information gathering on river system, normally undertaken by a team of experts.
b. Exchange opinion among agency representative, water managers, engineers and scientific experts to identify flow regimes; usually through a workshop.
c. Linking the environmental flow consideration with engineering considerations

Information gathering involves:

- identification of study area
- present condition of river for overall riverine habitat
- determination of importance of study area at local, regional and international scale.
- assessment of water quality
- biological surveys at selected points

- identification of building block method reaches and sites
- description of natural and present daily flow regime
- analysis of stage-discharge curve

The objective of the exchange of opinions among different experts is to identify and conclude on the recommended flows for the river. The experts should visit the site. Attention should be given to the flow features that are considered most important for maintaining or achieving the desired state of the river. These flow features are the building blocks which are constructed to create the in-stream flow requirements. Required flows are identified month-wise, considering low and high flows.

In final stage, recommended flow regimes are integrated with engineering concerns for the river. Two or three possible flow states with probable economic and social consequences are developed.

The building block method is costly for data collection and employment of experts. The final selection is dependent on professional judgment and experience.

5.3 Strategies for Water Resources Management

Strategies for water resources management can be broadly categorized into two classes:

(a) Demand management, and
(b) Supply management

5.3.1 Demand Side Management

5.3.1.1 Concept

Demand management options refer the actions that influence the use of water after the entry point. Demand side management is commonly implemented together with a water conservation program. Water conservation is generally accepted to mean "the minimization of loss or waste, the preservation, care and protection of water resources and the efficient and effective use of water." Demand side management involves a broad range of measures that aim to increase the efficiency of water use. Demand management may be defined as the adaptation and implementation of a strategy by a water institution to influence the water demand and usage in order to meet any of the following objectives: economic efficiency, social development, social equity, environmental protection, sustainability of water supply and services and political acceptability.

5.3.1.2 Different Approaches of Demand Management

The measures for demand management can include the following:

- conservation-oriented rate structures
- leak reduction program

5.3 Strategies for Water Resources Management

– landscaping with drought tolerant species
– water savings irrigation practices in agriculture
– water reuse
– pressure management
– water audit
– public awareness campaigns.

The demand side management may be pursued for various reasons, including postponding construction of new water or wastewater infrastructure; decreasing operation and maintenance costs; and environmental impacts of increasing withdrawals.

Demand-managed strategies applied around the world today can be grouped into four categories based on the approach employed:

(i) technological
(ii) economic
(iii) institutional and
(iv) behavioral

In the urban sector, techniques include the following:

– escalating block rate tariffs
– promotion of water-wise industries
– water auditing
– water loss management
– retrofitting with water-saving devices
– informative billing
– water-wise gardening, and
– public awareness

In the agricultural water sector, techniques include the following:

– reduction/removing of pricing subsidies
– efficient/water-saving irrigation scheduling
  - alternate furrow irrigation
  - alternate wetting and drying of rice field
  - soil drying during ripening of crop
  - deficit irrigation
  - adopting efficient irrigation method
– Soil water conservation measures
  - Reducing evaporation
  - mulching by various elements
  - using super absorbents

- selecting crops with high yields per unit water consumed
- appropriate cropping pattern
- adopting efficient irrigation devices/technologies
- land leveling
- water-wise tilling/field preparation
- water-wise crop cultivation method (e.g., direct seeding)
- soil-crop-weather management

  - cultivation of short duration cultivars
  - cultivation of low water demand crops (e.g., pulses, wheat)
  - analysis of long-term weather data (especially rainfall and temperature) and planning crop accordingly
  - priming of seed to mitigate low soil moisture at sowing

- Ponding rainwater at rice field by high levees (bunds)
- Storing rainwater in farm pond, lake, etc., and using in dry period

Urban Sector

Metering and Pricing

Metering and pricing are generally considered to be the building blocks of a demand side management. Metering may be more widespread in the industrial and institutional sectors. These categories may have fewer but larger water users, than in the residential and commercial sectors. In the short term, the introduction of metering can impact for water. However, ultimately metering must be used in combination with appropriate pricing structures, to provide an incentive for customers to reduce water use. In any given municipality, several rate structures may be used for different sectors (commercial, residential, industrial, institutional, etc.). Along with various rate structures, the increasing block-rate pricing structure (also termed as tiered water prices) may be an explicitly water conserving rate structure: when water use reaches a certain threshold (the boundary between "blocks"), the price per unit goes up. Tiered pricing provides an incentive for the farmers to choose efficient combinations of irrigation methods and management levels. However, to remain effective, such a structure must be keyed to inflation so that the amount charged for water is consistent with the real cost of providing the service. Declining block rate often is thought to encourage higher levels of water use; as the price per unit of water used goes down in steps or blocks, the amount used increases.

Water Ordinance (by Law)

State or municipal ordinance may promote water conservation and efficient use of water. Water rate ordinance simply authorizes various rate structures in place. Other ordinances (i.e., plumbing fixture, mandatory fixture retrofit, restrictions on specific users, and others) may also be employed. Regional municipalities or agricultural areas may have lawn watering ordinances in place, compared to others.

5.3 Strategies for Water Resources Management

Operational and Maintenance Measures Directing Reduced Water Loss

A wide range of operational and maintenance (O and M) measures directed at reducing water losses and consumption may be employed. Several of these measures, including leak detection, repair of water distribution lines, may be part of normal infrastructure maintenance. Other measures, such as installing new meters on unmetered accounts, reservoir renovation, water pressure reduction, etc. may be aimed at reducing water loss.

Plumbing Fixtures and Devices in a Voluntary Program

Motivation for voluntary retrofitting of plumbing fixtures (to more water-efficient ones), or the distribution of subsidized or free water-saving devices should be done. Efficient toilets and other devices can save a substantial amount of water in homes. The American Water Works Association (1999) reported that in a typical single family home, which has no water conservation fixtures, toilet use accounts for 27.7% (or 20.1 US gallons per capita per day). By installing water-efficient ultra low-flush toilets that use 1.6 US gallon per flush, toilet use declined to 19.3% or 9.6 US gallon per capita per day.

Agricultural Sector

Removing of Pricing Subsidies

Withdrawing of subsidies from water pumping (in the form of subsidies in diesels, electricity, etc.) will certainly make awareness regarding its efficient use, minimizing conveyance loss, and conservation measures.

Water-Saving Irrigation Scheduling

- *Alternate furrow irrigation*: Experimental results suggest that water use can be decreased by almost 33% by irrigating alternate furrows instead of every furrow. Most of the water savings, however, occur on the lower part of the field.
- *Alternate wetting and drying of rice fields*: In rice cultivation, instead of continuous ponding of the field to a certain depth, irrigating after 3–5 days of disappearance of ponded water saves about 30–35% water without reduction in yield.
- *Soil drying during ripening of crop*: In most crops (especially in cereals), we are interested to produce higher grain yield but not the straw yield. Soil drying during the grain-filling period of rice and wheat enhance early senescence. The grain-filling period may be shortened under such a condition, but a faster rate of grain-filling and enhanced mobilization of stored carbohydrate from vegetative parts to grain minimize the effect on yield. Thus, water demand can be minimized without reduction in yield.

Adopting efficient irrigation method: On-farm water use efficiency can be improved by moving to a more efficient irrigation system. Sprinkler and drip irrigation can save noneffective water loss. Minimization of water loss during land preparation in wetland culture of rice (substitution by dry-seeded rice) leads to lower total water requirement.

Deficit irrigation: Omitting irrigation at less sensitive growth stages of plants (with respect to water deficit) minimizes irrigation requirement without significant yield reduction. Research at ICARDA (Zhang and Oweis, 1999) has shown that applying only 50% of full supplemental irrigation (SI) requirements causes a yield reduction of only 10–15%. Assuming that under limited water resources only 50% of the full irrigation required by the farm would be available (i.e., 4,440 m^3 for a 4 ha field), the deficit irrigation was compared with other options by Zhang and Oweis (1999). They showed that a farmer having a 4 ha farm would on average produce 33% more grain from his farm if he adopted deficit irrigation for the whole area, than if full irrigation was applied to part of the area. The deficit irrigation increased the benefit by over 50% compared with that of farmers' usual practice of overirrigation.

In rice cultivation, instead of maintaining 3–5 cm ponded water, irrigating after 3–4 days of subsidence of ponded water (also termed as alternate wetting (ponding) and drying (to saturation or field capacity)) leads to 20–30% water saving without significant yield reduction. Deficit irrigation facilitates the use of applied and stored (within root zone) water more efficiently, and increases WP (Ali et al., 2007). Other measures of increasing water productivity may also be practiced (Ali and Talukder, 2008).

Water Conservation Measures

Water conservation can be done by adopting the following principles:

- reducing evaporation
- mulching
- using super-absorbents

Reducing Evaporation

From Water Surfaces Evaporation from lakes, reservoirs, or other water surfaces varies from about 2 m/yr for dry, hot climates to 1 m/yr or less for humid, cool climates. In the 1950s and 1960s, considerable research was done to reduce evaporation from open bodies of water by covering them with monomolecular layers of hexadecanol or octodecanol. While evaporation reductions of about 60% have been achieved under ideal conditions, actual reductions were much lower, and the use of monomolecular films to reduce evaporation from free water surfaces has found no practical application. Instead, more success has been obtained with floating objects in small reservoirs. Floating sheets of foam rubber have been successfully used. Evaporation reductions of close to 100% have been obtained with such covers.

Evaporation from open water surfaces can also be reduced by reducing the area of the water surface. For small surface storage facilities, this can be achieved by storing the water in deep, small reservoirs instead of in shallow, large reservoirs. For larger facilities, several ponds or compartmentalized ponds have to be available. When water levels in the ponds begin to drop, water is then transferred between ponds or between compartments so that only one or a few deep ponds are kept full while the others are dry, thus minimizing the water surface area per unit volume of water stored.

If the ponds are unlined, the effect of water depth on seepage loss from the pond must be taken into account. From a hydraulic standpoint, increasing the water depth would increase seepage the most.

From Crop Field Evaporation from soil is reduced by dryland farming techniques that are aimed at conserving water in the root zone during the fallow season for use by the crop in the next growing season. The main strategies are weed control, tillage, and leaving the stubble or other crop residue in the field during the dry or fallow season. Weed control prevents transpiration losses. Tillage is primarily needed on heavy soils that may crack during fallow and lose water by evaporation through the cracks. The purpose of the tillage, then is to close the cracks. Sands and other light-textured soils that do not crack are "self-mulching" and do not need tillage. Leaving the stubble or crop residue on the field during fallow periods reduces evaporation losses from the soil by lowering soil temperature and reducing wind velocities close to the soil surface.

In the Northern Great Plains of the United States, these dryland farming techniques reduce evaporation losses by about half the annual precipitation. Thus, if the precipitation is 38 cm/yr as in eastern Colorado, dryland farming techniques conserve about 19 cm of water per year.

Finally, evaporation of water from soil surfaces can be reduced by reducing the extent of wet areas from which water evaporates. In irrigated fields with incomplete crop covers (row crops in the beginning of the growing season, vineyards, orchards), evaporation from soil can be reduced by irrigating only the areas near the plants and leaving the rest of the soil (surface or subsurface). This can be accomplished, for example, with drip irrigation systems (surface or subsurface).

Mulching Mulching with crop residues during the summer fallow can increase soil water retention. Sauer et al. (1996) found that the presence of crop residue on the surface reduced soil water evaporation by 34–50%. Straw mulching can be easily implemented by local farmers and can be extended in the regional scale because material is most easily accessible, is available at low cost, and does not contaminate the soil.

Water Loss Minimization

Water loss in the conveyance system can be reduced through canal lining. A range of materials are available for that purpose. On-farm water loss (seepage through the borders of the plot) can be reduced through minimizing holes and proper maintenance of the borders with sufficient height. Thus, actual need of irrigation water would be reduced.

Adopting Efficient Irrigation Method/Technologies

This can be achieved through the following measures:

Improving Irrigation Efficiency If crop irrigation is practiced in areas with dry climates, much of the water use in those areas is for agriculture. Considering world average, about 75% of the total water use is for crop irrigation. Most of the irrigation systems are surface or gravity systems, which typically have efficiencies of 60–70%. This means that 60–70% of the water applied to the field is used for evapotranspiration by the crop, while 30–40% is "lost" from the conveyance system, by surface runoff from the lower end of the field, and by deep percolation of water that moves downward through the root zone.

Increased irrigation efficiencies allow farmers to irrigate fields with less water, which is an economical benefit. In addition, increased irrigation efficiencies generally mean better water management practices which, in turn, often give higher crop yields. Thus, increasing field irrigation efficiencies also saves water by increasing the crop production, thus allowing more crops to be produced with less water.

Field irrigation efficiencies of gravity systems can be increased by better management of surface irrigation systems (changing rate and/or duration of water application), modifying surface irrigation systems (changing the length or slope of the field, including using zero slope or level basins), or by converting to sprinkler or drip irrigation systems where infiltration rates and water distribution patterns are controlled by the irrigation system and not by the soil. Surface irrigation systems often can be designed and managed to obtain irrigation efficiencies of 80–90%. Thus, it is not always necessary to use sprinkler or drip irrigation systems when high irrigation efficiencies are desired.

Irrigation Scheduling As with increased field irrigation efficiencies, improved scheduling of irrigation conserves water only if runoff and/or deep percolation from the irrigated fields cannot be reused. Scheduling of irrigation can be based on soil water measurements (tensions and/or contents), or on estimates of daily evapotranspiration rates using climatological methods, evaporation pans, or lysimeter. Measurement of the plant water status through remotely sensed plant or crop canopy temperatures with infrared thermometers shows promise as a technique for scheduling irrigations. Better timing of irrigation could also increase crop yields per unit of evapotranspiration (for example, through less leaching of fertilizer), thus increasing crop water use efficiencies.

Water-Wise Cultivation Method

The wet-seeded or direct-seeded technique is an alternative to the transplanting method of rice crop establishment. This technique increases crop yield and water productivity, and reduces irrigation need.

5.3 Strategies for Water Resources Management

Soil-Crop-Weather Management

Manipulation of Seedling Age   In transplanted rice, seedlings 25–45 days old are normally used. Recent research results showed that up to 55–65-day-old seedlings can be used for Boro (Kharif) season (BINA, 2005). The older seedlings (55–65 days) could reduce crop duration up to 15 days. That means that total crop duration in the field (from transplanting to maturity) can be reduced by 30 days (15 days late transplanting +15 days early maturing), which obviously reduces the crop water requirement, and increases water productivity. Ali et al. (1992) obtained the highest yield with 60-day-old seedling for Aman (monsoon) season. Singh and Sharma (1993) and Paul (1994) observed insignificant yield difference for 30–60-day-old seedling for monsoon rice.

Priming or Soaking of Seed   The technique of seed priming, where the seed is soaked in water (usually 10–12 hrs), then surface dried and sown, has been shown to improve plant stands and provide benefits in terms of earlier maturity and increased seed yield in a range of crops (wheat, maize, lentil, chickpea, etc.) in rainfed, as well as irrigated crops grown on normal soils. This technique reduces the post-sowing or pre-sowing irrigation needs and saves water. Besides, priming had a significant positive effect on yield. Thus, the water productivity is increased. Kahlon et al. (1992) observed that soaking wheat for 24 or 48 h in water or pre-germinating seed sowing increased grain yield by 10.3, 16.3, and 21.2%, respectively, compared to sowing untreated seeds.

Crop Sowing Based on Weather Analysis or Forecast   With the probability analysis of long-term rainfall data or the use of the short- or medium-term weather forecast, dry spell can be avoided and thus the need of irrigation can be avoided. Chahel et al. (2007) observed that with the shifting of transplanting dates of rice at Punjab, India, from higher (mid-May) to lower (end of June onward) evaporative demand, there was an increase in grain yield, while there was a reduction in ET and irrigation water applied.

Changing Crops   Another method for reducing evapotranspiration in irrigated areas is to alter cropping patterns. In climates with hot summers and mild winters, summer crops can be minimized, and more winter crops (vegetables, flowers) can be grown. In addition, crops with lower water requirements can be introduced. Where there is some rainfall, dryland farming systems with supplemental irrigation (if necessary) can replace conventional irrigated agriculture.

Use of Anti-transpirants   Spraying plants with anti-transpirants may have some application for ornamental plants (lawns and shrubs) where production or fast growth is not important. For agricultural crops, however, a reduction in transpiration usually also means a reduction in yield. Thus, anti-transpirants generally are not feasible for reducing water use of agricultural crops.

Ponding Rain-Water at Crop Field by High Levees

During the wet spell of weather, water can be stored in the crop field by making high (~30 cm) levees around the plot, especially for rice crop. This stored water will serve the purpose of irrigation during long dry spell. This approach is very useful where the natural rainfall is uncertain and uneven, and supplemental irrigation is a must.

Increasing Soil Moisture Storage Capacity in the Crop Root Zone

Addition of Organic Matter Increasing water storage within the soil profile is necessary to increase plant available soil water. Increasing soil's organic matter (OM) has long been recognized as an effective way of improving soil physical and chemical conditions, and water-holding capacity. Increasing the OM content of sandy soils increases the available water storage capacity, permits greater infiltration, and increases root proliferation throughout the soil profile. The OM releases water slowly, resulting more efficient use of water, facilitating proper crop growth and thus increase in yield and water productivity.

Tillage and Subsoiling Active deep roots help to reduce drainage losses and abstract soil water. But in most cases, the subsoils are not suitable for root extension or inhibit root extension due to hard layer (often called plow pan). Tillage breaks surface soil crust. This leads to increased water storage by increased infiltration and increased root area. Sub-soiling or deep tillage breaks plow pan, and thus facilitate root expansion (and hence increased root area) and more soil moisture abstraction.

Other Management Factors

Among the management factors for more productive farming systems are the use of improved or suitable crop rotation, sowing dates, crop density, raised bed and the role of previous crops.

Public Awareness Approaches

Public awareness and education have been identified as key aspects of any demand side management plan in many places around the world. The most frequently used public awareness approaches directed to encourage water conservation are print media (brochures), information packages, media information, public lectures and meetings, working with local school, low water use landscape demonstration, etc. Demonstration of real cost savings to consumers may be an effective means to reduce water loss.

Other Measures

These include working with large water users to conserve water, water conservation plans or strategies, water audit, and working with other organizations or groups to promote water conservation.

5.3 Strategies for Water Resources Management

Discussion

There may be considerable variations in the water savings among the demand side measures. The appropriateness of different types of demand side measures may be a function of factors related to value of water, locality, financial capacity and habitat of the farmers, etc. Any combination of the measures or awareness campaign (suitable for the site concern) may be implemented. The effectiveness of some measures is known to depend on the presence of others. For instance, while metering can lead to a short-term 30–50% reduction in demand, water consumption can return to previous levels, if metering is not combined with pricing rate structures that tie the cost to the amount used. The long-term effectiveness of some measures can decline if certain considerations are not accounted for implementing them. For example, while increasing block rates have the potential to be effective conservation measures, to remain effective over the long term they must be keyed to inflation. Demographic factors such as per capita income should also be considered in setting rates. The nominal price of water would have to be raised annually by the rate of inflation plus the rate of change in real income simply to maintain constant rather than increasing water use.

The water conservation(WC)/demand management (DM) principles should be integrated fully into water supply planning, i.e., water potentially produced through increased efficiency, and decreased losses should be considered alongside other options at the beginning of supply-planning processes. Although the policy and legal framework for implementing WC/DM has been established in many countries, very few measures have been put in place. Water management entities should establish specific targets/standards for water use efficiency and allowable loss for each water sector and develop strategies to achieve those targets.

5.3.1.3 Obstacles to Implement Demand Management

Large capital projects result in substantial income for construction contractors, consulting firms, equipment and material suppliers, lending institutions, and government and other organizations. Historically, an extensive industry has developed around the glove attachment to supply-side solutions, while demand and conservation solutions have not attracted the same level of attention. Recently, in many countries, the private sector has become increasingly attracted to water supply in the urban sector, which could introduce the profit motive to implementing WC/DM measures.

Water produced through WC/DM can be 65–80% less expensive than water developed through new infrastructure. Despite the obvious economic benefits of water conservation and demand management, the WC/DM approach always cannot be implemented due to obstacles. It is necessary to identify and acknowledge these constraints in order to develop strategies to address them. The following are some of the obstacles and constraints:

(a) Financial/economic
  - Certain water conservation/demand (WC/DM) management measures depend on financial outlay by end users, who may not have adequate resources

- Water is allocated to consumers irrespective of economic value or efficiency of use
- The relatively low price of water, particularly in the agricultural sector
- Water institutions own water supply infrastructures
- Lack of funding or disproportionate funding for supply side measures at the expense of WC/DM.

One of the greatest obstacles to implement WC/DM measures is the initial capital cost. The initial costs for more efficient devices and technologies for the urban or agricultural sector can be substantial. This is despite the fact that costs for demand-side measures are significantly less than cost for supply-side measures. To build a dam, the government typically secures a loan and contracts private companies to construct it. Implementing WC/DM measures in the agricultural sector, on the other hand, would require individual farmers to purchase and install new devices or systems. Similarly, in the urban sector, individual homes or businessmen must purchase and install new devices or systems to achieve savings. Many cities and irrigation districts around the world have found that it makes economic sense to provide incentives such as rebates to encourage adoption of new technologies.

Another economic obstacle to efficient water use is subsidized water tariffs. The vast majority of agricultural water users, and to a certain extent of urban users enjoy subsidized water tariffs. Economic tools (increasing tariff or treating water as a economic good) are often the most powerful in reducing inefficient or wasteful use.

(b) Technical/institutional

- lack of adequate knowledge of the cause of growth in demand
- current planning practices choose the cheapest solution without regard to operating costs
- lack of understanding of the consumer and water usage patterns
- lack of cooperation among local authorities
- lack of cooperation among water services institutions
- officials and industry sectors protect their personal interests
- ignorance of WC/DM when promoting new infrastructure

Implementation of a properly designed and planned WC/DM program will fail without a strong policy and legislative foundation.

(c) Public perception

- Supply side management options appear easier to implement
- Water conservation measures are perceived only as drought relief strategies
- Lack of understanding of principles, scope, and potential of demand management
- Fears that water conservation will result in reduced service levels and reduced crop production in the agricultural sector
- Demand management strategies are often incorrectly perceived as punitive measures

5.3 Strategies for Water Resources Management

Changing habits, equipment, crops, and irrigation equipment/patterns is difficult for anybody including farmers and households, since they are unfamiliar with new technologies and uncertain of the benefits of making the change. A couple of methods have been used in the developed world to speed up this acceptance. They include a great deal of education and financial incentives from government and local supply authority.

> **Box 5.1 Potential for Demand Management in the Urban and Agricultural Sector of Bangladesh**
>
> Many areas and opportunities exist to make better use of demand side management measures. Water use in the urban and agricultural sectors of Bangladesh are generally highly inefficient, with waste/inefficiencies of up to 40 and 50%, respectively. The agricultural water sector holds even greater potential for savings than the urban sector because it uses three times as much water and is more inefficient than urban use.
>
> Flood irrigation, which achieves only 50–60% efficiency, is used in almost all irrigated lands. An increase in the efficiency of only 20% in urban and agricultural water use would save millions of dollars each year. Demonstration of real cost savings to consumers and the development of specific goals and objectives for demand side management programs are two important steps needed to overcome the challenges.

5.3.2 Supply Side Management

Supply management options refer to the actions that affect the quantity and quality of water at the entry point to the distribution system.

5.3.2.1 Approaches of Supply Management

Supply management options include the following measures:

(i) Development of new supply source
   i. surface water (rubber dam)
   ii. groundwater

(ii) Use of nonconventional water with appropriate strategic measures
   - saline water/poor quality water
     o with mixing/blending ratio
     o application of poor quality water at nonsensitive growth stages

- irrigation effluent
- drainage outflow

(iii) Rainwater harvesting at farm pond, lakes
(iv) Augmenting natural recharge
(v) Inducing artificial recharge
(vi) Sharing of international rivers
(vii) Network rehabilitation
(viii) Reallocation through volumetric constraints
(ix) Import of virtual water
(x) Use of super absorbents
(xi) Implementing dual water supply system
(xii) Implementing dual water use – aquaculture in rice field

5.3.2.2 Description of Different Approaches

Development of New Supply Source

Water is available on the planet both on the surface and under the surface of the earth. The water on the surface is termed "surface water", and the water under the surface is termed "groundwater." Surface water and groundwater are both important sources not only for human use but also for ecological systems.

Surface Water

Surface water is renewable, usually within few months or a year, while groundwater is also renewable, as it takes several months or a year. Development of surface water source includes construction of new dams and reservoirs, excavation of ponds, rehabilitation of natural water bodies such as lakes, rivers, and other lowlands. Rubber dam, a flexible and technically and economically sound technology, has been increasingly used in many parts of the world. But in all cases, environmental and ecological perspectives should be taken into consideration. Details about the rubber dam has been described in Chapter 11, Volume 1.

Groundwater

The agricultural sector is the largest consumer of water. For proper development of groundwater, a systematic survey and determination of aquifer characteristics is essential. Groundwater should be withdrawn based on safe yield concept. Safe yield of each aquifer should be determined, and groundwater should be exploited accordingly.

Safe Yield

Intensive groundwater development in many parts of the world has caused rapid decrease of groundwater level. This has resulted in aquifer depletion, water quality

5.3 Strategies for Water Resources Management

degradation, and many other problems. Under natural conditions before groundwater development, aquifers are in a state of dynamic equilibrium. The discharge by new installed wells must be balanced by an increase in recharge of the aquifer, or by a decrease in the old natural discharge, or by loss of storage in the aquifer, or by a combination of the above. Thus, to avoid an adverse effect of overexploitation of groundwater, development of groundwater should be based on safe yield.

Different researchers defined safe yield from different perspectives. In essence, safe yield is the quantity of water that can be harnessed from an aquifer annually without producing an adverse or undesirable result. The undesirable results may include the depletion of the groundwater reserves, degradation of water quality, intrusion of water of undesirable quality (e.g., saline water intrusion), violation of existing water rights, deterioration of economic advantages of pumping (i.e., increasing pumping cost), excessive depletion or reduction of stream flow by induced infiltration (or recharge), loss of wetlands and riparian ecosystem, land subsidence, etc. Often, a misperception among many hydrogeologists and water resources engineers and managers is that the development of groundwater is "safe" if the rate of withdrawal does not exceed the rate of recharge. Even with a pumping rate smaller than the natural recharge (so called "safe yield"), pumping may cause induced recharge or decreased discharge. The induced recharge may cause depletion of stream flow, and residual discharge may not be sufficient to maintain the ecosystem. In the 1980s, the concept of sustainability emerged, which has removed the problem of safe yield determination.

Sustainable Yield

Sustainability is a goal for the long-term welfare of both humans and the environment. Sustainable development must meet the needs of the present without compromising the ability of future generations to meet their own needs (World Commission on Environment and Development, 1987). Sustainability refers to renewable natural resources; therefore, sustainability implies renewability. Since groundwater is neither completely renewable nor completely nonrenewable, it begs the question of how much groundwater pumping is sustainable. In principle, sustainable yield is that which is in agreement with sustainable development. Alley et al. (1999) defined groundwater sustainability as development and use of groundwater in a manner that can be maintained for an infinite time without causing unacceptable environmental, economic, or social consequences. Furthermore, groundwater sustainability must be defined within the context of the complete hydrological system and with a long-term perspective to the groundwater resources management.

Basin sustainable yield can be defined by the following water balance equation (Kalf and Woolley, 2005):

$$P_s = R_0 + \Delta R_0 - D_R \quad (5.2)$$

where P_s is the sustainable pumping rate, R_0 is the natural recharge, ΔR_0 is the increased recharge induced by pumping, and D_R is the residual discharge.

If the total basin inflow, I_s, is defined as the sum of natural recharge and induced recharge by pumping, Eq. (5.2) can be rewritten as

$$P_s = I_s - D_R \tag{5.3}$$

where P_s is the basin sustainable yield.

The basin sustainable yield should be a compromised pumping rate, which can be sustained by groundwater recharge and will not cause unacceptable environmental, economic, or social consequences. Therefore, the basin sustainable yield should satisfy the following conditions (Zhou, 2009):

(a) It is a sustainable pumping rate defined by water balance equation (Eq. 5.2). However, the water balance equation is only a necessary condition but not an absolute condition. Other constraints must be satisfied.
(b) Environmental constraints require considering groundwater as a part of an integral water and ecological system. Pumping capture should not cause the excessive depletion of surface water and the excessive reduction of groundwater discharge to spring, rivers and wetlands. The cone of depression induced by pumping should not cause the intrusion of undesirable quality water, land subsidence, and the damage of groundwater dependent terrestrial ecosystems.
(c) Economic constraints require maximizing groundwater development to fulfill water demand for irrigation and industrial use.
(d) Social constraints require safe access of good quality groundwater for drinking water supply and equitable distribution of shared groundwater resources by all.

Maimore (2004) proposed a practical approach to define the sustainable yield. The approach includes considerations of spatial and temporal aspects, conceptual water balance, influence of boundaries, water demand and supply, and the stakeholder involvement.

Assessments of sustainable yield must reach beyond hydrogeology to encompass the interdisciplinary synthesis of surface water hydrology, ecology, geology, and climatology. In addition, since groundwater is a resource held in common, sustainable yield assessments must consider the socioeconomic context (Hardin, 1968). In general, different communities will have different perceptions of what constitutes an acceptable rate of groundwater withdrawal, and these perceptions may vary over time.

From the above discussion, it is clear that the basin sustainable yield cannot be simply calculated as a single value using the water balance equation (Eqs. 5.1 or 5.2). How much groundwater is available for use depends on how changes in recharge and discharge affect the surrounding environment and the acceptable trade off between groundwater use and these changes. Achieving this trade off in the long term is a central theme in the evolving concept of sustainability.

5.3 Strategies for Water Resources Management

Adverse Effect of Groundwater Overabstraction (or Groundwater Mining)

If the groundwater withdrawal is not based on safe yield or greater than the natural recharge (for long-term consideration), groundwater mining (or overdraft) occurs, which causes numerous adverse effects or environmental consequences:

- Failure of suction-mode pumps
- Reduction in stream flow
- Drying up of natural water bodies (e.g., ditches, ponds, canals), thus affecting the ecosystem
- Shallow-rooted trees become endangered
- Degradation of water quality
- Intrusion of polluted water (brackish or saline water) in coastal areas
- Disruption of ecosystem
- Land subsidence

If the groundwater becomes contaminated (such as arsenic contamination in West Bengal of India and Bangladesh), it will be a catastrophic event and troublesome to purify. Environmental consequences of such exploitation may become slowly critical in most cases. But the situation may be irreversible if care is not taken beforehand.

Fresh-Water Aquifer Underlying Saline Groundwater

In most natural saline zones, a relatively fresh water aquifer underlay the saline shallow aquifer (Rashid, 2008; Asghar et al., 2002). If such an aquifer is found, it could be used as a source of irrigation supply. Other management practices may be needed along with irrigation, depending on the salinity level of the water and the salt tolerance of the desired crop.

Use of Nonconventional Water with Appropriate Measures

Poor quality water or saline surface/ground water may be used to irrigate crops with some special measures (described below) and thus can be regarded as a vital source of water in water scarce areas.

Mixing Saline Water with Fresh Water

Irrigable water can be increased by mixing highly saline water with fresh water (good quality or low salinity water) to lower the salinity to acceptable/tolerable limits. Mixing does not reduce the total solute content but reduces the solute concentration due to dilution.

The salinity of the mixed water or the mixing ratio can be obtained by using the following equation (Adapted from Ayers and Westcot, 1985):

$$C_X = (C_i \times r_i) + (C_j \times r_j) \tag{5.4}$$

where

C_X = salt concentration of mixed water, mg/l
C_i = salt concentration of first category of water, mg/l
r_i = proportion of first category of water
C_j = salt concentration of second category of water
r_j = proportion of second category of water
and $r_i + r_j = 1$

Alternate Irrigation of Saline and Fresh Water

Fresh and saline water can be applied in alternate sequence (fresh-saline-fresh-saline) to increase irrigable water and minimize salinity hazard. In such a practice, some sort of stress hardening is developed in plants for salinity stress.

Irrigation with Saline Water at Less Sensitive Growth Stages

Where fresh water is unavailable or very costly, fresh water should be applied when it is essential. All growth stages of crops are not equally responsive to salinity stress. Hence, irrigation with fresh water at most sensitive growth stage(s), and irrigation with saline water (or a mix) at relatively insensitive stage(s) can facilitate reasonable plant growth and yield.

Rainwater Harvesting at Farm Ponds, Lakes

The source of all water is rain. The harvesting of rainwater simply involves the collection of water from surface on which rain falls and subsequently storing this water for later use. As a definition, it can be defined as the collection of rainfall or rainfall runoff for agricultural production, or domestic purposes, or livestock watering. In areas where there is inadequate groundwater supply, or surface resources are either lacking or insufficient, rainwater harvesting offers an ideal solution. Countries like Germany, Japan, United States, and Singapore are adopting rainwater harvesting. In urban areas, rainwater harvesting from the building roof and storing it underground can serve the need of about 40% of the demand of the urban people during rainy season. Productivity of rainfed agriculture can be improved through supplemental irrigation via water harvesting. Rainwater harvesting in existing local ponds, and using them for dry season cropping (particularly low water demanding crops, such as pulses, oilseed, wheat), seems a prospect in dry areas. Ali and Rahman (2009) found that 20:1 ratio of land area to pond area is sufficient to cultivate low-water demanding dryland crops (other than rice) in saline area. The details of water harvesting procedures have been discussed in Chapter 11 (*Water Conservation and Harvesting*), Volume 1.

5.3 Strategies for Water Resources Management

Increasing Groundwater Storage Through Recharge

Water can be stored above ground behind dams, or in underground aquifers. Underground storage can be enhanced by increasing the wetted area (width) of streams, using weirs or dams, or constructing levees in the streambed or flood plain. Also, groundwater recharge can be enhanced by constructing off-channel infiltration basins. Often, some form of upstream surface storage is needed to store short-term floods or other peak runoffs in the stream system. This water is then released at a slower rate from the reservoirs to allow infiltration through the downstream recharge system, so that eventually all the water is stored underground for future use.

Artificial Recharge

Recharge can be augmented through artificial arrangement. The best time for artificial recharge of groundwater is during the wet years when there is a water surplus. Artificial groundwater recharge through structures may be the most effective means for increasing groundwater potential in the rapidly declining groundwater areas. Artificial recharge includes several ways to recharge, from low to high cost involvement. A detailed procedure of artificial recharge has been described in Chapter 11 (*Water Conservation and Harvesting*), Volume 1.

Box 5.2 Water Supply Problems in Bangladesh

- Surface water is contaminated, needs treatment for drinking use
- Lowering of groundwater table
- Arsenic contamination of groundwater
- Excessive iron in groundwater at some places
- Salinity problem in the coastal areas
- Source problem of water in hilly areas
- Surface water is unavailable (diminishes) during dry months. Most of the river's source is outside the country

---

Sharing of International Rivers

More than 300 major river basins across national boundaries exist worldwide, which comprise about half of the total land of the world. The large river basins are shared by several countries. International treaty should be accomplished and employed to share the water among the countries. Good relations should be established among the countries in terms of cultural, trade, and other levels. If needed, some sort of sacrifice in other sectors may be allowed to hold the good relations and get a share of the water. Bilateral and multilateral environmental agreements should be accomplished and employed to deal with trans-boundary freshwater issues.

For almost all trans-boundary rivers, potentials of basin-wise arrangements is ignored as the riparian countries limit their planning and management considerations from a narrow nationalistic view. This attitude is also prevalent as a river flows from one state/province to another, even, within one country. However, resolution of interstate/interprovince disputes in India, United States, and Australia offer examples to be followed to overcome such stumbling blocks. There are many good examples where trans-boundary water disputes have been resolved. Involvement of a third party may facilitate resolution of such disputes, if it is between two unequal parties. Israel, Jordan, and Palestine have resolved their dispute over common water, despite all odds.

Reducing Quality Degradation

Degradation of quality to the point where water can no longer be used for its intended purpose or is no longer suitable for beneficial use in general is a form of water loss. Such losses occur when fresh water is polluted or when fresh water is discharged into salt water or seeps down to saline aquifers from where it no longer can be separately recovered. Where water is scarce, and water conservation is a necessity, such losses should be minimized. This indirectly increases the available source.

Cloud Seeding

Cloud seeding is a technique for artificial rainfall. Depending on local conditions, the economic aspects of cloud seeding can be quiet favorable. Clouds moving over an area without producing rain is a form of water loss for that area. Orographic storms are much more productive for seeding than convective storms. In western United States, seeding of orographic storms is expected to increase precipitation from those storms by 10–15% (the percentage increase is higher in dry years than in wet years). The preferred seeding technique is with ground-generated silver iodide crystals. Seeding from airplanes is much more expensive. Since only about 5% of the water in orographic storm clouds falls on the ground as natural precipitation, a 10–15% increase in precipitation due to cloud seeding would still leave plenty of water in the clouds for areas downwind from the seeded areas.

Network Rehabilitation

Initiation of rehabilitation projects for municipal supply network would save water. This option would be the easiest to implement and would bring overwhelming social acceptance. Similarly, reducing conveyance loss in irrigation systems provides a large savings and hence an increase in supply at the user point.

Reallocation Through Volumetric Constraints

The process of reallocating water from agriculture to cover deficits in household and industrial uses, or lower value crops to higher value crops may be an option to manage the scarce water resource.

5.3 Strategies for Water Resources Management

Import of Virtual Water

Where water is very scarce, the food or "virtual water" can be imported. In such situation, food imports may be invariably cheap below actual production costs. Gaining access to secure "virtual water" in the global system can be achieved by developing a diverse and strong economy. Alternatively, in oil-rich economies such access can be secured through oil revenues.

Superabsorbents for Water Capture

Superabsorbents are compounds which absorb and hold water in a form readily available to plants, slowly releasing this moisture in response to the change in water concentration in the surrounding medium. Their main function is to reduce moisture stress and transplant shock, thus creating optimum moisture conditions for plant growth. The water-holding capacity of the various superabsorbents available ranges from 30 to 1,000 times their own weight. In landscape construction, superabsorbents can be used as establishment aids for trees, shrubs, lawns and instant application. Details regarding superabsorbents are described in Chapter 11, Volume 1.

In water-short areas, there is usually no single solution for solving problems of inadequate water supplies. Rather, a broad approach is needed, saving water, using water more efficiently, and reusing water wherever possible. Only then can limited water supplies in arid and semiarid regions be effectively managed.

5.3.2.3 Implementation of Supply Management Options

In general, large-scale groundwater development requires the construction of a great number of wells and the use of a considerable amount of energy to extract groundwater. Low-income developing countries face considerable difficulties in implementing a groundwater development plan. In selecting the technologies to be used, a number of factors have to be taken into account which go beyond the physical merits of the technologies themselves, that is, human and financial resources availability, compatibility with cultural habits, level of dependence on foreign imports, etc.

Surface water schemes generally involve one or a small number of water storage and/or diversion facilities. Such a centralized structure can be easily monitored by a public, semipublic, or private organization. The availability of the resource is known at all times with reasonable accuracy, and its handling is almost permanently under control.

In groundwater development, two types of technologies may be considered for access to and extraction of groundwater: intermediate and modern technologies.

Intermediate technologies:

- the open (dug) well, lined with concrete down to the water level and fitted with a column of concrete rings below it

– the borehole, drilled by hand using simple machines. Such boreholes are equipped with simple hand or foot pump
– a variety of modernized traditional water extraction devices using metal instead of wood

Modern technologies:

– excavation machines for large diameter wells
– motorized drilling rigs of various types for borehole drilling
– motorized pumps of various sizes and capacities

Modern technologies allow rapid execution and development of resources compared to intermediate technology. In some cases, intermediate technologies are inefficient and difficult for the extraction of large yields. Choice of technology should be based on the above-mentioned criteria. As a guide, intermediate technologies are to be used when local conditions, mainly depth of water table and type of geological formations, allow for their use and when modest or moderate water yields are sought. Modern technologies are to be used if rapid and/or large-scale development is needed and if difficult hydrogeological conditions or lack of manpower do not allow for simpler, less expensive methods.

Large-scale groundwater schemes involve great numbers of extraction sites. The resource is invisible, its monitoring is complex. The operation of wells located on private properties is in most cases beyond any control. The scattering of the sites also creates specific problems. Most of the large-scale groundwater schemes fall into the following categories taken separately or in combination:

- High-yield well fields equipped with powerful pumping stations for urban or industrial supplies – less commonly for irrigation schemes
- Individual wells equipped with motorized pumps delivering moderate yields (5–10 to 30–50 l/s) for irrigation or industrial needs including tourism.
- Open wells or tube wells for village or rangeland supply and small-scale irrigation from which water is extracted using hand pump or wind energy. Nowadays, solar pumps are used in some places.

5.3.3 Integrated Water Resources Management

5.3.3.1 Meaning of Integrated Water Resources Management

Many different uses of water resources are interdependent. For example, high irrigation demands and polluted drainage flows from agriculture mean less freshwater for drinking or industrial use; contaminated municipal and industrial wastewater pollutes rivers and threatens ecosystems; if water has to be left in a river to protect fisheries and ecosystem, less can be diverted to grow crops. Integrated management means that all the different uses of water resources are considered together. Water allocations and management decisions consider the effects of each use on the others.

5.3 Strategies for Water Resources Management

Integrated water resources management is therefore a systematic process for the sustainable development, allocation, and monitoring of water resource use in the context of social, economic, and environmental objectives. The integrated approach to management of water resources necessitates coordination of the range of human activities which create the demand for water, determine land uses, and generate waterborne waste products. The principle also recognizes the catchment area or river basin as the logical unit for water resources management.

Since water sustains life, effective management of water resources demands a holistic approach, linking social and economic development with protection of natural ecosystems (environmental). Effective management links land and water uses across the whole of a catchment area or groundwater aquifer. Integrated water resources management (IWRM) balances the views and goals of affected groups, geographical regions, purpose of water management, and protects the water supplies for natural and ecological systems. IWRM implies more coordinated decision making across sectors (Fig. 5.1).

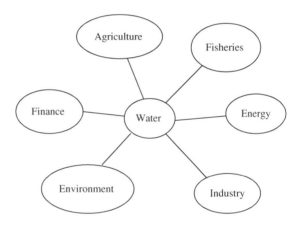

Fig. 5.1 Relevant sectors of water for coordinated decision making

In summary, integrated water resources management is entrenched in sectoral interests and requires that the water resource is managed holistically for the benefits of all. It offers a guiding conceptual framework with a goal of sustainable management and development of water resources.

Water is a subject in which everyone is a stakeholder. Real participation only takes place when stakeholders are part of the decision-making process. The type of participation will depend upon the spatial scale relevant to particular water management and investment decisions. A participatory approach is the best means for achieving long-lasting consensus and common agreement.

5.3.3.2 Approaches, Barriers, and Problems of Integrated Water Resources Management

Each country has its priority development and economic goals set according to environmental, social, economic, and political realities. Problems and constraints arise in each water use area, but the willingness and ability to address these issues in a coordinated way is affected by the governance structure of water. Recognizing the interrelated nature of different sources of water and also the impacts of the different water uses is a major step to the introduction of integrated water resources management.

> **Box 5.3 IWRM Issues in Bangladesh**
>
> IWRM issues in Bangladesh are as follows:
>
> (a) Systematic consideration of the various dimensions of water
>
> - surface water
> - groundwater
> - rainfall
> - quantity and quality
>
> (b) Interaction between water, land, and environmental encompassing such as
>
> - floodplain management
> - erosion control
> - preservation of wetland and fish habitat
> - irrigation and drainage
> - recreational use of water
> - nonpoint source of pollution
>
> (c) Interrelationship between water and social and economic development

Resolution of the many issues in the management of water resources calls for action on several fronts – technical, economic, social, legal, educational, and, not least, political. Under the existing conditions of institutional and legislative framework in most of the countries worldwide, implementing IWRM is likely to require reform at most of the stages in the planning and management cycle. An overall plan is required to envisage how the transformation can be achieved, and this is likely to begin with a new water policy to reflect the principles of sustainable management of water resources. To put the policy into practice is likely to require the reform of water law and water institutions. The institutional arrangements are needed to enable the following:

– The functioning of a consortium of stakeholders involved in decision making, with representation of all sections of society, and a good gender balance
– Organizational structures at basin and sun-basin levels to enable decision making at the lowest appropriate level

- Water resources management based on hydrological boundaries
- Government to coordinate the national management of water resources across water use sectors

Barriers in implementing IWRM include the following:

- Ignorance of interrelation between biophysical and socioeconomic aspects of a system
- Inadequate information and data on how water is used in agriculture
- Lack of understanding/awareness of IWRM principles and practices
- Demographic pressures
- Lack of political will and financial resource
- Incompleteness in water management policy and legal and regulatory frameworks

Problems with IWRM include the following:

- Institutional requirement will be huge.
- Can disrupt the existing system and therefore meet with a lot of frustration and negativity of the officials within the various departments.

5.4 Sustainability Issues in Water Resource Management

Water sustains life. It is our duty to sustain all sources of water. Scarcity and misuse of fresh water pose a serious and growing threat to sustainable development and protection of the environment. Human health and welfare, food security, industrial development, and the ecosystems on which they depend, are all at risk, unless water resources are managed more effectively at present and beyond what they have been in the past.

5.4.1 Concept of Sustainability

Sustainability rests on the principle that we must meet the needs of the present without compromising the ability of future generations. Sustainable water management and development is defined as a set of actions to secure the present functions of water without jeopardizing the interests of future generations. Sustainable management implies managing for the long term. Water resources systems that are able to satisfy to the extent possible the changing demands placed on them over time, without system degradation, can be called "sustainable.".

Sustainability, as defined in the Brundtland Commission's report "Our Common Future" (WCED, 1987), focuses on meeting the needs of both current and future generations. Development is sustainable if

> ... it meets the needs of the present without compromising the ability of future generations to meet their own needs.

Since the Brundtland report of 1987, sustainable development has become the subject of discussions and debates throughout the world. While the world "sustainability" can mean different things to different people, all would agree that it includes a consideration of the future. The Brundtland Commission was concerned about how our actions today will affect "the ability of future generations to meet their needs." Now, the question arises "what will those needs be?" We today can only guess as to what they may be. We can also argue over whether or not it is appropriate to try to meet present or future needs if they overstress the system designed to meet them. Another question arises, "over what time and space scales should we do it?" How do we allocate over time and space our renewable resources, i.e., the waters that exist in many deep groundwater aquifers that are not being replenished by nature? To preserve nonrenewable resources now for the use of our descendants in the future, in the interests of sustainability, would imply that those resources should never be consumed as long as there is a future. If permanent preservation seems unreasonable, then how much of a nonrenewable resource might be consumed, and when? It raises the question, "does everything need to be sustained?" If not, just what should? And over what spatial and temporal scales should sustainability considerations apply? Obviously we do enhance the welfare of future generations by preserving or enhancing the current state of our natural environmental resources and ecological systems. The debate over the definition of sustainability is among those who differ over just what should be sustainable and how to achieve it.

ASCE (1998) Defined Sustainability in Water Resources as:

> Sustainable water resource systems are those designed and managed to fully contribute to the objectives of society, now and in the future, while maintaining their ecological, environmental, and hydrological integrity.

Sustainable water resource systems are those designed and operated in ways that make them more adaptive, robust, and resilient to the changes in the natural system due to geo-morphological process, changes in the engineered components due to aging and improved technology, changes in the demands or desires due to a changing society, and even changes in the supply of water, possibly due to a changing climate. In the face of changes and uncertain impacts, an evolving and adaptive strategy is a necessary condition of sustainable water resources management. Adaptive management is a process of adjusting management actions and directions, as appropriate, in light of new information on the current and likely future condition of our total environment and on our progress toward meeting our goals and objectives.

Sustainable issues are not new issues, nor is sustainability a new concept. Yet the current interest in sustainable water resources management clearly comes from a realization that some of the activities that we are performing today on this earth, could cause irreversible damage of the ecosystem. This damage may have adverse effect not only our own lives but also the lives of those who follow us.

5.4.2 Scales of Sustainability

If we maintain too broad an interpretation of sustainable development, it becomes difficult to determine progress toward achieving it. In particular, concern only with the sustainability of large river basins could overlook the unique attributes of particular local watershed economics, environments, ecosystems, resource substitution, and human health. On the other hand, not every hectare of land or every reach of every stream in watershed need be sustainable or self-sufficient. Even at riverbasin or regional levels, it may not be possible to meet the "needs" or demands of even the current generations, if those needs or demands are greater than what can be obtained on a continuing basis at acceptable economic, environmental, and social costs. This highlights the need to consider the appropriate spatial scales when applying sustainability criteria to specific water resources systems.

We also need to consider the appropriate temporal scales when considering the sustainability of specific river basin water resource systems. The achievement of higher levels of water resource system sustainability does not imply that there will never be periods of time in the future in which the levels of welfare derived from those systems decreases. Given the variations in natural water supplies – the fact that floods and droughts do occur – it is impossible, or at least very costly, to design and operate water resources systems that will never "fail." During periods of "failure" the economic benefits may in fact depend on these events. One of the challenges of measuring sustainability is to identify the appropriate temporal scales in which those measurements should be made.

Sustainability measures provide ways by which we can quantify relative levels of sustainability. This can be defined in a number of ways. One way is to express relative levels of sustainability as separate or weighted combinations of statistical measures of various criteria that contribute to human welfare. These welfare criteria can be economic, environmental, ecological, and social. For many criteria, the time duration as well as the extent may be important.

5.4.3 Achieving Sustainability

In order to achieve sustainable development and protection of the environment in the context of water management, the Dublin statement identified four principles to guide action at local, national, and international levels. Each of these principles – an integrated approach, a participatory approach, explicit involvement of women, water as an interconnected package – is the basis for sustainable water management and development in the current twenty-first century. At least three themes fundamental in Agenda 21, publication from the Earth Summit in Rio de Janeiro, are directly relevant to water management and development. First, it was argued that planning should be well informed, systematic, and rigorous. Second, it was mentioned that more attention should be given to equity issues in the future. And third, Agenda 21 consistently advocates a need for greater decentralization in planning, management and development as one way of ensuring that decision making occurs at an appropriate level. This third theme highlights the significance of empowering local communities in water management, and other types of resource and environmental management.

Water resources management is the vector sum of a progression of legislation, policies, regulations, engineering practices and institutional traditions. Changes in how we manage our water resources are often motivated by each flood, drought, environmental disaster, or human health threat. All of these changes could be directed toward what we today might label sustainable development. Added to that are numerous cross-cutting and overlapping "programs" and "initiatives" designed to achieve sustainability in many areas of water resources and environmental protection.

From all of this activity, how do we determine the "right path," the "correct strategy" or the "optimal future?" What is or should be sustainable and what need not be sustainable? How do we account for the inevitable and profound effects of future technological developments, which may mitigate many of the adverse effects of current unsustainable practices? With the exception of the loss of species, what other resources are vulnerable to irreversible decision? Of all the possible strategies, which is the "best" for a location, is best for a state or country? What are the goals, and who should decide them? How would a water resource project, or watershed development strategy that is planned under sustainable development principle compare to one planned under the best of current planning and evaluation practices? Would there be a difference if the planning is for growth versus development? Is there a fundamental distinction between sustainable development strategies in "growing economics" versus the "developed economics?" There are many more questions like these that need answering, before we can say that we understand the ramifications of sustainable development and begin to truly implement sustainable development principles and goals.

The sustainable development paradigm seems to be advocating not so much for a different planning paradigm but rather for an extended set of evaluation factors – different criteria and weights on objectives to reflect a perceived shift in public preferences. One can formulate all the visionary, creative features for sustainable development that can be accommodated by computers and human imagination. Sorting through these possible features for water management requires a planning framework and a replicable set of evaluation criteria that should be common to all resource management agencies.

In many situations, the overall goals of conserving environmental and natural resources and alleviating poverty and economic injustice are compatible and mutually reinforcing. However, there will always be conflicting views on how these goals can be met. Trade-offs will have to be made among the conflicting views and objectives. The challenge for political leaders and professional resource managers is to make the best of situations that will require difficult decisions and choices if sustainable water resources management is to be achieved.

It is clear that there are many unanswered questions related to the sustainable development and management of any renewable or nonrenewable water resources

5.4 Sustainability Issues in Water Resource Management   177

system. No manager of water resources has the luxury of waiting until all these questions are answered. But those involved in managing the resources can still work toward increasingly sustainable levels of development and management. This includes learning how to get more from our resources and how to produce less waste that degrades these resources. New ideas and new technology will have to be developed to achieve increased economically efficient recycling and the use of recycled materials. Management approaches that are more nonstructural and compatible with the environmental and ecological life support systems must be identified. Better ways of planning, developing, upgrading, maintaining, and paying for the infrastructure that permits effective and efficient resource management and provides needed services must also be defined. For water resource managers, considerations of sustainability challenge is to develop and use better methods for explicitly considering the possible needs and expectations of future generations along with our own. We must develop and use better methods of identifying development paths that keep more options open for future populations to meet their own and their descendants' needs and expectations.

Finally we must create better ways of identifying and quantifying the amounts and distribution of benefits and costs when considering trade-offs in resource use and consumption among current and future generations as well as among different populations within a given generation. In striving for sustainable development of a river basin's water and related land resources, the effectiveness of any mechanism devised to realize that goal depends ultimately on the quality of the individuals entrusted with pursuing it. Engineers, economists, ecologists, planners, and professionals must be involved, but they can be only part of that involvement. Professionals must work within the social infrastructure of a community or region. Successful collaboration with an informed and involved public can lead to more socially compatible uses of resources and to more creative, appropriate, and hence sustainable uses of technology for addressing a community's or region's water resource problems or needs.

5.4.4 Strategies to Achieve Sustainability

The following strategies may be useful toward achieving a sustainable water management system:

5.4.4.1 Integrated Management of Water

Given the linkage between surface and subsurface water-related activities, the whole river basin should be considered by water-development policies. River basin authorities should be established with authority over inter-sectoral allocation of water, enforcement of water quality standards, arbitration in disputes, and compensation procedures. Water policies should encompass groundwater, surface water, and direct use of rainfall.

5.4.4.2 Participation of Users and Stakeholders

Governments should ensure that stakeholders are able to participate in the allocation and management of water resources by (a) identifying and establishing water rights for all users, with regard to economic, environmental, social and cultural considerations, (b) establishing compensation procedures, and (c) using impact assessment to identify livelihood rights and ecological considerations.

5.4.4.3 Environmental Protection

Both to guide development and to monitor changes, catchment-wide mapping of "hot spots" or areas of special problems or vulnerability regarding water quality should be undertaken, and sources of major pollutants should be identified and published. Water systems with special ecological or social significance should be protected and preserved. Training of water professionals should explicitly include attention to environmental dimensions and water conservation.

5.4.4.4 Research into the Impacts of Water Policy

Because the impacts of water policies and policy changes are not well understood, structural research on impacts is needed in order to move away from reliance on anecdotal evidence. Particular attention should be given to the equity effects of policies and actions, since they can injure individuals or groups even while providing benefits to a wider community. Also, any externally defined system of water rights must take into account the perceptions and customs of local users, if it is to have legitimacy.

5.4.4.5 Capacity Building for Integrated Management

The establishment of river basin authorities will require the education and training of people regarding coordination, planning of systems for monitoring and evaluation, sensitization to new responsibilities, integrating use of land and water, use of modelling techniques, implementation of guidelines, and procedures for stakeholder participation. Politicians and policy makers will need to become more aware regarding water scarcity and water needs, equity of access to water, water rights, sustainability, service improvement, enhancing the performance of water systems and system participation.

5.5 Conflicts in Water Resources Management

5.5.1 Meaning of Conflict

Conflict is often perceived as a process rather than an end of a disagreement or collision of interests, ideas or principles. Within such a process, conflict may develop from a difference in perception into a sharp disagreement and then into a fight or struggle, or war.

Conflicts in the water management of international river basins can be perceived as interaction of management issues in the following interdependent processes: integrated water resources management, international cooperation, and conflict management processes. Conflicts may appear in many forms as part of the integrated water resources management process, at different scales in the context of international cooperation, and in varying intensity in terms of conflict management.

5.5.2 Water Conflicts in the Integrated Water Resources Management Process

Water is life, and water management has thus become more complex in accordance with the increasing complexity of the economic and social development process. As integrated water resources management is itself a process, conflicts in water management evolve with the scope and intensity of the interaction between human beings and nature, among individuals, and between communities. Conflicts in the integrated water resources management process can therefore be seen from different perspectives: environmental, economic, and social or political.

5.5.2.1 Social Conflicts of Water Use

Water has long been perceived as a social good, and interaction between human beings and nature has, until recently, been based mostly on the sectoral perception of water resources ecosystems. This has resulted in various forms of water conflicts, which reflect different perceptions from sectoral needs for water or from different concepts of water-use priority in the process of social and economic development. This kind of conflict is called "social conflicts of water management." In terms of an integrated water resources management process, social conflicts of water management form the most important obstacle to the achievements of water-use efficiency of a water resource ecosystem.

5.5.2.2 Economic Conflicts

Apart from satisfying basic human needs and health, water resources are essential for food production, energy and the restoration and maintenance of ecosystems, and for social and economic development in general. While agriculture accounts for a major part of the global freshwater use and is necessary to ensure food security, the high economic growth expected in the developing countries call for better value-added utilization of water resources. It is imperative that freshwater resources development, use, management and protection be planned in an integrated manner, taking into account both the short- and long-term needs of the social dimension and the stability and sustainability of the social and economic development process.

Competition in limited water resources increasingly occurs among agriculture, rural, urban, industrial, and environmental uses. At the same time, disparity in the economic conditions between the urban and rural areas in a country and among countries continues to increase. Questions of economic efficiency in water use will eventually grow and assume greater significance in conflict management in water resources development. Effective use of market mechanism could contribute to conflict prevention in water management by making use of increased opportunities and incentives to develop, transfer, and use a resource in ways that would benefit all parties. On the other hand, the inability to integrate water resources management into the economic and social development process will lead to the aggravation of conflicts in water management. In order to avoid them, it is necessary to create conditions for an efficient environment for the economic use of water, including a well-defined legal and institutional framework for water utilization and conditions for a fair and equitable sharing of beneficial use of the water resources.

5.5.2.3 Legal Conflict

Application of the integrated water resources management concept to international river basins usually faces the most difficult obstacle: the legal context of water use. From an ecosystem point of view, the legal aspect of international river basins is the main source of inefficiency and conflicts in water management. These conflicts reflect the magnitude of problems in the legal aspects of water resources management, which may come from issues related to the allocation of water resources within a country or the management or sharing of water among the riparian countries of an international river basin.

5.5.2.4 Water Conflicts in Perspective

Water management in the current twenty-first century differs significantly from the traditional approach. With the increasing impact of the globalization of economic development and global changes, a variety of studies and research work have been undertaken to seek ways and means to deal with that impact and seek solutions to possible conflicts arising from these changes. Different scenarios under the impact of global climate change resulting from greenhouse effects have been studied for many international river basins, and various management concepts, including virtual water concept and privatization of the economy on water resources. In this context, conflicts in water management will therefore assume new dimensions: futuristic perspectives.

5.5.3 Scales of Conflicts in Water Management

Water conflicts may have a wide range of scales which usually reflect the true scales of water management problems in water utilization resulting from water shortage, water-related disasters and water pollution. Water conflict may have a larger scale resulting from different perceptions of needs, such as ecosystem needs for environmental protection, economic opportunities from water resources development, social equity and future demands for water. Large-scale water conflict may result from perceptions of local natural phenomena which affect the interests of the parties owing to the lack of information or communication. In general, water conflict may be in three different levels:

- global
- regional and
- upstream–downstream

5.5.3.1 Global Scale of Water Conflicts

The world population has tripled in the past 100 years, but water use for human purposes has increased sixfold. About half of all available freshwater was being used for human ends, the trend continued to increase, and there is a water crisis today. The poor management of water resources and the increasing disparity in the economic and social conditions between areas, countries, and regions were the root cause of the global scale of water conflicts. The global scale water conflicts results from the following three issues:

– lack of accessible water
– increasing environmental concern and
– economic value of water

Lack of Accessible Water

According to recent WHO/UNESCO estimates, the total volume of accessible water is less than 0.3% of the global water resources. The lack of accessible water is caused by the shortage of water in terms of both quantity and quality. Human behavior contributed a great deal to the worsening situation of accessibility to water in view of the fact that in many countries, both developing and developed, current pathways for water use are often not sustainable.

Increasing Environmental Concerns

Water resources ecosystems are increasingly recognized for the environmental goods and services they provide through healthy catchments. This trend has built up into global concerns for water quality protection and biodiversity conservation through a series of conservation programs, including the establishment of natural reserves and protected areas. Various other global measures for environmental conservation have also been introduced, among which was the introduction and application of the environmental assessment and management standards under the ISO 1400 series for development investments. In many instances, these environmental movements have contributed a great deal to the rehabilitation of rivers, lakes, and wetlands and improvement in the integrated management of catchments. It is well-recognized that, in spite of strenuous national and international efforts, the contamination of water bodies continues to pose a major threat to the sustainable development of water resources and environmental security.

Economic Value of Water

The scarcity of freshwater in the global system has been recognized as a major global concern with respect to food security but also to the well-being of humankind. In terms of demand management, it is necessary not only to conserve water but also to make the best of the available volume of water so as to improve the quality of life. Water is therefore seen not only as a social good but is increasingly recognized as having economic value. While the concept of assessing economic value to water resources is increasingly accepted, it is recognized as a threat to social security and, most importantly, to the basic need for universal access to drinking water, especially for the poor, in many developing countries.

5.5.3.2 Regional Scale of Water Conflicts

Water conflicts on a regional scale may come from three main categories: differences in water resources endowment, trans-boundary pollution and disputes in the management of international river basins. While water conflicts in the first two categories are less common, those in the third category are more frequent.

The different levels of water resources endowment have always been the main reason for the disparity in the distribution of population among areas of a country or a region. People tend to settle in areas with rich water resources where urban settlements continue to grow. Within a country, various measures, including economic incentives and development action, can be taken to minimize the impact of the differences in water resources endowment on the social and economic conditions between regions, although such measures may lead to other issues in water management. For an international region, options are usually limited and the discrepancy in resources can easily be conceived as better opportunities for development. This perception could lead, on the one hand, to explosive political issues, such as illegal activities to make use of the rich water resources, including the diversion of water resources or illegal fishing, and, on the other hand, to undue pressure to share those resources by using different control measures based on monopolistic market opportunities or transit control.

Trans-boundary pollution is becoming more and more frequent with higher levels of development intensity. Among the prominent issues of trans-boundary pollution are those related to acid rain caused by industrial development or the construction of major coal-fired thermal power plants. Intensive agricultural development and different industrial waste disposal schemes may lead to severe pollution of groundwater aquifers that extend beyond national boundaries.

With respect to the third category of conflicts on the regional scale, there are more than 300 major river basins and a number of major groundwater aquifers across national boundaries. These river basins comprise about half of the total land area of the world, which includes more than 60% of the area on the continents of Africa, Asia, and South America. Most of the river basins are areas of potential conflict, especially in the large river basins and those shared by several countries.

Box 5.4 Examples of Water Conflicts on the Regional Scale

Indian Subcontinent

Since the partition of the Indian subcontinent between India and Pakistan in 1947, longstanding conflicts over the Indus River became an international issue between the two countries overnight. With the help from the World Bank, negotiation over water issues between the two countries began in 1952. The Indus was divided between the two countries, with India receiving the three eastern and Pakistan the three western tributaries. The division deprived Pakistan of the original source of water for its irrigation system. In compensation, India paid for new canals to bring water from the rivers allocated to Pakistan and a consortium of countries financed the construction of storage dams to ensure Pakistan a reliable supply. At a price, the treaty defused a major source of potential conflict and allowed each country to develop its share of the basin's water.

Bangladesh, which gained its independence from Pakistan in 1971 with the aid of Indian army, was threatening to cancel its Treaty of friendship with its former liberator because of conflicts over water. Most of the rivers in Bangladesh flow from India. A major diversion from the Ganges river just a few miles from the Bangladesh border reduce water supply and increased salinity levels, threatening the livelihood of millions of Bangladesh. In 1996, a treaty between governments of India and Bangladesh was signed on sharing of the Ganges water at Farakka. This treaty provides a firm foundation for long-term development cooperation between the two countries.

The Middle East

Throughout the Middle East, the natural facts of water control, consumption and demand interact to form a complex hydro-political web. The allocation of the region's three major river basins, the Nile, the Euphrates-Tigris and the Jordan, nascent sources of tension and potential sources of conflict. Turkish relations with both Iraq and Syria are strained over Turkey's South East Anatolia Project. Egypt was concerned about the water development activities of the upstream users of the Nile. Of all the Middle East river basins, however, it is the Jordan River that hosts the most fraught and inflammable dispute. With the completion of the Ataturk Dam in 1990, Turkey is in a pivotal position to influence downstream flow of the river. Potentially, the dam could benefit all countries within the basin by reducing the variability of the river's natural flow.

5.5.3.3 Upstream and Downstream Relationship

The relationship between upstream and downstream states is usually the principal root cause of water conflicts in the management of international river basins. While water is a resource flowing from one place to another, it carries the impact of human intervention between places. Furthermore, the variability in time of water quantity adds complexity in the management of international water resources and confusion to the perception of changes from human intervention. This is typically true when a river flows between areas of different climatic conditions, such as dry and wet regions of a river basin. Depending on the relative importance of the impact of the changes on the economic and social conditions, water conflicts may develop from a bilateral issue into a river basin problem.

5.5.4 Analysis of Causes of Conflicts in Water Management

Conflicts in water management can be seen from the point of view of a management system which consists of three key spheres:

- Water
- Economic
- Political

In view of the complexity of the social and economic development processes, conflict management forms a prominent feature in the efficiency and productivity of the water sphere. In this sphere, conflicts are often affected by problems in the economic and political sphere. In essence, problem in any one sphere may lead to conflicts or disputes in the other two spheres.

Freshwater stocks are finite in global quantity, uneven, and highly restricted in terms of accessibility. A scenario based on current demographic and consumption trends shows that rapidly accelerating pressure will lead to significant global-scale problems. About 40% of the world's population live in shared fresh water basins, crossing political, ethical, and other boundaries. Multilateral environmental agreements constitute only one form of legal instrument for dealing with trans-boundary freshwater issues, and others must also be promoted and employed at various levels.

The economic and political factors are treated as separate driving forces. Although these factors have a strong interaction with the key factors affecting the water sphere directly, they may originate independently from the water sphere. Often, the problems in the economic and political spheres are caused by the lack of detailed information on good management of water resources or by differences in perception of a fair and equitable share of the water resources.

5.6 Impact of Climate Change on Water Resource

The potential impact of global climate change is one of the least addressed factors in water resources planning. Dealing with the potential threat of global warming has increased the complexity of water resources planning in arid and semiarid regions. Very large uncertainties are associated with both the magnitude and rate of change of climate associated with any future global warming.

5.6.1 Issues on Water Resources in Connection to Climate Change

In dealing with the impacts of climate change on water resources, there are essentially three issues to be determined:

(i) the future availability of water (water supply)
(ii) future demand of water (water demand)
(iii) the consequences of both of these on the environment

The possible impacts of climate change on water resources are described below:

5.6.1.1 Impact on Water Supply

Climate change has impacts on water resources and subsequently on the sustainability of our environment. Global warming would accelerate the hydrologic cycle, increasing both precipitation and evapotranspiration rates (Waggoner, 1990). The excessive (especially high intensity) rain can cause flash flood. Melting of polar ice caused by increase in temperature and thermal expansion may causes sea level rise. On balance, renewable water supply is likely to decline under global warming, but uncertainties as to the likelihood of global warming and its regional hydrologic implications present a dilemma for water planners.

5.6.1.2 Impact on Agricultural Water Demand

Since about 70% of the world water use is for irrigation, a change in climate affecting water resources will have the greatest impact upon this sector. Irrigation water use is affected primarily by potential temperature changes. Ali and Adham (2007) showed that if all other elements of weather remain constant, a 10% increase in maximum temperature would result in 7.3% increase in crop evapotranspiration demand. Higher temperature implies greater vegetative water consumption, particularly through evapotranspiration. The present percentage of water consumption may change when climate changes occur. Consequently, recharge to groundwater basins and surface runoff will decrease. Precipitation changes are not envisaged to change the water demand of irrigated crops in many parts of the world (e.g., Jordan, Bangladesh, India), as most of the irrigated crops are grown in the dry season, where no significant rainfall occurs.

5.6.1.3 Impact on Municipal, Industrial Use

Climate changes due to increased CO_2 and trace gases may affect the water supply for municipal and industrial uses. Also such changes may have an impact on water demand.

The secondary impacts of climate are as follows:

- flood
- drought
- coastal flooding
- salinity
- cyclone
- cold spell
- heat wave

Drought limits surface irrigation potential in drought-vulnerable areas and challenge food sufficiency programs of the country. Drought and flood affect the following aspects: quality of life, food security, economic development, and ecosystem sustainability.

According to the Intergovernmental Panel on Climate Change (IPCC), a body set up in 1988 to investigate global warming, agricultural impacts could be significant and water resources could be altered with relatively small climate changes causing large water resource problems in drought-prone areas. This implies that for regions which already experience water supply problems, it is prudent to include a global climate change dimension in planning studies. Long-term water supply planning, incorporating economic and regulatory instruments, needs to include an assessment of the sensitivity of the water resources to climate change in addition to forecasts of future water demand.

5.6.2 Adaptation Alternatives to the Climate Change

Potential measures to cope with the possible climate change are as follows:

(i) Structural
- increase freshwater supply by constructing dam, reservoir, etc.
- strengthen flood mitigation measures. Raise embankment to stop overtopping during flood

5.6 Impact of Climate Change on Water Resource

- rehabilitation of drainage system
- restrict salinity intrusion by constructing regulators

(ii) Nonstructural

- develop new water resources technology
- enhance drought protection strategies
- conservation measures and incentives for stakeholders
- educate the people
- demand management
- supply management
- proper selection of crop(s)/cropping pattern
- rain water harvesting
- import of virtual water
- land use zoning

5.6.2.1 New Technology

The current methods and information may not be sufficient for accurately assessing the impacts of climate change. New methodologies for better understanding the impacts of climate change are necessary, along with more sophisticated hydrologic model which can reflect the local hydrologic characteristics and spatial and temporal variations. Scientists should undertake program for development of low-cost water technologies.

5.6.2.2 Strengthen Flood Mitigation Measures

Floods often cause damage to life and property. With the climate change, the stream flow during the wet period may increase, and hence flood mitigation strategies should be further examined.

5.6.2.3 Enhance Drought Protection Strategies

Although stream flow may increase, the risk of water deficits may also increase in the future due to increased agricultural water demand. Thus, drought protection strategies should be enhanced.

5.6.2.4 Conservation Measure

In the agricultural sector, programs which encourage cropping patterns, conservation subsidies, and financial assistance in high technology water use in irrigation may be adopted and implemented. In the household sector, water conservation programs such as the use of water-saving devices could save a significant amount of water.

5.6.2.5 Educate the People

The success of water resources development and conservation relies on the inhabitants on the locality. The public should be aware of the impacts of climate change on water resources. Water resources professionals should contribute by helping governmental agencies to achieve their goals.

5.7 Challenges in Water Resources Management

5.7.1 Risk and Uncertainties

Water resource management is concerned with how the available and/or harvestable supply of water can be better allocated to fulfill the required demand. Thus, the water resource management measures involve a certain degree of uncertainties and risk because both water supplies and demands have inherent components of randomness. Some of the factors for this randomness can be attributed to the spatial and temporal variability of supply and/or demand.

5.7.2 International/Intra-national (Upstream–Downstream) Issues

The critical factors affecting international water management problem are the temporal characteristics associated with the objectives of the upstream and downstream countries. In some cases, the intra-national, that is, for upstream and downstream sides may face similar problems. Very often, the objectives of upstream and downstream sides differ from each other. For example, the objective of upstream country is to maximize the production of hydro-electric power while the objective of downstream side is to maximize their utilization of water for irrigation. This situation may lead to a major international conflict over the water.

5.7.3 Quality Degradation Due to Continuous Pumping of Groundwater

Groundwater is not an independent entity but is only one of the phases in which water exists in its overall hydrologic cycle. If the groundwater becomes contaminated (such as Arsenic contamination in West Bengal of India and Bangladesh), it will be a catastrophic event, and troublesome to purify. Mining or overexploitation of groundwater has already altered groundwater geochemistry possibly contributing to high levels of arsenic and increasing concentrations of fluorides (Serageldin, 1999). Environmental consequences of such exploitation may become critical slowly in most cases. But the situation may be irreversible if care is not taken beforehand.

5.7.4 Lowering of WT and Increase in Cost of Pumping

Generally, groundwater does not recycle as fast as the surface water, with rates of groundwater turnover varying from years to millennia, depending on aquifer location, type, depth, properties, and connectivity. Due to continual withdrawal of water, water table is declining gradually, both for rural and urban aquifers. This situation is widespread in many parts of the world. Besides, the recent declining trend is much higher than that of the past. The rate would be higher in the near future. This would be due to higher economic growth and increase in income and consequent change in habitat, demand in industry, and possible climate change.

Excessive pumping can lead to groundwater depletion, where groundwater is extracted at a rate faster that it can be replenished. Unregulated groundwater use leads to the "Tragedy of the Commons," with the eventual depletion of the resource and ruin to all. The cost of development and cost per unit quantity would be higher. The effects of excessive groundwater development tend to become apparent gradually, with time often measured in decades. To minimize abrupt decline of groundwater and thus the anticipated catastrophic events, the dependency on groundwater must be reduced and be used in a sustainable manner.

Relevant Journals

– Water Resources
– Water Resources Management
– Water Resources Research (American Geographical Union)
– Water International
– Journal of Water Resources Planning and Management (ASCE, USA)
– Water Resources Bulletin (American Water Works Association)
– Advances in Water Resources (Elsevier)
– Water and Environment Journal
– International Journal of Water Resources Development (Butterworths Scientific Ltd, UK)
– International Journal of Water
– Agricultural Water Management
– Journal of Irrigation and Drainage Division, ASCE
– Transactions of the American Society of Agricultural Engineers
– Journal of the Institution of Civil Engineers, London
– Ground Water (American Water Works Association)
– Ground Water Monitoring & Remediation
– Journal of Water Chemistry and Technology
– Journal of Hydrology (Elsevier)
– Irrigation and Drainage System
– Irrigation Science
– Journal of Applied Irrigation Science (Germany)
– Hydrological Sciences Journal (Blackwell, UK)
– Environmental and Resource Economics

Questions

(1) What do you mean by water resources management?
(2) "Water is an economic good?" Explain.
(3) What are the basic requirements of management of a resource?
(4) Discuss in brief, the procedure for estimating demand of water for agriculture and other sectors.
(5) Narrate the issues and principles of estimating water supply in a basin.
(6) What do you mean by "Environmental flow requirement?"
(7) What are the principal approaches of managing water resource?
(8) Discuss the various means/options of demand side management of water?
(9) Briefly explain the obstacles and potentials of demand side management in your province/state.
(10) Discuss in brief the various options of supply side management of water resource.
(11) What is sustainable development? Is sustainable development of water resource possible in your area?
(12) Discuss the strategies to achieve sustainability in water resource.
(13) Explain in brief, the impact of climate change on water resource.
(14) What adaptive alternatives do you suggest in water resource management to cope with climate change?
(15) What are the forthcoming challenges the mankind will face with respect to water resources? What preventive/remedial measures do you suggest against the challenges?

References

Ali MH, Adham AKM (2007) Impact of climate change on crop water demand and its implication on water resources planning – Bangladesh perspectives. J Agrometeorol 9(1):20–25

Ali MH, Hoque MR, Hassan AA, Khair MA (2007) Effects of deficit irrigation on wheat yield, water productivity and economic return. Agric Water Manage 92:151–161

Ali MH, Rahman MA (2009) Study of natural pond as a source of rain-water harvest and integrated salinity management under saline agriculture. In: Annual Report 2008–09, Bangladesh Institute of Nuclear Agriculture, Mymensingh, Bangladesh

Ali MY, Rahman MM, Rahman MM (1992) Effect of seedling age and transplanting time on late planted a man rice. Bangladesh J Training Dev 5(2):75–83

Ali MH, Talukder MSU (2008) Increasing water productivity in crop production – a synthesis. Agric Water Manage 95:1201–1213

Alley WM, Reilly TE, Franke OL (1999) Sustainability of groundwater resources. US Geol Surv Circ 1186:79

American Water Works Association (1999) Residential water use summary. Researched and compiled under contract by John Olaf Nelson Water Resources Management. http://www.waterwiser.org, Accessed 25 Aug 1999

ASCE Task Committee for Sustainable Criteria (1998) Sustainable criteria for water resource systems. ASCE Division of Water Resources Planning and Management, ASCE Press, Reston, VA, p 253

References

Asgar MN, Prathapar SA, Shafique MS (2002) Extracting relatively fresh groundwater from aquifers underlain by salty groundwater. Agric Water Manage 52(2,3):119–137

Ayers RS, Westcot DW (1985) Water quality for agriculture. FAO Irrigation and Drainage Paper 29 Rev. 1. FAO, Rome, p 174

BINA (Bangladesh Institute of Nuclear Agriculture) (2005) Effect of seedling age on yield and yield components of Binadhan-6. Annual Report for 2002–03, Bangladesh Institute of Nuclear Agriculture, Mymensingh, Bangladesh, pp 297–298

Chahal GBS, Sood A, Jalota SK, Choudhury BU, Sharma PK (2007) Yield, evapotranspiration and water productivity of rice-wheat system in Punjab (India) as influenced by transplanting date of rice and weather parameters. Agric Water Manage 88:14–22

Griffin RW (1997) Management, 5th edn. A.I.T.B.S.P and D, India

Hardin G (1968) The tragedy of the commons. Science 162:1143–1148

Kahlon PS, Dhaliwal HS, Sharma SK, Randhawa AS (1992) Effect of pre-sowing seed-soaking on yield of wheat under late-sown irrigated condition. Indian J Agril Sci 62(4):276–277

Kalf FRP, Woolley DR (2005) Applicability and methodology of determining sustainable yield in groundwater systems. Hydrogeol J 13:295–312

Maimore M (2004) Defining and managing sustainable yield. Ground Water 42(6):809–814

Paul SR (1994) Effect of age of seedlings and dates of transplanting on grain yield of sali rice in Assam. Ann Agric Res 15(1):126–128

Rashid MA (2008) Fresh water investigation for irrigation in coastal saline areas of Bangladesh. In: Proceedings of the Paper Meet 2008, Agricultural Engineering Division, The Institutions of Bangladesh, pp 1–8

Saleh AFM, Nishat A (1989) Review of assessment of groundwater resources in Bangladesh: quantity and quality. J Irrig Eng Rural Planning 17:48–59

Sauer TJ, Hatfield JL, Prueger JH (1996) Corn residue age and placement effects on evaporation and thermal regime. Soil Sci Soc Am J 60:1558–1564

Seckler D (1996) The new era of water resources management. Research Report 1, International Irrigation Management Institute (IIMI), Colombo, Sri Lanka

Serageldin I (1999) Looking ahead: water, life and the environment in the twenty-first century. Water Resour Dev 15(1/2):17–28

Singh KN, Sharma DK (1993) Effect of seedling age and nitrogen levels on yield of rice on a sodic soil. Field Crops Res 31:3–4

Waggoner PE (1990) Climate change and U.S. water resources. Wiley, New York, NY

WCED (World Commission on Environment and Development) (1987) Our common future. Oxford University Press, New York, NY

World Bank (1997) Water allocation mechanisms – principles and examples. World bank, Agric Natural Resource Department, Washington, DC

Zhang H, Oweis T (1999) Water-yield relations and optimal irrigation scheduling of wheat in the Mediterranean region. Agric Water Manage 38:195–211

Zhou Y (2009) A critical review of groundwater budget myth, safe yield and sustainability. J Hydrol 370:207–221

Chapter 6
Land and Watershed Management

Contents

6.1	Concepts and Scale Consideration	194
6.2	Background and Issues Related to Watershed Management	195
	6.2.1 Water Scarcity	196
	6.2.2 Floods, Landslides, and Torrents	196
	6.2.3 Water Pollution	196
	6.2.4 Population Pressure and Land Shrinkage	196
6.3	Fundamental Aspects of Watershed Management	197
	6.3.1 Elements of Watershed	197
	6.3.2 How the Watershed Functions	198
	6.3.3 Factors Affecting Watershed Functions	198
	6.3.4 Importance of Watershed Management	198
	6.3.5 Addressing/Naming a Watershed	198
6.4	Land Grading in Watershed	199
	6.4.1 Concept, Purpose, and Applicability	199
	6.4.2 Precision Grading	200
	6.4.3 Factors Affecting Land Grading and Development	201
	6.4.4 Activities and Design Considerations in Land Grading	203
	6.4.5 Methods of Land Grading and Estimating Earthwork Volume	205
6.5	Runoff and Sediment Yield from Watershed	212
	6.5.1 Runoff and Erosion Processes	212
	6.5.2 Factors Affecting Runoff	213
	6.5.3 Runoff Volume Estimation	214
	6.5.4 Factors Affecting Soil Erosion	219
	6.5.5 Sediment Yield and Its Estimation	221
	6.5.6 Sample Workout Problems on Sediment Yield Estimation	223
	6.5.7 Erosion and Sedimentation Control	225
	6.5.8 Modeling Runoff and Sediment Yield	226
6.6	Watershed Management	227
	6.6.1 Problem Identification	227
	6.6.2 Components of Watershed Management	228

	6.6.3	Watershed Planning and Management	229
	6.6.4	Tools for Watershed Protection	230
	6.6.5	Land Use Planning	230
	6.6.6	Structural Management	230
	6.6.7	Pond Management	231
	6.6.8	Regulatory Authority	231
	6.6.9	Community-Based Approach to Watershed Management	231
	6.6.10	Land Use Planning and Practices	234
	6.6.11	Strategies for Sustainable Watershed Management	235
6.7		Watershed Restoration and Wetland Management	236
	6.7.1	Watershed Restoration	236
	6.7.2	Drinking Water Systems Using Surface Water	236
	6.7.3	Wetland Management in a Watershed	237
6.8		Addressing the Climate Change in Watershed Management	238
	6.8.1	Groundwater Focus	238
Relevant Journals			238
Relevant FAO Papers/Reports			238
Questions			239
References			239

A watershed is the geographic area where all water running off the land drains to a common outlet. Watershed management activities can be considered at the state, river basin, individual watershed level or regional scale. For improving watershed protection and restoration, it is a prerequisite to know how agricultural systems influence soil and water resources.

For managing watershed effectively, it is necessary to identify and address land-use practices and other human activities that pollute local water resources or otherwise alter watershed functions. To develop sustainable programs, land and water must be managed together and an interdisciplinary approach is needed. Watershed management recognizes that the water quality of our streams, lakes, and estuaries results from the interaction of upstream features. It unites social, economic, and environmental concerns and the cumulative effects of site-specific actions on rangelands, forests, agricultural lands, and rural communities.

6.1 Concepts and Scale Consideration

We all live in a watershed – the area that drains to a common waterway, such as a stream, lake, estuary, wetland, aquifer, or even the ocean – and our individual actions can directly affect it. A watershed is the geographic area where all water running off the land drains to a given stream, river, lake, wetland, or coastal water. Simply speaking, a watershed is the area of land where all of the water that is under it drains off or goes into the same place.

Watershed can also be viewed as that area of land (or a bounded hydrologic system), within which all living things are inextricably linked by their common water course and where, as humans settled, simple logic demanded that they become part of a community.

Everyone is an integral part of the watershed in which he or she lives. We often think of rivers as simply water flowing through a channel, but river systems are complex and intimately connected to and affected by the characteristics of their surrounding watersheds – the land that water flows over and under on its way to the river. Many human activities that occur on the land, such as agriculture, transportation, mining, and construction, affect our river systems and how they function.

Watershed management activities can be considered at the state, river basin, individual watershed level or regional scale. Using watersheds, we can take a broader view of the environment, which is complicated and interconnected with our activities across local and regional scales.

An understanding of natural resource responses to agricultural activities at regional and watershed scales is necessary for successful and efficient management of the resources. For improving watershed protection and restoration, it is a prerequisite to know how agricultural systems influence soil and water resources. We can best understand the overall health of aquatic systems on a watershed basis.

6.2 Background and Issues Related to Watershed Management

Watersheds have been viewed as useful systems for planning and implementing natural resource and agricultural development for many centuries. Recognition of the importance of watersheds can be traced back to some of the earliest civilizations; ancient Chinese proverbs state that "Whoever rules the mountain also rules the river," and "Green mountains yield clean and steady water."

Expanding human populations and their increasing demands for natural resources have led to exploitation and degradation of land and water resources. Revenga et al. (1998), in an assessment of 145 watersheds globally, emphasized that expanding human demands for resources have intensified watershed degradation, with the result that some of the watersheds with the greatest biological production are becoming the most seriously degraded. Development projects and programs by all types of organizations (national governments, multinational and bilateral agencies, nongovernmental organizations (NGOs), etc.) have proliferated in response to these problems.

Current and expanding scarcities of land and water resources, and the human response to these scarcities, threaten sustainable development and represent paramount environmental issues for the twenty-first century. An added concern is to develop means of coping with the extremes and uncertainty of weather patterns, such as the 1997–1998 El Nino effect that resulted in severe droughts in some parts of the world and record flooding elsewhere. Watershed management provides both a framework and a pragmatic approach for applying technologies to cope with these issues, which are discussed below.

6.2.1 Water Scarcity

Water scarcity has been widely called the top global issue of concern in developed and developing countries alike. By 2025, it is estimated that between 46 and 52 countries, with an aggregate population of about 3 billion people, will suffer from water scarcity. Coping with water scarcity is compounded by soil degradation, groundwater depletion, water pollution, and the high costs of developing new water supplies or transferring water from water rich to water poor areas. Through watershed management, we can recognize both the opportunities and limitations of water yield enhancement through vegetative and structural measures.

6.2.2 Floods, Landslides, and Torrents

Floods, landslides, and torrents result in billions of dollars being spent each year globally for flood prevention, flood forecasting, and hillslope stabilization. Yet the cost of lives and property damage due to floods, landslides, and debris flows are staggering. The impact of these naturally occurring phenomena are exacerbated by human encroachment on floodplains and other hazardous areas, which is often the result of land scarcity discussed below. In many parts of the world there has been an overreliance on structural solutions (dams, levees, channel structures, etc.) in river basins, along floodplains, and in areas susceptible to debris torrents, all of which impart a false sense of security to those living in hazardous areas. In addition, the replacement of natural wetlands, riparian systems, and floodplains with urban and agricultural systems can cumulatively add to downstream problems. A watershed perspective brings these cumulative effects and linkages into focus, but the ability to develop solutions requires that we have the appropriate policy and institutional support.

6.2.3 Water Pollution

Point and nonpoint water pollution continue to plague many parts of the world, threatening the health of humans, compounding water scarcity issues noted above, and adversely impacting aquatic ecosystems, with subsequent implications for fish and wildlife. Best Management Practices (BMPs) and related technologies of watershed management have the advantage of stopping nonpoint pollution at its source.

6.2.4 Population Pressure and Land Shrinkage

Scarcity of land and natural resources results from a shrinking arable land due to expanding populations of humans and livestock. Land degradation resulting from cultivation, grazing, and deforestation of marginally productive lands compounds the effects of land scarcity. These are often steep areas with shallow soils that experience accelerated surface and gully erosion, soil mass movement, and increased sediment and storm flow damage to downstream communities. In the tropics, it is estimated that about 0.5 ha of farmland is needed to feed one person (Pimental et al. 1995). Lal (1997) indicated that by the year 2025, 45 countries in the tropics will have less than 0.1 ha of arable land per capita. Globally, of the 8.7 billion ha of agricultural land, forest, woodland, and rangelands, over 22% has been degraded since mid-century, with 3.5% being severely degraded (Scherr and Yadav, 1996). Deforestation continues to gain worldwide attention with most of the concern expressed in terms of lost biodiversity; of equal importance are the implications of deforestation on watershed functions.

From overall consideration of the watershed, the challenges include the following.

- high population growth
- increasing land use intensity
- highly degraded or threatened river systems
- no legal protection for in-stream flow needs
- watersheds that are driven by glacially fed rivers and then flow into the driest part of the province
- significant allocation to irrigation
- an extensive plumbing systems of canals and dams
- a highly uncertain water supply and
- emerging water conflicts.

6.3 Fundamental Aspects of Watershed Management

6.3.1 Elements of Watershed

Watershed management recognizes that the water quality of our streams, lakes, and estuaries results from the interaction of upstream features. It unites social, economic, and environmental concerns and the cumulative effects of site-specific actions on rangelands, forests, agricultural lands, and rural communities.

Wetlands are important elements of a watershed because they serve as the link between land and water resources. Oceans, coasts, and estuaries provide critical natural habitat and recreational areas for our nations.

The term "watershed processes" connotes consideration of distributed systems with processes which are neither uniform in space nor constant in time. Watershed processes also suggest processes such as mass flux (water, sediment, or contaminant) relative to a specified contributing area.

6.3.2 How the Watershed Functions

Because they convey the water that runs over the land and into the ground, watersheds provide many vital ecological and hydrological functions. Hydrologically, watersheds collect water from rainfall and snowmelt, storing some of this precipitation in wetlands, soils, trees, and other vegetation, and underground in aquifers. The floodplain along the banks of a river also serves as an important storage site for water during periods of heavy runoff. These natural storage sites help eliminate contaminants as suspended particles settle out and as water infiltrates into the soil where biological and chemical reactions can break down impurities. Some of this stored water eventually flows into streams, rivers, and lakes as runoff.

Ecologically, watersheds provide critical habitat for many plant and animal species, as well as transport paths for sediment, nutrients, minerals, and a variety of chemicals. Watersheds also provide water to human communities for drinking, cleaning, recreation, navigation, hydroelectric power, and manufacturing.

6.3.3 Factors Affecting Watershed Functions

Like all organisms, humans are an integral part of the watersheds in which they live. Therefore, human activities, both in the water and on the land, can have a great impact on the watershed functions described previously. The creation of buildings, parking lots, and roads; the draining of wetlands; mining; deforestation; and agricultural activities can all alter the quality and quantity of water that flows over and infiltrates into the ground. These changes can alter watershed functions by eliminating critical water storage sites (e.g., wetlands and floodplains) and by contributing additional sediments and chemicals to runoff. Human activities can also eliminate critical natural habitat sites, thereby limiting biodiversity in the watershed.

6.3.4 Importance of Watershed Management

People and wildlife require healthy watersheds. Everyone lives in a watershed and each person's actions in turn affect their neighbors and the land and water "downstream." For managing water effectively, the most appropriate land unit is the watershed.

Activities of all land uses within watersheds impact the water quality of down gradient water bodies. Point and nonpoint sources of pollution in a watershed contribute nutrients, bacteria, and chemical contaminants to waterways. Watershed management encompasses all the activities aimed at identifying sources and minimizing contaminants to a water body from its watershed.

6.3.5 Addressing/Naming a Watershed

We may refer to watersheds by their proper name as well as by a grouping of numbers. This set of numbers is called the watershed's Hydrologic Unit Code (HUC), also known as the watershed address. The HUC can range from 2 to 16 digits in length.

Watersheds are delineated by United States Geological Survey (USGS) using a nationwide system based on surface hydrologic features. This system divides the country into 21 regions, 222 subregions, 352 accounting units, and 2,262 cataloguing units. A hierarchical hydrologic unit code (HUC) consisting of 2 digits for each level in the hydrologic unit system is used to identify any hydrologic area. The 6-digit accounting units and the 8-digit cataloguing units are generally referred to as basin and subbasin. There are many states that have defined down to 16-digit HUCs.

6.4 Land Grading in Watershed

6.4.1 Concept, Purpose, and Applicability

6.4.1.1 Concept

Land is the important basic resource that supports the production of all agricultural commodities, including livestock, which is dependent on land to produce the grain and forage consumed. *Land grading* is the manipulation of the land surface from its natural condition. It is reshaping or restructuring the existing land surface for planned grades as determined by engineering survey, evaluation, and layout. "Planned grade" is depended upon the planned activities such as irrigation (in target method) and/or subsequent drainage, erosion control, specific crop cultivation (e.g., horticultural crop *or* cereals such as rice). *Land improvement* consists of betterments, site preparation and site improvements that ready the land for its intended use.

Altering the surface of agricultural crop land is a common practice used primarily to improve the drainage of water from a field or to increase the efficiency of surface irrigation. There are several common names for this practice, including land smoothing, land leveling, land grading, and land forming. Although each of these practices generally involves altering the land surface, there are distinct differences in the methods used to level the surface of the land (ASAE, 1998).

Generally, there are two basic types of practices that alter the surface of agricultural land: land smoothing and land grading. Land smoothing involves shaping the land to remove irregular, uneven, mounded, broken, and jagged surfaces without using surveying information. This operation would typically be performed by a tractor pulling a land leveler or other type of smoothing implement. Land grading is the operation of shaping the surface of land to predetermined grades so that each row or surface slopes to a drain, or is configured for efficient irrigation water application.

Land grading also involves smoothing the land surface to eliminate minor depressions and irregularities, but without changing the general topography. Irregular micro-topography can cause severe crop loss.

6.4.1.2 Purpose

The purpose of grading is to provide more suitable topography for facilitating crop culture, irrigation, surface runoff control, and to minimize soil erosion and sedimentation. By land grading, the land surface is formed to the predetermined grades so that each row or surface slopes to a field drain.

Land grading permits uniform and efficient application of irrigation water without excessive erosion and at the same time provides for adequate surface drainage. One of the objectives for grading or leveling agricultural fields is to improve the efficiency of surface irrigation. Surface irrigation or gravity flow methods range from furrow irrigation, whereby the field is irrigated by allowing irrigation water to move across the field by flowing down the row furrows, to flood irrigation as used in rice production, whereby several centimeters of irrigation water is maintained on the field throughout the growing season. A plane surface (longitudinal and transverse slopes) is easiest to manage and maintain. Land grading can reduce the number of field drains; thus more land is available for use.

6.4.1.3 Essential Conditions and Applicability

All lands to be graded for irrigation should be suitable for use as irrigated land and for the proposed methods of water application. Water supplies and the delivery system should be sufficient to make irrigation practical for the crops to be grown and the irrigation water application methods to be used.

Land grading practices are applicable at sites with uneven or steep topography or easily erodible soils, because it stabilizes slopes and decreases runoff velocity. It is necessary where grading to a planned elevation is necessary and practical for the proposed development of a site, and for proper operation of sedimentation-control practices. Grading activities should maintain existing drainage patterns as much as possible.

6.4.2 Precision Grading

Precision grading is a land improvement that can increase the productivity and value of agricultural crop land by improving field drainage or increasing the efficiency of surface irrigation. In precision grading, instruments (Scapers) having GPS/Laser-controlled cut-and-fill work facility are used. The cost of precision grading represents a long-term investment in farm.

The efficiency of surface irrigation methods can be improved substantially by precision grading. A field leveled to a precise grade will ensure furrow irrigation water will travel the entire length of row and not be blocked or diverted by depressions in the field. For flood irrigation, precision grading can allow for a reduction in the minimum depth of water that must be maintained on the field to cover the entire field. By increasing surface irrigation efficiency, irrigation costs can also be reduced.

Precision grading land for the purpose of improving surface irrigation efficiency is applicable to a wide variety of crop production situations. Approximately one half of the total irrigated agricultural acreage in the United States is irrigated by surface or gravity flow methods. Major crops utilizing gravity flow irrigation systems include corn, soybeans, rice, hay, cotton and orchards. Gravity flow systems are also used on crops ranging from wheat and barley to tomatoes and berries. A large number of farms also use surface irrigation on pasture acreage. The efficiency of surface irrigation on all of these crops could be improved with precision land grading.

Since precision grading constitutes a long-term investment in the land, it will require several years of production to recover the costs invested. Therefore, land chosen for precision grading should be land that has a high probability of remaining in production for several years into the future.

6.4.3 Factors Affecting Land Grading and Development

Successful land planning and grading requires knowledge of engineering, agricultural, and architectural branches. The following factors should be considered for selecting and designing land grading:

6.4.3.1 Topography

The primary design determinant is the form of the land, its slope characteristics, storm-water runoff, and surface drainage patterns. Topographic analysis reveals where favorable views exist.

Excavation and fill materials to be required for or obtained from such structures as ditches, pads, and roadways should be planned for as a part of the overall grading job. The cut-to-fill ratio will normally be between 1.30 and 1.50 to allow for losses due to compaction, hauling, and undercutting.

6.4.3.2 Soil, Hydrologic, and Geologic Conditions

Soil and subsoil condition affect grading and planting options. Subsoil bore test, analysis, and geologic mapping provide important information regarding soil compaction, rock strata, drainability, erosivity (erosion susceptibility), water table (aquifer) position, etc. All these factors should be considered with respect to intended objective, since soil and hydrology must be compatible with the target vegetation.

Soils should be deep enough so that, after the needed grading work is done, an adequate, usable root zone remains over most of the field that will permit satisfactory crop production with proper conservation measures. Limited areas with shallower soils may be graded to provide adequate drainage, irrigation grades or a better field arrangement.

6.4.3.3 Climate and Microclimate

Competent design should always take into account the sunshine, temperature, shade, precipitation, wind, etc.

6.4.3.4 Indigenous Vegetation and Wildlife Habitats

Faced with vegetated or natural sites, the designer should advocate use of native plant species; as the animals that inhabit or depends on a site's natural features.

6.4.3.5 Existing Regulatory Context

In addition to other factors, the existing ordinance and regulations (of the State/Province) governing land use, site development, construction and conservation must be considered for land development and planning, water development, drainage, etc.

6.4.3.6 Sustainability Issues

Engineers and architects should concern with the sustainability issues of the resources and environments of the surroundings the land to be developed and graded.

All grading work for drainage or irrigation should be planned as an integral part of an overall farm system to conserve soil and water resources. Boundaries, elevations and direction of slope of individual field grading jobs should be such that the requirements of all adjacent areas in the farm unit can be met. Designs for the area being graded should include plans for removing excess irrigation and storm runoff water from the fields.

6.4.3.7 Proposed Plan of Use

The finished degree of smoothness or slope depends on the irrigation method, cropping, anticipated depth to the water table and water quality. As the slope or gradient increases, less precise grading is needed.

6.4.3.8 Field Size

Under certain circumstances (especially in a mechanized farming system), the grading costs are related to the field size. The larger the farm, the lower the costs. As the farm becomes smaller and irrigation runs shorter; labor requirement increases, machinery operating cost increases, and the proportion of unproductive land increases.

6.4.3.9 Cost and Benefit Factors

Volume of earthwork to be needed should be included in estimates of total land grading costs. Unit costs for grading vary with the depth of cuts, length of haul, how smooth the surface must be, etc. There is a trade-off between the cost of land grading and the benefits to be achieved, such as efficient water use. If water use efficiency will inevitably be low, there is no justification in achieving more than a smoothing of the land in the direction of the slope.

In deciding whether or not to make the investment in precision grading of agricultural fields on a farm, certain cost factors should be considered. The first consideration is whether the producer should purchase the laser leveling and dirt moving equipment and do the work himself or whether the work be hired out to someone else on a custom-hired basis. The second cost consideration is determining how many years of crop production will be required to recover the investment in precision grading costs.

The total amount of acreage on the farm to be leveled is a critical component in this decision of whether to invest in precision grading equipment. If only a small amount of acreage is planned to be graded, for example 50 ha, the producer may choose to hire out the work rather than purchase the equipment for such a small amount of acreage. However, if a large amount of acreage is to be graded over a multiyear period, for example 300 ha graded over a 5-year period, it would probably be more economical for the producer to purchase the laser leveling and dirt moving equipment and perform the work himself.

The number of years of crop production required to recover investment in precision grading costs can be estimated by comparing the precision grading costs per hectare with the increased net returns per hectare resulting from increased production or the reduction in cultivation, irrigation, and other production costs resulting from increased field efficiency. An estimate of the number of years required to recover precision grading costs can be determined by dividing the precision grading costs per hectare by the increased annual net returns per hectare.

6.4.4 Activities and Design Considerations in Land Grading

A grading plan should be prepared that establishes and includes the following:

- Identification of areas of the site to be graded
- An estimate of the land grading requirements
- How drainage patterns will be directed
- How runoff velocities will affect receiving waters
- Information regarding when earthwork will start and stop
- The degree and length of finished slopes
- Where and how excess material will be disposed of (or where borrow materials will be obtained if needed)
- Berms, diversions, and other storm water practices that require excavation and filling should be incorporated into the grading plan.

Before grading activities begin, decisions must be made regarding the steepness of cut- and fill-slopes and how the slopes will be protected from runoff, stabilized and maintained.

The degree of slope acceptable for irrigation development and therefore the cost of land leveling depends on the anticipated method of irrigation, intensity and amount of rainfall, susceptibility of the soil to erosion, and planned cropping system.

In the United States, gravity irrigation on slopes greater than about 12% is seldom practiced. With sprinkler or drip systems, limitations on slope due to an erosion hazard or the operation of farm machinery are important. Slopes of 20% are currently considered the maximum acceptable in the United States for cultivated crops irrigated by sprinklers. In areas that experience severe thunderstorms, the maximum usable slope may be less. Land devoted to dense cover crop or grass may permit irrigation of steeper slopes than for row or field crops.

Although excessive slope is the most frequent problem, lack of slope may also be a limitation. Excessive flatness may result in higher grading costs to increase the slope and achieve the smooth uniform surface necessary for uniform distribution of irrigation water. Extremely gentle gradients may make irrigation of slowly permeable soils difficult because standing water induces scaling and waterlogging. Very permeable soils and extremely flat topography may prevent uniform irrigation without excessive deep percolation and water use. On the other hand, very flat land provides an opportunity to use really efficient surface irrigation methods such as basin and border strip, where soils are suitable.

6.4.4.1 Furrow Grades

Land graded for irrigation with subsequent drainage will have a slope in the row direction between 0.1 and 0.5% on deep alluvial soils. There should be no reverse grade in the row direction. The most desirable surface is a plane. Fields graded to minimum slopes will require more maintenance of grade than steeper slopes.

Design grades on claypan soils may have furrow grades up to 1.0% to avoid exposing large areas of subsoil in cut areas. Special residue management may be necessary to minimize erosion where furrow grades exceed 0.5%.

6.4.4.2 Cross Slope

Cross slope (slope perpendicular to row slope) is permitted in order to reduce cut volume or to establish the "plane of best fit." Cross slopes must be such that "breakthroughs" from both irrigation water and runoff from rainfall are held to a minimum. Terraces should be used on cross slopes of 2% or more. Where terraces are necessary on fields to be irrigated, rows should be parallel to terraces. Land forming may be necessary between terraces to eliminate reverse grade in irrigated rows.

6.4.4.3 Maximum Length of Runs for Irrigation

Maximum length of runs for irrigation should be limited by furrow flow rates available, furrow cross-sectional area, erosion hazard to the furrow, and water intake characteristics of the soil. Erosion hazard is a function of soil texture, crop residue and slope. A frequently used guide to maximum furrow flow rates is $Q = 10/s$, where s is furrow slope in percent and Q in gallons per min per furrow.

6.4.4.4 Other Considerations

- Land grading is a major source of sedimentation and must be carefully planned and carried out.
- The use of phasing, natural buffers, mulching, and temporary and permanent seeding should be the primary methods of addressing erosion control for land grading projects.
- Fall and winter erosion control measures must be upgraded and refined to protect the site from spring runoff and snowmelt.

In estimating the cost of grading, the field boundaries and the type of surface irrigation (small basins, large basins, furrow, and border strip lengths, etc.) must be determined. There is an interrelationship between the irrigation field size and the amount of land grading required. Where grading will cause damage by exposing subsurface horizons and hardpans, alternative development options should be considered.

6.4.5 Methods of Land Grading and Estimating Earthwork Volume

6.4.5.1 Planning and Early Surveying

Grading plan and installation should be based upon adequate surveys and soil investigations. In the plan, show disturbed areas, cuts, fills, and finished elevations of the surface to be graded. All practices necessary for controlling erosion on the graded site and minimizing sedimentation downstream should be included in the plan. Such practices may include the following:

- Vegetation, vegetated- and lined-waterways
- Grade-stabilization structures
- Surface and subsurface drains

There are four aspects of topography that have special bearing on land leveling and grading for surface irrigation: (1) slope, (2) microrelief, (3) macrorelief, and (4) cover. The land classifier and/or surveyor must achieve competence in distinguishing and evaluating those topographic features that are significant. Considerable experience is required to achieve acceptable accuracy in estimating the costs of leveling from field observations. Topographic maps do not always give sufficient information for accurate assessments. Guidance and training may be provided by an experienced agricultural engineer engaged in detailed layout studies. Detailed farm layouts of representative areas showing the costs of land grading can provide the best guidelines. If done properly, evaluation of the topography based on experience and field layout studies is adequate for most planning studies.

The important considerations in the early surveys are as follows:

(a) the topographic features of the land that influence the flow of water by gravity or the elevation and distance to which water must be pumped
(b) the depths of barriers that can act as obstructions to the constructing of canals, drains, and other structures or affect grading and land leveling operations
(c) the presence of unstable subsurface materials that may lead to subsidence problems
(d) the permeabilities of soils on which canals and drains will be constructed and the associated losses of water for unlined or lined channels
(e) the substratum condition as it affects the installation of permanent structures such as diversion weirs, storage reservoirs.
(f) soil conditions for installing field and main drainage (i.e., depth to barrier, nature of barrier, etc.)
(g) the location of dug-wells or tube-wells in respect not only to water but also to the land that will be irrigated to obtain the best advantages in terms of energy-saving and topography
(h) the size and shape of potential management units or fields
(i) the positioning of bunds or levees according to topography and changes in soil texture or other land characteristics, thus improving the efficiency of water use and productivity
(j) the assessment of basin sizes, furrow lengths, etc. in relation to the earthmoving costs and the acceptable slopes and microrelief after grading
(k) the matching of water supply and demand and the scheduling of water in terms of frequency, rate, and duration of application. The design of the canal or pipe networks to the field and the engineering costs depend on any one or all of these factors.

In rehabilitation schemes, quite different assessments may be required depending on whether the scheme is in a rice area, or in an arid, or semiarid area subject to waterlogging and salinity problems. Other categories also occur in the intermediate rainfall zones.

In the rice land situation, rehabilitation often involves upgrading the primary, secondary, and tertiary water supply networks or the installation of improved water control structures (diversion weirs, measuring devices, storage structures, etc.). The land evaluator may be called upon to evaluate land suitabilities relating to the improvement of these engineering works.

In the rehabilitation of saline, sodic, and waterlogged land in arid and semi-arid areas, surveys are generally required for the engineering works, especially topographic surveys and groundwater level surveys for the proper location of irrigation and drainage channels. If very high construction costs are implicated, the land suitability class of the associated land may be downgraded accordingly.

Lot Benching

Lot benching is the grading of lots within a subdivision so that the runoff from each lot is directed to a stable outlet rather than to an adjacent lot. This practice is most applicable in subdivision developments on hilly or rolling topography. Lot benching will reduce the slope and length of slope of disturbed areas, thereby reducing the erosion potential. This practice also establishes drainage patterns on individual lots within a subdivision at the time of rough grading.

Lot benching can be very effective for controlling erosion on hilly developments. By reducing slope lengths and the steepness of slopes, the potential for erosion is lowered. The amount of benefit derived from this practice depends upon the steepness of the slopes and the erodibility of soils on the site.

6.4.5.2 Cutting and Filling

This operation is typically performed by tractors pulling dirt buckets or scrapers that pick up soil in high points in a field and deposit it in low points in the field. Dirt scraping operations are controlled by laser equipment that enables the slope of a field to be cut to a specific grade. In some cases, the final operations of precision grading would be to use a land leveler to smooth the graded field surface.

Several land grading methods are considered to be conservation practices by the Natural Resources Conservation Service of the US Department of Agriculture. Land smoothing (NRCS, code 466) is the practice of removing irregularities on the land surface by use of special equipment. This practice is classified as rough grading and does not require the use of a complete grid survey or other soil engineering data. The purpose of this practice is to improve surface drainage, provide for more effective use of precipitation, obtain more uniform planting depths, provide for more uniform cultivation, improve equipment operation and efficiency, and facilitate contour cultivation. Precision land forming (NRCS, code 462) is the practice of reshaping the surface of land to planned grades. All land-forming operations under this practice are performed on the basis of detailed engineering survey and layout. The purpose of this practice is similar to that of land smoothing, primarily focused on improving surface drainage. Irrigation land leveling (NRCS, code 464) is a conservation practice very similar to precision land forming. The major distinction is that the practice of irrigation land leveling is performed for the primary purpose of increasing surface irrigation efficiency.

Specific soil engineering practices related to the design and construction of surface drainage systems on agricultural lands can be used to improve field drainage. These practices are designed to improve the construction and maintenance of agricultural surface drainage systems, which are adapted to modern farm mechanization.

6.4.5.3 Construction Guidelines

Land to be graded should be cleared of brush and excessive crop residue, trash, or vegetative material. Grading should not be attempted when soil moisture exceeds that permitting normal tillage or plowing.

Bring the land to design grade or grades in accordance with a detailed plan showing cuts, fills, and grades. Fills of more than 15 cm should be built up by spreading the soil in successive layers. Disk or chisel the field surface after scoop work is completed and before final land leveling. Leaving undisturbed temporary and permanent buffer zones (i.e., vegetated buffer strips) in the grading operation may provide an effective and low-cost erosion-control measure that will help reduce runoff velocity and volume and off-site sedimentation.

Finish work with a land plane so that the field is free from depressions that would cause ponding of water. The land plane should be operated over the field three times: once at a 45° angle to the direction of the rows; once at a right angle to the direction of the rows; and finally in the direction of the rows. Field checking to determine compliance with design grades should have a maximum tolerance of ±3.0 cm at any grid point, with no reverse grade. Instruments (scrapers) are now available for heavy excavation and GPS/Laser controlled cut and fill work. Finishing scrapers should be used to level land, precision finishing, and land-forming work.

6.4.5.4 Maintenance Considerations

During the first year after grading, normally cut areas swell and fill areas settle. This may require minor cuts-and-fills and additional land leveling.

All graded areas and supporting erosion and sediment control practices should be periodically checked, especially after heavy rainfalls. All sediment should be removed from diversions or other storm water conveyances promptly. If washouts or breaks occur, they should be repaired immediately. Prompt maintenance of small-scale eroded areas is essential to prevent these areas from becoming significant gullies.

6.4.5.5 Earth Work Volume Estimation

The accuracy of the earthwork quantities estimated is directly related to the completeness of the surveying work performed. Excavation and fill materials required for or obtained from such structures as ditches shall be planned as a part of the overall leveling job.

"Cut-Fill" or "Volumetrics" is the calculation or analysis of landform volumes. A "Cut" volume is defined as the volume of material, which is excavated below existing site levels (Fig. 6.1). "Fill" volume is the volume of material, which is mounded above existing site levels. Negative volume values indicate areas that have to be filled, positive volume values indicate regions that have to be cut.

Before volumes can be calculated, the areas of excavation (either horizontal or vertical) must be established. Horizontal areas are associated with cross-sections cut

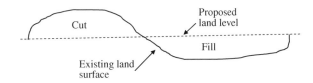

Fig. 6.1 Schematic of showing cut and fill on a site

6.4 Land Grading in Watershed

through the earthwork volume and vertical surfaces associated with Digital Terrain Model (DTM) volumes. Horizontal areas usually refer to the area extent of the excavation and horizontal areas enclosed by contour elevation lines. These areas are defined by a series of discreet points along their boundaries. Areas are calculated by connecting these points in a series of continuous triangles that extend across the area. Given the northing and easting of each of the three vertices of each triangle, and the lengths of each of the three sides of the triangles, each triangle's area can be calculated as follows:

$$A = \sqrt{s(s-a)(s-b)(s-c)} \tag{6.1}$$

where

A = the area of the triangular area (m^2)
$a, b, c,$ = the lengths of the three sides of the triangle (m)
$s = (a+b+c)/2$

There are several ways of calculating earth and soil volumes.

Depth Area Method

Depth Area Method (DAM) is the simplest method. In this method, volume of soil is obtained by multiplying the thickness of the strata to be excavated by the surface area of the strata. This can be done with reasonable accuracy only for strata that is consistently thick and whose area extent is known. It is perfectly suitable for estimating the amount of topsoil to be stripped at a consistent depth (usually 15 cm). It is also applicable for estimating the volume of regular (square or rectangular with vertical sideslopes) excavations of a consistent depth below a relatively flat surface. Volumes are calculated as follows:

$$V = T \times A \tag{6.2}$$

where

V = volume (m^3)
A = surficial slope area (m^2)
T = thickness of strata or even cut (m)

Total volume, $V_T = \sum V_i$

Grid Method

The Grid Method extends DAM to an excavation of varying depths. The excavation volume is obtained by applying a grid to the excavation area. The grids can be staked

Fig. 6.2 Schematic of grid method

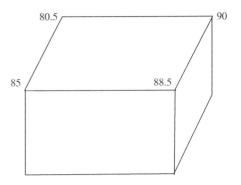

to squares of 3, 6, 15, 30, or more meter depending on the project size and the accuracy desired. For each grid square, final elevations are established for each corner of every grid square (Fig. 6.2). These are subtracted from the existing elevations at the same location to determine the depth of cut or height of fill at each corner. For each grid square an average of the depths/heights of the four corners is multiplied by the area of the square to determine the volume of earthwork associated with the grid area. The total earthwork volume for the project is calculated by adding the volumes of each grid square in the excavation area. Volumes are calculated as follows:

$$V = ((D_1 + D_2 + D_3 + D_4)/4) \times A \tag{6.3}$$

where

V = volume (m^3)
A = area of the grid square (m^2)
D = depth of cut/fill at each grid corner (m)

Total volume, $V_T = \sum V_i$

End Area Method

The End Area Method (EAM) utilizes the areas of parallel cross-sections at regular intervals through the proposed earthwork volume. The cross-sections can be spaced at intervals of 10, 20, 50, or 100 m depending on the size of the site and the required accuracy. They are aligned perpendicular to a baseline that extends the entire length of the excavation area (Fig. 6.3). There are several types of cross-sections, which can be drawn by hand or generated by CADD. For flat terrain or level excavation, a level section is suitable. Irregular sections are used for most excavations in rough terrain. Cross-sectional areas are calculated with either the triangular area method described above (if the cross-sections are geometrically simple) or by the Length Interval Method for more complicated cross-sections. Unit volumes are calculated as follows:

6.4 Land Grading in Watershed

Fig. 6.3 Schematic of end area method

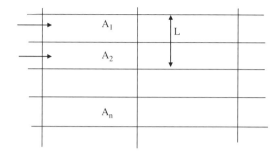

$$V = L \times [(A_1 + A_2)/2] \quad (6.4)$$

where

V = volume (m³)
A = area of the grid square (m²)
L = distance between cross-section along the baseline (m)

Total volume, $V_T = \sum V_i$

Prismoidal Formula

The Prismoidal Formula (PF) allows for greater accuracy than EAM. It is especially useful when the ground is not uniform or significantly irregular between cross-sections. PF adds an additional cross-sectional area midway between the two cross-sections defining the volume being calculated. Note that this cross-section is calculated separately and is not an average between the two end areas. Volumes are calculated as follows:

$$V = L \times [(A_1 + (4 \times A_m) + A_2)/6] \quad (6.5)$$

where

V = volume (m³)
A_1, A_2 = areas of the adjacent cross-sections (m²)
A_m = area of the midway cross-section (m²)
L = distance between cross-section along the baseline (m)

Contour Area Method

The Contour Area Method (CAM) uses the area of the excavation elevation contour lines to determine volumes. From a topographic map of the site, the areas enclosed by regular contour intervals are measured (Fig. 6.4). This area measurement can

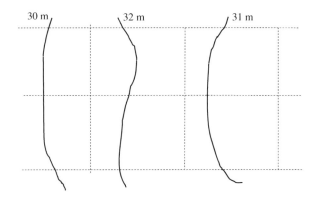

Fig. 6.4 Schematic of contour method

be done by hand with a planimeter, electronically by a digitizer, or directly with a CADD program. If the horizontal areas enclosed by each contour line are large relative to the elevation difference between the two contour elevations, averaging the two areas and multiplying the average by the height difference can determine volumes. However, for relatively small earthworks (like spoil piles and borrow areas), volumes can be calculated based on the formula for the volume of a truncated pyramid:

$$V = \frac{h}{3}\left[(B_1 + B_2 + \sqrt{(B_1 \times B_2)}\right] \tag{6.6}$$

where

V = volume (m^3)
B_1, B_2 = areas of the contour elevation lines (m^2)
h = elevation difference between the contour elevations (m)

6.5 Runoff and Sediment Yield from Watershed

6.5.1 Runoff and Erosion Processes

Surface runoff is the result of precipitation and is the amount of water which appears in the stream channel network during and after precipitation. Surface runoff, as direct flow of water over the soil surface and in small, definable channels, is termed overland flow. Overland flow is not necessarily sheet flow, although it may be under idealized conditions and on a sufficiently small scale. It consists of flow to, into, and within small concentrated flow channels or rills. Overland flow is thus sheet flow on the interrill areas and channel flow in the many small rills. For surface runoff to be classified as overland flow, it must be that the mean flux per unit width of the flow area cross-section is proportional to the storage in an incremental area. When surface flow cannot be hydrologically or hydraulically treated as overland flow, it is channel flow. Again, these distinctions are somewhat arbitrary and difficult to describe quantitatively, but they are useful, conceptually and mathematically.

Soil particles are detached when the impact of raindrops or the erosive force of flowing water is in excess of the ability of the soil to resist erosion. Sediment particles are transported by raindrop splash and by overland flow. Deposition of soil particles occurs when the weight of the particle exceeds the forces tending to move it. This condition is often expressed as sediment load exceeding sediment transport capacity.

Particles detached in the interrill areas move to the rills by splash mechanisms and as a result of suspension and siltation in overland flow. Thus, their detachment and movement is independent (except for morphological features of rill and channel systems controlling length and slope of interrill areas) of processes in rill and stream channels. The converse, however, is definitely not true; the amount and rate of water and sediment delivered to the rills determine rill erosion rates, sediment transport capacity in rills, and rate of sediment deposition.

6.5.2 Factors Affecting Runoff

A range of factors like climate (especially rainfall), soil, topography, and land-use determine rates of runoff and the total volume.

6.5.2.1 Soil Type/Infiltration Rate

The soil type and the infiltration rate influence surface runoff. If the infiltration rate is rapid, there will be little surface runoff.

6.5.2.2 Rainfall

The rainfall intensity and duration primarily contribute to peak runoff rate and total volume. High intensity rainfall for a short duration (but higher than the time of concentration) will result in a higher peak runoff rate. On the other hand, uniform and moderate intensity rain for prolonged period will yield higher runoff volume.

If the climate is dry, more water will evaporate, hence less will be available for contributing to runoff.

6.5.2.3 Vegetation

Vegetation retards the velocity of flow, thus increases infiltration opportunity. As a result the runoff and sediment transport will be decreased.

6.5.2.4 Topography

Slope length, steepness, and shape are the topographic characteristics that most influence peak runoff rate and total runoff volume. Sloppy lands facilitate quick response time to peak runoff. As the infiltration opportunity time is decreased with increasing slope, the total runoff volume is also increased.

6.5.2.5 Surface Depressions, Ponds, and Other Natural Water Storage

If the watershed contains much land depressions, ponds, and other detention storage, the total runoff volume will certainly be decreased, and the time of concentration will be increased.

6.5.2.6 Mulches and Crop Residues

Mulches and crop residues slow down the velocity of flow, thus allowing more opportunity time for infiltration, and hence decrease the runoff volume.

6.5.2.7 Land Use

Land use is an important factor affecting peak runoff rate and runoff volume. Vegetative and canopy pattern, surface roughness, mechanical soil disturbance, amount of biomass in the surface and top layer, etc. affect water velocity, retention, infiltration, and runoff rate and volume.

6.5.2.8 Soil Cover Management

Cover management practices affect both the forces applied to the soil by erosive agents and the susceptibility of the soil to detachment. For a given land use like cropland, important features include the crops that are grown, the type of tillage system such as clean, reduced, or no till. Important features on a construction site include whether or not the land is bare, mulch has been applied, the slope has been recently reseeded, or the soil material is a cut or fill. Important features on range and reclaimed land include the native or seeded vegetation and degree of ecological maturity.

6.5.2.9 Other Management Practices

Support practices include ridging (e.g., contouring), vegetative strips and barriers (e.g., buffer strips, strip cropping, fabric fence, gravel bags), runoff interceptors (e.g., terraces, diversions), and small impoundments (e.g., sediment basins, impoundment terraces).

6.5.3 Runoff Volume Estimation

Two general methods are available to compute runoff on small upland areas. The first method is based on models such as Richards' equation or various approximations to it called infiltration equations. This method uses precipitation data as a

6.5 Runoff and Sediment Yield from Watershed

function of time, together with an infiltration equation to separate rainfall rate data (intensity) into the amount entering the soil (infiltration), and the amount which moves over the soil surface (runoff as overland flow).

The second method which is used to compute runoff on small upland areas is based on rainfall depth alone or on rainfall depth and statistics representing rainfall intensity to compute runoff volume. Given runoff volume, other procedures are used to estimate peak rate of runoff or the runoff hydrograph. The USDA Soil Conservation Service runoff curve number procedure is the best known and widely used model of this type.

6.5.3.1 The Rational Method

Peak Runoff from Single Storm Event

The rational method of estimating peak flow on small watershed (< 12 km^2) is based on the assumption that for storms of uniform intensity, distributed evenly over the basin, the maximum rate of runoff occurs when the entire basin area is contributing at the outlet, and this rate of runoff is equal to a percentage of the rainfall intensity.

Runoff from an area from a storm event can be calculated using Rational method as

$$Q = \frac{CIA}{360} \qquad (6.7)$$

where

Q = runoff (m^3/s)
C = runoff coefficient (dimensionless), representing the ratio of peak runoff rate to average rainfall intensity
I = rainfall intensity (mm/h)
A = area contributing runoff to the point of consideration (ha)
360 = factor from unit conversion

The runoff coefficient varies with the surface condition. For given storm events of same rainfall intensity, nonconstant coefficients of runoff are obtained. The variation is due to the different antecedent moisture conditions. For loose soil (high infiltration) its value is low (0.3–0.6) whereas for hard surface its value may be up to 0.8.

Design Peak Flow

This method is used to estimate peak flow for designing small structures like culverts. The peak flow rate resulting from a storm with an average recurrence interval (ARI) of Y years is calculated using the following formula:

$$Q = \frac{C_y \times I_y \times A}{360} \tag{6.8}$$

where

Q = peak flow rate resulting from storm ARI of Y Years (m³/s)
C_y = runoff coefficient for design event having an ARI of Y years (dimensionless)
A = area of catchment (ha)
I_y = rainfall intensity (mm/h) corresponding to a particular storm duration and ARI. The duration is set equal to a sub-catchment time of concentration.

The following guidelines are provided for the use of the Rational Method. The applicable average recurrence interval, runoff coefficient, area of catchment, and design average rainfall intensity should be determined as discussed below.

Design Average Recurrence Interval (ARI)

The hydraulic design for the total drainage system (the underground pipeline plus the surface drainage system) should cater for 100-year ARI flood flows.

Runoff Coefficient (C)

Runoff coefficients provide for the relationship between runoff and rainfall volumes and make provision for the intermixing of pervious and impervious surfaces. For urban storm-runoff estimation, C is about 0.8–0.9, whereas, for agricultural fields, it is about 0.4–0.7, depending on the size and pattern of the catchment or drainage basin.

Time of Concentration (T_c)

The time of concentration at a particular location is generally the time required for runoff to travel by the longest available flow path to that location.

In many cases, however, a "partial area" effect occurs through the lower part of the catchment, where flows are higher than those calculated for the entire catchment, because the time of concentration is lower and the design rainfall intensity is higher.

6.5.3.2 SCS Method

The Soil Conservation Service (SCS) [recent "Natural Resources Conservation Service"] method is an empirical equation. The SCS runoff equation is

$$Q = \frac{(P - I_a)^2}{(P - I_a) + S} \tag{6.9}$$

where

Q = runoff (mm)
P = rainfall (mm)
I_a = initial abstraction (mm)
S = potential maximum retention after runoff begins (mm)

The I_a is approximated as (USDA-SCS, 1985)

$$I_a = 0.2S$$
$$\text{Thus, } Q = \frac{(P - 0.2S)^2}{P + 0.8S} \tag{6.10}$$

The potential retention (S) can range from zero on smooth, impervious surface to infinity in deep gravel. For convenience, the "S-values" were converted to runoff curve numbers (CN) by the following transformation.

$$\text{CN} = \frac{1,000}{10 + S} \tag{6.11}$$

$$\text{Or, } S = \frac{1,000}{\text{CN}} - 10$$

The curve number is based on the area's hydrologic soil group, land use, treatment, and hydrologic condition.

From the above equation, it is revealed that for $S = 0$, $\text{CN} = 100$, and when S approaches infinity, CN approaches zero. But for practical applications, the CN values are almost limited to the range of 40–98. The CN values for most US watersheds are tabulated in Chapter 9 of NEH-4 (National Engineering Handbook, Section 4) (USDA-SCS, 1985; USDA-SCS, 1993). The amount of runoff can be found if the rainfall amount and curve number is known. The runoff curve number, an index/indication of runoff producing potential, represent the combined hydrological effect of soil type, land use, agricultural land treatment, hydrologic condition, and antecedent soil moisture. For agricultural land (grassland and well-developed cropfield), the curve number for low-, moderate-, well-, and high-runoff potential soils usually range 40–68, 60–79, 65–86, and 80–89. Here, low-, moderate-, well-, and high-runoff potential soil indicate high infiltration rate and well-drained; moderate infiltration rate and moderate well to well-drained; slow infiltration rate and impediment of downward water movement; and, very slow infiltration rate and having claypan/clay-layer at or near surface; respectively.

Weighted curve number for an area can be obtained as

$$\text{CN}_w = \frac{\sum (\text{CN}_i \times A_i)}{\sum A_i} \tag{6.12}$$

The basic assumption of the SCS curve number method is that, for a single storm, the ratio of actual soil retention after runoff begins to potential maximum retention is equal to the ratio of direct runoff to available rainfall.

Although this is an accepted method for runoff estimation, studies have indicated that it should be evaluated and adapted to regional agro-climatic condition. The curve number tables should be used as guidelines and actual curve numbers and their empirical relationships should be determined based on local and regional data. Curve number method does not account for the antecedent moisture condition – less runoff under dry conditions and more runoff under wet conditions.

6.5.3.3 Sample Workout Problems on Runoff Estimation

Example 6.1

A field at East-Central Illinois was seeded with pasture. The soil of the field is silt-loam. The magnitude of a storm event was 70 mm. The potential maximum retention after runoff begins was approximated at 25 mm. Estimate the storm runoff from the field.

Solution

We know, runoff estimates by SCS method, $Q = \dfrac{(P - 0.2S)^2}{P + 0.8S}$

Here,
$P = 70$ mm
$S = 25$ mm

Putting the values in above equation, $Q = \dfrac{(70 - 0.2 \times 25)^2}{70 + 0.8 \times 25} = 46.9$ mm (Ans.)

Example 6.2

A rice field in North China Plain consists of 12 ha clay-loam soil having moderate organic matter. The topography of the basin is flat, and each unit plot is surrounded by 10 cm levee. The rainfall recorded at a nearby weather station for a particular storm event was 120 mm. Estimate the runoff from the field.

Solution

We know runoff $Q = \dfrac{(P - 0.2S)^2}{P + 0.8S}$

Here,
$P = 120$ mm
For the prevailing field condition, Assuming $S = 11$ cm $= 110$ mm

Thus, $Q = \dfrac{(120 - 0.2 \times 110)^2}{120 + 0.8 \times 110} = 46$ mm

Runoff volume, $V =$ Area $\times$ runoff depth

Given, $A = 12$ ha $= 120{,}000$ m^2
Runoff depth $= 46$ mm $= 0.046$ m
Thus $V = 120{,}000 \times 0.046 = 5{,}250$ m^3 (Ans.)

Example 6.3

A watershed of 1,000 ha area consists of clay-loam soil. The area is covered by pasture, wheat, and corn, and the general field slope toward the drainage outlet is about 0.5%. A 3-h storm of 25 mm/h occurs over the watershed. The anticipant moisture in the area was near at field capacity. Compute the peak flow rate at the outlet of the watershed

Solution

We know, using Rational method, peak discharge,
$Q = CIA / 360$, m^3/s
Here,
$I = 25$ mm/h
$A = 1{,}000$ ha
For the prevailing field condition, assuming $C = 0.65$
Putting the above values, $Q = (0.65 \times 25 \times 1{,}000)/360 = 45.13 m^3/s$ (Ans.)

6.5.4 Factors Affecting Soil Erosion

The major factors affecting soil erosion rates are climate (mainly rainfall), soil, topography, and land-use pattern. The mechanisms of erosion hazard of the factors are described below:

6.5.4.1 Climate

The most important climatic variable is rainfall erosivity, which is related to rainfall amount (how much it rains) and intensity (how hard it rains). Another important climatic variable is temperature because temperature and precipitation together determine the longevity of biological materials like crop residue and applied mulch used to control erosion.

Seasonal variations in wind, temperature, humidity, and rainfall may create more ideal conditions for erosion.

6.5.4.2 Soils

Soils vary in their inherent erodibility. Erodibility, the property to break down soil structure, is dependent on soil composition and texture. Soils with high erodibility require less energy to detach soil particles. Soil erodibility for disturbed soils is high. On the other hand, permanent stable soils have low erodibility.

6.5.4.3 Topography

Slope length, steepness, and shape are the topographic characteristics that most affect rill and interrill erosion. Agriculture on slopes greater than 3% increases the risk of soil erosion. Steeper and longer slopes generate runoff with more velocity and energy to erode and transport more sediment.

6.5.4.4 Land Use

Land use is the single most important factor affecting rill and interrill erosion because the type of land use and land condition are features that can be most easily changed to reduce excessive erosion.

6.5.4.5 Vegetative Cover

Vegetation shields soils from the impact of raindrops and traps suspended sediment from runoff.

6.5.4.6 Soil Cover Management/Cultural Practices

Cover management practices affect both the forces applied to the soil by erosive agents and the susceptibility of the soil to detachment. For a given land use like cropland, important features include the crops that are grown, the type of tillage system such as clean, reduced, or no till. Important features on a construction site are as follows: whether or not the land is bare, the soil material is a cut or fill, mulch has been applied, or the slope has been recently reseeded. Important features on range and reclaimed land include the native or seeded vegetation, and degree of ecological maturity.

6.5.4.7 Other Management Practices

Other support practices include ridging (e.g., contouring), vegetative strips and barriers (e.g., buffer strips, strip cropping, fabric fence, gravel bags), runoff interceptors (e.g., terraces, diversions), and small impoundments (e.g., sediment basins, impoundment terraces). These practices reduce erosion primarily by reducing the erosivity of surface runoff and by causing deposition.

Soil erosion is usually caused by the impact force of raindrops and by the sheer force of runoff flowing in rills and streams. Raindrops falling on bare or sparsely vegetated soil detach soil particles, runoff in the form of sheet flow along the ground, picks up and carries these particles to surface waters. As the runoff gains velocity and concentration, it detaches more soil particles, cuts deeper rills and gullies into the surface of the soil, and adds to its own sediment load. The further the runoff runs uncontrolled, the greater its erosive force and the greater the resulting damage.

The fate of eroded material within a watershed is influenced by hydrologic, topographic, vegetative, and groundcover characteristics.

6.5.5 Sediment Yield and Its Estimation

6.5.5.1 Concept

Sediment yield is defined as the amount of sediment per unit area removed from a watershed by flowing water during a specified period of time. Sediment yield affects rates of soil development and influences the recovery of disturbed surfaces downslope from source areas in desert landscapes. Sediment yield is strongly affected by surficial materials, topography, rainfall seasonality, and vegetation cover and can be increased by soil disturbance, which often occurs as the result of land use.

6.5.5.2 Mechanism

Sediment yield from upland areas is simply the final and net result of detachment, transport, and deposition processes occurring from the watershed down to the point of interest where sediment yield information is needed. Depending upon the scale of investigation and definition of the problem, this point of interest can be the edge of a farm field, delivery point to a stream channel, a position on a hillslope, a property boundary at a construction site, watershed outlet or some other location dependent upon topography. In any event, sediment yield at the point of interest is determined by the occurrence of physical processes of sediment detachment, transport, and deposition at all positions in the contributing watershed area above the point of interest. When the energy in the stream (containing sediment) dissipates to a level that can no longer support the transport of the sediment, the sediment falls out of the water column and deposits.

6.5.5.3 Estimation

(A) Delivery Ratio Method

Sediment yield is often computed based on the use of a delivery ratio defined as the change per unit area from the source to the point of interest. The delivery ratio (D in percent) is often expressed as

$$D = 100 \ Y/T \qquad (6.13)$$

where Y is the total sediment yield at the downstream point of interest, and T is the total material eroded (gross erosion) on the watershed area above the point of interest. Values of Y and T are given in units of mass per unit area per unit time (e.g., T/A/yr).

(B) Empirical Equations for Estimating Sediment Yield

The Universal Soil Loss Equation (USLE)

The most widely used and successful model to predict soil loss from upland areas is the Universal Soil Loss Equation (USLE). The USLE is an erosion model designed to predict the long-term average soil losses in runoff from specific field areas in specific cropping and management systems. The procedure of USLE is founded on an empirical soil loss equation that is believed to be applicable wherever numerical values of its factors are available.

The USLE was originally derived and presented in English units. Conversion to SI units was accomplished thereafter. The USLE is

$$A = R \times K \times \text{LS} \times C \times P \tag{6.14}$$

where

A = Soil loss per unit area per unit time (t/ha/yr)
R = rainfall and runoff factor (MJ-mm/ha-h-y)
K = soil erodibility factor (MJ-mm/ha-h-y)
LS = slope length and steepness factor (–)
C = cover and management factor (–)
P = support practice factor (–)

Rainfall and Runoff Factor (R)   The factor R is computed as the product of rainfall storm energy (E) and the maximum 30-min rainfall intensity (I_{30}). The product term (EI) is described by Wischeier and Smith (1978) as a statistical interaction term that reflects how total energy and peak intensity are combined in each particular storm.

It indicates how particle detachment is combined with transport capacity. An average annual value of R is about 100–90,000 SI unit.

An approximate equation to estimate R is (Lane et al., 2010)

$$R = 0.417 \, P^{2.17} \tag{6.15}$$

where R is in MJ-mm/ha-h-y, and P is the 2-yr, 6-h rainfall amount in millimeters.

Soil Erodibility Factor (K)   The K is the soil loss rate per erosion index unit for a specified soil as measured on a unit plot. Unit plot is defined as 22.1 m length of uniform 9% slope continuously clean-tilled fallow condition. The value of K for agricultural soil usually ranges from 0.013 to 0.053 MJ-mm/ha-h-y.

Slope length and steepness factor (LS)   The factor LS is the ratio of soil loss per unit area of a field slope to that from a unit plot. The value of LS normally ranges from 0.2 to 6.0.

6.5 Runoff and Sediment Yield from Watershed

Cover and Management Factor (C) The factor C is the ratio of soil loss from an area with specified cover and management to that from an identical area in tilled and continuous fallow. It is a measure of the combined effects of all cover and management and is primarily affected by crop stage period, canopy cover, applied mulch, residue mulch, tillage, and crop residuals. The value of C usually ranges from 0.01 to 1.4.

Support Practice Factor (P) The factor P is the ratio of soil loss with a specific support practice to the soil loss on a unit plot. For croplands, the value of P for contouring is about 0.6–0.9, for strip-cropping 0.3–0.9, for contour-farmed and terraced fields 0.05–0.9.

Modified Forms of USLE Williams (1975) replace the R factor in USLE by a runoff factor and interpret the other USLE factors on a watershed-wide basis (referred to as MUSLE). Thus, MUSLE is a watershed, rather than an upland, sediment yield model. Onstad and Foster (1975) modified the R factor in USLE to allow individual storm estimation of upland soil loss. Other factors of USLE retain their original meaning and interpretation.

6.5.6 Sample Workout Problems on Sediment Yield Estimation

Example 6.4

A farmer plowed his field in spring and planted grain corn. The soil is sandy loam with medium organic matter content. The climate of the area is semi-humid. The field is 400 m long with a 5% slope. Contour farming was performed in the field. Estimate the soil loss from the field. Assume standard value of the variables, if needed.

Solution

We know soil loss, $A = R \times K \times LS \times C \times P$ (t/ha/yr)
We get rainfall and runoff factor, R, which can be expressed as

$$R = 0.417\, P^{2.17}$$

where P is the 2-yr, 6-h rainfall amount.
For the given semi-humid area, assume $P = 50$ mm
Then, $R = 0.417 \times [50]^{2.17} = 2{,}027$

Assumed values of other factors (along with their conceptual judgment) are summarized below:

Factor	Relevant field condition	Typical values	Assumed value
K	Soil erodibility factor: The soil is loam	0.013–0.053	0.03
LS	Slope length and steepness factor: 5% slope, 400 m long	0.2–6.0	4.0
C	Cover and management factor: Crop is corn, moderate coverage	0.01–1.4	0.5
P	Support practice factor: Contour farming	0.6–0.9	0.6

Putting the values,

$$A = 2{,}027 \times 0.03 \times 4.0 \times 0.5 \times 0.6 = 72.9 \text{ t/ha/yr (Ans.)}$$

Example 6.5

Mr. John Mitchell has a 100 ha watershed in Melbourne. The watershed has 20 ha of good pasture in silty-clay soil of 2% slope (CN = 65), 30 ha of pasture on silt-loam soil of 0.5% slope (CN = 70) and remaining 50 ha of wheat on loam soil of 1% slope (CN = 85). Determine the weighted curve number for the watershed.

Solution

We know

$$CN_w = \frac{\sum (CN_i \times A_i)}{A}$$

The data are arranged in tabular form as follows:

Sl no.	Area, A (ha)	Curve no., CN	$A \times$ CN
1	20	65	1,300
2	30	70	2,100
3	50	85	4,250
	$\sum A = 100$		$\sum (A \times CN) = 7{,}650$

Putting the values in the above equation, $CN = 7{,}650/100 = 76.5 \approx 77$ (Ans.)

Example 6.6

In a humid sub-tropic climatic zone in India, a farmer has sown winter wheat. The subsequent crop in the field is broadcasted (dry-seeded) Aman rice. The soil is silt-loam having a moderate organic matter. The average grade of the field is 1% and the maximum length of run of individual plot is 200 m. No additional mulch is applied in the field. The 2-yr 6-h rainfall for the area is 40 mm. Estimate the soil loss from the field.

Solution

We know soil loss, $A = R \times K \times LS \times C \times P$ (t/ha/yr)
We get rainfall and runoff factor,

$$R = 0.417 \, P^{2.17}$$

where P is the 2-yr, 6-h rainfall amount.
Given, $P = 40$ mm
Then, $R = 0.417 \times [40]^{2.17} = 1,249$

Assumed values of other factors (along with their conceptual judgment and typical values) are summarized below:

Factor	Relevant field condition	Typical values	Assumed value
K (Soil erodibility factor)	The soil is silt-loam	0.013–0.053	0.02
LS (Slope length and steepness factor)	Slope length 200 m, slope 1%	0.2–6.0	1.0
C (Cover and management factor)	Crops are wheat and rice, dense coverage	0.01–1.4	0.2
P (Support practice factor)	No mulching	0.3–0.9	0.4

Putting the values,

$$A = 1,249 \times 0.02 \times 1.0 \times 0.2 \times 0.4 = 1.99 \text{ t/ha/yr (Ans.)}$$

6.5.7 Erosion and Sedimentation Control

Sedimentation involves the following geologic processes: erosion, transportation, and deposition. These are natural geologic phenomena. However, land development and other man-made activities accelerate the process. Excessive sediment loads result in turbid waters and heavy deposition over the substrate. Sediment-laden waters affect human activity through degradation of water quality. Consequently, minimizing the occurrence of erosion and effective control of sediment transport is imperative.

Principles of erosion and sedimentation control are based on minimizing the effects of the soil and climatological factors (described earlier) that accelerate the erosion process. Any single strategy may not be able to control those factors nor can they all be performed at every site. However, adoption/integration of as many control measures as possible may provide the most effective erosion and sedimentation control. The principles or strategies for erosion and sedimentation control are summarized below:

(a) Reduce longitudinal slope (<3%) and length of run for furrows or borders
(b) Reduce the velocity of runoff water (e.g., by maintaining vegetative cover, preserving vegetative buffer strip around the lower perimeter of the land disturbance)
(c) Divert storm water runoff
(d) Practice minimum tillage or conservation tillage in erosive soil
(e) Use buffer strip, strip cropping, and gravel bags
(f) Provide runoff interceptors (e.g., terraces)
(g) Construct siltation basin/sediment basin at different points of the drainage channel
(h) Apply organic/crop residues in the field which is prone to erosion
(i) Select land-use pattern (e.g., vegetative cover with crops) which minimizes rill and gully erosion
(j) Increase the organic matter of soil to increase shear strength, thereby decrease erosivity.

6.5.8 Modeling Runoff and Sediment Yield

Upland processes and processes in individual channel segments are combined through the channel network and interchannel areas to influence runoff and sediment yield from watersheds. In addition to the complex relationships on upland areas in stream channels, processes affecting watershed runoff and sediment yield include interactions (e.g., channel junctions and backwater) as well as land use, soil and cover characteristics, and other factors varying over the drainage area. The state-of-the-art in hydrology and erosion/sedimentation is such that runoff and sediment yield from a watershed cannot be described adequately or predicted without resorting to the use of indices, fitted parameters, and the application of judgment and experience.

Erosion models play critical roles in soil and water resources conservation and nonpoint source pollution assessments, including sediment load assessment and inventory, conservation planning and design for sediment control, and for the advancement of scientific understanding. The two primary types of erosion models are process-based models and empirically based models. Process-based (physically based) models mathematically describe the erosion processes of detachment, transport, and deposition, and through the solutions of the equations describing those processes provide estimates of soil loss and sediment yields from specified land surface areas. Empirical models relate management and environmental factors directly to soil loss and/or sediment yields through statistical relationships.

Several process-based models have been developed to predict runoff volume and sediment yield from watershed.

6.5.8.1 SWAT

"SWAT" stands for "Soil and Water Assessment Tool." SWAT is a continuous-time model developed at the USDA-ARS to predict the impact of land management practices on water, and sediment yields in large (basin scale) complex watersheds with varying soils, land use and management conditions over long periods of time (> 1 year). SWAT provides large-scale assessment of the status of surface water hydrology in the watershed and is useful for ungauged watersheds, and watersheds with limited data.

6.5.8.2 KINEROS

"KINEROS" stands for "Kinematic Runoff and Erosion Model." KINEROS is a physically based model designed to simulate runoff and erosion from single storm events in small watersheds less than about 100 km^2.

6.5.8.3 AGWA

"AGWA" stands for "Automated Geospatial Watershed Assessment Tool." AGWA is a GIS (Geographic Information System)-based hydrologic modeling tool that is designed for performing relative assessments (change analysis) resulting from land cover/use change (Burns et al. 2005).

6.5.8.4 RUSLE2

RUSLE2 is Revised Universal Soil Loss Equation (USDA-ARS). RUSLE2 includes several components. One major RUSLE2 component is the computer program that solves the many mathematical equations used by RUSLE2. A very important part of the RUSLE2 computer program is its interface that connects the user to RUSLE2. Another major component of RUSLE2 is its database, which is a large collection of input data values. The user selects entries from the database to describe site-specific field conditions. The other major component of RUSLE2 is the mathematical equations, scientific knowledge, and technical judgment on which RUSLE2 is scientifically based.

6.6 Watershed Management

6.6.1 Problem Identification

As human populations expand, and demands upon natural resources increase, the need to manage the environments in which people live becomes not only more important but also more difficult. Land and water management is especially critical as the use of upstream watersheds can drastically affect large numbers of people living in downstream watersheds.

It is necessary to identify and address land-use practices and other human activities that pollute local water resources or otherwise alter watershed functions.

Box 6.1 Watershed Pollution and Control-Sample Lesson from Ohio

Since passage of the federal Clean Water Act in 1972 and the Safe Drinking Water Act in 1974, great progress has been made in reducing the amount of pollutants discharged into Ohio's waters from point sources such as wastewater treatment plants and industries. But as point sources of pollution were reduced, other forms of pollution, called nonpoint source or diffuse pollution, came to the forefront. Nonpoint source pollution results from human land-use practices such as agriculture, mining, forestry, home septic systems, and contaminated runoff from urban landscapes. Now these nonpoint sources of pollution, combined with the physical destruction of aquatic habitat, are the major remaining sources of impairment of Ohio's rivers and lakes.

6.6.2 Components of Watershed Management

Watershed management consists of those coordinated human activities which aimed at controlling, enhancing, or restoring watershed functions. Among various areas of watershed management are improved land management, water harvesting and storage structures, improved agricultural equipments, integrated nutrient management, vermi-composting, nuclear polyhedrosis virus (NPV) production, improved cropping systems, and soil conservation measures.

Each state is required to develop the limit for the maximum amount of a specific pollutant (tolerable limit) that a water body can accommodate without causing the water body to become unable to serve its beneficial use (MDL), for all water bodies listed "impared." The MDL process is just one component of watershed management. Effective watershed management is an ongoing process that must be flexible enough to adapt to the unique characteristics of different watersheds as well as changing circumstances within a single watershed. It results in reduction of contaminants within watersheds and improvement of water quality.

6.6.3 Watershed Planning and Management

Watershed planning and management comprise an approach to protect water quality and quantity that focuses on a whole watershed. This is a departure from the traditional approach of managing individual wastewater discharges, and is necessary due to the nature of polluted runoff, which in most watersheds is the biggest contributor to water pollution. Polluted runoff is caused by a variety of land use activities, including development, transportation, agriculture and forestry, and may originate anywhere in the watershed. Due to its diffuse nature, polluted runoff cannot be effectively managed through regulatory programs alone.

Watershed planning and management involve a number of activities: targeting priority problems in a watershed, promoting a high level of involvement by interested and affected parties, developing solutions to problems through the use of the expertise and authority of multiple agencies and organizations, and measuring success through monitoring and other data gathering. Watershed management activities may take place at the state, river basin, or individual watershed level. Many issues are best addressed at the individual watershed level. For example, identifying sources of pollution that are carried by storm water to a lake is best carried out by people working within that lake watershed. Other issues are more appropriate at the basin level, such as determining appropriate discharge limits for wastewater licenses within the basin. Still others may best be operated at the state level, such as the operation of a statewide permit program.

Of course any planning process would benefit from the inclusion of both watershed and social data. Land use planners are commonly faced with a challenging mix of resource management, residential, habitat and aesthetic values, and issues.

A process for collecting and combining available watershed and social data with the goal of improving land use decision making in rural regions is necessary. The process should integrate watershed assessment, public participation, and land use planning concepts in an effort to provide tools to maintain rural community quality of life while conserving the water, the land, and other resource values. The strategy should include characteristics of a community plan, a watershed plan, and a community "visioning" document. It should be combined directly with a land use planning process.

Watershed management plans generally include the following elements:

(i) Definition of the area of concern, the purpose of the plan, and who was involved in developing the plan.
(ii) Description of the physical, ecological, and social characteristics of the watershed and the communities within its boundaries.
(iii) Description of the problems that affect watershed functions.
(iv) Identification of responsible parties and of planned activities for addressing identified problems and responsible parties.
(v) Explanation of how progress will be measured once implementation of a plan begins.
(vi) Evaluation of the success.

6.6.4 Tools for Watershed Protection

Watershed protection mechanism should provide links to a number of resources, including maps of public drinking water systems that use surface water and the county-based "Watershed Connections."

Geographic Information Systems (GIS) tools have been used in making watershed maps and predicting land use impacts from development, fertilizer, and pesticides.

New and versatile tools are necessary to help watershed coordinators to more effectively plan and deal with barriers to watershed restoration, in addition to strengthening the network of agencies, organizations, and local planning groups.

Farmers can contribute to watershed protection and community health by continuing to improve and implement best management practices. The farm assessment program should offer easy-to-conduct self-assessment tools for farmstead and field practices, soil monitoring, and livestock and pasture management. Citizens can take advantage of such programs to conduct an environmental assessment of the home and property.

Watershed related county-based publication is useful for providing information on local watersheds and water quality. It can be used by educators, county government, plan commissions, and citizen groups who want an overview of local water-related issues and resources.

6.6.5 Land Use Planning

Land use planning may be linked with watershed planning at the local level. Statewide educational program may be launched by project basis. The project should be designed to empower communities to prevent and solve natural resource problems resulting from changing land use in growing watersheds and to empower local officials to incorporate watershed protection measures into comprehensive land use plans.

The impacts of land use change on water resources may be simulated using computer models that can be used to determine short- or long-term impacts of urbanization and other land use change. The model results should be interpreted or translated in such a way that is understandable to general people.

6.6.6 Structural Management

Emphasis on structural solutions to water storage and flooding problems has given way to a new approach that recognizes the multitude of functions watersheds provide and the need to meet multiple objectives such as flood prevention, erosion control, wildlife habitat, and provision of recreation. There has also been increasing awareness that watershed management is not solely the responsibility of government agencies and conservancy districts.

6.6.7 Pond Management

In addition to beautifying the landscape, ponds provide important ecosystem services such as storm water management, habitat for aquatic life, and ecosystem health and stability. Proper pond management can prevent problems and ensure a healthy functioning pond habitat.

6.6.8 Regulatory Authority

Central or State control regulatory authority is needed to control land-use practices that alter aquatic habitat and cause nonpoint source pollution. For most effective results, government agency representatives, public officials, educators, scientists, concerned citizens, and other private interests may join together to identify and address land-use practices and other human activities that pollute local water resources or otherwise alter watershed functions, and materialize the solution strategies.

6.6.9 Community-Based Approach to Watershed Management

Community-based watershed management is an approach to water-resource protection that enables individuals, groups, and institutions with a stake in management outcomes (often called stakeholders) to participate in identifying and addressing local issues that affect or are affected by watershed functions.

Proponents of community-based watershed management involving local stakeholders results in more locally relevant solutions that take into account each community's unique social, economic, and environmental conditions and values. Stakeholder participation is also thought to create a sense of local ownership of identified problems and solutions, thus ensuring long-term support for resulting management plans.

Some key stakeholders may include those people who have the authority to make land-use decisions, such as individual landowners, farmers, and local government officials. Other stakeholders may include representatives from environmental and community groups, schools, national/state EPA, Department of Natural Resources, the local Soil and Water Conservation Districts, and University Extension departments.

6.6.9.1 Characteristics of Community-Based Watershed Management

Changing Roles and Relationships

As local communities participate more actively in watershed management, the roles and relationships of resource managers and stakeholders will change. Traditionally, resource managers were viewed as experts who were uniquely qualified to identify and implement watershed management strategies. But community-based watershed management recognizes that all stakeholders have a critical role to play in the management planning process. Resource managers and other stakeholders can contribute in many different ways, but all must work collaboratively to understand and address watershed issues when a community-based approach is used.

Whole-System Perspective

Watershed management is not a single strategy but is a general approach to water-resource protection that recognizes the interconnectedness of all the physical and biological components of the landscape, including human communities. A community-based approach considers not only the physical characteristics of a watershed, but it also takes into account the social and economic factors associated with watershed issues. The goal of community-based watershed management is to protect and restore watershed functions while considering the variety of social and economic benefits of those functions.

Integration of Scientific Information and Societal Values

Watershed management decisions should be based on sound scientific information, both in terms of identifying problems and selecting options for addressing those problems. However, resource managers have learned that management decisions that are based on scientific evidence alone often fail in the long run because they conflict with a community's economic or other social values. Community-based approaches to watershed management attempt to incorporate a broad range of values in the management process by involving representatives from a diverse cross-section of the community throughout the management planning process. In some cases, by involving diverse interests early on, many conflicts can be resolved during the planning process, thereby avoiding more costly battles once plans are put into action.

Adaptive Management Style

Addressing environmental, social, and economic issues at the watershed scale is complex, and often there is a high level of uncertainty regarding the outcomes of management decisions. Effective community-based watershed management entails an experimental approach to management in the sense that participants must be prepared to learn from their mistakes and to adapt their management strategies to changing conditions. In many ways, watershed management planning is never complete, because as old issues are resolved, new ones arise. For this reason, the long-term commitment of the stakeholders involved in a community-based watershed-management project is critical to its success.

6.6.9.2 Challenges Associated with Community-Based Watershed Management

Community-based watershed management is neither easy nor always effective at protecting or restoring watershed functions. Some of the challenges to face in adopting a community-based approach include the following:

(i) Watersheds may cover thousands of acres of public and privately owned land. Developing even a basic understanding of how human activities affect watershed functions is a major undertaking.
(ii) Some key stakeholders may lack the time, motivation, skills, or resources to participate effectively throughout the management planning process.
(iii) Resource management professionals may be reluctant to give up their role as experts and to share authority with lay persons regarding resource management issues.
(iv) Conflicts between stakeholders over management goals and the means to accomplishing those goals are inevitable, and resource management professionals are often ill-prepared to facilitate constructive dialogue to resolve these conflicts.
(v) Community-based approaches require time and resources to generate interest and to build relationships between stakeholders. Funding agencies and stakeholders may grow impatient with the lack of observable outcomes.

6.6.9.3 Keys to Success of Community-Based Watershed Management

There is no easy formula for successful community-based watershed management. However, experience from efforts around the globe suggests that several key factors, such as those listed here, are common to many successful projects:

i. Involve stakeholders in the management planning process in a way that is meaningful to them and that allows them to use their particular skills and knowledge most effectively.
ii. Do not be discouraged if some stakeholders choose not to participate initially. Begin by educating and informing key audiences about the values of the watershed to the community, the watershed management process, and specific actions they can take to get involved.
iii. Determine the appropriate scale for addressing watershed problems. Actions aimed at changing land-use practices are easiest to implement at the local level and become more difficult to manage on a larger scale.
iv. View the watershed management plan as a starting point and not the end product. Be prepared to adapt the plan as conditions change and groups learn from their mistakes.
v. Make management decisions, when possible, based on a consensus of a broad range of stakeholders. Efforts to resolve conflicts before management decisions are made may bring dividends in the long run.

vi. Focus on desired outcomes (e.g., clean water), which can often be more helpful and motivating for participants than emphasizing problems and who is causing them.

6.6.10 Land Use Planning and Practices

Land use and management activities within and adjacent to a particular region have the potential to affect water quality and other aquatic and terrestrial resources. Integrated watershed management has become to be recognized internationally as an important holistic approach to natural resources management, which seeks to promote the concept of sustainable development. In that context, sustainable land use forms an overall planning framework, whilst sound land-use planning concepts, together with the adoption of appropriate land-use practices, provide key guidelines for land and water resources development and management, which should be undertaken with the integrated objectives of reducing natural disasters, boosting productivity, and achieving sustainable development

Sound land-use planning methods and practices can be developed from an end-use standpoint, such as social and economic development of national or regional planning, or from a sectoral point of view, i.e., in the context of development planning for various sectors such as agriculture, forestry, mining, and water resources. There are strong linkages in planning between the two viewpoints; national and sectoral, between the two levels; national and regional, and among the various sectors. Integrated land-use planning aims to address these linkages. The important elements of these linkages include the management system, financial resources, institutional, and legal frameworks and community participation.

Planting a vegetated buffer between one's house or landscape and the lake or stream is a very positive step that one can take to protect water quality. Vegetated buffer strips are a proven means of controlling erosion and other sources of nonpoint source pollution. Several options are available when designing a buffer: a natural buffer, an enhanced buffer, or a landscaped buffer.

A natural buffer is the simplest and least expensive of the three options. To develop this requires only a decision on your part about the size of the vegetated strip you wish to have, a commitment to stop mowing the area, and the patience to allow plant material to become established and grow. Plants establish themselves in succession, and it will probably be several years before shrubs and trees become rooted and thrive. Advantages of this option are that the native plants that do become established are tough and resilient and a natural part of the lake ecosystem with no need of investing funds.

An integrated approach that stresses both the importance of participatory planning and the institutional and technical constraints and opportunities is therefore necessary. The institutional and technical context for managing watersheds and river basins, including the involvement of both the public and private sectors, is also appreciated.

Some other measures/techniques useful in watershed management are

- broad bed and furrow (BBF) system as a solution to waterlogging in vertisols.
- adopting a successful watershed and demonstrating the benefits of integrated watershed management in terms of increased crop yields, reduced use of pesticides and chemical fertilizers, increased ground water levels and conserving water and soil.
- local tools for protecting wetlands.
- awareness on the direct and indirect impacts of urbanization on wetlands.
- adapting watershed tools to protect wetlands.

6.6.11 Strategies for Sustainable Watershed Management

In the past, water resource, forestry, and agricultural projects were often developed with little regard to watershed management and upstream–downstream linkages. Furthermore, the role of local people and the importance of changing land use practices by those people are critical factors in achieving successful programs. Commonsense tells us that to develop sustainable programs, land and water must be managed together and that an interdisciplinary approach is needed. Now the question, "Are we moving in that direction?" There are some indicators that this may be happening. People who are trained and educated in watershed management are assuming leadership positions in many countries. Such movements indicate that policies and institutions that support integrated watershed management are emerging. Furthermore, the emergence of citizen-based watershed organizations in the United States and other countries recognizes on one hand, that a watershed management approach is relevant, but on the other hand, existing governmental institutions are not fulfilling the role of watershed management.

The following measures are also needed for sustainable watershed management:

(a) Interdisciplinary approaches to project design are needed that integrate the technical and human dimensions of watershed management. This requires an understanding of cultures and traditional land use practices. Watershed planning has historically relied upon engineering and technical expertise but has been deficient in socioeconomic aspects, resulting in less than optimal outcomes and a diminished flow of benefits beyond the termination of projects.

(b) Socioeconomic research and participatory techniques need to be incorporated early in the conceptual design and planning stages of projects. Without coincident local participation, top-down approaches alone often have inconsistent and unpredicted results, even though they may be technologically sound. Bringing in local participation, and socioeconomic specialists later on when problems arise may be too late, places undue responsibility on those not responsible for original project design. Participatory monitoring and evaluation methods should be used throughout the project cycle.

(c) Before utilizing subsidies or cash-for-work incentives, other means of providing incentives should be considered. Negative externalities can result when projects rely on subsidies; such economic strategies that may not fit because of cultural and economic differences between donor agencies and receptor countries.
(d) Both environmental and socioeconomic monitoring are needed throughout implementation and following project completion to assist in informed decision making.
(e) Project design and planning should consider scale and topography aspects in coping with upstream–downstream interactions and cumulative watershed effects. Small-scale projects with clearly defined watershed management objectives have a greater chance of demonstrating positive outcomes that can lead to long-term programs in contrast to large, ambitious, and complex projects that are difficult to manage and administer.
(f) Administrative and institutional structures should be developed that recognize watershed boundaries, without becoming overly complex. Flexibility in planning and management is essential.
(g) Regional training and networking programs at all levels should be promoted, building upon existing networks. Long-term funding support for technical professionals, managers, and policy makers should receive the same attention as operational field projects. Through expanded training programs, including training of trainers, diffusion of technology occurs and the continuity of positive project outcomes can be enhanced.

6.7 Watershed Restoration and Wetland Management

6.7.1 Watershed Restoration

It is required that the authority of each State or Province will conduct water quality assessments to determine whether its streams, lakes, and estuaries are sufficiently "healthy" to meet their designated uses, i.e., drinking, irrigation, fishing, or recreation. A water body that does not meet its designated use is defined as "impaired." Effective planning and long-term change in impaired watersheds requires citizen participation in many stages of the process. Engaging stakeholders in the watershed management process (including watershed scale planning and implementation) may result in changing attitudes and behaviors that reduce contamination throughout watersheds and consequently improve water quality. Detail descriptions about the water quality of stream, river, and estuaries are described in Chapter 6 of Volume 1.

6.7.2 Drinking Water Systems Using Surface Water

Community water systems that use surface water can benefit by developing a watershed protection plan to protect their water supply from current and future contamination. To help communities, watershed protection plan maps should be developed of all public water supply watersheds.

6.7.3 Wetland Management in a Watershed

Policies should be developed to protect and manage the wetlands in the watershed (Fig. 6.5). Wetland is necessary to help keep groundwater recharged.

Fig. 6.5 Wetland

In watershed management program, focus should be made covering the following aspects:

(1) *Water quality improvement from management practices in agricultural watersheds.*

Under this category, measurement should be taken on the transport of nitrate, phosphorus, and fecal indicator bacteria in watersheds. This information is related to the terrain, soils, and agricultural practices within the watersheds. Methods should be developed to identify areas where wetlands, buffers, and other conservation practices can provide water quality benefits.

(2) *Biological buffers for improving water quality in agricultural landscapes.*

Riparian buffers are largely effective in reducing transport of nutrients, pesticides and sediments by filtering the runoff water before it enters a stream. Subsurface nutrient uptake by buffer system may be examined.

(3) *Site- and time-specific crop, tillage, and nutrient management for sustainable agro-ecosystems.*

The long-term yield responses to soils, nitrogen fertilizer, and water should be monitored. This information may allow us to determine management zones within production fields, which would allow farmers to improve their site-specific farming systems.

(4) *Integration of research information into a decision support system for resource conservation and water quality.*

Conservation planning tools to be used by national resource conservation department, and environmental and economic goals should be evaluated. Models may be used to simulate environmental outcomes resulting from combinations of soils and management practices.

6.8 Addressing the Climate Change in Watershed Management

The impacts of climate change on the management of watershed should be focused. There are considerable challenges in watershed management, including in-stream flow needing gaps, changes in agricultural water demand, changes in precipitation, changes in river flows, reduced water availability, etc.

6.8.1 Groundwater Focus

The principle that groundwater and surface water quality must be preserved is part of the "Water for Life" strategy. The long-term objective is to understand the state of the quality and quantity of groundwater supply.

Relevant Journals

- Landscape and Ecological Engineering
- Journal of Hydrology
- Hydrology Journal
- Hydrological Sciences Journal
- Water Resources Update
- Journal of Soil and Water Conservation
- Journal of Hydraulic Engineering
- Journal of American Water Resources Association
- Environment, Development, and Sustainability

Relevant FAO Papers/Reports

- FAO Land and Water Bulletins 9 (Land-Water Linkages in Rural Watershed)
- FAO Soils Bulletins 13 (Land Degradation)

References 239

- FAO Soils Bulletins 57 (Soil and Water Conservation in Semi-arid Areas)
- FAO Soils Bulletins 78 (Conservation Agriculture – Case Studies in Latin America and Africa)
- FAO Irri. & Drainage Paper 55 (Control of Water Pollution from Agriculture)

Questions

(1) What do you mean by watershed? What are the elements of a watershed?
(2) Why is watershed management important?
(2) Briefly discuss the issues related to watershed management.
(3) Discuss the factors affecting watershed functions.
(4) What do you mean by "land grading" and "land improvement?" What is precision grading?
(5) What are the factors influencing land grading and development?
(6) Write down the activities and design considerations in land grading.
(7) What points should you consider in the early surveys?
(8) What is lot benching?
(9) What are the considerations in construction work? What are the maintenance considerations?
(10) Discuss in brief the various methods for determining earth-work volume.
(11) Discuss the runoff and sediment generation processes.
(12) Discuss the factors affecting runoff rate and volume.
(13) Briefly describe the various methods of runoff volume estimation.
(14) Discuss the factors affecting soil erosion
(15) What is sediment yield? Describe the Universal Soil Loss Equation for estimating sediment yield
(16) How can the erosion and sediment generation be controlled?
(17) Briefly explain the modeling principles of runoff and sediment yield. Name five models for estimating runoff and sediment yield.
(18) Explain the principles of watershed management.
(19) Briefly explain the measures to be taken for watershed protection.
(20) Discuss the climate change issues to be considered in watershed.
(21) Discuss the principles of sustainable watershed management.
(22) How can the wetland in a watershed be managed?
(23) Discuss the impacts of climate change on watershed management.

References

ASAE (American Society of Agricultural Engineers) (1998) Soil and water terminology. ASAE Standards, 1998, Standards Engineering Practices Data ASAE S526.1:936–952

Lal R (1997) Soils of the tropics and their management for plantation forestry. In: Nambier EKS, Brown AG (eds) Management of soil, nutrients and water in tropical plantation forests, AIC Conference, Sydney, pp 97–121

Lane LJ, Hakonson TE, Foster GR (2010) Watershed erosion and sediment yield affecting contaminant transport. http://eisner.tucson.ars.ag.gov/hillslopeerosionmodel/pdfFiles/ct_formatted.pdf. Accessed 6 May 2010

Onstad CA, Foster GR (1975) Erosion modeling on a watershed. Trans ASAE 18(2):288–292

Pimental D, Harvey C et al (1995) Environmental and economic costs of soil erosion and conservation benefits. Science 267:1117–1122

Revenga C, Murray S, Abramvitz J, Hammoud A (1998) Watersheds of the world: ecological value and vulnerability. World Watch Institute, Washington, DC

Scherr SJ, Yadav S (1996) Land degradation in the developing world: implications for food, agriculture, and the environment. Discussion Paper 14, International Food Policy Research Institute, Washington, DC

USDA-SCS (1985) National engineering handbook, Section 4 – hydrology. USDA-SCS, Washington, DC

Williams JR (1975) Sediment yield prediction with universal equation using runoff energy factor. In: Present and prospective technology for predicting sediment yields and sources, ARS-S-04. USDA, Agric. Res. Serv., Washington DC, pp 244–252

Wischeier WH, Smith DD (1978) Predicting rainfall erosion losses, a guide to conservation planning. Agriculture Handbook No. 537. USDA Science and Education Administration, Washington, DC

Chapter 7
Pollution of Water Resources from Agricultural Fields and Its Control

Contents

7.1 Pollution Sources	242
7.1.1 Point Sources	242
7.1.2 Nonpoint Sources	242
7.2 Types of Pollutants/Solutes	243
7.2.1 Reactive Solute	243
7.2.2 Nonreactive Solute	243
7.3 Extent of Agricultural Pollution	243
7.3.1 Major Pollutant Ions	243
7.3.2 Some Relevant Terminologies	244
7.3.3 Factors Affecting Solute Contamination	244
7.3.4 Mode of Pollution by Nitrate and Pesticides	247
7.3.5 Hazard of Nitrate (NO_3–N) Pollution	248
7.3.6 Impact of Agricultural Pollutants on Surface Water Body and Ecosystem	248
7.4 Solute Transport Processes in Soil	250
7.4.1 Transport of Solute Through Soil	250
7.4.2 Basic Solute Transport Processes	251
7.4.3 Convection-Dispersion Equation	254
7.4.4 Governing Equation for Solute Transport Through Homogeneous Media	254
7.4.5 One-Dimensional Solute Transport with Nitrification Chain	256
7.4.6 Water Flow and Solute Transport in Heterogeneous Media	257
7.5 Measurement of Solute Transport Parameters	258
7.5.1 Different Parameters	258
7.5.2 Breakthrough Curve and Breakthrough Experiment	259
7.6 Estimation of Solute Load (Pollution) from Agricultural Field	261
7.6.1 Sampling from Controlled Lysimeter Box	261
7.6.2 Sampling from Crop Field	261
7.6.3 Determination of Solute Concentration	262
7.7 Control of Solute Leaching from Agricultural and Other Sources	265
7.7.1 Irrigation Management	265

    7.7.2 Nitrogen Management 265
    7.7.3 Cultural Management/Other Forms
           of Management 266
7.8 Models in Estimating Solute Transport from Agricultural and Other Sources 266
Relevant Journals 267
Questions 267
References 269

Chemicals from agricultural field and other sources frequently enter the soil, subsoil, and aquifer. This may happen either by normal management practices or by accident, and the resulting chemical residues pose hazards to the environment and ecosystem. Whether we are using fertilizer or other pollutants, it is useful to know how fast it moves.

Surface and groundwater pollution caused by chemicals from agricultural (e.g., fertilizers, pesticides, insecticides, herbicides) and industrial sources have caused public concern for decades. Chemicals migrating from municipal disposal sites also represent environmental hazards. The same is true for radionuclides emanating from energy waste disposal facilities. Extensively and, specifically, intensively cropped areas are major sources of groundwater recharge. Excessive use of chemical fertilizers can lead to pollution of surface and groundwater if associated with high rainfall or irrigation, shallow rooted crops, and sandy soils. To preserve the groundwater and reduce economic losses for the farmers, estimation of solute/chemical leaching below the root zone and prevention of such leaching is crucial.

7.1 Pollution Sources

To check or mitigate the adverse effects of pollution, it is necessary to understand the sources, the characteristics, and the interaction processes of the pollutants with the environment.

The sources of pollution can be broadly divided into two classes:

(i) Point source
(ii) Nonpoint source (or diffuse source)

7.1.1 Point Sources

Point source of pollution is a type of pollution that can be traced to a single source, such as pipes, wells or ditches, industrial or metropolitan outlet, and sewage treatment plants. It has a clearly definable point of entry into the waterways.

7.1.2 Nonpoint Sources

Nonpoint source (also termed diffuse source) is the source that cannot be traced to a single point but rather distributed throughout the system. Diffuse pollution loads are transient and highly variable hydrological phenomena related to the use or misuse of land, they may also result from atmospheric decomposition of local or distinct origin. Agricultural practices are the major nonpoint source of groundwater and/or surface water pollution.

7.2 Types of Pollutants/Solutes

Solutes are chemicals that are dissolved in water. Solutes are of two types:

7.2.1 Reactive Solute

Reactive solute is a solute that reacts with soil during transportation. As a result, it does move with the same velocity of water. Examples are K and Ca.

7.2.2 Nonreactive Solute

Nonreactive solute is a solute that does not react with soil or other chemicals during transportation. This type of solute moves with the same velocity of water. Examples are Cl and Br.

7.3 Extent of Agricultural Pollution

7.3.1 Major Pollutant Ions

The growing needs for agricultural commodities due to the ever-increasing population dictate intensification of agricultural activities through the application of irrigation, chemical fertilizers, and pesticides. Application of fertilizer as an agricultural practice is a major contributor to the total load of pollutants in the soil. Despite increased agricultural productivity, the deleterious effects of these chemical fertilizers on the surface and groundwater quality are of alarming concern throughout the world. Many commercial chemicals are used extensively to provide plants with the primary nutrients like nitrogen (N), phosphorous (P), potassium (K), and sulphur (S). Leaching of these nutrients as deep percolation loss from the root zone of agricultural crops continue to endanger the long-term groundwater quality in many rural areas of the world. The fate of nitrogen in the soil is of major concern because of the potential hazard for N, applied in excess of the natural decomposing capacity of the soil, to contaminate shallow and deep aquifers.

Contamination of groundwater by nitrate has been a problem in many areas of the world for many years and continues to occur. During the last two decades, many cases of groundwater contamination from extensive fertilizer applications in watersheds utilized for agricultural purposes have been reported worldwide. The major ion concentrations in such cases of groundwater pollution are nitrate (NO_3^-), sulfate (SO_4^{-2}), potassium (K^+), and phosphate (PO_4^{-3}).

Nitrate contamination of groundwater is considered one of the most serious problems worldwide because of its diffuse type and the difficulties in its control and management. There are several sources of nitrogen (N) in soil, including on-site domestic waste disposal systems (e.g., septic tanks and cesspools), dairies, animal feedlots, and irrigation with sewage effluent. The main source of nitrate contamination is agriculture, especially the intensive application of nitrogen fertilizers. A survey of wells constructed by the "U.S. Geological Survey" Department in 1985 found that more than 10% of the 2732 wells sampled in California were contaminated by nitrates at concentration of the Federal drinking water standard of 10 mg/l nitrate-nitrogen (nitrate-N) (USGS, 1985). Pollutants from municipal and industrial waste water include chloride (Cl^{-1}), lead (Pb), cadmium (Cd), zinc (Zn), molybdenum (Mo), cobalt (Co), etc.

7.3.2 Some Relevant Terminologies

Contaminant/pollutant: The substance that makes pollution/contamination is called contaminant/pollutant.

Contamination/pollution: Generally, the presence of any undesired substance(s) in an environmental component makes that component polluted/contaminated. Scientifically, the presence of any undesired substance(s) in any environmental component in amount(s) more than its recommended/permissible limit makes that component polluted/contaminated. The process or activity that renders the environmental component polluted/contaminated is called environmental pollution.

Environment: Environment of an individual is everything surrounding the same individual and their inter- and intra-relationship. For example, a man's environment is everything around the man. Among biotic components (i.e., all living beings) are animals, plants, microbes etc.; and among abiotic components (i.e., all physical entries) are air, water, food, weather, etc.; and the relationship between /among the components.

Maximum permissible limit: "Maximum permissible limit" is the concentration of a pollutant (normally expressed in mg/kg or ppm) in a substance or environment (such as drinking water, air, food commodities, animal feed) to be legally permitted by the national authority.

Persistent pollutants: The pollutants which persist or exist in the environment for a long time. This type of pollutant degrades very slowly. They have long half-life. Half-life is the time span within which the pollutant will degrade by 50%.

7.3.3 Factors Affecting Solute Contamination

The initiation of irrigation may result in large quantities of water moving through the soil profile. Concerns about subsurface water quality grow as more water moves through the soil profile because advective transport of pollutants may also increase. It has been found that in irrigated agriculture, the timing, and the method of irrigation and fertilizer application are the key factors affecting the leaching of chemical fertilizers (Sarkar and Ali, 2010). In addition to fertilizer rate, factors such as tillage system, cropping system, soil type, and environmental conditions can play important roles in determining the amount of solute/contaminants leached into groundwater. The groundwater contamination is more likely where soils are well-drained, and irrigation is necessary for crop growth and where nitrogen is applied above recommended rates. There is a direct relationship between large NO_3–N losses, excess nitrogen inputs and inefficient irrigation management.

The major factors affecting nitrate and/or other solute leaching are as follows.

(a) *Excess use of fertilizer*: Use of chemical fertilizer in excess of crop requirement results in disproportionate leaching. The use of N-fertilizer at rates higher than the rate of uptake by the plant increases the potential for increased nitrate leaching. If nitrogen fertilizer is applied when temperature is not high enough to support rapid plant growth, is at risk of leaching. Applying excess nutrients fertilizers will directly affect subsurface water quality especially for NO_3–N, which is highly mobile.

(b) *Inconsistence with irrigation water*: If nitrogen fertilizer is applied without consideration of the water regulation, a high concentration of nitrate appears in drain outlet water. Downward movement of ammonia and nitrate from fertilizers in rain and irrigation water are observed.

(c) *Drier condition & residue of fertilizer at crop harvest*: Under dryland culture or under limited irrigation, more solute/nutrient retains in the soil (Fig. 7.1). Nitrate leaching strongly depends on the amount of N remaining in the soil at the harvest of the crop. Nitrogen uptake by plants and nitrogen mineralization both increase under irrigation and could mitigate the increased nitrate leaching.

(d) *Irrigation schedule and fertilizer type*: Nitrate leaching could be attributed to both irrigation schedule and fertilizer type. Slow-releasing fertilizer has low risk of leaching that of fast releasing fertilizer. The effect of advection is higher than that of diffusion in the nitrate transfer phenomena when the rate of irrigation is increased.

(e) *Crop root length & density*: Absorption of nutrient or solute largely depends on the extent and length of root zone. If the root length is higher, there is greater opportunity time to absorb the solute. Inclusion of deep-rooted crops, such as alfalfa, in the rotation could reduce the amount of residual N capable of leaching below the root zone.

(f) *Crop type & growth rate*: The uptake rate of water and solute depends on the vegetative growth pattern of the crop grown, which is a function of crop type and growth rate.

(g) *Soil type and organic matter content*: Decomposition and mineralization of organic residues and other nitrogenous fertilizers depend on the organic matter content, microorganisms, and physical and bio-physical properties of the soil concerned. A decrease in organic matter lower total mineralization. Sandy soil

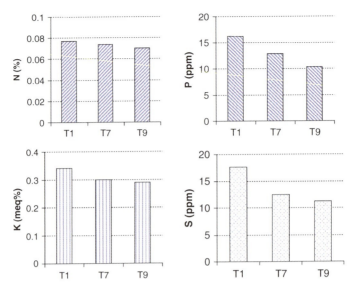

Fig. 7.1 Nutrient retention in soil at harvest of wheat crop under no irrigation (T1), two irrigation (T7), and four irrigation (T9) frequency (after Ali, 2008)

facilitates quick percolation of water and, thus, greater chance of transport of solutes.

(h) *Crop rotation*: Crop rotation may influence the uptake, decomposition, and/or addition of nitrogen in soil and, thus, on the leaching of nitrate.

(i) *Irrigation or rainfall*: In dryland agriculture, the increase in soil moisture that results from irrigation or rainfall dissolves excess nitrate (NO_3–N) present in the soil profile, and makes it more susceptible to leaching. Higher moisture contents will also raise microbial activity including mineralization. The increase in mineralization rate directly affects nutrient leaching. The initiation of irrigation caused a flash of NO_3–N to the shallow groundwater.

(j) *Management factors*: Management factors such as tillage number, mulching, application of herbicide, etc. influence on the decomposition and mobility of organic and inorganic nitrogen downwards. A more intensive tillage practice under irrigated agriculture could increase the mineralization rate. Increased mineralization rates contribute to the elevated NO_3–N concentrations in the subsurface water.

Once released into the subsurface environment, industrial and agricultural chemicals are generally subjected to a large number of simultaneous physical, chemical, and biological processes, including sorption-desorption, volatilization, photolysis, and biodegradation, as well as their kinetics. The extent of degradation, sorption, and volatilization largely determines the persistence of a pollutant in the subsurface. The fate of organic chemicals in soils is known to be strongly affected by the kinetics of biological degradation.

The literature suggests that nitrate leaching from the crop field may vary from 30 to 170 kg-N/ha/year depending on the crop, irrigation, fertilizer use (type & amount), crop rotation, and management factors.

7.3.4 Mode of Pollution by Nitrate and Pesticides

Nitrogen is essential to plant growth and increases yield. Under ideal conditions, only the fertilizer that can be used by the plant would be applied, leaving no residual to move below the root zone. However, in most cases, not all the applied nitrogen is assimilated by the plants, allowing some to move below the root zone. Nitrogen in the soil that is not returned to the atmosphere in the form of nitrogen gas or ammonia is generally converted to the nitrate form by bacteria. Nitrate is very mobile, and if there is sufficient water in the soil, it can move readily through the soil profile. Among the nutrients leached or allowed to run off, N is the most abundant and is of major concern as the source of ground and surface water pollution.

The pesticides that are applied for preventing, destroying, repelling, mitigating, or controlling any insect, fungus, bacterial organism or other plant or animal pest; deposit in or on a site of plant, soil, or water. Then through different processes move and enter into other environmental components. Based on the characteristics of the component and characteristics of the pesticides such movement, their fate and impact on the environmental components become evident. The persistent pesticides persist in the environmental component for a long time and move intact from one component to another even at long distances (Fig. 7.2).

In the past, most pesticides were regarded as involatile, but now volatilization is increasingly recognized as being an important process affecting the fate of pesticides in field soils. Another process affecting pesticide fate and transport is the relative reactivity of solutes in the sorbed and solution phases. Several processes such as gaseous and liquid phase molecular diffusion, and convective-dispersive transport,

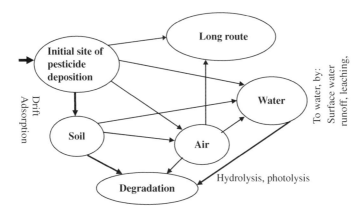

Fig. 7.2 Schematic of fate of pesticide residues in environment

act only on solutes that are not adsorbed. Degradation of organic compounds likely occurs mainly, or even exclusively, in the liquid phase. On the other side, radioactive decay takes place equally in the solution and adsorbed phases, while other reactions or transformations may occur only or primarily in the sorbed phase.

Pollution of surface water from storm runoff and its management has been described in Chapter 3 (*Water*), Volume. 1.

7.3.5 Hazard of Nitrate (NO_3-N) Pollution

There is sufficient evidence that high nitrate contamination in drinking water can cause serious health problems. Thus, acute and chronic health effects of contaminated water due to agricultural activities are serious threats to human beings. Nitrate may cause methemoglobinemia (blue baby syndrome) in infants (Stone et al., 1997) when it is above the maximum contamination level of 10 mg/l. Additionally, nitrate interaction with other dietary substances may cause health problems in humans (Maidson and Brunett, 1985).

7.3.6 Impact of Agricultural Pollutants on Surface Water Body and Ecosystem

7.3.6.1 Impact of Eutrophication

The process of nutrient over-enrichment of natural water – "eutrophication" is a threat for sustainable water resource management. Modern agriculture is dependent on irrigation, chemical fertilizers and pesticides. Irrigated agriculture accounts for over two-thirds of the world's current freshwater use. In some cases excessive fertilizer is used, i.e., more than the plant's absorbing capacity. As a consequence, the river water and lakes have become enriched with nitrogen and phosphorous nutrients which in turn degraded the water environment.

The eutrophication has devastating impacts on fisheries, biodiversity, and other activities in the aquatic ecosystem. Generally the growth of floating algae in the aquatic ecosystem depends on the availability of light, nutrients, and densities of grazing species, but they start to grow very fast in the nutrient enriched water. Subsequently, the huge biomass of algal bloom produced due to their rapid growth requires enormous amount of oxygen from the water column to be decomposed and consequently the oxygen concentration becomes too low for fish and invertebrates to survive. Fisheries resources may be increased to a certain level of eutrophication but low oxygen concentration reduces fish stock. High phosphorus concentration in river water may harm the plant communities and modifies the biological balance. Food chains in the aquatic ecosystem may be altered.

Eutrophication can be minimized by planting vegetation at the river bank for reduction of erosion and nutrient effluent, managing the amount and timing of fertilizer application, controlling runoff and facilitating biological activities to disintegrate harmful chemical compounds.

7.3.6.2 Impact of Pesticides and Other Chemicals

The persistent pollutants/agrochemicals degrade slightly, mostly by abiotic processes like photolysis, hydrolysis, and oxidation. In the aquatic system, they undergo bio-magnification through the food chain. During the movement through different environmental components, the pesticides seriously affect the living organisms encountered on the pathway. They are biodegraded very slowly and at minimum by the biotic processes/organisms. Reversely, the nonpersistent pesticides biodegrade very quickly by the biotic processes including microbes, and they do not bio-magnify the food chain.

The pollutants/agrochemicals, particularly the extremely and highly hazardous ones cause different hazards to the organisms in their respective habitats ranging up to death. A list of such hazards to different organisms is provided below:

Hazards to fisheries:

- reduction of fish population
- reduced reproduction
- species endangered

Human health hazards:

- skin disease
- headache
- eye irritation
- respiratory problems
- immune system disorder
- reduced longevity
- reproductive disorder
- deformed fetus
- tumor
- cancer
- death

Hazards on wild life:

- reduction of wild population
- reproduction failure of wild species
- species endangered

Reduction in biodiversity
Hazards to apiculture/honeybee

All of the above effects may be direct or indirect, via secondary contamination or food chain. The long-term ecological impacts of increased rates of agricultural nitrogen and phosphorus input will depend on the levels to which these nutrients accumulate in various nonagricultural ecosystems. These levels are uncertain because of the complexities of the global biogeochemistry of nitrogen and phosphorus. These nutrients accumulate in a variety of forms in many different sinks (such as arable soil organic matter, atmospheric nitrous oxide, groundwater, freshwater, and marine ecosystems and their sediments) after agricultural application.

7.4 Solute Transport Processes in Soil

For minimizing the pollution of soil, water and the environment, we should know the processes involved in the transport, the ability of soils to transmit chemicals in soil solution, and techniques to remove them from the solution.

7.4.1 Transport of Solute Through Soil

Transport processes in soils, particularly the movement of water and solutes, play a vital role in the provision of suitable conditions for plant growth, and in the replenishment and quality of groundwater supplies. Our understanding of these processes has developed largely from experimental studies on simple uniform porous materials, such as sands, that have led to the widespread acceptance of Darcy's law and Richards' equation for soil-water flow and of the dispersion equation for solute movement. These equations imply that soils can be considered as continua and effectively assume that the water and solutes pervade the whole volume rather than being contained in the complex network of pores in which the velocity of flow and the solute concentration vary from point to point. They adequately describe the macroscopic water and solute movement in simple porous materials. However, field soils are commonly more complex, often having a bimodal pore structure with a network of macropores separating aggregates of soil particles that form regions of micropores. The transport behavior in these aggregated soils, especially when unsaturated, can be very different from that in the simple porous materials in which the theory has been developed.

The way in which water and solutes move in aggregated soils depends on the mode of saturation of the pore space that is made up of the micropore region within the aggregates and the macropores surrounding them. When both regions are saturated, a hydraulic head gradient causes water to flow preferentially in the macropores with little flow within the aggregates, so that movement of solutes into or out of the aggregates is mainly by diffusion caused by the difference between the solute concentrations of the water in the two regions. The movement of water in the micropore region within the aggregates can be considered to behave as if in a continuum and can be described by Darcy's law. With water moving into the aggregates

7.4 Solute Transport Processes in Soil

during the wetting up, there is convective movement of solutes, so that the dispersion equation (with boundary conditions imposed by conditions in the macropores) must be used to describe the solute movement. When the velocity becomes very small, the solute movement then approximates to a diffusion process and can be calculated from Fick's law. Thus, the salts which is transporting toward the centers of the aggregates (micropore regions), redistribute eventually to a uniform equilibrium condition throughout the aggregates. When the macropores are empty, the aggregates become almost isolated so that redistribution of water and solutes occurs only within the aggregates.

The bimodal soil structure of aggregated soils has a profound effect on bulk soil hydraulic conductivity and on soil hydrology. Such structure occurs when cracks and fissures are produced during natural soil shrinkage on drying or during mechanical drainage operations, and also as a result of tillage. The presence of interconnecting macropores assists drainage and promotes rapid leaching of solutes when the macropores are full of water, but inhibits these processes when empty.

7.4.2 Basic Solute Transport Processes

Solutes are transported through three basic processes:

i. Convection
ii. Diffusion
iii. Dispersion

Solutes can exist in all three phases of liquid, solid, and gaseous. The decay and production processes can be different in each phase. Solutes are transported by convection and dispersion in the liquid phase, and by diffusion in the gaseous phase.

7.4.2.1 Convective Solute Transport

Convective solute transport refers to the movement of dissolved solutes through the soil with the flowing water. In the absence of sorption and dispersion, the solutes move at the same average velocity of water. Under such a condition, the total mass flow of solute is related to the law that governs the movement of water through the soil. Thus, the convective flux of solute passing through a unit area of soil can be expressed as

$$q_c = qC \qquad (7.1)$$

where

q_c = convective flux density of solute (kg/m²/s)
q = Darcian water flux *or* volumetric flux density (m³ of water/m²/s)
C = concentration of solute in water (kg/m³ of water)

7.4.2.2 Diffusive Solute Transport

Diffusion is the process of moving fluids from higher concentration to lower concentration. Ionic and molecular species in solute move from areas of higher concentration to lower concentration. Transfer of mass of a chemical species is caused by gradient in concentration. Molecular diffusion in solutions is controlled by Fick's first and second laws. It is assumed that the rate of transfer of solute by diffusion through a unit area of a section of soil is proportional to the gradient in concentration normal to the section. The diffusive mass flux of solute through a unit area is

$$q_d = -\theta D_c \frac{\partial C}{\partial x} \qquad (7.2)$$

where

q_d = diffusive flux of solute (kg/m²/s)
θ = total volumetric water content of soil (m³ water/m³ soil)
D_c = chemical diffusion coefficient of solute in pure water (m²/s)
C = concentration of solute in water (kg/m³ of water)

The negative sign in the equation arises because diffusion occurs in the direction opposite to that of increasing concentration.

Fick's first law: Steady state flux of solute is given by

$$F = -D \frac{\partial C}{\partial x} \qquad (7.3)$$

where

F = flux of solute under steady state (mass/area/time)
D = Diffusivity coefficient (1×10^{-9} to 2×10^{-9} m²/s)
C = Solute concentration (mass/volume)
$\frac{\partial C}{\partial x}$ = Concentration gradient

Fick's second law: Rate of change of solute concentration over time is given by

$$\frac{\partial C}{\partial t} = D \frac{\partial^2 C}{\partial x^2} \qquad (7.4)$$

where

$\frac{\partial C}{\partial t}$ = Change of solute concentration with time

7.4.2.3 Dispersive Solute Transport

In porous media, solute dispersion is caused by two mechanisms: (i) molecular diffusion and (ii) hydrodynamic dispersion. Hydrodynamic dispersion is explained by the tortuous nature of the convective stream lines resulting from microscopic fluctuation of the advection velocity.

The size and shape of the pores in soil differ, which cause variations in the velocity of water through the pores. Also, the velocity is faster at the center of a pore than near the periphery. The complexity of the pore system causes mixing of the soil solution along the direction of flow, and hence dispersion of the solute. Dispersion is a velocity-dependent process. It occurs only during the flow of water. Sometimes it is termed mechanical dispersion. The mechanical dispersion is described in a similar way as chemical diffusion using Fick's law as

$$q_m = -\theta D_m \frac{\partial C}{\partial x} \qquad (7.5)$$

where

q_m = dispersive flux of solute (kg/m²/s)
θ = total volumetric water content of soil (m³ water/m³ soil)
D_m = mechanical dispersion coefficient of solute in pure water (m²/s)
C = concentration of solute in water (kg/m³ of water)

The transport of solutes in a porous medium is governed by the combined effect of convection (the average solute particle velocity), diffusion, and dispersion. Due to similarity between the chemical diffusion and mechanical dispersion, the coefficients D_c and D_m are assumed to be additive. Thus,

$$D = D_c + D_m$$

where D is the longitudinal hydrodynamic dispersion coefficient (m²/s)

Usually, D is simply referred to as the dispersion coefficient. It is also termed as apparent diffusion coefficient or the diffusion-dispersion coefficient.

The average solute particle velocity defines the centroid of the solute plume at a given time or the average arrival time of solutes at a given depth. For a homogeneous porous medium, steady-state water flow and an inert (nonreactive) solute, the average solute particle velocity equals q/θ. The solute dispersion quantifies the dispersion of the solute plume around the centroid at a certain time or the dispersion of the solute breakthrough around the average arrival time at a certain depth.

7.4.3 Convection-Dispersion Equation

The movement of solute in porous media is commonly described by the convection-dispersion equation (CDE) (Bear, 1972). For a steady state, one-dimensional transport of a reactive solute through a uniform soil, the CDE can be written as

$$R\theta \frac{\partial C}{\partial t} = D \frac{\partial^2 C}{\partial z^2} - v \frac{\partial C}{\partial z} \tag{7.6}$$

where
  R = retardation factor (dimensionless)
  D = hydrodynamic dispersion coefficient (m^2/s)
  C = solute concentration (mol/s)
  θ = water content (volumetric)
  v = pore water velocity (m/s)
  z = depth (m)

The dispersion coefficient D is often assumed to be linearly related to the pore water velocity, $D = \lambda v$, where λ is the dispersivity (m).

The CDE describes solute transport as the sum of the average convection with flow, and the hydrodynamic dispersion.

7.4.3.1 Assumptions in CDE

The CDE is based on three basic assumptions:

– Continuous porous media
– Average flow velocity
– Fick's first law for solute dispersion

7.4.3.2 Drawbacks of CDE

The most important drawbacks of CDE when it is used to simulate transport, can be attributed to non-Fickian behavior of dispersive transport as well as the apparent scale dependence of the dispersivity.

The solute dynamics in the root zone profile can be described by different mathematical equations under different situations.

7.4.4 Governing Equation for Solute Transport Through Homogeneous Media

7.4.4.1 Two-Dimensional Equation

The partial differential equations governing two-dimensional nonequilibrium chemical transport of solutes involved in a sequential first-order decay chain during

7.4 Solute Transport Processes in Soil

transient water flow in a variably saturated rigid porous medium are (Simunek and van Genuchten, 1995) as follows:

$$\frac{\partial \theta c_1}{\partial t} + \frac{\partial \rho s_1}{\partial t} + \frac{\partial a_v g_1}{\partial t} = \frac{\partial}{\partial x_i}\left(\theta D_{ij,1}^w \frac{\partial c_1}{\partial x_j}\right) + \frac{\partial}{\partial x_i}\left(a_v D_{ij,1}^g \frac{\partial g_1}{\partial x_j}\right) - \frac{\partial q_i c_1}{\partial x_i} - Sc_{r,1} -$$
$$(\mu_{w,1} + \mu'_{w,1})\theta c_1 - (\mu_{s,1} + \mu'_{s,1})\rho s_1 - (\mu_{g,1} + \mu'_{g,1})a_v g_1 + \gamma_{w,1}\theta + \gamma_{s,1}\rho + \gamma_{g,1}a_v \quad (7.7)$$

$$\frac{\partial \theta c_k}{\partial t} + \frac{\partial \rho s_k}{\partial t} + \frac{\partial a_v g_k}{\partial t} = \frac{\partial}{\partial x_i}\left(\theta D_{ij,1}^w \frac{\partial c_k}{\partial x_j}\right) + \frac{\partial}{\partial x_i}\left(a_v D_{ij,k}^g \frac{\partial g_k}{\partial x_j}\right) - \frac{\partial q_i c_k}{\partial x_i} - Sc_{r,k} -$$
$$(\mu_{w,k} + \mu'_{w,k})\theta c_k - (\mu_{s,k} + \mu'_{s,k})\rho s_k - (\mu_{g,k} + \mu'_{g,k})a_v g_k + \mu'_{w,k-1}\theta c_{k-1} +$$
$$\mu'_{s,k-1}\rho s_{k-1} + \mu'_{g,k-1}a_v g_{k-1} + \gamma_{w,k}\theta + \gamma_{s,k}\rho + \gamma_{g,k}a_v$$

$$k\varepsilon(2, n_s) \quad (7.8)$$

where c, s, and g are solute concentrations in the liquid [ML^{-3}], solid [MM^{-3}], and gaseous [ML^{-3}] phases, respectively; q_i is the ith component of the volumetric flux density [LT^{-1}]; μ_w, μ_s, and μ_g are first-order rate constants for solutes in the liquid, solid, and gas phases [T^{-1}], respectively; μ_w', μ_s', and μ_g' are similar first-order rate constants providing connections between individual chain species; γ_w, γ_s, and γ_g are zero-order rate constants for liquid [ML^{-3}T^{-1}], solid [T^{-1}], and gas phases [ML^{-3}T^{-1}], respectively; ρ is the soil bulk density [ML^{-3}], a_v is the air content [L^{-3}L^{-3}], S is the sink term in the water flow, c_r is the concentration of the sink term [ML^{-3}], D_y^w is the dispersion coefficient tensor [L^{2}T^{-1}] for the liquid phase, and D_y^g is the diffusion coefficient tensor [L^{2}T^{-1}] for the gas phase. The subscript w, s, and g correspond with the liquid, solid, and gas phases, respectively; while the subscript k represents the kth chain number, and n_s is the number of solutes involved in the chain reaction.

7.4.4.2 One-Dimensional Transport

Solute transport equation for a homogeneous, isotropic porous medium during steady-state unidirectional groundwater flow reduces to the following:

$$D_T \frac{\partial^2 c}{\partial x^2} + D_L \frac{\partial^2 c}{\partial z^2} - v\frac{\partial c}{\partial z} - \lambda Rc = R\frac{\partial c}{\partial t} \quad (7.9)$$

where λ is a first-order degradation constant, D_L and D_T are the longitudinal and transverse dispersion coefficients, respectively; v is the average pore water velocity (q_z/θ) in the flow direction, R is the solute retardation factor, and z and x are the spatial coordinates parallel and perpendicular to the direction of flow. The initial solute-free medium is subjected to a solute source, c_0, of unit concentration. The source covers a length $2a$ along the inlet boundary at $z=0$ and is located symmetrically about the coordinate $x = 0$. The transport region of interest is the half-plane ($z \geq 0; -\infty \leq x \leq \infty$). The boundary condition may be written as

$$c(x, 0, t) = c_0 \qquad -a \leq x \leq a$$

$$c(x, 0, t) = 0 \qquad \text{other values of } x$$

$$\lim_{z \to \infty} \frac{\partial c}{\partial z} = 0 \qquad (7.10)$$

$$\lim_{x \to \pm\infty} \frac{\partial c}{\partial x} = 0$$

7.4.5 One-Dimensional Solute Transport with Nitrification Chain

One special group of degradation reactions involves decay chains in which solutes are subject to sequential (or consecutive) decay reaction. Problems of solute transport involving sequential first-order decay reactions frequently occur in soil and groundwater systems. Examples are the migration of various radionuclides, the simultaneous movement of interacting nitrogen species, organic phosphate transport, and the transport of certain pesticides and their metabolities.

For solute transport in a homogeneous, isotropic porous medium during steady-state unidirectional groundwater flow, the solute transport Eqs. (3.1) and (3.2) reduces to

$$R_1 \frac{\partial c_1}{\partial t} = D \frac{\partial^2 c_1}{\partial x^2} - v \frac{\partial c_1}{\partial x} - \mu_1 R_1 c_1 \qquad (7.11)$$

$$R_1 \frac{\partial c_1}{\partial t} = D \frac{\partial^2 c_1}{\partial x^2} - v \frac{\partial c_1}{\partial x} + \mu_{i-1} R_{i-1} c_{i-1} - \mu_i R_i c_i \qquad i = 2, 3 \qquad (7.12)$$

where μ is a first-order degradation constant, D is the dispersion coefficients, v is the average pore water velocity (q_x/θ) in the flow direction, x is the spatial coordinate in the direction of flow, and where it is assumed that three solutes participate in the decay chain. For this specific three-species nitrification chain:

$$NH_4^- \to NO_2^- \to NO_3^-$$

The boundary condition may be written as

$$\left[-D \frac{\partial c_i}{\partial x} + v c_1 \right] = v c_{0,1}(0, t)$$

$$\left[-D \frac{\partial c_1}{\partial x} + v c_i \right] = 0 \qquad i = 2, 3 \qquad (7.13)$$

7.4 Solute Transport Processes in Soil

$$\lim_{x \to \infty} \frac{\partial c_i}{\partial x} = 0 \quad i = 1, 2, 3$$

7.4.6 Water Flow and Solute Transport in Heterogeneous Media

In macroporous heterogeneous soil, the classical water flow theory based on Buckingham-Darcy law may not adequately describe the infiltration and redistribution of water. Although these macropores may comprise only a small fraction of the total soil volume, they can have a profound effect on the rate of infiltration and redistribution of water. As a result, pollutants dissolved in water can reach deeper soil layers and ground water tables much faster as could be expected assuming homogeneous flow in the matrix.

Under such conditions, the two-domain concept or double-, dual-, bi(multi)modal-porosity models, with macropores as a second domain next to the less permeable micropore region, are nowadays widely accepted for modeling water flow and solute transport in heterogeneous soils. Both regions are treated as continua and the continuum approach is used to establish the flow and transport equations in each region. Both equations are coupled by means of an exchange term accounting for the mass transfer of water and solutes between both regions.

7.4.6.1 Dual Porosity Model for Solute Transport

In the mobile-immobile solute transport model, the soil water volume is split up into a "mobile" and an "immobile" region. Solute transport in the mobile region is described by the convective-dispersion equation, and solute exchange between both regions is modeled by a first-order kinetic diffusion process:

$$\theta\beta \frac{\partial C_m}{\partial t} + \theta(1-\beta)\frac{\partial C_{im}}{\partial t} = \theta\beta D_m \frac{\partial^2 C_m}{\partial z^2} - \theta\beta v_m \frac{\partial C_m}{\partial z} \quad (7.14)$$

$$\theta(1-\beta)\frac{\partial C_{im}}{\partial t} = \alpha(C_m - C_{im}) \quad (7.15)$$

where β is the ratio of the mobile water content, θ_m, versus the total water content; $1-\beta$ is the ratio of the immobile water content θ_{im} versus θ; C_m and C_{im} are the concentrations in the mobile and immobile phases; D_m is the dispersion coefficient for the mobile region, α is the first-order mass exchange rate; and v_m the average pore-water velocity in the mobile liquid phase.

7.5 Measurement of Solute Transport Parameters

7.5.1 Different Parameters

Study of mobility of contaminants in an aquifer is an important issue for the proper remediation of contaminated groundwater. Transport of contaminants in aquifer depends largely upon the aquifer and their chemical properties. The aquifer property can be defined as the medium or material property which consequently is related to the velocity and dispersion of contaminants. These intrinsic factors arising from the physical aspects of the aquifer materials finally construct two important solute transportation parameters, i.e., advection and dispersion coefficients. Analysis of solute transport parameters therefore is essential for prediction of the fate of contaminants.

The main parameters of the solute transport equations are as follows: advection coefficient (V), hydrodynamic dispersion coefficient (D), retardation factor (R), first-order rate constant, zero-order rate constant, partition coefficient, pore water velocity (v), dispersivity, and length (α).

Dispersivity: It indicates the spreading of a solute across the "front" of average bulk flow. It is a geometrical proportionality constant of a media. It is determined by the pore-size distribution of the channels involved in transmission of a bulk solution.

Retardation factor (R): It is a unique medium property. It is a dimensionless parameter reflecting the retarding effect of adsorption on solute transport. It is expressed as

$$R = 1 + \rho K_d/\theta,$$

where ρ is the bulk density, K_d is partition coefficient, θ is volumetric water content.

First-order rate constant: In first-order kinetics (a form of reaction process), the reaction rate is proportional to the concentrations of the component reactants. The proportional constant is termed as first-order rate constant (or degradation constant, in degradation reaction).

Zero-order rate constant: In zero-order kinetics, the constant reaction rate is termed as zero-order rate constant. Zero-order kinetics refers to the reaction process in which the reaction rate is independent of the concentrations of the reactants.

Partition coefficient: It is a proportionality constant in an equilibrium adsorption isotherm. It is determined as the ratio of adsorbed content to the concentration in an aqueous solution for the linear equilibrium isotherm.

Hydrodynamic dispersion: It refers to the spreading of a solute beyond the region expected due to the flow alone. It is the sum of hydraulic dispersion and molecular diffusion.

Pore water velocity (v): It is the average velocity through soil pores. It is determined as: $v = q/\theta$, where v is the pore water velocity, q is the Darcy velocity, θ is the water content.

7.5 Measurement of Solute Transport Parameters

Dispersivity length (α): Generally it increases with the travel distance of solute. The discrepancy between laboratory and field test value of α is termed as scale effect of dispersion.

Solute transport parameters can be obtained by column studies and breakthrough curve (BTC) experiments in soils and well-tracer tests in aquifers.

7.5.2 Breakthrough Curve and Breakthrough Experiment

7.5.2.1 Breakthrough Curve

Breakthrough curves are the graphical presentation of relative solute concentration (outflow/inflow) of water passing through a soil column against the corresponding time of measurement or the pore volume. The curve is used to derive solute transport parameters.

Instead of relative solute concentration, the normalized concentrations are also used in constructing breakthrough curves. The pore volume, also called the dimensionless time, is the total volume of water in the column of soil at a particular water content.

The shape of the breakthrough curve (See Fig.7.3) varies with water flux and solute concentration (input rate).

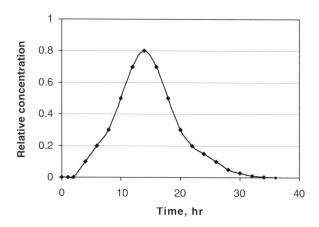

Fig. 7.3 A typical breakthrough curve

7.5.2.2 Breakthrough Experiment

The breakthrough experiments may be conducted by applying a solution of salt to the column of soil as a pulse or in a step. For the pulse-type experiment, a small volume of salt solution is spread uniformly over the surface of the soils in the column instantaneously during steady water flow. The salt solution moves through the soil column with the moving water and causes changes in the concentration of solute in the soil column. The changing concentrations are monitored over time until the whole solution moves out of the column. The solute concentration is monitored using TDR or automated technique.

In the step-type experiment, salt solution is continuously added to the soil column at a constant rate. The concentration of salt in the soil column increases gradually and reaches a constant value when the whole initial soil-water phase is replaced by the flowing solution. The increasing concentrations are monitored over time to construct the BTCs.

The pulse-type experiment gives a better insight into the physical processes of the transport system from the falling segment of BTCs. It also needs small quantities of salt solution. But for unsaturated condition, a pulse experiment is time consuming. On the other hand, the step-type experiment needs several steps of application of solute to characterize the soil properly. This needs large quantity of salt solution. It is time consuming to attain a constant concentration of salt in the experimental soil under field condition.

7.5.2.3 Analysis of Breakthrough Curve

Breakthrough curves (BTCs) are analyzed to determine the parameters governing the transport of solute involved in the transport process. A number of mathematical models predict the transport of solutes through the vadose zone. The methods used to analyze the breakthrough curve are as follows:

- analytical method
- numerical method
- quasi-analytical method
- method of moment
- curve-fitting method

The most widely used method to determine the solute transport parameters is to fit the transport models to breakthrough curves measured in the laboratory or field. Minimizing the squared deviations (least-square technique) is an accurate and reliable method for fitting the observed BTCs to the transport models. Different models are usually based on the analytical solutions of the one-dimensional convection-dispersion equation, usually for steady flow of water through soil. The method of time moment is applied to determine the solute transport parameters directly from the measured breakthrough data. It is applicable to linear processes only.

In least-square technique, the BTCs measured are compared with those precdicted by the analytical solutions based on assumed values of parameters. If the measured and estimated BTCs agree well, the values of the parameters used for the prediction are regarded as reliable.

7.6 Estimation of Solute Load (Pollution) from Agricultural Field

Estimates of diffusive-source pollution loads over different temporal and spatial scales are required to evaluate the severity of the pollution problems and to determine appropriate management measures.

7.6.1 Sampling from Controlled Lysimeter Box

Solute leaching from the agricultural field can be estimated by growing the crop in the controlled lysimeter (drainage type) and collecting the free drainage outflow at the bottom of the lysimeter. To obtain a realistic estimate, the soil and other conditions in the lysimeter should resemble those of the field. The outflow tap is kept closed but only during drainage collection. Drainage collection may be accomplished at various stages of the crop, or at predetermined dates from fertilizer application. The solute load in a particular drainage collection is

$$M_s = V_d \times C_s \qquad (7.16)$$

where

M_s = mass of solute leached in the particular drainage event (mg)
V_d = volume of drainage outflow (l)
C_s = concentration of solute in the drainage sample (mg/l)

Seasonal or yearly amount of solute is obtained by summing up the individual mass of solute for each drainage event.

7.6.2 Sampling from Crop Field

Samples of percolated water from the crop field can be collected by installing water sampler and using vacuum extractor (vacuum hand pump). Soil water sampler consists of porous ceramic cup. The samplers are installed at various depths (30–250 cm) and intensities depending on the specific objectives. Sampling may be done at various stages of the crop or at predetermined dates since fertilizer application. At the predetermined /scheduled date, vacuum is created in the sampler and one has to wait for several hours to obtain an appreciable amount of water. Then the sample is collected through the pump. A schematic view of the ceramic cup, vacuum suction pump, and sampling arrangement is shown in Fig. 7.4.

If water enters the cup, the pressure in the dial of the suction pump falls down. In case of loose fitting of the arrangement, the pressure may also fall down. In such a case, the suction is created repeatedly. If soil-water is low, the suction may need to prolong for 24 h to get an appreciable amount of sample.

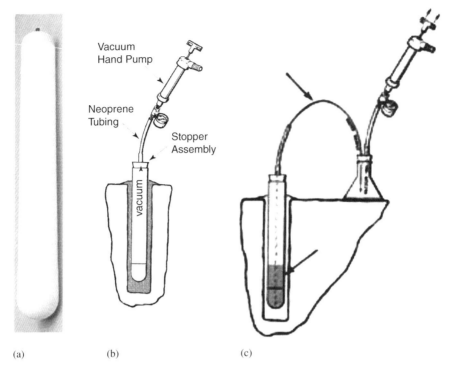

Fig. 7.4 Water sampling equipments (**a**) ceramic cup, (**b**) vacuum suction pump, and (**c**) water sampling arrangement (Courtesy: Soil Moisture Equipment Co.)

Collection of water sample following the crop period is also necessary to monitor the residual nutrients, specially during rainy/wet periods.

To determine the total solute leaching during a particular period, determination of solute concentration and correct estimation of leaching amount or deep percolation is a prerequisite. Duration of a particular leaching event may be fixed depending on the date of sample collection such as 5 days prior to 5 days post-sampling date. Leaching amount may be estimated using field water balance equation or other reliable techniques (e.g., tensiometer readings, TDR, or Neutron moisture meter data).

Determination of basic infiltration rate and hydraulic conductivity particularly *in situ* condition, and determination of soil organic matter is needed to interpret the solute transport data.

7.6.3 Determination of Solute Concentration

It is beyond the scope of this book to discuss various methods for determining different solutes/pollutants in the water and/or soil sample. The readers are referred to the relevant analytical books/handbooks, such as *Standard Methods for Examination*

7.6 Estimation of Solute Load (Pollution) from Agricultural Field

of Waters & Wastewaters (Rand et al., 1976 and later editions, APHA Pub.); *Soil, Plant, Water and Fertilizer Analysis* (Gupta, 2007); *FAO Soils Bulletin 10* (Dewis and Freitas, 1970); *FAO Soils Bulletin 31* (Rhoades and Merrill, 1976), *USDA Handbook 60*.

7.6.3.1 Solute Transport Study by Tracer

Evidence showed that the diffuse (nonpoint) source of pollution plays a major role in the degradation of the quality of our soil and groundwater systems. Transport processes governing soil and groundwater contamination by a diffuse source can be studied by means of a tracer (e.g., chloride and bromide ion (KCl, KBr), tritium).

7.6.3.2 Sample Examples on Solute Transport Problem

Example 7.1

In a lysimeter (4m × 2m), maize crop was grown. The leaching amount was collected and the concentration of NO_3–N of the respective leachate for the growing season and the fallow period are given below. Find out the total amount of NO_3–N leached down during the period.

DAE	20	40	60	80	100	120	140	160	180	200
Drainage (l)	0.8	0.6	0.5	0.7	0.7	0.6	3	2.5	3	4
Conc. of NO_3–N in drainage water (mg/l)	4	5	4.5	3	3	4	6	5	4.5	4

DAE is the days after seedling emergence.

Solution

We know, solute amount, $M_s = V_d \times C_s$
= drainage volume (l) × concentration of solute in the drainage water (mg/l)
For the 20 DAE, NO_3–N = 0.8 × 4 = 3.2 mg
Thus the whole calculation follows as

DAE	20	40	60	80	100	120	140	160	180	200
Drainage (l)	0.8	0.6	0.5	0.7	0.7	0.6	3	2.5	3	4
Conc. of NO_3–N in drainage water (mg/l)	4	5	4.5	3	3	4	6	5	4.5	4
Total NO_3–N, mg	3.2	3	2.25	2.1	2.1	2.4	18	12.5	13.5	16

Total NO_3–N for 8 m² area during the whole period = 75.05 mg
Thus, total NO_3–N per hectare of land = 0.094 kg

Example 7.2

In an experimental field of rice crop, water sampler was installed at 120 cm depth to collect the deep drainage and analyzed at laboratory for NO_3–N concentration. Nitrogenous fertilizer was applied in two splits. Water samples were collected at 15-day intervals and deep drainage was estimated using field water balance approach. The sampling was done during the crop season and continued for a month after the crop harvest. The following data were gathered:

DAT	15	30	45	60	75	90	105	120	135	150
Estimated drainage (mm)	30	25	20	18	22	20	20	24	30	31
Conc. of NO_3–N in water (mg/l)	3	2.5	4.2	2.2	4	2.3	2.7	4	5	4

Determine the amount of NO_3–N leached per hectare during the period.

Solution

Drainage volume from a particular drainage event, say 15 DAT, per hectare = (30/1,000) m × 10,000 m^2 = 300 m^3

Then, solute amount, $M_s = V_d \times C_s = (300 \times 10^6)$ cm^3 × $(3/1,000)/1,000$ g/cm^3
= $(300 \times 10^6) \times (3/10^6) = 900$ g

DAT	15	30	45	60	75	90	105	120	135	150
Estimated drainage (mm)	30	25	20	18	22	20	20	24	30	31
Drainage vol., m^3/ha	300	250	200	180	220	200	200	240	300	310
Conc. of NO_3–N in water (mg/l)	3	2.5	4.2	2.2	4	2.3	2.7	4	5	4
Solute (NO_3–N) amount (g)	900	625	840	396	880	460	540	960	1,500	1,240

Thus, total NO_3–N leached = 8,341 g/ha
= 8.34 kg/ha

7.7 Control of Solute Leaching from Agricultural and Other Sources

The mass of solute leaching is directly related to the percolation volume and solute concentration of the leachate. Thus, both water and nitrogen fertilizer (nitrogen) management is important in controlling nitrate leaching.

The application of the following guiding principles minimizes nitrate leaching:

(1) Nitrogen use in excess of crop requirement results in disproportionate leaching
(2) Nitrogen application when temperature is not high enough to support rapid plant growth is at risk from leaching
(3) Irrigation should be used to enhance nitrogen recovery by crops, and
(4) Accumulation of nitrate in soil by mineralization after harvest should be minimized

7.7.1 Irrigation Management

Poor irrigation management aggravates the pollution effects of fertilizers. Over irrigation should be avoided. Traditional irrigation applications can be altered to enhance nutrient and water uptake while minimizing leaching below the root zone. Increasing irrigation efficiency could reasonably be accomplished by changing from a furrow or basin to a drip or sprinkler system which has potentially high irrigation efficiency. Nitrogen leaching is less for the alternate-furrow irrigation. Contamination of groundwater with nitrate is attributed to deep percolation of water containing the chemical. Thus proper irrigation design, scheduling irrigation according to available soil-water depletion (implementing zero runoff irrigation), and improving efficiency of water application and irrigation uniformity can reduce deep percolation, resulting in reduced nitrate losses to a certain extent. A combination of sprinkler irrigation and N fertigation significantly reduces N leaching with only minor reductions in crop yield. Matching irrigation amount to crop ET also will reduce N loading to the groundwater.

7.7.2 Nitrogen Management

Nitrogen (N) management practices for reduction of nitrate losses include the following:

(i) *Correct nitrogen application (applying fertilizer based on plant species need to meet yield goals)*: Proper N application for realistic yield goals is probably the best method for controlling nitrate leaching
(ii) *Split application of N (reduce the availability and thus loss)*: Applying small amount of N fertilizer early in the growing season, use tissue analysis to schedule additional fertilizer N as the season progresses.

(iii) *Soil nitrate test for N side-/top-dressing requirements*: Soil test for N offers the possibility of reducing the extent of over-fertilizing with N.
(iv) *Use of slow-releasing fertilizer*: Fertilizers that release N at the slower rate meet the crop demand and reduce the chance of leaching.
(v) *Appropriate method of application (e.g., deep placement, fertigation)*: Fertilizer N can be applied in the irrigation water (fertigation). Side-dress application for most N fertilizer (in the humid regions or under irrigation in the drier climate) can reduce the extent of nitrate pollution.
(vi) *Accumulation of nitrate in soil by mineralization after harvest should be minimized*: Cover crops to take up nitrate remaining in the soil at the end of the growing season can reduce the extent of nitrate pollution.
(vii) *Some deficit irrigation (~40%)*: Some allowable deficit irrigation can reduce or eliminate percolation, and thus reduce N leaching. Matching irrigation amount to crop evapotranspiration (ET) will also reduce N loading to the groundwater.

7.7.3 Cultural Management/Other Forms of Management

(i) *Cover crop*: Some amount of solutes (N) remains in the soil at harvest of the crop. After harvest, where the land remains fallow, if a cover crop is grown, it will absorb the available nutrients and thus will reduce the chance of leaching.
(ii) *Zero tillage/minimum tillage*: In some instances, where zero tillage or minimum tillage is possible, it will reduce percolation rate and thereby increase the opportunity time, which will reduce the leaching of solutes.

Within the state-of-the art, the fertilizer and irrigation application can be altered to enhance nutrient and water uptake, while minimizing leaching below the root zone. Thus, appropriate management of nitrogen fertilizer, irrigation, and other management practices can reduce nitrate leaching to a large extent.

7.8 Models in Estimating Solute Transport from Agricultural and Other Sources

Nowadays, computer simulation models have become a useful tool in understanding the transport of water and solutes through the soil into the groundwater. Models are useful tools for integrating different processes involved in nitrogen transport in soil and can be used in forecasting how a system will behave without actually making measurements in the physical system. Models are particularly useful describing and predicting transport processes, simulating conditions which are economically or technically impossible to carry out by field experiments.

Some of the available & commonly used models for NO_3–N leaching are as follows:

- NLEAP
- SOIL
- SOILN
- SWMS_2D
- DRAINMOD-N
- LEACHN
- MACRO
- HYDRUS-2D (two-dimensional water and solute transport model)
- IRRSCH (irrigation scheduling model)
- Opus (transport of nonpoint source pollution)

Relevant Journals

- Water, Air and Soil Pollution – An International Journal of Environmental Pollution
- Paddy and Water Environment
- Journal of Contaminant Hydrology
- Water Quality and Ecosystem Modeling
- Environmental Monitoring and Management
- Ecosystem, Ecotoxicology, Environmental Management, Environmental Monitoring and Assessment
- Vedos Zone Hydrology
- Water Quality, Exposure and Health
- Journal of Hydrology (Amsterdam)
- Water Resources Research
- Agricultural Water Management
- Agronomy Journal
- Irrigation Science

Questions

(1) Describe in brief the sources of pollution from agricultural field.
(2) Write a short note on the following:
    Pollutant, pollution, persistent pollutants, Maximum permissible limit
(3) Explain the factors influencing solute leaching from crop field.
(4) Explain the way by which the nitrate and/or pesticides pollute water and soil.
(5) What are the impacts of agricultural pollutants on water resource and the ecosystem?
(6) Describe the mechanisms/pathway of solute transport thorough soil.
(7) Explain in brief, the basic solute transport processes.

(8) Write down the governing equations for solute transport under the following conditions:

(a) Two-dimensional transport through homogeneous media
(b) One-dimensional transport through homogeneous media
(c) One-dimensional transport with nitrification chain
(d) Dual porosity flow in heterogeneous media

(9) Define convection-dispersion equation (CDE) for reactive solute through a uniform soil.
(10) Define the following: Dispersivity, Retardation factor, First-order rate constant, Zero-order rate constant, Hydrodynamic dispersion, Dispersivity length.
(11) What is a breakthrough curve? Draw a typical breakthrough curve.
(12) Explain the procedure for estimating solute load from agricultural fields.
(13) Write down the guiding principles for minimizing nitrate leaching from agricultural fields.
(14) Describe the procedures/measures for minimizing nitrate leaching from agricultural fields.
(15) Name some models for solute transport study along with their salient features.
(16) In a lysimeter (8m × 4m), wheat crop was grown. The leaching amount was collected, and the concentration of NO_3–N of the respective leachate for the growing season and the fallow period are given below. Find out the total amount of NO_3–N leached down during the period.

DAE	20	40	60	80	100	120	140	160	180	200
Drainage (l)	1.0	2.9	3.8	1.6	1.0	2.5	3.5	1.5	4.0	5.0
Conc. of NO_3–N in drainage water (mg/l)	3.0	4.5	3.5	2.8	2.9	3.0	4.0	4.5	5.5	3.0

DAE = days after seedling emergence.

(17) In an experimental field of rice crop, nitrogenous fertilizer was applied in two splits. Water sampler was installed at 150 cm depth to collect the deep drainage. Water samples were collected at 15-day interval and the collected samples were analyzed at laboratory for NO_3–N concentration. Deep drainage was estimated using field water balance approach. The sampling was done during the crop season and continued for a month after the crop harvest.

The following data were gathered:

DAT	15	30	45	60	75	90	105	120	135	150
Estimated drainage (mm)	25	35	30	28	32	27	19	21	25	29
Conc. of NO_3–N in water (mg/l)	3.5	3.5	3.2	2.0	4.0	3.3	2.9	4.1	4.0	3.5

Determine the amount of NO_3–N leached per hectare during the period

References

Ali MH (2008) Deficit irrigation for wheat cultivation under limited water supply condition. Dissertation.Com, Boca Raton, USA, p 183

Bear J (1972) Dynamics of fluids in porous media. American Elsevier Publishing Co, New York, NY

Dewis J, Freitas F (1970) Physical and chemical methods of soil and water analysis. FAO Soils Bulletin 10. FAO, Rome, 275p

Gupta PK (2007) Soil, plant, water and fertilizer analysis. AGROBIOS, India

Maidson RJ, Brunett JO (1985) Overview of the occurrence of nitrate in groundwater of the United States. U.S. Geological Survey Water-Supply Paper No. 2275

Rand MC, Greenberg AE, Taras MJ (1976) Standard methods for the examination of water and wastewater. Washington, DC, American Public Health Association

Rhoades JD, Merrill SD (1976) Assessing the suitability of water for irrigation: theoretical and empirical approaches. In: Prognosis of salinity and alkalinity. Soils Bulletin 31. FAO, Rome, pp 69–109

Sarkar AA, Ali MH (2010) Evaluation of different water management practices for water saving, nitrate leaching and rice yield. A Research report, BINA/Ag. Engg. Division-8, Agricultural Engineering Division, Bangladesh Institute of Nuclear Agriculture, p 28

Simunek J, van Genuchten MTh (1995) Numerical model for simulating multiple solute transport in variably saturated soils. In: Wrobel LC, Latinopoulos P (eds) Proceedings of the water pollution III. Modeling, management, and prediction. Computation Mechanics Publication, Ashurst Lodge, Ashurst, Southampton, pp 21–30

Stone KC; Hunt PG, Johnson MH, Tatheny TA (1997) Groundwater nitrate-N concentrations on an eastern coastal plains watershed. Paper No. 97–2152, ASAE Meeting Presentation

USGS (United States Geological Survey) (1985) National water summary 1984, hydrological events, selected water-quality trends, and ground-water resources. U.S. Geological Survey Water Supply Paper 2275

Chapter 8
Management of Salt-Affected Soils

Contents

8.1	Extent of Salinity and Sodicity Problem		272
8.2	Development of Soil Salinity and Sodicity		273
	8.2.1	Causes of Salinity Development	273
	8.2.2	Factors Affecting Salinity	277
	8.2.3	Mechanism of Salinity Hazard	278
	8.2.4	Salt Balance at Farm Level	278
8.3	Diagnosis and Characteristics of Saline and Sodic Soils		279
	8.3.1	Classification and Characteristics of Salt-Affected Soils	279
	8.3.2	Some Relevant Terminologies and Conversion Factors	282
	8.3.3	Diagnosis of Salinity and Sodicity	285
	8.3.4	Salinity Mapping and Classification	290
8.4	Impact of Salinity and Sodicity		293
	8.4.1	Impact of Salinity on Soil and Crop Production	293
	8.4.2	Impact of Sodicity on Soil and Plant Growth	294
8.5	Crop Tolerance to Soil Salinity and Effect of Salinity on Yield		295
	8.5.1	Factors Influencing Tolerance to Crop	295
	8.5.2	Relative Salt Tolerance of Crops	297
	8.5.3	Use of Saline Water for Crop Production	298
	8.5.4	Yield Reduction Due to Salinity	299
	8.5.5	Sample Examples	300
8.6	Management/Amelioration of Saline Soil		301
	8.6.1	Principles and Approaches of Salinity Management	301
	8.6.2	Description of Salinity Management Options	302
8.7	Management of Sodic and Saline-Sodic Soils		317
	8.7.1	Management of Sodic Soil	317
	8.7.2	Management of Saline-Sodic Soil	319
8.8	Models/Tools in Salinity Management		320
8.9	Challenges and Needs		323
Relevant Journals			323

Relevant FAO Papers/Reports 323
FAO Soils Bulletins 324
Questions 324
References 325

Salinity is an important land degradation problem. Salinity is more widely known and refers to the amount of soluble salt in a soil. Sodicity refers to the amount of sodium in soils. The consequences of salinity have detrimental effects on plant growth and final yield damage, reduction of water quality for users, sedimentation problems, and soil erosion. When crops are too strongly affected by the amounts of salts, they disrupt the uptake of water into roots and interfere with the uptake of competitive nutrients.

Effective management of salinity or sodicity problem requires correct diagnosis of the problem. Understanding the soil-salinization processes and limitations to crop production is the key to improvements in crop productivity related to salinity and sodicity. Although several treatments and management practices can reduce salt levels in the soil, there are some situations where it is either impossible or too costly to attain desirably low soil-salinity levels. In some cases, the only viable management option is to plant salt tolerant crops, or management of farming systems to minimize their impacts. Use of saline water in agriculture requires some changes from standard irrigation practices. This chapter discusses all of the issues.

8.1 Extent of Salinity and Sodicity Problem

Land and water are the two most important natural resources for agricultural development. Land degradation due to soil and water salinity and waterlogging is threatening the sustainable use of the resources. Saline and sodic (alkali) soils can significantly reduce the value and productivity of affected land. Excess salt ions in the soil or from irrigation water can cause salinity and limit agricultural production. Saline soils are found almost throughout the world. Soil salinization has been identified as a major process of land degradation. Based on the FAO/UNESCO Soil Map of the World, the total area of saline soils is about 394 million hectare and that of sodic soil is about 434 million hectare (ha). These are not necessarily arable but cover all salt-affected lands at global level. Out of the current 232 million ha of irrigated land, 45 million ha are salt-affected soils (about 19.5%) (Ghassami et al., 1995); and out of the 1,500 million ha of dryland agriculture, about 32 million ha (2.1%) are salt-affected at varying degrees.

The majority of human-induced land salinization (secondary salinization) in the world is associated with irrigation. Countries suffering the effects of secondary salinization include Australia, United States (particularly the states of Montana, North Dakota, and South Dakota), Canada (the prairie provinces of Manitoba, Alberta, and Saskatchewan), Thailand, South Africa, Turkey, India, Pakistan, and Argentina.

8.2 Development of Soil Salinity and Sodicity

8.2.1 Causes of Salinity Development

Soil salinization is the accumulation of free salts in soil to such an extent that it leads to degradation of soils and vegetation. The salts originate from the natural weathering of minerals or from fossil salt deposits left from ancient sea beds. Salts accumulate in the soil of arid climates as irrigation water or groundwater seepage evaporates, leaving minerals behind. Irrigation water often contains salts picked up as water moves across the landscape, or the salts may come from human-induced sources such as municipal runoff or water treatment. As water is diverted in a basin, salt levels increase as the water is consumed by transpiration or evaporation.

Salinity problems are caused from the accumulation of soluble salts in the root zone. The problems include high total salts, excess exchangeable sodium, or both. Generally, it is a natural process, and may results from the following:

- high levels of salt in the soils
- irrigation with saline water
- shallow saline groundwater
- landscape features that allow salts to become mobile (movement of water table)
- climatic trends that favor accumulation of salts
- land-use practice and rainfall pattern
- man-made activity
- urban area
- seepage salting

8.2.1.1 Salts in Soil

Salt is a natural element of soils and water. Originally salts came from the weathering of rocks that contain salt. In some areas (for example in Australia), salinity is an inherent situation (enormous amounts of salts are stored in the soils). However, human practices have increased the salinity of top soils by bringing salt to the surface through disrupting natural water cycles, by allowing excess recharging of groundwater and accumulation through concentration. Sodicity is generally dependent on the parent material from which a soil is formed and is found on older soils where there has been sufficient weathering of clay minerals to cause a dominance of sodium. The most common salts in soil include chlorides, sulfates, carbonates and sometimes nitrates of calcium, magnesium, sodium, and potassium. The chemical composition, common name, and equivalent weight of common salts are given in Table 8.1.

8.2.1.2 Salinity from Irrigation Water

Salinity problems can also occur on irrigated land, particularly when irrigation water quality is marginal or worse. It has been estimated that slightly more than one-fourth

Table 8.1 Common salts in saline area and their equivalent weight

Salt name	Chemical formula/ composition	Common name	Equivalent weight (g)	Cation (+)	Anion (−)
Sodium chloride	NaCl	Table salt	58.45	Sodium	Chloride
Sodium sulfate	Na_2SO_4	Glauber's salt	71.03	Sodium	Sulfate
Magnesium sulfate	$MgSO_4$	Epsom salt	60.18	Magnesium	Sulfate
Sodium bicarbonate	$NaHCO_3$	Baking soda	84.01	Sodium	Bicarbonate
Sodium carbonate	Na_2CO_3	Sal soda	53.00	Sodium	Carbonate
Calcium sulfate (with 2 molecules of water)	$CaSO_4.2H_2O$	Gypsum	86.09	Calcium	Sulfate
Calcium carbonate	$CaCO_3$	Calcite (lime)	50	Calcium	Carbonate

of irrigated farmland in the United States is affected by soil salinity. Salinity from irrigation can occur over time wherever irrigation occurs, since almost all water (other than natural rainfall) contains some dissolved salts. When the plants use the water, the salts are left behind in the soil and eventually begin to accumulate.

8.2.1.3 Shallow Saline Groundwater

Salinity from drylands can occur when the water table is between two to three meters from the surface of the soil and the groundwater is saline (which is true in many areas). The saline water is raised by capillary action and salts are concentrated on the surface resulting from evaporation of water (salts are left behind). In some circumstances, the process is favored by land use practices allowing more rainwater to enter the aquifer than it could accommodate. For example, the clearing of trees for agriculture is a major reason for drylands in some areas, since deep rooting of trees has been replaced by shallow rooting of annual crops.

There are two vectors acting on the salt-infected water, the upward pull from evaporation and capillary action and the downward force of infiltration. Whenever the net flow is up, a saline soil will result. Any factor that increases downward infiltration in a recharge area or any practice that increases evaporation and decreases downward percolation in a discharge area will increase the potential for having a saline soil.

Since soil salinity makes it more difficult for plants to absorb soil moisture, these salts must be leached out of the plant root zone by applying additional water. This, in turn, can lead to rising water tables, requiring drainage to keep the saline groundwater out of the root zone. If the water table rises too high, then natural soil evaporation will begin to draw the salts back upward into the soil profile. The problem is accelerated when too much water is added too quickly due to inefficient water use such as over-irrigation, applying more than is required for leaching, using bad estimates of evapotranspiration and poor system design, and is also greatly increased by poor drainage and use of saline water for irrigating agricultural crops.

8.2.1.4 Landscape Feature

In the landscape, soil salinity develops as excess water from well-drained recharge zones moves to and collects in imperfectly to poorly drained discharge zones. The buildup of excess water brings dissolved salts into the root zone of the discharge area (Fig. 8.1).

Fig. 8.1 Salinity development in discharge area

Salinity is more widely known and refers to the amount of soluble salt in a soil. Unlike sodicity, movements of water influence salinity. Hence, salinity has been related to clearing and irrigation development and results from changes of land use and water movements in landscapes.

8.2.1.5 Climatic Condition

Soil salinity and related problems generally occur in arid or semiarid climates where rainfall is insufficient to leach soluble salts from the soil (or where surface or internal soil drainage is restricted). In humid regions, salt problems are less likely because rainfall is sufficient to leach soluble salts from the soil, but even in higher rainfall areas, salinity problems occur. In some areas with high water tables, problems may occur with surface evaporation leaving salts to accumulate. Rainfall patterns can influence the spread and severity of saline soil. Low rainfall and high evaporative demand can stimulate capillary rise and hence high salt accumulation at the surface.

Under irrigated conditions in arid and semiarid climates, the buildup of salinity in soils is inevitable. The severity and rapidity of buildup depends on a number of interacting factors such as the amount of dissolved salt in the irrigation water and the local climate.

8.2.1.6 Man-Made Activity

Man-made activity can stimulate the salinization process. One of the best examples of excess salination was observed in Egypt in 1970 when the Aswan High Dam was built. The change in the level of groundwater before the construction had enabled soil erosion, which led to high concentration of salts in the water table. After the construction, the continuous high level of the water table led to the salination of the arable land.

8.2.1.7 Salinity from Urban Area

Salinity in urban areas often results from the combination of irrigation and groundwater processes. Cities are often located on drylands, leaving the rich soils for agriculture. Irrigation is also now common in cities (gardens and recreation areas).

8.2.1.8 Seepage Salting/Salty Groundwater Discharge

Saline seeps (or seepage salting) can form where salty groundwater discharges at the ground surface. This is most commonly found in basaltic (open downs) areas. The occurrence of Black Tea Tree (*Melaleuca bracteata*) is a useful guide to seepage areas (both fresh and salt water).

Salts occur naturally in many bedrock deposits and in some deposits on top of the bedrock. Groundwater flowing through these deposits dissolves and transports the salts. Under certain conditions, groundwater discharges at the soil surface. When the water evaporates, the salts are left behind. Over time, the salts accumulate in the groundwater discharge area, forming a saline seep. A white salt crust forms where the salt concentration is very high.

Salt accumulates by water entering the soil at a "recharge area"; this water flows through the soil profile and into aquifers in the bedrock. The water flows through these aquifers accumulating salts into solution, as the water flows through areas that have high concentrations of salt, the salt concentration in the water increases. Eventually, due to bedrock formation, the water in the aquifer is forced close to the soil surface and the water table is elevated. There are different mechanisms that cause an elevated water table. Once the water table is within 2 m of the soil surface, it is possible for the salt infected water to creep up to the surface by capillary action. The location where the water creeps to the surface is called the "discharge area." This upward flow of water, accompanied by evaporation, leaves high concentrations of salt on or near the soil surface.

Factors Affecting the Formation of Saline Seeps

- Generally, recharge areas occur in upper slope positions, while discharge areas occur in lower slope positions.
- Periods of high precipitation and irrigation canal seepage can increase the risk of seep.
- Growing high-moisture-use crops, like alfalfa, in recharge areas can reduce the flow of water to discharge areas.
- Summer-fallow in recharge areas can increase the risk of saline seep formation because there are no crops to take up moisture from deep in the soil.

8.2 Development of Soil Salinity and Sodicity

Knowing the source of salt problems can be helpful in diagnosing and managing salt-affected fields.

8.2.2 Factors Affecting Salinity

The main factors affecting the intensity and extent of salinity are as follows:

- soil moisture content
- soil type
- depth to saline water table
- climate
- irrigation method
- irrigation frequency

8.2.2.1 Soil Moisture

Soil salinity can be difficult to notice from one season to the next because it is influenced by moisture conditions. In wet years, there is sufficient leaching and dissolving of salts so that they are not visible on the soil surface and some crop growth may be possible. However, the excess water received in wet years contributes to the overall salinity problem over time. In dry years, increased evaporation dries out the soil and draws salts up to the soil surface, producing white crusts of salt. In dry years, producers become more concerned with salinity because salts are highly visible and little-to-no crop growth occurs in the affected areas.

Salt-affected soils can occur locally (only a few square meters in size, scattered over a given landscape) or regionally (large areas several hectares in size). Depending on moisture conditions, these areas can increase in size or intensify in salt concentration.

8.2.2.2 Depth to Water Table

Overall outcomes are primarily dependent on the movement, salt content, and depth of groundwater.

8.2.2.3 Method of Irrigation

Irrigation method and frequency have significant effects on growth and yield of crop and also on salinity hazard. Drip irrigation gives the greatest advantages when saline water is used. Sprinkler irrigation may cause leaf burn on sensitive crops with saline water. Damage may be reduced by night irrigation and by irrigating continually rather than intermittently.

8.2.2.4 Soil Factor

Soil properties such as fertility, texture and structure play a role in altering the salinity response function. At high soil fertility levels, three will be a larger yield reduction per unit increase in salinity than under low fertility. On the contrary, the soils having balance nutrients can minimize the effect of salinity than the unbalanced one.

The effects of soil texture and structure are revealed through their influence on the infiltration capacity. For the same evapotranspiration rate, a sandy soil will lose proportionately more water than a clay soil, resulting in a more rapid increase of soil solution concentration. If good irrigation practices are followed, the sandy soil will have to irrigate more frequently, thereby reducing the damage which may be caused by increased concentration. The infiltration capacity of a field is an extremely variable parameter and is a function of soil texture and structure. This variability is largely responsible for the variability of soil salinity.

8.2.2.5 Climate

Three elements of climate – temperature, humidity, and rainfall – may influence salinity response, out of which temperature is the most critical. High temperatures increase the stress level to which a crop is exposed. This may be due to increased transpiration rate or because of the effect of temperature on the biochemical transformation in the leaf. The increase in the stress level results in changes in the salinity response function.

High atmospheric humidity tends to decrease to some extent the crop stress level, thus reducing salinity damage. High rainfall dilutes the salt concentration and increases leaching, thus reducing the salinity level.

8.2.3 Mechanism of Salinity Hazard

It is interesting to know what causes the problem with plant growth when salt is present in the soil solution. When plants take in water, nutrients are also present in the water and are taken up. Plants naturally have salt present in their rooting systems which pull water into the plant from the difference in osmotic pressure. Salt in the soil solution decreases the osmotic potential of the system and slows or even stops the uptake of water. As the difference in concentration decreases, the osmotic potential decreases. When the concentration of salt in the soil increases and approaches that of the plant attempting to grow, the osmotic potential decreases. As the osmotic potential decreases, the movement of soil solution into the plant decreases. Salt-sensitive plants basically perish from water deprivation. The plant will express symptoms of drought even though the soil is saturated with water. The water is present but is unavailable to the plant.

8.2.4 Salt Balance at Farm Level

From the principle of conservation of mass,

Total salt inflow into the soil system − total salt outflow from the soil
= change in salt content in soil

8.3 Diagnosis and Characteristics of Saline and Sodic Soils

If the amount of salt inflow in the soil system is greater than the salt outflow (or removal), then excess salt will build up, and adversely affect crop growth. Salt in the soil system comes from soil mineral, irrigation water, rainwater, and capillary flux. The salt is removed from the system by physical removal (scraping), leaching/deep drainage water, surface drainage water, and by the plants. Mathematically, it can be expressed as

$$(M_s + V_i C_i + V_{rw} C_{rw} + U C_u) - (P_r + V_d C_d + V_{sd} C_{sd} + V_{pw} C_{pw}) = 0 \quad (8.1)$$

where

M_s = amount of salt dissolved from soil mineral of a specified (or unit) soil volume, mg

V_i, V_{rw}, U are the volume of water added to the same volume of soil by irrigation, rainfall, and upward flux (capillary rise), respectively, liter

C_i, C_{rw}, C_i are the concentration of salt in irrigation water, rainwater, and upflow, respectively, mg/l

P_r = amount of salt removed from surface by physical means, mg

V_d, V_{sd}, V_{pw} are the volume of water removed by deep drainage, surface drainage, and plant uptake, respectively, liter

C_d, C_{sd}, C_{pw} are the concentrations of salt in deep drainage, surface drainage, and plant uptake water, respectively, mg/l.

8.3 Diagnosis and Characteristics of Saline and Sodic Soils

8.3.1 Classification and Characteristics of Salt-Affected Soils

Salt-affected soils may be divided into three groups depending on the amounts and kinds of total soluble salts present (estimated by electrical conductivity, EC), soil pH, and exchangeable sodium percentage (ESP) (Table 8.2):

Table 8.2 Generalized classification of salt-affected soils

Soil class	Criteria		
	EC (dS/m)	ESP	pH
Saline soil	>2	<15	<8.5
Sodic (alkali) soil	<2	>15	>8.5
Saline-sodic soil	>2	>15	<8.5

Note: Normal soil – EC < 2 dS/m, ESP < 15

Saline soil is affected by too much total dissolved salts, sodic soil is affected by too much sodium, and saline-sodic soils is affected by an excess of both soluble salts and sodium. It is fairly common for both sodic and saline conditions to occur together in many soils.

8.3.1.1 Saline Soil

Salinity refers to the amount of salts in soils. Saline soils are high in soluble salts but low in sodium. When a solution extracted from saturated soil is 2.0 dS/m or greater, the soil is saline. This soil is commonly referred to as white-alkali soil due to the formation of white crust on soil surface.

Characteristics:

- Electrical conductivity (EC) is greater than 2 dS/m, showing higher soluble salt concentration
- Exchangeable sodium percentage (ESP) is less than 15 (or, sodium absorption ratio, SAR value is less than 13), indicating low level of exchangeable sodium
- The principal cations are Ca^{++} and Mg^{++} with smaller amounts of Na^+ and K^+ ions
- The Cl^- and SO_4^- are the dominant anions with lesser amounts of HCO_3^-
- The pH value of this soil is generally less than 8.5.

Saline soils are formed from the accumulation of salts. There are many different types of salt and they vary in ability to create saline soils. Basically, the more soluble a salt is, the more it can contribute in forming saline soil. Some common salts are listed in Table 8.3 along with their corresponding solubility. Gypsum and lime are present in some saline soils but their low solubility indicates they are not as damaging as other salts such as Glaubers's and Epsom salts.

Table 8.3 Some common salts and their solubility in water

Salt	Chemical formula	Solubility (g/l)
Calcium carbonate (lime)	$CaCO_3$	0.01
Calcium sulfate (gypsum)	$CaSO_4$	2
Magnesium sulfate (epsom salt)	$MgSO_4$	300
Sodium sulfate (Glauber's salt)	Na_2SO_4	160

Saline soils often are in normal physical condition with good structure and permeability. They are characterized by irregular plant growth and salty white crusts on the soil surface. These salts are mostly sulfates and/or chlorides of calcium and magnesium. In many cases, it is possible to see the salts in the soil when they precipitate out of solution. Salts are not always visible. For example, lime is quite difficult to see in the soil. Soil can easily be tested for lime by applying a dilute acid such as HCl to the soil; fizzing and bubbling indicates its presence. Soil salinity is a limitation where plant growth is reduced due to the presence of soluble salts in soil which holds water more tightly than the ability of plants to extract water from the soil. As a result, many plants exhibit symptoms of doughtiness, but the soil is often relatively moist.

Movements of water influence salinity. Hence, salinity has been related to clearing and irrigation development and results from changes of land use and water movements in landscapes. Saline soils are the easiest of the salt-affected soils to reclaim if good quality water is available and the site is well drained.

8.3.1.2 Sodic Soil

Sodicity refers to the amount of sodium in soils. Sodic soils are low in soluble salts but high in sodium. This soil is also referred to as "black alkali soil."

Characteristics:

- Exchangeable sodium percentage (ESP) is greater than 15 (or, sodium absorption ratio, SAR value is greater than 13), reflecting greater amount of exchangeable sodium.
- Electrical conductivity (EC) is less than 2 dS/m, indicating lower amounts of soluble salts.
- The pH value is generally greater than 8.5.
- The dominant cation is Na^{++} with smaller amounts of Ca^{++}, Mg^{++} Na^+, and K^+ ions.
- The HCO_3^- and CO_3^- are the dominant anions with smaller amounts of Cl^- and SO_4^-.

Sodic soils have exchangeable sodium percentages (ESP) of more than 15. This means that sodium occupies more than 15% of the soil's cation exchange capacity (CEC). High *ESP* is manifested by a dispersed clay system, which greatly reduce both air and water entry into the soil system. Thus, sodic soils have restricted water movement, are easy to get stuck in when wet, form lumpy seed beds and often have unfavorable pH for crop growth. They can be reclaimed, but it may be slow and expensive due to the lack of a stable soil structure, which slows water drainage.

Sodic soil develops through a process whereby sodium ions build up in preference to other soil cations (particularly calcium) on the exchange complex of the soil. Increases in soil pH and decreases in calcium and magnesium usually accompany this process. If sodium salts are the dominant type of salts present, a relatively small amount of sodium salts can negatively affect soil structure and create a sodic soil condition but may not necessarily have high electrical conductivities.

If a soil is highly sodic, a brownish-black crust sometimes forms on the surface due to dispersion of soil organic matter. By the time darkened crusts are visible on the soil surface, the problem is severe, and plant growth and soil quality is significantly impacted. Dispersion of soil particles often results in crusting and poor emergence. Plants growing on sodic soils may appear stunted and often show a burning or drying of tissue at the leaf edges, progressing inward between veins.

8.3.1.3 Saline-Sodic Soil

These soils contain large amounts of total soluble salts and greater than 15% exchangeable sodium. The pH is generally less than 8.5. Physical properties of these soils are good as long as an excess of soluble salts is present.

Characteristics:

- Electrical conductivity (EC) is greater than 2 dS/m, showing higher soluble salt concentration.
- Exchangeable sodium percentage (ESP) is greater than 15 or, sodium absorption ratio, SAR value is greater than 13, indicating high level of exchangeable sodium.
- The pH value of this soil is generally less than 8.5.

Dispersion will occur in soils having excess sodium and relatively low Ca and Mg. As a result of clay dispersion, soils will have poor physical properties. This results in a massive or puddled soil with low water infiltration, poor tilth, and surface soil crust formation.

8.3.2 Some Relevant Terminologies and Conversion Factors

Salts: Soluble mineral substances present in soil, rocks, groundwater, and surface water. The salts most commonly occurring in soil are common salt (for example, sodium chloride – NaCl), gypsum, and lime.

Ions: Charged elements that join to form different salts, e.g., sodium ions (Na^+) and chloride ions (Cl^-) form sodium chloride (NaCl) or common table salt.

Soluble salts: Major dissolved inorganic solutes.

Salt loads: The amount of salt generated in the dryland area.

Salt scalds: Salty subsoils that have been exposed by wind erosion.

Saline groundwater: Groundwater containing dissolved salts.

Dryland salinity: All areas of salinity where irrigation is not present.

Ion specific effects: Effect of different ions, e.g., chloride (Cl^-), sodium (Na^+), Iron (Fe), or boron (B) on plants.

Salinity control regions: For the management of salinity, a State may be divided into several regions. The boundaries may be drawn up taking into account salinity provinces, catchment boundaries and other administrative boundaries.

Salinity mitigation: Any activity which reduces the salinity problem. For example, re-vegetation, improved cropping practices, reducing fallow and cultivation, irrigation with freshwater, planting salt tolerant species, and so on.

Salt tolerant: Species able to withstand at high levels of salinity.

Halophytes: Plants which live in salt-affected soils (for example, Saltbush).

8.3 Diagnosis and Characteristics of Saline and Sodic Soils

Sodicity: The presence of a high proportion of sodium ions relative to other cations bound to clay particles.

Calcareous: A soil containing significant amounts of naturally occurring calcium carbonate, which fizzes when dilute acid is added.

Alkalinity: Soils having alkaline reaction, pH >7.0. Problems usually appear at pH >7.8 as nutrient deficiencies.

Gypsum: Calcium sulfate ($CaSO_4 \cdot 2H_2O$) used to supply calcium and sulfur to improve sodic and saline-sodic soils.

Leaching: The flushing of salts from soil by the downward movement of water.

Leaching fraction: The proportion of the water entering the soil that soil conditions will allow to pass below the root-zone.

8.3.2.1 Electrical Conductivity (EC)

Electrical conductivity measures the ability of a water sample to conduct electricity. This relates to the amount of total soluble salts (TSS) in the water sample. Pure water has very low conductivity. As TSS increases, water becomes more conductive. Although different dissolved substances affect conductivity differently, the average TSS = $0.66 \times$ EC. The SI unit for EC is deci-Siemens per meter (dS/m). However, milli-mhos per centimeter (mmhos/cm) and micro-mhos per centimeter (μmhos/cm) are frequently used.

EC is a measure of salinity resulting from all the ions dissolved in a water sample or saturated paste. This includes negatively charged ions (e.g., Cl^-, NO_3^-) and positively charged ions (e.g., Ca^{2+}, Na^+).

8.3.2.2 Total Soluble Salts (TSS) or Total Dissolved Salts (TDS)

TSS refers to the total amount of salt dissolved in the soil extract expressed in parts per million (ppm). The salts include substances that form common table salt (sodium and chloride) as well as calcium, magnesium, potassium, nitrate, sulfate, and carbonates.

Cation Exchange Capacity (CEC): The amount of exchangeable cations that a soil can adsorb at a specific pH, expressed as milli-equivalents per 100 g of soil.

8.3.2.3 Exchangeable Sodium Percentage (ESP)

ESP is a measure of soil sodicity or sodium hazard. ESP is the sodium fraction adsorbed on soil particles expressed as a percentage of cation exchange capacity. It is calculated by dividing the exchangeable sodium over the Cation Exchange Capacity (or CEC) and multiplying the product by 100, i.e.,

$$ESP = (\text{exchangeable sodium/cation exchange capacity}) \times 100$$

8.3.2.4 Sodium Adsorption Ratio (SAR)

It is another method for measuring sodicity. It is a relation between soluble sodium and soluble divalent cations (normally Ca and Mg) in a soil-water solution, which can be used to predict the exchangeable sodium fraction of soil equilibrated with a given soil-water solution.

The concentration of sodium ions relative to calcium and magnesium ions in the soil is called the sodium adsorption ratio (SAR). It is calculated as

$$\text{SAR} = \frac{\text{Na}^+}{\sqrt{\frac{\text{Ca}^{++} + \text{Mg}^{++}}{2}}} \tag{8.2}$$

where the concentrations are expressed in millequivalents per liter (meq/l) and are obtained from a saturated paste soil extract. By definition, SAR has unit of $(\text{meq/l})^{1/2}$, but following conventional usage, SAR values in the text will not include this unit.

To convert ppm or mg/l into meq/l, divide the concentration value by the respective equivalent weight. For Na^+, Ca^{2+}, and Mg^{2+}; it is 23, 20, and 12.2, respectively.

Soil water extracts with SAR values >13 are indicative of a soil with a sodium problem. Even at SAR values >8, there are instances when relatively high concentrations of Na relative to Ca and Mg results in dispersion of clay particles, soil structural breakdown, and soil pore blockage which reduces infiltration rates and increases erosion potential.

8.3.2.5 Salinity

The amount of dissolved salts in soil or water. It is usually described as the concentration of dissolved salt in milligrams per liter (mg/l) and measured as the capacity of a solution to carry an electric current (EC).

8.3.2.6 Salinization

The accumulation of salts in the root zone and on the soil surface is termed salinization. It may be a result of rising water tables and the upward movement of saline groundwater. Salinization causes destruction or reduction of vegetation, soil structure breakdown, erosion, silting up of streams, and pollution of water resources.

8.3.2.7 Salinity and Osmotic Potential

Salinity is correlated to osmotic potential, which is the primary cause of plant damage, and death. Osmotic potential causes dissolved constituents in soil to try to retain water, so plants have to compete with salt for water. The presence of excessive salts in soils causes plants to prematurely suffer drought stress even though substantial water may be present in the soil. Osmotic potential is a direct result of the combined concentrations of dissolved Na, Ca, K, and Mg cations, and Cl^{-1}, SO_4^{-2}, HCO_3^{-1}, and CO_3^{-2} anions which are common constituents in salty water.

8.3.2.8 Unit of EC

The SI unit of conductivity is "Siemens" (symbol "S") per meter. The equivalent non-SI unit is "mho" and, 1 mho = 1 Siemens. Thus,

1 mmho/cm = 1 dS/m
(one milli-mho per centimeter = One deci-Siemen per meter)

8.3.2.9 Unit Conversion

Conductivity to mmol per liter:
mmol/l = 10 × EC (EC in dS/m), for irrigation water and soil extracts in the range 0.1–5 dS/m.
Conductivity to osmotic pressure in bars:
OP = 0.36 × EC (EC in dS/m), for soil extracts in the range of 3–30 dS/m.
Conductivity to mg/l:
mg/l = 0.64 × EC (in μmhos/cm)
= 640 × EC (in dS/m), for waters and soil extracts having conductivity up to 5 dS/m
mg/l = 800 × EC (in dS/m), for waters and soil extracts having conductivity >5 dS/m
mmol/l (chemical analysis) to mg/l:
Multiply mmol/l for each ion by its molar weight and obtain the sum.
Ions to EC:
Total sum of cations and anions (meq/l) = 10 × EC (dS/m), 0.1 < EC < 5.0 dS/m.

8.3.3 Diagnosis of Salinity and Sodicity

Proper diagnosis is critical to any successful problem correction. The first step toward management of any salt-affected soil is an assessment of the soil including the soil profile. Effective management of salinity or sodicity problem requires correct diagnosis of the problem. A salt-alkali soil test should be established to see whether the soil is saline, sodic, saline-sodic, or not affected by salts. An examination of the soil profile along with soil survey information will help determine soil permeability characteristics, which are important in leaching salts. In some cases, it may be necessary to facilitate drainage using tile drains or open ditches to allow successful reclamation.

Diagnosis of salt-affected soil can be performed by

- field observation, and/or
- laboratory examination

8.3.3.1 Diagnosis by Field Observation

Salinity may be a local problem, so a site visit is recommended regardless of the availability of detailed soil information. Some visual symptoms can be used to help diagnose these problems, but ultimately soil and irrigation water analyses are the best ways to make an accurate diagnosis. When salinity is suspected to be caused by a high water table, you may be able to measure groundwater depth by boring holes with an auger. If free water collects in holes less than 1.5–2 m deep, a drainage problem is indicated.

Field Symptoms

There are many visual indicators of salinity in affected areas. The absence of crop or poor crop in seeded areas can be a good indicator that salts are present. Another indicator of soil salinity is the presence of a "Bathtub ring" around sloughs or depressions; this is an area around a slough where it is easy for salts to accumulate.

Check for poor crop growth, light gray, or white colors on soil surface, areas that take longer to dry and growth of salt-tolerant weeds (foxtail barley, kochia, Russian thistle, etc.). Vegetation is a good indicator of the location of these soils. Dawson Gum (*Eucalyptus cambageana*), False Sandalwood (*Eremophila mitchellii*), and Poplar Box (*Eucalyptus populnea*) are the main trees found on sodic soils. Plants growing in saline soils may appear water stressed. In some cases, a white crust is visible on a saline soil surface (Fig. 8.2).

Salt-affected soils may inhibit seed germination and cause irregular emergence of crop seedlings, particularly in sensitive crops such as beans or onions. Plants sprinkler irrigated with saline water often show symptoms of leaf burn, particularly on young foliage. If a soil is highly sodic, a brownish-black crust sometimes forms on the surface due to dispersion of soil organic matter. By the time darkened crusts are visible on the soil surface.

High soil pH sometimes may confuse with the symptom of salinity or sodicity, as yellow stripes on middle-to- upper leaves or dark green or purple coloring of the lower leaves and stems can be signs of high soil pH. Symptoms of salinity, sodicity, and high soil pH are given in Table 8.4.

8.3.3.2 Diagnosis by Laboratory Determination

Soil testing labs typically evaluate pH and EC (electrical conductivity) as part of a routine program. If the pH is high (>8.5), sodium adsorption ratio (SAR) or ESP should also be calculated. Proper soil sampling depth to diagnose saline conditions

8.3 Diagnosis and Characteristics of Saline and Sodic Soils

Fig. 8.2 View of salinity affected crop fields: (**a**) salt appears at the surface (*top one*), (**b**) rice plants are partially damaged due to saline water (*bottom one*)

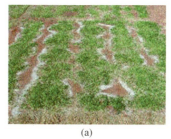

(a)

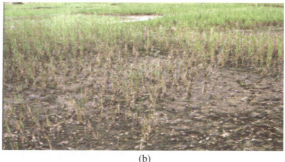

(b)

Table 8.4 Potential symptoms of salinity, sodicity and high soil pH

Soil problem	Potential symptoms
Saline soil	• White crust on soil surface
	• Water stressed plants
	• Leaf tip burn
Sodic soil	• Dark powdery residue on soil surface
	• Poor drainage, crusting or hardsetting
	• Low infiltration rate; high runoff and erosion
	• Stunted plants with leaf margins burned
Saline-sodic soil	• Generally, same symptoms as in saline soil
High soil pH	• Stunted yellow plants
	• Dark green to purplish plants

depends upon cropping system and the nature of the situation. However, crop salinity thresholds are based on the average salinity level of the active root zone. If the site in question is irrigated, water sample should be collected for analysis as well. High levels of salts and sodium may come from irrigation water, a high water table, manure or fertilizer inputs, or from the soil parent material. To manage the problem effectively, you need to know the source of the salts. Although 2.0 dS/m is used as a threshold EC to define saline soils, many crops may show symptoms and reduced yields at much lower ECs (1.5–2.0 dS/m).

Other information, including soil texture, cation exchange capacity, type of clays present, calcium carbonate content, organic matter, depth to groundwater, and soil profile information, will help in planning a reclamation program.

Tests frequently employed in diagnosing salinity/sodicity problems are summarized below:

- pH
- Electrical Conductivity (EC)
- Sodium Adsorption Ratio (SAR)
- Exchangeable Sodium Percentage (ESP)
- Cation Exchange Capacity (CEC)
- Lime Estimate
- Total Dissolved Solids (TDS – water only)
- Anions and cations: e.g., Ca^{2+}, Mg^{2+}, Na^+, Cl^-, SO_4^{2-}, CO_3^{2-}, HCO_3^-
- Available gypsum ($CaSO_4 \cdot 2H_2O$) and gypsum requirement
- Soil texture estimate
- EC of Irrigation water

Soil Sampling

The soil samples should represent the salt-affected areas only, instead of the entire field. The top 10 cm or plow depth should be sampled if the salt accumulation is induced by irrigation or other farming practices. On the other hand, samples from 0 to 15 cm and 30 cm increment below the surface need to be collected to evaluate the extent of brine contamination. A similar sample from nearby normal soil is often useful as a benchmark for comparison. Multiple subsamples (15–20 cores with a soil probe) are needed to make a representative composite.

Composite soil sampling may not provide an accurate measurement of the overall salinity level of a field. To assess a suspicious area of a field for salinity, take soil samples to 0.6 m from the affected area and an adjacent non-affected area. In many cases, comparison of soil samples from the affected area and surrounding normal appearing areas will be beneficial in diagnosing the problem. If you wish to map an entire field for its salinity status, there are indirect measurements using specialized equipment that can be used.

To determine the source of salinity, dig a pit in the soil of both the affected and unaffected areas, check for salt particles and check for carbonates using dilute hydrochloric acid (HCl). Since soluble salts are more mobile than carbonates, this can be used to determine the net direction of water movement. Observation wells and piezometers may be installed to identify recharge and discharge areas.

In general, when sampling for salinity or sodicity problems, collect a composite sample of several soil cores from the active root zone or the zone you plan to manage, concentrating on the areas that appear most impacted. High pH, salt or sodium levels are rarely uniformly distributed across fields. Areas of the field suspected of these problems should be mapped and sampled separately to fully understand the severity of the problem. Be sure to tell the laboratory that you suspect a problem and ask that they analyze for SAR or ESP and perform a gypsum test, if necessary. Tentative test required for different problems are summarized in Table 8.5.

8.3 Diagnosis and Characteristics of Saline and Sodic Soils

Table 8.5 Test required for different problem identification

Tentative problem	Test required
Suspect a salinity or sodicity problem	Soil EC, pH, and SAR
Suspect poor quality irrigation water	TDS, EC, SAR, Na^+, Cl^-, B, HCO_3^-
A salinity problem exists and want to monitor or calculate leaching requirement	Soil EC, irrigation water EC, consider spatial mapping of field EC
A sodicity problem exists and want to calculate gypsum requirement	CEC, ESP and/or SAR, lime estimate

8.3.3.3 Determination of Intensity of Salinity and Sodicity Hazard

Irrigators should determine the potential salinity and sodium hazard of their water and soil before planning crop cultivation or any successful management planning. The salt content of a soil can be estimated by measuring the electrical conductivity (EC) of the saturated soil extract.

Estimation of Salinity

Salinity is estimated by measuring electrical conductivity, EC. The ability of a solution to carry current is called electrical conductivity. The electrical conductivity of a solution is proportional to its soluble salt content (In an aqueous solution, the EC increases with the concentration of ions and hence the concentrations of total dissolved salts). The more salts in the sample, the greater is its electrical conductivity. Similarly, the lower the EC reading, the lower the salinity. The severity of salinity or sodicity is based on the value of EC or ESP, as described in the preceding section.

Measuring EC

Take a sample of soil, add enough water (salt-free) to the soil sample to completely saturate it, stir it for some time, and extract water from the saturated soil using a vacuum pump. Then measure the electrical conductivity (EC) of the saturation extracts using an EC meter or other type of instrument. In the absence of vacuum pump, some recent devices (handy EC meters, under different brand/commercial name) allow inserting and measure EC directly in the soil saturated paste.

Available hand-held devices are much quicker and easier than sending away a soil sample to a laboratory.

Measuring/Estimating ESP, SAR

Estimation of ESP

A second important measure is the amount of exchangeable sodium, determined by extracting the soil with 1 M ammonium acetate and measuring the amount of sodium in the extract. The results of this extraction must be corrected for soluble sodium measured but not exchangeable. Once this correction is made, the results are expressed as percent exchangeable sodium.

Estimation of SAR

After determining the ionic concentrations of Na, Mg, and Ca; *SAR* value can be calculated using standard formula as described in an earlier section.

Interpretation of the Results

The general interpretation of the results is summarized in Tables 8.6 and 8.7.

Table 8.6 Interpretation of salinity hazard based on EC value

EC of saturation extract (dS/m)	Salt rank	Interpretation
0–2	Low	Very little chance of injury on all plants
2–4	Medium	Sensitive plants and seedlings of other plants may show injury
4–8	High	Most non-tolerant plants will show injury
8–16	Excessive	Salt tolerant plants will grow, other crops will show severe injury
>16	Very excessive	Very few plants will tolerate and grow

Table 8.7 Interpretation of sodicity hazard based on Na, SAR, ESP, and Ca:Na ratio

Parameter	Value	Alkali rank	Interpretation
Sodium percent (%)	0–10	Low	No adverse effect on soil
	>10	Excessive	Soil dispersion results in poor soil physical condition and plant growth is restricted
SAR	> 13		Indicates sodic soil
ESP	>15		Indicates sodic soil
Ca:Na	< 10:1		Sodium may begin to cause soil structural problems

8.3.4 Salinity Mapping and Classification

8.3.4.1 Salinity Mapping

The purpose of classifying and mapping salinity is to identify various types and extent to quantify the issue and move toward recommended management practices. The maps provide an excellent extension tool to local agents and let farmers know the extent of salinity issues beyond their property. Analyses of the data allow policy

8.3 Diagnosis and Characteristics of Saline and Sodic Soils

makers to design and target programs specific to the extent and the nature of the problem.

Salinity mapping assists in

- targeting salinity control activities
- focusing on the areas with the best opportunity for cost-effective control
- creating a database with saline areas specified by the legal land location
- generating the color-coded salinity maps from the database
- calculating the areas affected by each salinity type
- applications at the provincial and farm levels
- creating regional and provincial salinity maps and database
- monitoring changes in salinity from provincial to farm level
- targeting provincial soil salinity control programs

The salinity databases have a number of applications. Most importantly, the municipality can use the information to target its salinity control programs. Because the information is in digital format, it is simple to calculate statistics on the areas affected by the different salinity types. These statistics allow the municipality to focus on those saline areas which can be easily controlled with low-cost methods.

8.3.4.2 Salinity Classification

The classification is helpful in salinity control programs because the cause of the salinity determines which control practice is appropriate. Salinity may be classified based on hydrogeology, surface water flow, geology, topography, and soils:

A: Classification based on causes/sources of salinity
B: Classification based on the development mode
C: Classification based on salt content or EC

A: Classification Based on Causes/Sources of Salinity

There are two different kinds of salinity visible around the world:

1. Groundwater fed salinity or Seasonal salinity
2. Saline soil or Permanent salinity

Groundwater Fed Salinity or Seasonal Salinity

In this category, salinity is groundwater fed with the rising and lowering of groundwater. This happens mainly in low-rainfall area (dry area) and to seasonally dry soil, caused by capillary rise of saline shallow underground water. Example is: the eastern part of North Dakota, USA, which become saline in the dry part of the year.

In wet years or wet seasons, there is sufficient leaching and dissolving of salts so that they are not visible on the soil surface and some crop growth may be possible. However, the excess water received in wet years may contribute to the overall salinity problem over time. In dry years, increased evaporation dries out the soil and draws salts up to the soil surface, producing white crusts of salt. In dry years, producers become more concerned with salinity because salts are highly visible and no crop growth occurs.

Saline Soil or Permanent Salinity

In this case, the soil is saline by formation, i.e., geologically formatted from saline rocks, having deep saline profile, or land deposited from sea loads. For example, coastal area (southern part) of Bangladesh. During wet season, crops are also grown here, due to dilution of salt in surface soil. But if the rainfall is low, crops are affected by salinity.

B: Classification Based on the Development Mode

Based on the mode of development, salinity can be categorized as

1. Primary salinity
2. Secondary salinity

Primary Salinity

This kind of salinity is developed naturally. It results from the long term continuous discharge of saline groundwater. Saline soils due to primary salinity often have high EC values. These soils are not suited to crop production without special management. The best course of action for primary salinity is to leave the affected area in its natural state. If the land has been tilled, salt-tolerant vegetation should be established.

Secondary Salinity

This kind of salinity is human induced, and is the result of human activities that have changed the local water movement patterns of an area. Soils that were previously nonsaline have become saline due to changes in saline groundwater discharge.

Saline soils due to secondary salinity may have lower EC values and may be improved with management. In order to optimize production in saline, discharge areas, water must be utilized in the adjacent, nonsaline recharge areas. This will decrease the movement of excess water from recharge areas to discharge areas.

C: Classification Based on Salt Content or EC

Sensitive crops may exhibit negative effects of salinity at levels <2 dS/m. An EC of 4 was considered in the past as a general salinity rating for traditional annual crops (wheat, canola) which are not significantly affected by soil salinity levels below 4 dS/m. The rating systems of salinity with greater detail are as follows:

i. Nonsaline (0–2 dS/m)
ii. Slightly saline (2–4 dS/m)
iii. Weakly saline (4–8 dS/m)
iv. Moderately saline (8–15 dS/m)
v. Strongly saline (>15 dS/m)

Soil salinity is typically described and characterized in terms of concentrations of soluble salts. While a general classification scheme based upon soil EC and SAR has been accepted, be aware that salinity problems occur on a continuum rather than at a given threshold. Appropriate management strategies should be evaluated from the following and other measurements:

Saline = high salt content (EC > 2 dS/m)
Sodic = high sodium content (SAR > 13)
Saline-sodic = high salt, high sodium
Basic = high pH (problems usually start at pH > 7.8)

8.4 Impact of Salinity and Sodicity

8.4.1 Impact of Salinity on Soil and Crop Production

The consequences of salinity are detrimental effects on plant growth and final yield damage, damage to infrastructure (roads, bricks, corrosion of pipes and cables), reduction of water quality for users, sedimentation problems and soil erosion. When crops are too strongly affected by the amounts of salts, it disrupts the uptake of water into roots and interferes with the uptake of competitive nutrients. Salinity causes damage to the soil structure, including dispersion, erosion, and waterlogging. Salinity reduces plant available water content (PAWC) and root depth for dryland agriculture.

The salinity of the soil water affects swelling. Clay swelling and dispersion are the two mechanisms which account for changes in hydraulic properties and soil structure. Swelling that occurs within a fixed soil volume reduces pore radii, thereby reducing both saturated and unsaturated hydraulic conductivity. Swelling results in aggregate breakdown, or slaking, and clay particles movement, which in turn leads to blockage of conducting pores. Clay swelling occurs because clay particles imbibe water to lower the exchangeable cation concentration near the negatively charged surfaces of the clay. Divalent calcium ions are more strongly absorbed to clay surfaces than monovalent sodium. Consequently, calcium clays swell les than sodium clays.

Salinity has five main impacts on plant nutrition:

(i) The first is the osmotic effect. As salts accumulate in soil, the soil-solution osmotic pressure increases. When this happens, the amount of water available for plant uptake decreases and plants exhibit poor growth and wilting even though the soil is not dry. Thus, salts in the soil water solution can reduce evapotranspiration by making soil water less available for plant root extraction. When the salt concentration in the soil solution exceeds the salt concentration inside plant roots, this causes water to move out of the roots. In extreme cases plants can be left severely dehydrated.

(ii) The second impact of salinity on plant nutrition is a specific ion effect which has two outcomes. Saline soil can cause toxic accumulation of ions – sodium and chloride, inside plant cells, which hinder vital physiological processes. Salinity causes reduced root growth due to osmotic pressure and toxicity. Thus, high salt concentrations limit the plant root's ability to take up water and nutrients, which restricts crop growth and reduces yields.

(iii) The third is the interfering uptake of competitive nutrients. Some ions have an antagonistic effect on the uptake of other ions or elements. Table 8.8 shows some of the specific ion effects that can occur in saline soil.

Table 8.8 Specific ion effect

Ions accumulating in plant cells	Ions inhibited from entering plant
Ca^{2+}	K^+
Na^+	Zn^{2+}
Cl^-	NO_3^-
SO_4^{2-}	PO_4^{3-}

(iv) The fourth impact of salinity on poor nutrition of crop is the slow mineralization of organic nitrogen in saline soil.

(v) The fifth impact of salinity is that it inhibits plant growth as a result of stomatal closer, which reduces the CO_2 to O_2 ratio in the leaves and inhibits CO_2 fixation. As a result, the rate of leaf elongation, enlargement and cell division is reduced.

The availability of phosphorous in saline soil depends on the pH and the degree of salinity of the soil. In general, toxicity of iron may develop in saline soil; and deficiency of phosphorus and zinc may occur. The soil may become acidic. From various research findings, it is revealed that for a particular salinity level, the effect of a single salt is different from that of the multiple salts.

8.4.2 Impact of Sodicity on Soil and Plant Growth

Sodicity in soils has a strong influence on the soil structure of the layer in which it is present. Sodic soils are low in total salts but high in exchangeable sodium.

The combination of high levels of sodium and low total salts tends to disperse soil particles, making sodic soils of poor tilth. These soils are sticky when wet, nearly impermeable to water and have a slick look. As they dry, they become dense, hard, cloddy, and crusty. This dense layer is often impermeable to water and plant roots. Dispersion occurs when the clay particles swell strongly and separate from each other on wetting. In addition, scalding can occur when the topsoil is eroded and sodic subsoil is exposed to the surface, reducing plant available water content and increasing erodibility.

Crop growth and development problems on sodic soils can be nutritional (sodium accumulation by plants), associated with poor soil physical conditions, or both. Plants on sodic soils usually show a burning or drying of tissue at leaf edges, progressing inward between veins. General stunting is also common. Crops differ in their ability to tolerate sodic soil, but if sodium levels are high enough, all crops can be affected. Problems of sodic soil include loss of grazing, salty water supplies, death of trees, erosion, siltation, and so on.

Sodic soils may impact plant growth by

i. specific toxicity to sodium sensitive plants,
ii. calcium deficiencies or nutrient imbalances caused by excessive exchangeable sodium
iii. high pH, and
iv. dispersion of soil particles, resulting in poor physical conditions in the soil.

If a soil is highly sodic, a brownish-black crust sometimes forms on the surface due to dispersion of soil organic matter. By the time darkened crusts are visible on the soil surface, the problem is severe, and plant growth and soil quality is significantly impacted. Dispersion of soil particles often results in crusting and poor emergence of seed.

8.5 Crop Tolerance to Soil Salinity and Effect of Salinity on Yield

Excessive soil salinity reduces the yield of many crops. This ranges from a slight crop loss to complete crop failure, depending on the type of crop and the severity of the salinity problem. Although several treatments and management practices can reduce salt levels in the soil, there are some situations where it is either impossible or too costly to attain desirably low soil-salinity levels. In some cases, the only viable management option is to plant salt tolerant crops.

8.5.1 Factors Influencing Tolerance to Crop

Tolerance of plants to soil salinity is not a fixed characteristic of each species or a variety, but may vary with the growth stage and environmental conditions. The factors influencing tolerance to crop are

- Variety or species
- Growth stage
- Environment
- Salinity-stress history or sequence of salinity stress
- Rootstocks
- Soil type
- Moisture condition

8.5.1.1 Varietal Differences in Salt Tolerance

Crops differ in the ability to tolerate salinity, but if levels are high enough (> 16 dS/m), only tolerant plants will survive. Differences in varietal tolerance to salinity and other adverse soil conditions have been known to exist for decades but it is only in the latest decades that serious efforts have been initiated to exploit the genetic potential of salt-tolerant crop varieties through different breeding programs. Wheat, soybean and barley are genetically moderately tolerant (~EC = 8 dS/m) to salinity (Ortiz-Moasterio et al., 2002; Ali and Rahman, 2008a). Some wheat cultivars can produce reasonable yield at irrigation salinity of 14 dS/m (Ali and Rahman, 2008a).

In recent years intensive efforts have been made at the International Rice Research Institute at Los Banos in the Philippines to breed varieties for tolerance to various adverse soil conditions and many advanced lines in IRRI's breeding program show tolerance for one or more adverse soil factors. It is apparent that breeding crop varieties tolerant to salinity offers significant opportunities for better management of areas where salinity is a perpetual problem.

8.5.1.2 Growth Stage

Although some crops seem to tolerate salinity during seed germination as well as during later growth stages, germination failures are most commonly responsible for poor and spotty stands and bare spots in otherwise cultivated fields. Frequently this is not the result of crops being especially sensitive during germination, but rather is caused by exceptionally high salt concentration in the shallow surface zone where seeds are planted. These high salt concentrations result from the salt that is left behind as the upward moving water is evaporated near the soil surface.

Most plants are more sensitive to salinity during germination than at any other growth stage. However, there are large variations in the sensitivity of germinating seeds to salinity. Beans and sugarbeet are more sensitive to salts at germination than are alfalfa and barley.

Maas and Hoffmann (1977) reviewed data on tolerance of crops in relation to growth stage and showed that the tolerance pattern of barley, wheat and maize was nearly the same as that of rice. Sugarbeet and safflower, on the other hand, were sensitive during germination while the tolerance of soybean could either increase or decrease between germination and maturity depending on the crop variety.

8.5.1.3 Environmental Factors

Climatic conditions greatly influence plant responses to salinity. Normally, the crop species show greater tolerance to salinity at the location where the environment is cold and humid than at the location where the environment is hot and dry. In some parts of India, rice is grown during both the rainy season (kharif) and the dry season (rabi). The yield reduction with increasing salinity is much more in the dry than in the wet season. This is because the dry climate accelerates evaporation, which accumulates salts at the surface. In addition, the higher transpiration loss from the plant reduces plant-water status (but can not recover fully), which hampers plant growth.

8.5.1.4 Rootstocks and Salinity Tolerance

Most fruit crops are more sensitive to salinity than are field, forage or vegetable crops. Grapes, citrus, stone fruits, berries and avocados are all relatively sensitive to salinity. However, certain stone-fruits, citrus and avocado rootstocks differ in their ability to absorb and transport sodium and chloride ions and have, therefore, different salt tolerance.

Most fruit crops are also sensitive to other toxic elements, particularly boron. This ion is present in most irrigation water and in saline soils. It is toxic to many plants at a concentration only slightly in excess of that required for optimum growth. Small quantities of boron absorbed by the roots are accumulated by the leaves and values above 250 ppm result in typical leaf burns.

8.5.1.5 Soil Moisture

As the soil moisture deceases, the concentration of salt, and hence the salinity increases. The vice versa is true for increasing soil moisture. Thus the soil moisture affects the plant stand and growth.

8.5.1.6 Soil Type

For a given moisture content, soil-water potential is higher in clay soil than that of the sandy soil. Thus the plant can easily uptake the moisture from sandy soil compared to clay soil.

8.5.2 Relative Salt Tolerance of Crops

Some crops are more sensitive to salinity than others. Crops such as pulses, row crops and special crops are particularly sensitive to salinity. The salt tolerance of some crops changes with growth stages.

Generally, corn and grain sorghum are intermediate and wheat and alfalfa are more tolerant. Crested and tall wheatgrass and a few sorghum–sudan hybrids are

Table 8.9 Relative tolerance of annual field crops and forages

Electric conductivity (dS/m)	Annual crop	Forage crop
Nonsaline to slightly saline (0–4)	Soybeans, field beans, fababeans, peas, corn	Red clover, alsike, timothy
Moderately saline (4–8)	Canola, flax, mustard, wheat, oats	Reed canary, meadow fescue, intermediate wheat, crested wheatgrass, alfalfa, sweet clover
Severely saline (8–16)	Barley may grow but forages are more productive in severe salinity	Altai wild ryegrass, Russian wild grass, tall wheatgrass, salt meadow grass

very tolerant, able to grow on soils with exchangeable sodium percentages above 50%.

Sensitive crops, such as pinto beans, cannot be managed profitably in saline soils. Table 8.9 shows the relative salt tolerance of field, forage, and vegetable crops. The table shows the approximate soil salt content (expressed as the electrical conductivity of a saturated paste extract (EC_e) in dS/m at 25°C) where 0, 10, 25, and 50% yield decreases may be expected. Actual yield reductions will vary depending upon the crop variety and the climatic conditions during the growing season. Fruit crops may show greater yield variation because a large number of rootstocks and varieties are available. Also, stage of plant growth has a bearing on salt tolerance. Plants are usually most sensitive to salt during the emergence and early seedling stages. Tolerance usually increases as the crop develops. The salt tolerance values apply only from the late seedling stage through maturity, during the period of most rapid plant growth. Crops in each class are generally ranked in order of decreasing salt tolerance. Even though many field crops do not grow well in saline soil there are other cropping options, such as forages, that will grow well. Crop selection is a valuable tool in a salinity management program.

Sunflower is moderately sensitive to soil salinity, where it can tolerate salinity upto EC equals to 1.7 dS/m. Cotton can tolerate higher salinity levels than some other crops.

8.5.3 Use of Saline Water for Crop Production

Supplies of good quality irrigation water are expected to decrease in the future. Thus, irrigated agriculture faces the challenge of using less water, in many cases of poorer quality. The water having salinity and sodicity may require only minor modifications of existing irrigation and agronomic strategies in most cases, there will be some situations that require major changes in the crops grown, the method of water application, and the use of soil amendments.

8.5 Crop Tolerance to Soil Salinity and Effect of Salinity on Yield

Use of saline water in agriculture requires the following principal changes from standard irrigation practices:

(i) Selection of appropriate salt tolerant crops
(ii) Improvements in the water management, and in some cases, the adoption of advanced irrigation technology (such as drip, sprinkler, subirrigation)
(iii) Maintenance of soil physical properties to assure soil tilth and adequate soil permeability to meet crop water and leaching requirements.

Irrigation with saline water may cause some degree of salinization of the soil. This will cause a decrease in crop yield relative to yield under nonsaline conditions. Relative yields decrease with increasing salinity of the irrigation water.

8.5.4 Yield Reduction Due to Salinity

Relative yield (Y_r, in percent) at any given soil salinity (EC) above the threshold level can be calculated by the following straight line equation (adapted from Ayers and Westcot, 1985):

$$Y_r = 100 \times \frac{EC_0 - EC_e}{EC_0 - EC_{100}} \tag{8.3}$$

where

Y_r = Relative yield, in percent
EC_0 = EC of soil at zero yield (i.e., EC_e where $Y = 0$) (dS/m)
EC_e = electrical conductivity (EC) of the soil saturation extract (dS/m) in question
EC_{100} = the salinity threshold level above which the crop yield starts to decline (maximum soil salinity level at 100% yield potential, i.e., the upper EC_e where $Y_r = 100$)

The values of EC_{100} and EC_o for a given crop can be taken from the literature or appropriate figure or table.

The actual yield,

$$Y_{act} = Y_r \times Y_p \tag{8.4}$$

where Y_p is the potential yield under nonsaline condition.

A piece-wise linear model can be established from the principle of the Eq. (8.3):

$$Y_r = 100, \; 0 < EC_e \leq EC_t \tag{8.5a}$$

$$Y_r = S(EC_e - EC_t), \; EC_t < EC_e \leq EC_0 \tag{8.5b}$$

$$Y_r = 0, \; EC_e > EC_0 \tag{8.5c}$$

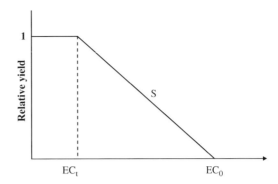

Fig. 8.3 Schematic representation of relative yield versus EC

where

Y_r = relative yield reduction (%)
EC_e = salinity of the saturation extract of field soil (dS/m)
EC_t = threshold salinity level of the soil above which yield starts to decline (maximum EC value for 100% yield potential) (dS/m)
EC_0 = salinity level of soil above which yield is zero (dS/m)
S = slope of the response function between EC_t and EC_0 (Fig. 8.3)
$= \dfrac{1}{EC_0 - EC_t}$

Yields of most crops are not significantly affected where salt levels are 0–2 dS/m. Generally, a level of 2–4 dS/m affects some crops (slightly tolerant). Levels of 4–5 dS/m affect many crops (moderately tolerant) and above 8 dS/m affect all but the very tolerant crops.

8.5.5 Sample Examples

Example 8.1

In a saline area, the EC of a wheat field during its growth period was found 7.0 dS/m. Estimate yield reduction due to salinity, if the wheat cultivar can maintain potential yield up to 4 dS/m, and the yield at EC>22 dS/m is zero.

Solution

Here, $EC_e = 7$ dS/m
$EC_0 = 22$ dS/m
$EC_{100} = 4$ dS/m
We get, relative yield, $Y_r = 100 \times \dfrac{EC_0 - EC_e}{EC_0 - EC_{100}}$
Putting the above values, $Y_r = 83.33\%$
Yield reduction $= (100 - 83.33)\% = 16.67\%$ (Ans.)

Example 8.2

Estimate the actual yield of a rice variety which is grown in a field having EC of soil saturate $(EC_e) = 10$ dS/m, critical salinity level above which the yield reduction starts $(EC_t) = 6$ dS/m, the salinity level at which yield becomes zero $= 20$ dS/m, and potential yield in a nonsaline situation $= 5.0$ t/ha.

Solution

We know, relative yield, $Y_r = 100 \times \dfrac{EC_0 - EC_e}{EC_0 - EC_{100}}$

Here, $EC_e = 10$ dS/m
$EC_0 = 20$ dS/m
$EC_{100} = 6$ dS/m

Thus, $Y_r = 100 \times \dfrac{20 - 10}{20 - 6} = 71.42\% = 0.7142$

$Y_{act} = Y_r \times Y_p = 0.7142 \times 5 = 3.57$ t/ha (Ans.)

8.6 Management/Amelioration of Saline Soil

The mechanisms of formation of saline soils are important but what is even more important is the management of the infected soil and how to slow the formation of salinity in these soils. Remediation of salt contaminated sites and managing high salt content soil for crop production are more effective if the specific nature of the salt in the soil is considered.

8.6.1 Principles and Approaches of Salinity Management

Strategies for management of saline soil lie mainly on the following principles:

1. Reduction of capillary rise of saline underground water
2. Increasing downward movement of fresh water (mainly rainwater) for washout of salts
3. Controlling saline water intrusion
4. Escape of high salinity period by proper crop calendar/crop adaptation
5. Introduce crop cultivars adaptable to developed salinity level

Salt-affected soils exist under a wide range of hydro-geological, physiographical conditions, soil types, rainfall and irrigation regimes, and different socioeconomic conditions. Therefore, a single technique or system may not be applicable to all areas and conditions. Several practices may be required to combine into an integrated system that functions satisfactorily under the prevailing production constraints and soil types to give together economic benefit on a sustainable basis.

The following approaches may be adopted to ameliorate the salinity effect:

- Removing surface crust/scraping
- Controlling saline water
- Hydraulic/engineering practices (e.g., salt leaching, drainage, artificial recharge)
- Chemical practices (e.g., reclamation/treatment of saline soil)
- Irrigation and water management practices
- Physical methods
- Retardation of saline water intrusion
- Biological reduction of salts
- Increasing water use of annual crops and pastures
- Other management practices
  Salinity avoidance, mulching/crop residue application, physical management, organic manuring, row orientation, crop choice/growing salt tolerant crops, appropriate/well adjusted fertilization
- Developing salt tolerant crops
- Policy formulation

8.6.2 Description of Salinity Management Options

8.6.2.1 Removing Surface Salts/Scraping

The total salts and sodium must be reduced before plants can grow normally. The only effective way to reduce salts in soil is to remove them. If the salts are accumulated at the surface and forms a thick layer, the simple and easiest way to remove them is to scrape them out (i.e., take off the layer). Soils having a shallow water table, or a highly impermeable profile; surface flushing of salts from soils that contain salt crusts at the surface may be practiced to ameliorate the saline soil.

8.6.2.2 Control of Saline Water

There are two areas of concern of saline soils; recharge and discharge areas. We should realize that salinity is a water problem in most cases. Excess water at the recharge area causes most salinity problems. Preventing the accumulation and resulting deep percolation of water to the bedrock is important. Excess water in recharge areas may arise as a result of man-made ponding, excess accumulation of snow, excessive summer fallowing, excess annual cropping, and decreased forage and perennial cropping. Control of water accumulation in recharge areas can be established by drainage. Care should be taken when attempting any type of drainage as it may result in causing salinity elsewhere.

There are also different management strategies for saline discharge sites. The goal of discharge management should not be to remove salts completely, rather decrease the salt concentration in the top 30 cm of the soil. Practicing direct seeding in these areas reduces evaporation and increases deep percolation of water. This

8.6 Management/Amelioration of Saline Soil

is achieved because the trash layer insulates the soil and consequently reduces evaporation. The trash layer also decreases water runoff which increases deep percolation.

8.6.2.3 Engineering Practices

There are several engineering approaches to mitigate salinity hazard and facilitate crop production. These include leaching, drainage, artificial recharge through tube well (known as "recharge well"), harvesting rain water at farm ponds and canals, etc.

(A) Leaching

Leaching consists of applying enough good quality water to thoroughly leach excess salts from the soil. Leaching is accomplished on a limited basis at key times (saline sensitive stages of crop) during the growing season, particularly when a grower may have high quality water available. There are two ways to manage saline soils using this approach:

(i) First, salts can be moved below the root zone by applying more water than the plant needs. This method is called the leaching requirement method.
(ii) The second method, where shallow water tables limit the use of leaching, combine the leaching requirement method with artificial drainage (Leaching plus artificial drainage). For proper management of salinity problem, the irrigator should monitor both the soil and irrigation water salinity.

For leaching, water should be added in sequential applications, allowing time for the soil to drain after each application. The quantity of water necessary for leaching varies with initial salt level, desired salt level, irrigation water salinity, and how the water is applied.

Calculation of Leaching Requirement

The fraction of applied or irrigation water which passes through the root zone is termed as leaching fraction (LF). Sometimes it is expressed as the percent of additional water needed above crop water requirements to wash out the salt. Mathematically, it can be expressed as

$$\text{LF} = \frac{D_d}{D_i} \tag{8.6}$$

where

D_d = depth of drainage water (cm)
D_i = depth of infiltrated or irrigated water (cm)

The term "Leaching fraction" and "Leaching requirement" is interchangeably used in the literature. Leaching requirement (LR) is the absolute amount (depth) of water required to leach out the salts below the crop root zone. But in practice, it is expressed as a fraction of net irrigation demand (or ET) of the crop.

Derivation of Equation for Total Irrigation Depth in Leaching Purpose

From the principle of mass balance of salt, we can write

$$V_1 C_1 = V_2 C_2 = \ldots$$

where V_1, V_2 are the volume of solution (water) containing salt for state 1 and 2 respectively, and C_1, C_2 are the concentration of salt in the solutions, respectively. Then we can write

$$\frac{V_1}{V_2} = \frac{C_2}{C_1}$$

For a fixed surface area, we can write

$$\frac{D_1}{D_2} = \frac{C_2}{C_1}$$

where D_1 and D_2 are the depths of the solutions (cm). If we express the salt concentration in terms of EC (electrical conductivity), then

$$\frac{D_1}{D_2} = \frac{EC_2}{EC_1} \tag{8.7}$$

For leaching purpose, the total (or gross) depth of irrigation (AW) consists of net irrigation amount, D_i (or evapo-transpirative demand, ET) plus drainage or leaching amount, D_d. That is,

$$AW = ET \text{ (or } D_i) + D_d$$

$$\text{Or, } D_d = AW - D_i$$

Thus, the Eq. (8.6) can be rewritten as

$$LF = \frac{D_d}{D_i} = \frac{AW - D_i}{D_i} = \frac{AW}{D_i} - 1 \tag{8.8}$$

$$\text{Or, } \frac{AW}{D_i} = 1 + LF$$

Or, $AW = D_i(1 + LF) = ET(1 + LF)$ [since $D_i = ET$]

Considering a field water application efficiency, E_f, the above equation becomes

$$AW = \frac{ET(1 + LF)}{E_f} \qquad (8.9)$$

This equation is more straightforward and versatile to calculate total irrigation depth (or water to be applied both for crop water demand and leaching). For practical purpose, the net irrigation demand or ET demand is calculated as the water required to bring the root zone soil to field capacity.

Equation proposed by Ayers and Westcot (1985) to calculate total depth of water for a particular season or period is

$$AW = \frac{ET}{1 - LR} \qquad (8.10)$$

where

AW = depth of applied water, cm/season
ET = total crop water demand, cm/season
LR = leaching requirement, expressed as fraction (leaching fraction)

The Eq. (8.9) has several merits over the Eq. (8.10):

- it has strong theoretical basis
- it is straightforward
- it is logical that the total applied depth will vary directly with the leaching fraction
- it includes field application efficiency, which is a logical demand

The leaching fraction can be calculated based on the salinity of applied water using the following equation (Rhoades, 1974):

$$LF = \frac{EC_w}{5EC_t - EC_w} \qquad (8.11)$$

where EC_w is salinity of the applied irrigation water, and EC_t is the threshold salinity (average soil salinity tolerated by the crop).

A leaching fraction can be calculated for sprinkler irrigated fields using the equation:

$$\% \text{ Leaching fraction} = \frac{EC_{water}}{2 \times EC_{max}} \times 100 \qquad (8.12)$$

In this equation, EC_{max} is the maximum soil EC wanted in the root zone.

In general, leaching fractions of 10–20% can be used depending on the degree of existing soil salinity and salinity of irrigation water.

Considerations in Leaching Method

(a) In situations where a grower has multiple water sources of varying quality, consider planned leaching events at key salinity stress periods for a given crop.

Most crops are highly sensitive to salinity stress in the germination and seedling stages. Once the crop grows past these stages, it can often tolerate and grow well in higher salinity conditions. Planned periodic leaching events might include a post-harvest irrigation to push salts below the root zone to prepare the soil (especially the seedbed/surface zone) for the following spring. Fall is the best time for a large, planned leaching event because nutrients have been drawn down. However, since each case is site-specific, examine the condition of the soil, groundwater, drainage, and irrigation system for a given field before developing a sound leaching plan.

(b) Apply the leaching fraction to coincide with periods of low soil N and residual pesticide.

(c) Surface irrigators should compare leaching requirement values to measurements of irrigation efficiency to determine if additional irrigation is needed. Adding more water to satisfy a leaching requirement reduces irrigation efficiency and may result in the loss of nutrients or pesticides and further dissolution of salts from the soil profile.

Other Aspects/Options in Leaching Calculation

There are several methods available for estimating leaching requirement (LR) based on different perspectives on how to estimate the average root zone salinity. Differences among methods can be significant, particularly if the root zone salinity is weighted for the amount of water uptake as Rhoades and Merrill (1976) proposed for high frequency irrigation. Leaching requirement obtained using this approach are considerably smaller than for the other methods, indicating that increasing irrigation frequency should be beneficial when irrigating with saline water.

According to Rhoades and Merrill (1976) and Ayers and Westcot (1985), the LR is based on a water uptake distribution of 40:30:20:10% for the first through fourth quarters of the root zone. LRs obtained using equation of Rhoades tends to be low for crops with low threshold salinities. However, the equation is quite useful for quick estimates of LR.

Different Perspectives on Calculating LF

Case 1: No drainage limitation

The depth of infiltrated water can be expressed as the product of average infiltration rate (I_f, cm/day), and irrigation time (t_i), as

$$D_i = I_f \times t_i \tag{8.13}$$

8.6 Management/Amelioration of Saline Soil

Substituting Eq. (8.13) into Eq. (8.6) yields

$$\text{LF} = 1 - \frac{\text{ET}}{I_f \times t_i} \tag{8.14}$$

The above relationship is applicable when no drainage limitation exists.

Case 2: Internal drainage is limiting

Assume that internal drainage is limiting for the prevailing field condition. Now, if the average drainage rate is R_d (cm/d) for the irrigation cycle t_c (days), LF can be expressed as

$$\text{LF} = \frac{D_d}{D_i} = \frac{R_d t_c}{\text{ET} + R_d t_c} \tag{8.15}$$

where ET is the evapotranspiration for the irrigation cycle (cm)

(B) Drainage

Where the subsoils are permeable, natural drainage may be sufficient. Otherwise, drainage system may be needed. Various types of drainage are used throughout the world, such as surface drainage, subsurface drainage, mole drainage, vertical drainage.

> *Surface drainage*: Ditches are provided on the surface so that excess water will run off before it enters the soil.
> *Subsurface drainage*: For the control of groundwater table to a certain depth (safe position for the crop), deep open ditches or tile drains or perforated plastic pipes are installed below ground level.
> *Mole drainage*: Shallow channels left to a bullet-shaped device are pulled throughout the soil. This acts as a supplementary drainage system connected to the main drainage system.
> *Vertical drainage*: pumping out excess water from tubewells when the deep horizons have an adequate hydraulic conductivity.

The depth and spacing of the drainage system should be based on soil type (subsoil strata) and local economic considerations.

Artificial Recharge of Rainwater to Aquifer Through Recharge Well

Through recharge tubewells, excess rainwater can be conveyed to aquifer. Large amount of water can be recharged through recharge well. As a result, the salinity of the aquifer water will be lower due to dilution and become within acceptable range. In flat topography, where aquifer with good transmissibility exists at shallow depth (in the first aquifer of 20–60 m depth), recharge structures with tubewells

Fig. 8.4 Schematic of a recharge well

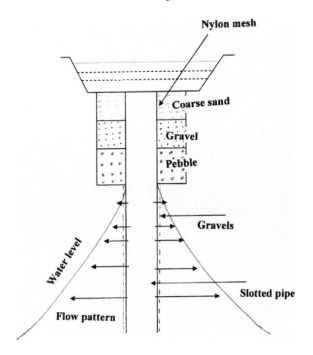

are often better choice than surface storage. In areas having alluvial aquifers with good transmissibility, hydraulic conductivity and specific yield, recharge tubewells improve water quality and availability more quickly than gradual percolation from percolation tank or check dams. Figure 8.4 illustrates the design of a recharge well. It is similar in construction to a discharge well (well for pumping).

Harvesting Rainwater at Farm Pond or Canal

The rain water can be harvested at the excavated farm pond or canal to cultivate dry-season crop, when the soil salinity is high and there is no source of fresh water. Ali and Rahman (2008b) found 20:1 ratio of land area to pond area is sufficient to cultivate low-water demanding dryland crops (other than rice) in saline area.

8.6.2.4 Chemical Practices (Reclamation/Treatment of Saline Soil)

These include chemical amendment and mineral fertilization. Saline soils are dominated by neutral soluble salts and at high salinities sodium chloride is most often the dominant salt although calcium and magnesium are present in sufficient amounts to meet the plant growth needs. Since sodium chloride is most often the dominant soluble salt, the *SAR* of the soil solution of saline soils is also high. Sometimes amendment (gypsum application) is necessary for the reclamation of salt-affected soil depending on the practical importance. Chemical amendment neutralizes exchangeable *Na* and *Na* carbonate. Leaching followed by amendment is needed for removal of salts derived from the reaction of the amendments. Gypsum, sulfur and sulfuric acid are commonly used.

One way to prevent excess salination would be the use of humic acids, especially in regions where too much irrigation was practiced. In soils with excess salts, humic acids can fix anions and cations and eliminate them from the root regions of the plants.

8.6.2.5 Irrigation and Water Management Practices

Water management is the key to successful salinity management. Frequent light watering, using subsurface drip or sprinklers, can help leach excess salts without allowing excess deep drainage to contribute to water table problems. Irrigation management also impact on nutrition of saline soil.

Pre-sowing (or Pre-plant) Irrigation

Most crop plants are more susceptible to salt injury during germination or in the early seedling stages. An early-season application of good quality water, designed to fill the root zone and leach salts from the upper 15–30 cm of soil, may provide good enough conditions for the crop to grow through its most injury-prone stages.

Irrigation Frequency Management

Salts are most efficiently leached from the soil profile under higher frequency irrigation (shorter irrigation intervals). Keeping soil moisture levels higher between irrigation events effectively dilutes salt concentrations in the root zone, thereby reducing the salinity hazard. Most surface irrigation systems (flood or furrow systems) cannot be controlled to apply less than 3 or 4 in. of water per application and are not generally suited to this method of salinity control. Sprinkler systems, particularly center-pivot and linear-move systems configured with low energy precision application (LEPA) nozzle packages, or properly spaced drop nozzles, and drip irrigation systems provide the best control to allow this type of irrigation management.

Drip Irrigation

The drip system provide for opportunities to enhance the use of saline waters in water scarce areas and even on saline soils. It results in considerable savings in irrigation water thus reducing the risks of secondary salinization.

Subsurface Irrigation

Subsurface drip irrigation pushes salts to the edge of the soil wetting front, reducing harmful effects on seedlings and plant roots.

Mixing/Blending of Saline and Fresh Water

Highly saline water can be mixed with good quality or low salinity water to lower the salinity to acceptable/tolerable limit. This approach is only possible where a relatively better quality source is available, and that the better quality supply is not enough to meet the demand. Mixing does not reduce the total solute content but reduces the solute concentration due to dilution.

The salinity of the mixed water or the mixing ratio can be obtained by using the following equation (Adapted from Ayers and Westcot, 1985):

$$C_m = (C_1 \times r_1) + (C_2 \times r_2) \tag{8.16}$$

where

C_m = concentration of mixed water
C_1 = concentration of first category of water
r_1 = proportion of first category of water
C_1 = concentration of 2nd category of water
r_1 = proportion of 2nd category of water
and, $r_1 + r_2 = 1$

The concentration can be expressed as mg/l, ppm, meq/l or, EC_w (dS/m), but must be consistent all throughout.

Alternate/Cyclic Irrigation with Saline and Fresh Water

Fresh and saline water can be applied at alternate sequence to minimize salinity hazard. In such a practice, the nonsaline soil may turn to saline, and a saline soil to more saline. Thus, any attempt to utilize saline water (even with conjunction with fresh water) calls for the leaching of accumulated salts, which may be furnished by natural rainfall, or applying additional irrigation water for leaching (i.e., LF).

Irrigation with Saline Water at Less Sensitive Growth Stages

All growth stages of crops are not equally sensitive to salinity stress. Irrigation with fresh water at sensitive stage(s) and irrigation with saline water (or a mix) at relatively insensitive stage(s) can facilitates optimum plant growth and yield.

Irrigation at Alternate Furrows

Alternate furrow irrigation may be desired for single-row bed systems. This is accomplished by irrigating every other furrow and leaving alternating furrows dry. Salts are pushed across the bed from the irrigated side of the furrow to the dry side. Care is needed to ensure enough water is applied to wet all the way across the bed to prevent buildup in the planted area.

8.6.2.6 Biological Reduction of Salts

In areas with negligible irrigation water or rainfall available for leaching, biological reduction of salts by harvest of high-salt accumulating aerial plant parts may be a strategy to reduce the salt hazard. This can be done by plant uptake and removal. A type of plant that is of particular interest for salinity affected areas is the saltbush, which is able to tolerate saline conditions and draws salt up into its leaves.

8.6.2.7 Other Management Options

Salinity Avoidance

Salinity can be escaped through cultivating/adopting crop cultivars during the period of low salinity potential of the soil, avoiding the high salinity period. Breeding program can be launched to develop such variety.

Mulching/Crop Residue Management

Crop residue at the soil surface reduces evaporative water losses, thereby limiting the upward movement of salt (from shallow, saline groundwater) into the root zone. Evaporation and thus, salt accumulation, tends to be greater in bare soils. Fields need to have 30–50% residue cover to significantly reduce evaporation. Under crop residue, soils remain wetter, allowing fall or winter precipitation to be more effective in leaching salts, particularly from the surface soil layers where damage to crop seedlings is most likely to occur. Plastic mulches used with drip irrigation effectively reduce salt concentration from evaporation.

In the soils showing poor water-transmission property, addition of amendments like sand, rice-husk or rice-straw may be used to improve leaching under limited water availability. The presence of subsurface drains may also help in increasing the leaching of salts in silty clay-loam soil.

Physical Management

Several mechanical methods may be used to improve infiltration and permeability in the surface layer and root zone and thus control saline conditions. These include land leveling, deep plowing, and special planting technique.

Land leveling is needed to achieve a more uniform application of water for better leaching and salinity control. Tillage is useful for soil permeability improvement. Deep tillage is most beneficial on stratified soils having impermeable layer. It loosens the soil aggregates, improves the physical condition of the layer, and increases air spaces and hydraulic conductivity. Special planting procedures that minimize salt accumulation around the seed include planting on sloping beds or raised furrows in single or double rows, which are helpful in getting better stands under saline conditions.

Organic Manuring

Incorporation of organic manure in the soil has two principal beneficial effects on saline soil: improvement of soil permeability and release of carbon dioxide and certain organic acids during decomposition. It also acts as a source of nutrients.

Application of Sand (Sanding)

For fine textured soils, sand may be added to the surface soil to increase permeability. As a result of such application, soil texture is permanently changed. If properly done, sanding can result in improved air and water movement, and enhance leaching by saline/sodic water. Sanding with plowing provides better results.

Row/Seed Bed Management

In addition to leaching salt below the root zone, salts can also be moved to areas away from the primary root zone with certain crop bedding and surface irrigation systems. The goal is to ensure the zones of salt accumulation stay away from germinating seeds and plant roots.

In ridge bed system, salts accumulate at the center of the bed, and the shoulders and corners are relatively salt-free (Fig. 8.5), thus seedling establishment and plant stand are better at the corner and sloping beds or shoulders compared to normal planting.

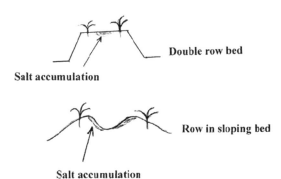

Fig. 8.5 Arrangement of row/seed bed for salinity management

Crop Choice/Growing Salt Tolerant Crop Species or Cultivar

Although several treatments and management practices can reduce salt levels in the soil, there are some situations where it is either impossible or too costly to attain desirably low soil salinity levels. In some cases, the only viable management option is to plant salt-tolerant crops. Sensitive crops, such as pinto beans, cannot be managed profitably in saline soils.

Selecting salt-tolerant crops may be needed in addition to managing soils. Crop selection can be a good management tool for moderately saline soils. Determination of appropriate crops for irrigation should be done on the bases of the salt tolerance of the crop and the salinity of the irrigation water. We have to realize that management practices, irrigation water quality, and environment also affect tolerance. Just as crops differ in tolerance to high salt concentrations, they also differ in their ability to withstand in high sodium concentrations.

Appropriate/Well Adjusted Fertilization

It is important to have good fertility management for the crops grown to have a better chance of successful stand/establishment. Well-adjusted fertilization could improve the yields of crops. There have been only a limited number of studies on the effect of salinity on the nutrition of crops in respect of micronutrients. A disturbed balance in the uptake and composition of major nutrients is bound to influence the plant composition of micronutrients. Besides the generally known toxic effects of boron, there is a need to understand better the behavior of *Fe*, *Mn*, *Zn*, *Cu*, etc., in relation to soil salinity particularly with a view to establishing limiting values – so far only developed for normal soils.

High salinity may interfere with the growth and activity of the soil's microbial population and thus indirectly affect the transformation of essential plant nutrients and their availability to plants.

In saline soils, the availability of *P* is more a function of plant root length and area (which is restricted due to salinity) and the negative effect of excess chlorides on *P* absorption by roots. Application of judicious quantities of P-fertilizers in saline soils helps to improve crop yields by directly providing phosphorus and by decreasing the absorption of toxic elements like *Cl*.

On moderately saline soils, the application of potassic fertilizers may increase the crop yields either by directly supplying *K* or by improving its balance with respect to *Na*, *Ca*, and *Mg*. However, under high salinity conditions it is difficult to exclude Na effectively from the plant by the use of *K*-fertilizers.

8.6.2.8 Developing/Cultivating Salt Tolerant Crops

Breeding program should be launched to develop/screen salt tolerant crop cultivars. International Rice Research Institute (IRRI) and other international/national institutes have developed several rice and other crop genotypes, which can tolerate salinity up to certain level (8–14 dS/m soil or irrigation water EC).

8.6.2.9 Increasing Water Use of Annual Crops and Pastures

In some areas, salinity is caused by change in vegetation. Native vegetation is cleared and replaced with annual crop and pasture species, which allow a larger proportion of rainfall to remain unused by plants and to enter the groundwater. As a result, groundwater tables have risen; bringing dissolved accumulated salts to the surface. In such areas, increasing the water use of existing annual crops and pastures may play an important rule to minimize salt accumulation at surface. But this practice alone may not sufficient.

8.6.2.10 Policy Formulation

Land-use planning should be done at the national level, indicating what crops should be grown at each soil and/or agro-ecological zone. Based on the salinity problem, suggested specific crop(s) must be grown and it should be monitored and materialized. This would enhance the best utilization of the land and optimize crop production.

8.6.2.11 Overall Discussion

Salinity problems may often more complex. Management of salt-affected lands for agricultural use is largely dependent on the water availability, climatic conditions, period of salinity, crop standing, and the availability of resources (capital, inputs). It may require a combination of agronomic and engineering management practices, depending on careful definition of the main production constraints and requirements based on a detailed, comprehensive investigation of soil characteristics, water monitoring (rainfall, irrigation water and water table), and a survey of local conditions including climate, crops, economic, social, political and cultural environment, and existing farming system. Proper management procedures (soil moisture, irrigation system uniformity and efficiency, local drainage, and the right choice of crops), combined with periodic soil tests, can improve the conditions (i.e., prolong the productivity) in salt-affected soils to a certain extent.

8.6.2.12 Sample Examples

Example 8.3

A pond water has the salinity level 2.0 dS/m. The groundwater salinity level at that area is 15 dS/m, which is marginal for wheat irrigation. Find out the mixing ratio of pond water to groundwater to lower the salinity to 8.0 dS/m.

Solution

Given,

 EC of pond water, $EC_p = 2.0$ dS/m
 EC of groundwater, $EC_g = 15$ dS/m

8.6 Management/Amelioration of Saline Soil

The final/target EC of the mixture, $EC_m = 8.0$
We know,

$$C_m = C_1 \times r_1 + C_2 \times r_2 \quad (A)$$

and, $r_1 + r_2 = 1$
Or, $r_2 = 1 - r_1$
Putting the values in Eq. (A), $8 = 2 \times r_1 + 15(1 - r_1)$
Or, $13\, r_1 = 7$
Thus, $r_1 = 0.53$
Then, $r_2 = (1 - r_1) = 0.47$
Thus the ratio would be 53:47 (Ans.).

Example 8.4

The salt concentrations of two solutions are 10 and 50 mg/l. Find the salt concentration of mixed water, if the proportions of the solutions are 70 and 30%, respectively.

Solution

We know, $C_x = C_1 \times r_1 + C_2 \times r_2$
Given,
Concentration of 1st solution, $C_1 = 10$ mg/l
Proportion of 1st solution, $r_1 = 70\% = 0.70$
Concentration of 2nd solution, $C_2 = 50$ mg/l
Proportion of 2nd solution, $r_2 = 30\% = 0.3$
Putting the above values, salt concentration of the mixed water, $C_x = (10 \times 0.7) + (50 \times 0.3) = 22$ mg/l (Ans.)

Example 8.5

A cereal crop having 120 days growing period has the following data from field condition:

Total ET: 80 cm
Irrigation cycle: 10 days
Irrigation time: 6 h
Average infiltration rate: 1.0 cm/h
EC of drainage water: 1.5 dS/m

Assuming no drainage limitation, calculate leaching fraction and permissible electrical conductivity of irrigation water.

Solution

Given, average infiltration rate, $I_f = 1.2$ cm/h
$= 28.4$ cm/day

Infiltration time $(t_i) = 6$ h $= 6/24$ day $= 0.25$ day
ET during infiltration cycle $= (80/120) \times 10 = 6.67$ cm
We know, leaching fraction (LF) $= 1 - (ET/I_f t_i) = 1 - 6.67/(28.8 \times 0.25)$
$= 0.074$

LF $= D_d/D_i = EC_{iw}/EC_{dw}$
Thus, $EC_{iw} = $ LF $\times EC_{dw} = 0.07 \times 1.5 = 0.111$ dS/m (Ans.)

Example 8.6

A wheat crop is planted in a silt loam soil and irrigated from a river water. The EC of the river water at the time of irrigation is 1.5 dS/m. The crop ET during the growing season is 76 cm. How much additional water must be applied for leaching?

Solution

Given,

$$EC_{iw} = 1.5 \text{ dS/m}$$

For wheat, at 100% yield potential, $EC_e = 6.0$
We know,

$$LR = EC_{iw}/(5\ EC_e - EC_{iw})$$

$$= 1.5/(5 \times 6.0 - 1.5)$$

$$= 0.0526$$

The required amount of water for both ET demand and leaching requirement:

(a) Using the relation, AW $=$ ET/(1$-$ LR)
$= 76/(1 - 0.0526)$
$= 80.22$ cm (Ans.)
(b) Using the relation, AW $=$ ET $(1+LF)/E_f$
and assuming $E_f = 90\%$, AW $= 76\ (1 + 0.056)/0.9 = 89.14$ mm (Ans.)

8.7 Management of Sodic and Saline-Sodic Soils

8.7.1 Management of Sodic Soil

Sodic soils require different management than saline soils. Excess sodium must first be replaced by another cation and then leached. Sodic soils are treated by replacing the sodium with calcium from a soluble source. Gypsum ($CaSO_4 \cdot 2H_2O$) is considered the cheapest soluble calcium source for reclamation of sodic soils. Soil application of gypsum should be applied according to the gypsum requirement of the soil. Technically, soil reclamation is possible but may not always economically feasible.

Chemical amendments available for reclamation are

(a) soluble calcium salts: Calcium chloride ($CaCl_2 \cdot 2H_2O$), Gypsum ($CaSO_4 \cdot 2H_2O$).
(b) Acids or acid formers: Elemental sulfur (S), sulfuric acid (H_2SO_4), Iron sulfate ($FeSO_4$).
(c) Insoluble calcium salts: Ground lime stone ($CaCO_3$).

On calcareous soils (soils with excess $CaCO_3$ present), elemental sulfur may be added to furnish calcium indirectly. Sulfur is oxidized to sulfuric acid which reacts with the calcium carbonate to form gypsum. Oxidation of elemental sulfur is slow, so this method may be of limited value.

Three processes are occurred for the reclamation of sodic soils:

(i) Sodium ion (Na^+) is replaced by calcium ion (Ca^{++}) on the exchange complex
(ii) Soil hydraulic conductivity increases
(iii) Sodium salts are leached from the soil system.

The use of gypsum to increase water infiltration is a common and an old practice. Surface application of gypsum reduces soil crusting thereby increases water infiltration and in turn, crop yield. Ayers and Westcot (1985) reported that for a potato crop, gypsum applied and disked into the soil at rates as high as 10 Mg/ha/yr resulted in greatly improved infiltration. Adding gypsum to the water at rates sufficient to increase the calcium concentration by 2–3 mmol/l was also effective. Machines are now available which make it relatively easy to apply the gypsum to the irrigation water.

8.7.1.1 Cation Exchange Reaction

(a) Reclamation by Gypsum (Na^+ Is Replaced by Ca^{++})

Ca^{++} replaces the Na^+ from the colloidal surface, and Na^+ with SO_4^- form Na_2SO_4 which is readily soluble in water and leach down or drain out from the field with irrigation water.

$$2NaX + CaSO_4 = CaX + Na_2SO_4 \quad \downarrow$$

(leaching to the soil system)

(b) Reclamation by Acid Formers or Acids

(i) Elemental S and sulfuric acid

Elemental sulfur (S) is oxidized into sulfur dioxide and then sulfur dioxide is formed into sulfuric acid.

$$2S + 3O_2 \rightarrow 2SO_2$$

(microbial oxidation)

$$2SO_2 + H_2O \rightarrow H_2SO_4$$

Sulfuric acid reacts with sodium compound as follows:

With the presence of lime:

$$H_2SO_4 + CaCO_3 = CaSO_4 + CO_2 + H_2O$$

$$2NaX + CaSO_4 = CaX + Na_2SO_4 \quad \downarrow$$

(leached)

Without lime:
In this case, sodium is replaced by hydrogen.

$$NaX + H_2SO_4 = HX + Na_2SO_4$$

(ii) Iron sulfate

$$FeSO_4 + H_2O = H_2SO_4 + FeO$$

With lime:

$$H_2SO_4 + CaCO_3 = CaSO_4 + CO_2 + H_2O$$

$$2NaX + CaSO_4 = CaX + Na_2SO_4 \quad \downarrow$$
(leached)

Without lime:

$$NaX + H_2SO_4 = 2HX + Na_2SO_4$$

8.7 Management of Sodic and Saline-Sodic Soils

As a thumb rule, reclamation of a foot (12 in.) depth of sodic soil on one acre requires 1.7 t of pure gypsum for each milli-equivalent of exchangeable sodium present per 100 g of soil. The rate of gypsum needed for reclamation may be determined by a laboratory test.

Once the gypsum is applied and incorporated, sufficient good quality water must be added to leach the displaced sodium beyond the root zone. Reclamation of sodic soils is slow because soil structure, once destroyed, is slow to improve. Growing a salt-tolerant crop in the early stages of reclamation and disking in crop residues adds organic matter which increases water infiltration and permeability, speeding up the reclamation process.

8.7.1.2 Sample Examples

Example 8.7

A sodic soil has an ESP of 35 and CEC of 22 meq/100 g soil. Calculate the amount of gypsum required to reduce the ESP to 15 at a depth of 40 cm.

Solution:

Change in ESP = 35 − 15 = 20
We know, ESP = (exchangeable Na/CEC) × 100
Ex. Na = (ESP×CEC)/100
    = (20 × 22)/100
    = 4.4 meq/100 g soil
We know gypsum required for 1 ha-30 cm soil to change 1 meq ex. Na/100 gm
    = 3.85 metric t (mt).
Thus, gypsum req. per hectare in this case, GR = (3.85 × 4.4/30) × 40 mt
    = 22.58 mt
Considering safety factor for field application,
Field requirement, FR = GR × safety factor for field application = 22.58 × 1.25 = 28.23 mt (Ans.)

8.7.2 Management of Saline-Sodic Soil

Saline-sodic soils are characterized by the occurrence of sodium (Na^+) to levels that can adversely affect several soil properties and growth of most crops. Some practices suited to saline soils can be used for saline-sodic soils.

In reclamation of saline-sodic soils, the leaching of excess soluble salts must be accompanied (or preceded) by the replacement of exchangeable sodium by calcium. If the excess salts are leached and calcium does not replace the exchangeable sodium, the soil will become sodic. Even nonirrigated sodic or saline-sodic soils show dramatic improvement with gypsum application. Gypsum (15 t/acre) applied to a saline-sodic soil in Kansas increased wheat yields an average 10 bushels per acre over a 5-year period in an area with 28 inch average annual rainfall.

Management of salt-affected soils requires a combination of agronomic practices depending on soil characteristics, water quality, and local conditions including climate, crops, economic, social, political and cultural environment, and existing farming systems. There is usually no single way to combat salt-problem in irrigated agriculture. However, several practices can be combined into an integrated system to achieve a satisfactory yield goal.

8.8 Models/Tools in Salinity Management

Salinity management requires an understanding of catchment data and processes. The aim of modeling and model use are to assist farmers, extension workers and policy makers in making on-farm investment decisions to manage salinity in the catchment. A number of models have been developed to predict long term behavior of groundwater, root-zone salinity index, desalinization of soil profile, quality of groundwater and drainage, efficient solute transport; and crop response models to simulate crop production. Some of the models are outlined below:

2CSalt Model

It was developed by Department of Environment and Resource Management, The State of Queensland (Australia). It is a model of a catchment surface and groundwater and salt balance, which considers effects of spatial land use.

BC2C Model

A model used to estimate the response time of stream flow and salt loads to land use change, developed by Department of Environment and Resource Management, The State of Queensland (Australia).

TARGET Model

It is a spreadsheet model. It was developed as a multi-enterprise, multi-period, whole-farm analysis tool with an emphasis on "what if" types of analysis. Financial inputs and production inputs can be readily varied on a yearly basis. It was developed in Australia.

SALT

Analyzes crop salt tolerance response data, developed by USDA.

WATSUIT

It was developed in US Salinity Laboratory. It predicts the salinity, sodicity, and toxic-solute concentration. It is a transient state model and is used for assessing water suitability for irrigation which can incorporate the specific influences of the many variables that can influence crop response to salinity, including, climatic, soil properties, water chemistry, irrigation and other management practices.

UNSATCHEM

It was developed in US Salinity Laboratory (USA). It is a one-dimensional solute transport model, which simulates variably saturated water flow, heat transport, carbon dioxide production and transport, solute transport and multi-component solute transport with major ion equilibrium and kinetic chemistry. UNSATCHEM package may be used to analyze water and solute movement in the unsaturated, partially saturated, or fully saturated porous media. Flow and transport can occur in the vertical, horizontal, or in an inclined direction. This package is a good tool to understand the chemistry of unsaturated zone in case of saline water use and development of analytical model to predict the changes in groundwater and soil quality.

SWMS-3D

Simulates water and solute movement in three-dimensional variably saturated media. It was developed by USDA Salinity Laboratory.

SIWATRE

It was developed in ILRI, the Netherlands, for simulation of water management system in arid regions (unsaturated flow model) which has the components as sub-model DESIGN for water allocation to the intakes of the major irrigation canal, sub-model WDUTY for estimation of water requirement at farm level, sub-model REUSE for the water losses to the atmosphere, and WATDIS sub-model for water distribution within the command.

CropSyst – Modified Version

CropSyst, a comprehensive crop growth/management model was modified for assessing crop response and water management in saline conditions. It was done at the Department of Biological Systems Engineering, Washington State University.

LEACHC (Version of LEACHM)

A computer model can be used for irrigation scheduling and assessing the impacts of various fresh and saline shallow water tables on soil salinity built up.

SIWM Model

SIWM (Salinity Impacts of Wetland Manipulation) model was developed in Australia. The model uses a daily water and salt balance approach in attempt to simulate wetland behavior.

SIMPACT Model

SIMPACT (Salinity Impacts) model assess the salinity impacts of actions that increase or decrease recharge to regional aquifers. The model was developed in Australia.

SWAM

The simulation model SWAM can be used in classifying the waters, assessing their suitability and evaluating management strategies for applying poor quality (saline) waters for crop production. It was developed by Singh et al. (1996), India.

SGMP Computer Model

It was developed in ILRI, the Netherlands, as a numerical groundwater simulation model to quantify the amount of recharge from the top system to the aquifer and its spatial variation and to assess its effects on water table depths.

SALTMOD Computer Model

It was developed in ILRI, the Netherlands, to predict long-term effects of groundwater conditions, water management options, average water table depth, salt concentration in the soil, groundwater use, drain and well water yields, dividing the soil-aquifer system into four resources surface reservoir, soil reservoir (root zone), an intermediate soil reservoir (vadose zone), and a deep reservoir (aquifer).

ENVIRO-GRO

It is an integrated water, salinity, and nitrogen model. It was developed by Pang and Letey (1998).

SWASALT/SWAP

It was a package on an extended version of SWATRE model. The depth and time of irrigation applied, quality of irrigation water used, soil type and initial soil quality can be modified and the effects on crop performance, soil salinization and desalinization process, soil water storage (excess/deficit) can be obtained from the model output.

8.9 Challenges and Needs

Salinity development and salt leaching are interacted by many factors. Evaluation of traditional concepts such as leaching requirement, leaching fraction, salt balance concept demands incorporation of a rapid, efficient, and economic way of monitoring changes in salinity during amelioration. New scientific information is needed to improve the economic gains of salinity amelioration/management because the future improvements in salinity management seem to be limited by economic rather than a lack of technological means. In addition to the existing salt-tolerant crop genotypes, research is needed to seek out or develop genotypes with increased tolerance to salinity and sodicity.

Relevant Journals

- Irrigation Science (Springer)
- Agricultural Water Management
- Agronomy Journal
- Soil Science Society of America Journal
- Natural Resources Research (Springer)
- Water Resources Management
- Field Crops Research
- Crop Science
- Plant and Soil

Relevant FAO Papers/Reports

- FAO Irri. and drainage Paper 57 (Soil salinity assessment: methods and interpretation of electrical conductivity measurement) [1999]

- FAO Irri. and drainage Paper 48 (The use of saline waters for crop production) [1993]
- FAO Irri. and drainage Paper 29 (Water quality for agriculture) [1976]

FAO Soils Bulletins

– FAO Soils Bulletin No. 39 (Salt-affected soils and their management) [1998]

Questions

(1) Narrate the extent of salinity problem. Why the process of soil salinization is important to know?
(2) Describe the process of soil salinization.
(3) What are the factors affecting salinity problem?
(4) Briefly describe the characteristics of salt-affected soils.
(5) Write a short note on: EC, TSS, ESP, CEC, SAR.
(6) Discuss the field and laboratory methods for diagnosis of saline and sodic soil.
(7) How will you determine the intensity of salinity and sodicity hazard?
(8) How will you interpret (a) the salinity hazard based on EC, (b) sodicity hazard based on Na, SAR, ESP and Ca:Na ratio?
(9) Classify salinity based on different approaches.
(10) What are the impacts of (a) salinity, and (b) sodicity on soil and plant growth and crop production?
(11) Describe the factors influencing tolerant to crop.
(12) What are the principles for management of saline soil?
(13) Briefly describe the different approaches to ameliorate the salinity effect.
(14) What is leaching fraction (LF)? What points should be considered in leaching process?
(15) Write down the mathematical formula for determining LF under different perspectives.
(16) Write down the mathematical formula for determining total water requirement for crop evapotranspiration and leaching.
(17) Discuss the management methods of (a) saline, (b) sodic, and (c) saline – sodic soils.
(18) Write down the names of some models/tools available for salinity management.
(19) A canal water has the salinity level 5.0 dS/m. The groundwater salinity level at that area is 20 dS/m, which is not suitable for irrigation. Find out the mixing ratio of canal water to groundwater to lower the salinity to 12.0 dS/m.
(20) The electrical conductivity (EC) of a maize field during its growth period was found 9.0 dS/m. Estimate yield reduction due to salinity, if the maize cultivar

can maintain potential yield up to 6 dS/m, and the yield at EC>16 dS/m is zero.
(21) Estimate the actual yield of a wheat genotype which is grown in a field having EC of soil saturate (ECe) = 12 dS/m, critical salinity level above which the yield reduction starts (EC$_t$) = 8 dS/m, the salinity level at which yield becomes zero (EC$_0$) = 20 dS/m, and potential yield in a nonsaline situation = 5.0 t/ha.
(22) A sodic soil has an ESP of 30 and CEC of 20 meq/100 g soil. Calculate the amount of gypsum required to reduce the ESP to 12 at a depth of 30 cm.

References

Ali MH, Rahman MA (2008a) Study on the effect of different levels of saline water on wheat yield. In: Annual report 2008–09, Bangladesh Institute of Nuclear Agriculture, Mymensingh, Bangladesh

Ali MH, Rahman MA (2008b) Study of natural pond as a source of rain-water harvest and integrated salinity management under saline agriculture. In: Annual report 2008–09, Bangladesh Institute of Nuclear Agriculture, Mymensingh, Bangladesh

Ayers RS, Westcot DW (1985) Water quality for agriculture. FAO Irrigation and Drainage Paper 29 Rev. 1. FAO, Rome, p 174

Ghassami F, Jakeman AJ, Nix HA (1995) Global salinization of land and water resources: human causes, extent and management. Centre for Resource and Environmental Studies, Australian National University, Canberra, pp 211–220

Maas EV, Hoffmann GJ (1977) Crop salt tolerance – current assessment. ASCE J Irr Drain Div 103(IR2):115–134

Ortiz-Moasterio JI, Hede AH, Pfeiffer WH, van Ginkel M (2002) Saline/sodic subsoil on triticale, durum wheat and bread wheat under irrigated conditions. Proceedings of the 5th international triticale symposium, Radzikow, Poland, Annex 30 June–5 July 2002

Pang XP, Letey J (1998) Development and evaluation of ENVIRO-GRO, an integrated water, salinity, and nitrogen model. Soil Sci Soc Am J 62:1418–1427

Rhoades JD (1974) Drainage for salinity control. In: van Schilfgaarde J (ed) Drainage and agriculture. Agronomy, vol 17. American Society of Agronomy, Madison, pp 433–461

Rhoades JD, Merrill SD (1976) Assessing the suitability of water for irrigation: theoretical and empirical approaches. In: Prognosis of salinity and alkalinity. Soils Bulletin 31. FAO, Rome, pp 69–109

Singh CS, Gupta SK, Ram S (1996) Assessment and management of poor quality waters for crop production: a simulation model (SWAM). Agric Water Manage 30:25–40

Chapter 9
Drainage of Agricultural Lands

Contents

9.1	Concepts and Benefits of Drainage		329
	9.1.1	Concepts	329
	9.1.2	Goal and Purpose of Drainage	329
	9.1.3	Effects of Poor Drainage on Soils and Plants	329
	9.1.4	Benefits from Drainage	330
	9.1.5	Types of Drainage	330
	9.1.6	Merits and Demerits of Deep Open and Buried Pipe Drains	332
	9.1.7	Difference Between Irrigation Channel and Drainage Channel	334
9.2	Physics of Land Drainage		334
	9.2.1	Soil Pore Space and Soil-Water Retention Behavior	334
	9.2.2	Some Relevant Terminologies	335
	9.2.3	Water Balance in a Drained Soil	338
	9.2.4	Sample Workout Problem	340
9.3	Theory of Water Movement Through Soil and Toward Drain		341
	9.3.1	Velocity of Flow in Porous Media	341
	9.3.2	Some Related Terminologies	341
	9.3.3	Resultant or Equivalent Hydraulic Conductivity of Layered Soil	342
	9.3.4	Laplace's Equation for Groundwater Flow	345
	9.3.5	Functional Form of Water-Table Position During Flow into Drain	346
	9.3.6	Theory of Groundwater Flow Toward Drain	346
	9.3.7	Sample Workout Problems	347
9.4	Design of Surface Drainage System		349
	9.4.1	Estimation of Design Surface Runoff	349
	9.4.2	Design Considerations and Layout of Surface Drainage System	349
	9.4.3	Hydraulic Design of Surface Drain	349
	9.4.4	Sample Work Out Problem	350
9.5	Equations/Models for Subsurface Drainage Design		351
	9.5.1	Steady-State Formula for Parallel Drain Spacing	351
	9.5.2	Formula for Irregular Drain System	355
	9.5.3	Determination of Drain Pipe Size	356
9.6	Design of Subsurface Drainage System		356

	9.6.1	Factors Affecting Spacing and Depth of Subsurface Drain	356
	9.6.2	Data Requirement for Subsurface Drainage Design	357
	9.6.3	Layout of Subsurface Drainage	357
	9.6.4	Principles, Steps, and Considerations in Subsurface Drainage Design	358
	9.6.5	Controlled Drainage System and Interceptor Drain	361
	9.6.6	Sample Workout Problems	362
9.7	Envelope Materials	365
	9.7.1	Need of Using Envelop Material Around Subsurface Drain	365
	9.7.2	Need of Proper Designing of Envelop Material	365
	9.7.3	Materials for Envelope	365
	9.7.4	Design of Drain Envelope	366
	9.7.5	Use of Particle Size Distribution Curve in Designing Envelop Material	367
	9.7.6	Drain Excavation and Envelope Placement	368
9.8	Models in Drainage Design and Management	368
	9.8.1	DRAINMOD	368
	9.8.2	CSUID Model	369
	9.8.3	EnDrain	369
9.9	Drainage Discharge Management: Disposal and Treatment	369
	9.9.1	Disposal Options	369
	9.9.2	Treatment of Drainage Water	370
9.10	Economic Considerations in Drainage Selection and Installation	371
9.11	Performance Evaluation of Subsurface Drainage	371
	9.11.1	Importance of Evaluation	371
	9.11.2	Evaluation System	372
9.12	Challenges and Needs in Drainage Design and Management	373
Relevant Journals	373
FAO/World Bank Papers	374
Questions	374
References	376

For the maximum growth of plants, it is essential to provide a root environment that is suitable for it. Water logging has an effect on the uptake of nutrients by plants. Drainage provides favorable condition for root growth, enhance organic matter decomposition, reduce salinity level above the drain, and maintain productivity of the land. For successful drainage design, the complex interaction of water, soil and crop in relation to quality of water must be well understood beforehand. The knowledge of drain material (drainage conveyance conduit) and placement technique of the drain (including selection of envelop material and its design with respect to soil bedding) are also of great importance. Judgment of alternate design options, and, monitoring and evaluation of the system after installation are demanded for successful outcome from the drainage projects. Disposal and treatment issues of drainage effluent are of concerns now-a-days. This chapter discusses all of the above points.

9.1 Concepts and Benefits of Drainage

9.1.1 Concepts

Drainage is the removal of excess water from the root zone area. The drainage problem is caused by excess of water either on the surface of the soil or in the root zone beneath the surface of the soil. The removal of water from the surface of the land and the control of the shallow groundwater table (referred as drainage) improves the soil as a medium for plant growth. The sources of excess of water may be precipitation, irrigation water, snowmelt, overland flow, artesian flow from deep aquifers, flood water from channels, or water applied for special purposes as leaching salts from the soil.

9.1.2 Goal and Purpose of Drainage

Drainage is used in both humid and arid areas. The main goal or objective of providing drainage is to increase production and to sustain yields over long periods of time.

The purpose of drainage is to provide a root environment that is suitable for the maximum growth of plants. The drainage serves the following or one of the following purposes:

- Remove extra water from the crop root zone
- Control salinity buildup in the crop root zone and soil profile.

9.1.3 Effects of Poor Drainage on Soils and Plants

Water that fills the soil pores not only displaces the air in the soil but also obstructs the gases which are given off by the roots to escape. The oxygen content in the soil is limited not only because of small amount of oxygen dissolved in the water but also because of extremely low through sub-soils.

After the dissolved O_2 in water-logged soil has been consumed, an aerobic decomposition of organic matters takes plane. This results in the production of reduced organic compounds and complex aldehydes. Mineral substances in the soil are altered from the oxidized state to a reduced state, and toxic concentration of ferrous and sulfide ions may develop within a few days after submergence of the soil.

Water logging generally leads to a deceleration in the rate of decomposition of organic matter. Thus nitrogen (N_2) tends to remain logged up in the organic residues. N_2, therefore, is often a limiting factor to plant growth on poorly drained soil.

Water logging has an effect on the uptake of nutrients by plants. This is shown by certain symptoms which develop under circumstances of water logging. These symptoms are yellowing, reddish, or stippled (dotted) appearance of the leaves.

In some sensitive plants, prolonged water logging damage the root system, and consequently the plants die due to lack of nutrient and water uptake.

9.1.4 Benefits from Drainage

9.1.4.1 Major Benefits

(a) Makes favorable soil condition for plant roots to grow
(b) Increases the availability of plant nutrients
(c) Reduces the hazard of toxic substances in the soil
(d) Helps in the decomposition of organic matter and thereby increases the availability of N_2.
(e) Extends the period of cultivation
(f) Provides a better environment for crop emergence and early growth
(g) Can reduce soil compaction

9.1.4.2 Additional Benefits

Drainage helps to get ride of the following problems:
   (i) Ponding of water during the summer may cause scalding (burn by hot water) of the crops.
  (ii) A health hazard is created by mosquitoes which breed in ponded field.
 (iii) A high water table prevents the soil to warm up readily in spring and thus affect seed germination.
 (iv) Plant diseases are more active under water-logged conditions. Fungus growth is particularly prevalent (most common).
  (v) Bearing capacity of water-logged soil is low. Thus, create problem with construction and compaction of soil by animals and machines.

9.1.5 Types of Drainage

Drainage may be either natural or artificial. In case of natural drainage system, water flows from the fields toward low-land, then to swamps or to lakes (in some places local name is "khal" or "bill") and then to rivers, through natural system. In some cases/soils, the natural drainage processes are sufficient for plant growth and production of agricultural crops, but in many other soils, artificial drainage is needed for efficient agricultural production.

Artificial drainage is divided into two broad classes: (i) surface, and (ii) subsurface. Some installations serve both purposes.

9.1.5.1 Surface Drainage

Surface drainage is the removal of water that collects on land surface. It is a system of drainage measures, such as open drains and land forming, to prevent ponding by diverting excess surface water to a collector drain.

9.1 Concepts and Benefits of Drainage

Fig. 9.1 Schematic view of surface drain

Many fields have depressions or low spots where water ponds. Surface drainage techniques such as land leveling, construction of shallow ditches or waterways can allow the water to leave the field rather than causing prolonged wet areas (Fig. 9.1). The water from field plots flows toward drain or shallow ditches through artificial down-slopes. Then shallow ditches discharge into larger and relatively deeper drain called collector drain. The collector drain is connected to the lake or other deep water bodies.

A surface drainage system always has two components:

– Open field drains to collect the ponding water and divert it to the collector drain
– Land forming to enhance the flow of water toward the field drains

9.1.5.2 Subsurface Drainage

Subsurface drainage is the removal of excess water from the soil profile. This is normally accomplished by buried pipe drains or deep open drain. Subsurface drainage increases the productivity of poorly drained soil by lowering the water table, and providing higher soil aeration.

Deep Open Drain

This type of drainage channel is deeper than the natural normal drainage channel (Fig. 9.2) and usually constructed with an excavator; often with a bulldozer or grader shaping the spoil. It is appropriate for

- small catchments
- where the soil is compact and stable

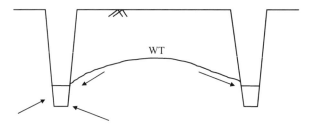

Fig. 9.2 Schematic of deep open drain

- where a suitable outlet is available to dispose of the quantity and quality of water collected
- where it can be used as alone or with other system (as outlet of normal surface drain)

Deep open drain has the potential in turning salt-affected land into productive. It directs surface and subsurface water away from the landscape, taking with it the salt that has accumulated in the soil.

Buried Pipe Drain

A network of perforated tubes or pipes is installed 0.8–1.5 m below the soil surface. The excess water from the root zone flows into the pipe through the openings (Fig. 9.3). The pipes convey the water to a collector drain. A gentle slope of the pipe (0.5–1.0%) is provided toward the collector drain. The drain pipe may be made of plastic, concrete, or clay. These tubes are commonly called "tiles" because they were originally made from short lengths of clay pipes known as "tiles." In clay and concrete pipes, water enters the pipes through the small spaces between the tiles, and drain away. In plastic pipes, water enters into the drain through perforation distributed over the entire length of the pipe. The buried pipe lowers the water table to the depth of the pipe or tiles over the course of time (several days).

Typical Length of Different Pipe Drain

    Plastic: 100–200 m long, 10–15 cm diameter
    Clay and concrete: 25–30 m long, 5–10 cm diameter.

9.1.6 Merits and Demerits of Deep Open and Buried Pipe Drains

There are some merits and demerits of each of the drain systems – deep open and buried pipe drain. They are summarized below:

Fig. 9.3 Buried pipe drain system (**a**) drain position (**b**) water-table position during operation

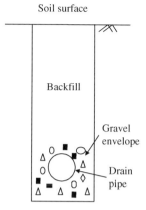

(a) Drain placement

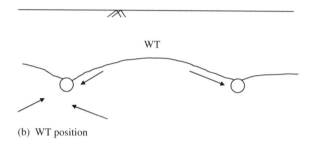

(b) WT position

Deep open drain	Buried (subsurface) pipe drain
1. Causes loss of cultivable land, that otherwise could be used for crops	1. Causes no loss of cultivable land
2. Restricts the use of machines	2. Does not restrict the use of machines
3. It requires a large number of bridges or culverts for road crossing and access to the field	3. Does not require
4. Requires frequent maintenance	4. Maintenance requirements are very limited
5. Cost of construction is lower compared to the cost of buried pipe system	5. Total cost of the system is higher compared to deep open drain due to material cost, equipment, and skilled manpower involved
6. Effective and/or economic life is short	6. Effective and/or economic life is long
7. Allows for removal of much larger volumes of water in a shorter time span than subsurface drainage	7. Allows for removal of lower volumes of water in a shorter time span

9.1.7 Difference Between Irrigation Channel and Drainage Channel

Drainage channel can be distinguished from the irrigation channel with the following characteristics:

Irrigation channel	Drainage channel
1. Irrigation channel network spreads from the source toward the field	1. Drainage channel networks convergence from the field toward the main drain
2. Irrigation channel is normally elevated from the plot or at least at equal level to the field plot	2. Drainage channel is certainly lower than the field plot
3. In some instances, irrigation channel may be used as drainage channel (e.g., channel at lower part of field can be used as drainage channel of upper part)	3. Drainage channel can not be used as irrigation channel without structural modification

9.2 Physics of Land Drainage

Drainage systems are engineering structures that remove water according to the principles of soil physics and hydraulics. To understand the mechanism how drainage influence the water balance in the soil and control the subsurface hydrology, we should understand the basic concepts regarding soil (pore space) water and their retention or release characteristics.

9.2.1 Soil Pore Space and Soil-Water Retention Behavior

The soil bulk volume consists of both solid and pore space (Fig. 9.4). The proportion of pore space in bulk volume depends primarily on soil texture. The typical range of pore space is 30–55%. For practical implications, soil pores can be classified as follows:

Pore type	Pore diameter
Micro-pore	<0.01 mm
Meso-pore	0.01–0.2 mm
Macro-pore	>0.2 mm

Water is held within the pore space by weaker capillary forces. Around the soil particles, water is held as "film" by stronger adsorptive forces. Within the saturation range of a soil, the classification of total water is illustrated in Fig. 9.5.

9.2 Physics of Land Drainage

Fig. 9.4 Schematic of solid and pore space in soil bulk volume

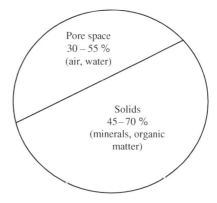

Fig. 9.5 Schematic representation of different types of water in soil

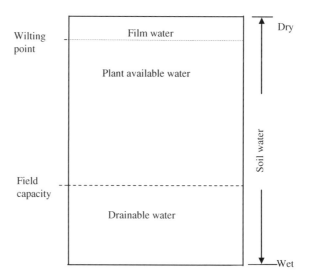

9.2.2 Some Relevant Terminologies

9.2.2.1 Void Ratio (*e*)

It is defined as the ratio of the volume of voids or pore spaces (V_p) to the volume of solid particles (V_s). That is, void ratio,

$$e = \frac{V_p}{V_s}$$

It represents the proportionate amount of pore volume and solid volume.

9.2.2.2 Porosity

It is the ratio of volume of voids or pores to the bulk or total soil volume (solid plus pores). That is,

$$\eta = \frac{V_p}{V} = \frac{V_p}{V_p + V_s}$$

It represents proportionate amount of pore volume and bulk volume.

9.2.2.3 Relation Between Porosity and Void Ratio

$$\eta = \frac{V_p}{V_p + V_s} = \frac{1}{1 + \frac{V_s}{V_p}} = \frac{1}{1 + \frac{1}{e}} = \frac{e}{1 + e}$$

or

$$\eta = \frac{e}{1 + e} \tag{9.1}$$

9.2.2.4 Moisture Concentration

The moisture concentration is defined as the volume of water per unit volume of soil.

9.2.2.5 Pore Water

Water that is held in the soil pores is termed as pore water.

9.2.2.6 Drainable Pore Space or Drainable Porosity (P_d)

Drainable pore space or drainable porosity is the air-filled pores present when the soil is drained to field capacity (i.e., after gravity drainage). Alternatively, the macro-pore spaces which releases or drains water due to gravity drainage is termed as drainable porosity. This can be estimated as

$$P_d\,(\%) = \text{soil porosity}\,(\%) - \text{soil moisture at field capacity}\,(\%)$$

It is influenced by soil texture and structure. Coarse-textured (sandy) soils have large drainable porosity, whereas fine-textured (clayey) soils have smaller drainable porosities (Table 9.1). This implies that for a certain amount of water drained, a sandy soil shows a smaller water-table drop than that of a clay soil.

9.2 Physics of Land Drainage

Table 9.1 Typical drainable porosity in different textured soils

Soil texture	Drainable porosity (%)
Sandy	20–30
Loam	12–18
Clayey (clays, clay loam, silty clays)	5–12

Physical meaning of drainable porosity (P_d)

By definition,

Drainable porosity (%) = (Volume of drainable pore spaces in total volume of soil) ×100/total volume of soil

Another way of expressing drainable porosity is the quantity of water drained (d_d, mm) from a given drop in water table (h), expressed as percentage. That is,

$$P_d = \frac{d_d}{h} \times 100 \qquad (9.2)$$

A P_d of 8% means that draining 8 mm of water lowers the water table by 100 mm.

9.2.2.7 Drainable Water

Drainable water is that water which can be drained from a saturated soil by gravity or free drainage. The amount of drainable water in the soil depends on the amount of "drainable pore space" or drainable porosity. In drainage system, drainable water is expressed in units of depth (meter or millimeter). Expressing drainable water in this way assumes that its depth applied to a unit area (i.e., square meter or hectare). The volume of water from this depth can be computed simply by multiplying the depth of drainable water by the area of drainage (area of interest), making sure to keep the units consistent.

In the field, the soil moisture is not constant but changing with time and varies throughout the soil profile – from ground surface to a particular depth. Soil closer to the water table is wetter than soil closer to the ground surface. This means that as moving up from the water table, the soil pores contain proportionately less water. The change in proportion of air-filled and water-filled pores between the ground surface and water table (after the downward flow of excess water) is illustrated by the curved line in Fig. 9.6. At some height above the water table, the soil moisture will have drained to field capacity. Poorly drained soils may have water tables at or very close to the soil surface for prolonged period of time. Under such condition the proportion of air-filled pores in the soil profile is very small, and that's why the soil lacks proper aeration to support plant growth.

Suppose a pot with soil have no holes at the bottom for water to escape. The pot is watered until water spills over the top. At this point the soil is saturated and no air in the soil pores. If a hole is then made at the bottom of the pot, the free or "drainable" water will drain out and the soil will be left at field capacity. In case of

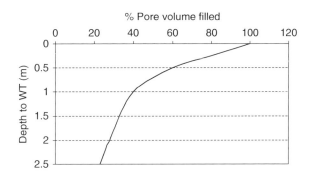

Fig. 9.6 Schematic of proportion of water-filled pores with depth to water table

subsurface drain, the excess water from the soil profile drains by the same process through the drain until the water table is lowered to the drain.

"Drainable water" can be measured directly from a predetermined drainage area following the above procedure. It can also be estimated indirectly from drainable porosity, as

$$\text{Drainable water (m}^3) = \text{drainable porosity (\%)} \times \text{drainable volume (m}^3)$$

i.e., $D_w = P_d \times D_v$.

9.2.2.8 Drainable Pore Volume Under Negative Pressure (or Suction)

The drainable pore volume at certain suction represents the volume of water that can be drained from a unit (or particular) volume of soil, when the soil-moisture pressure decreased from atmospheric pressure to some specific negative pressure.

9.2.3 Water Balance in a Drained Soil

In a crop-soil system, the term "water balance" describes the fate of precipitation and various components of water flow in and around the soil profile. Drainage affects soil-water, and thus other components of the water balance are also affected. Subsurface drainage influences the hydrology of heavily drained regions significantly and permanently, by substantially reducing surface runoff, shortening periods of surface pondage, and lowering of water table. Water balance on a soil profile with good natural drainage and in an artificially drained soil profile is depicted in Fig. 9.7

In the typical natural drainage system, precipitation (rainfall, snowmelt) (P), irrigation (IR) (if applied) are the major water input to the system affecting surface runoff (R), crop evapotranspiration (ET), deep percolation (DP), and changes in soil-water storage (S). Assuming that no water enters the soil from adjacent areas by horizontal flow, the water balance can be mathematically expressed as

9.2 Physics of Land Drainage

Fig. 9.7 Schematic of water balance of (**a**) artificial drained (*upper one*) and (**b**) natural drained soil

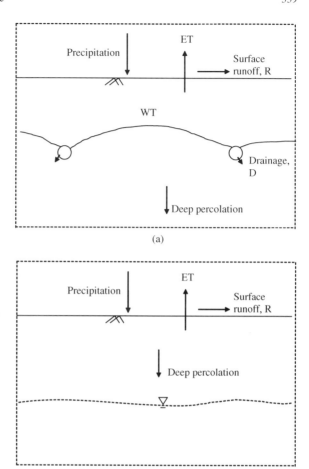

$$P + IR - R - ET - DP = S \qquad (9.3)$$

If the deep percolation continues, there is an opportunity for the water table to rise. It is evident from the water balance that the amount of deep percolation depends on the extent to which the precipitation and/or irrigation input to the soil is reduced by R, ET, and S.

In case of artificially drained soil profile, similar water balance equation holds true. However, the drainage flow becomes a major component of the water leaving the system. The water balance equation can be written as

$$P + IR = R + ET + DP + S + D \qquad (9.4)$$

As in case of natural drainage, the amount of drainage is dependent on how much precipitation is lost to R, ET, and the drainage capacity itself. In artificial drained soil, runoff component is greatly influenced due to change in air/water filled pores. After draining the water, the soil has more pore volume available for water infiltration during the next rain because of the larger volume of empty pores. Consequently, more infiltration and less runoff may occur with an artificially drained soil compared to a poorly drained (or with no artificial drained) soil.

The amount of infiltration in drained soil depends on many factors such as the nature and timing of the next rain, soil texture (pore space and its distribution), hydraulic conductivity, and depth of drain and spacing. The amount will be greater when the difference between the shallow (initial) and deep water level (final) is greater, and soil texture is coarser. Smaller rains of low intensity will reduce the total runoff rate, because proportionally more water will have an opportunity to infiltrate and pass through the drainage system. Smaller rains may cause surface runoff on the undrained soil and no surface runoff at all on the drained soil. However, if one or more rains occur before the drained soil has had time to drain the previous water adequately, water balance differences between the two soils will be lessened.

9.2.4 Sample Workout Problem

Examples 9.1

An agricultural soil contains 47% pore space, and the moisture content after gravity drainage is 39% (by volume). Find the void ratio, drainable porosity, and drainable water volume from a 20 m × 15 m plot having 1.0 m root zone depth.

Solution

$$
\begin{aligned}
\text{We know, void ratio} &= (\text{volume of void/volume of solid}) \\
&= \text{vol. of void}/(100 - \text{vol. of void}) \\
&= 47/(100 - 47) \\
&= 0.886
\end{aligned}
$$

$$
\begin{aligned}
\text{Drainable porosity} &= \text{total porosity} - \text{water content after gravity drainage} \\
&= 47 - 39\% \\
&= 8\%
\end{aligned}
$$

$$
\begin{aligned}
\text{Drainable water volume} &= \text{drainable porosity} \times \text{drainable soil volume} \\
&= (8/100) \times (20 \times 15 \times 1 \text{ m}^3) \\
&= 24 \text{ m}^3
\end{aligned}
$$

Ans.: Void ratio = 0.886, drainable porosity = 8%, drainable water volume = 24 m^3.

9.3 Theory of Water Movement Through Soil and Toward Drain

9.3.1 Velocity of Flow in Porous Media

Discharge velocity V is defined as the quantity (or volume) of water being discharged from a soil column per unit cross-sectional area of the column per unit time, i.e., $V = Q/A$.

The discharge velocity is useful when we are mainly concern with the amount of water flowing through a porous medium, but does not relate to the time of travel of this water. The flow can occur only through the pores, thus the velocity across any cross-section must be taken as some kind of average over the cross-section.

Let "m" be the ratio of the effective area of pore (A_p) to the total area (A) of the indicated section, then

$$m = \frac{A_p}{A}$$

or, $A_p = A \times m$
Then,

$$Q = AV = A_p V_{av} = mAV_{av} \qquad (9.5)$$

where V_{av} is an average velocity through the pores of the cross-section.

9.3.2 Some Related Terminologies

9.3.2.1 Drainage Intensity/Drainage Coefficient/Drainage Requirement

Drainage intensity *or* drainage coefficient is defined as the depth of water (mm) that is removed or to be removed per day (i.e., in 24 h) from a field for successful growth of crop or land amelioration. Its unit is mm/d (millimeter per day).

9.3.2.2 Drainage Density

It is the length of drain per unit area.
Drainage density, D_d (m/ha) = (total length of drain in an area, m)/(area of the land, ha)

9.3.2.3 Head

It is the energy (potential) of water at a point of interest, expressed as water height (m).

9.3.2.4 Water Table

Free water surface in soil, where soil-water pressure equals zero.

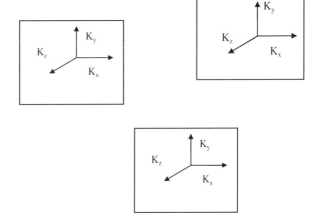

Fig. 9.8 Schematic of homogeneous and isotropic soil

9.3.2.5 Equipotential Line

It is the line of equal hydraulic head or potential.

9.3.2.6 Homogeneous and Isotropic Media

A media (here soil) is termed as homogeneous with respect to a certain property (say, hydraulic conductivity), if the property is same at every point of the media. For example, for homogeneous soil, $\frac{dK}{dS} = 0$. That is, the value of K does not change with space or distance.

A media is termed as isotropic with respect to a certain property, if the property is same in all directions at a particular point. That is, for isotropic soil, $\frac{dK}{dx} = \frac{dK}{dy} = \frac{dK}{dz}$.

Thus, a homogeneous soil may or may not be isotropic. Similarly, an isotropic soil may or may not be homogeneous.

A soil is termed as homogeneous and isotropic (with respect to a certain property, say, hydraulic conductivity), if the property (hydraulic conductivity) is same in all places in the media and also in all directions (Fig. 9.8). That is, for homogeneous and isotropic soil, $\frac{dK}{dS} = 0$ and $\frac{dK}{dx} = \frac{dK}{dy} = \frac{dK}{dz}$

9.3.3 Resultant or Equivalent Hydraulic Conductivity of Layered Soil

Theoretical derivation from Darcy's equation and experimental determination of soil hydraulic conductivity (K) has been described in Chapter 4 (*Soil*), Volume 1. Here, only the equivalent or resultant hydraulic conductivity will be discussed.

9.3.3.1 Resultant Horizontal Hydraulic Conductivity

Consider a soil column having n layers, and the depth (m) and horizontal hydraulic conductivity (m/d) of the layers are $d_1, d_2, d_3, \ldots d_n$ and $K_1, K_2, K_3, \ldots K_n$, respectively (Fig. 9.9). Assume that the hydraulic gradient in the horizontal direction is gradφ.

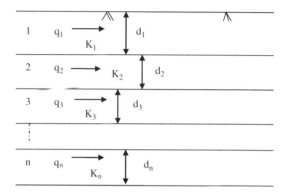

Fig. 9.9 Schematic of horizontal hydraulic conductivity

The total depth of the layers, $D = d_1 + d_2 + d_3 + \ldots d_n$

Flow through unit width of layer d_1 under gradφ hydraulic gradient, $q_1 = A_1 V_1 = (1 \times d_1) \times (K_1 \text{ grad}\varphi) = K_1 d_1 \text{ grad}\varphi$

Similarly,

Flow through "d_2", $q_2 = K_2 d_2 \text{ grad}\varphi$
Flow through "d_3", $q_3 = K_3 d_3 \text{ grad}\varphi$
Flow through "d_n", $q_n = K_n d_n \text{ grad}\varphi$

Total flow through the layers,

$$Q = \sum_{i=1}^{n} q_i = (K_1 d_1 + K_2 d_2 + K_3 d_3 + \cdots + K_n d_n)\text{grad}\varphi$$

or,

$$Q = \sum K_i d_i \text{ grad}\varphi \tag{9.6}$$

Now, we assume that the equivalent horizontal hydraulic conductivity of the whole column (layers) is K_H (m/d). Then,

$$Q = D \times K_H \text{grad}\varphi = \sum d_i \times K_H \text{ grad}\varphi \tag{9.7}$$

where D is the depth of soil column.
Equating the Eqs. (9.6) and (9.7), we get

$$K_H = \frac{\sum K_i d_i}{\sum d_i} \tag{9.8}$$

9.3.3.2 Resultant Vertical Hydraulic Conductivity

Consider a soil column having *n* layers, and with different thickness and hydraulic conductivity for each layer (Fig. 9.10). Now, we are interested to know the resultant (or equivalent) hydraulic conductivity for the soil column.

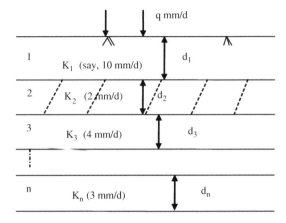

Fig. 9.10 Schematic of vertical soil layer with differential hydraulic conductivity

For explanation purpose, assume that layer-2 is relatively impervious than the other layers (sample values are given within the parenthesis). Assume that constant input, q (rainfall or irrigation rate), is higher than the lowest conductivity value (K_{min}) of the layers (here K_2). Although the layer-3 is relatively more pervious than the layer-2, the resultant flux or hydraulic conductivity will be limited by this layer. That is, if $q \geq K_{min}$, the unit flux or flux density will be controlled by the K_{min}. This is because, the K, Darcy's proportionality constant, or the hydraulic conductivity of the media, is the flux density (or in short "flux") under unit hydraulic gradient ($m^3/m^2/d$), not the flow velocity (m/d). In some text books, it is erroneously treated and expressed as flow velocity.

If the supply flux (q) is smaller than the K_{min}, the resultant vertical conductivity (K_V) will be limited by the q. Thus,

$$\begin{aligned} K_V &= K_{min}, \text{ if } q \geq K_{min} \\ &= q, \text{ if } q < K_{min} \end{aligned} \tag{9.9}$$

In reference to Fig. 9.10, at the top of layer-2 (i.e., at the bottom of layer-1), positive pressure will exist, since the incoming flux is higher than the outgoing flux. In contrast, at the top of layer-3 (i.e., at the bottom of layer-2), negative pressure will exist, as the outgoing flux capacity is higher than the incoming flux.

9.3.3.3 Resultant Conductivity of Horizontal and Vertical Direction

Resultant conductivity of horizontal hydraulic conductivity (K_H) and vertical hydraulic conductivity (K_V) can be obtained as

$$K' = \sqrt{K_H \times K_V} \qquad (9.10)$$

9.3.4 Laplace's Equation for Groundwater Flow

Groundwater flow can be described by Laplace's equation. Laplace's equation combines Darcy's equation and equation describing mass continuity.

In general form, Darcy's equation can be written as

$$q = V \times 1 = K_i \times 1 = K\frac{dh}{dS} \qquad (9.11)$$

where K is the hydraulic conductivity, i is the hydraulic gradient, dh is the head difference with the distance dS.

For a three-dimensional system, it can be written as

$$V_x = K\frac{dh}{dx}, \quad V_y = K\frac{dh}{dy}, \quad V_z = K\frac{dh}{dz}$$

For steady-state condition (no change in storage), mass continuity equation can be written as

$$\frac{dV_x}{dx} + \frac{dV_y}{dy} + \frac{dV_z}{dz} = 0$$

Putting the values of V_x, V_y, and V_z, we get

$$K_x\frac{d^2h}{dx^2} + K_y\frac{d^2h}{dy^2} + K_z\frac{d^2h}{dz^2} = 0 \qquad (9.12)$$

For homogeneous and isotropic soil system, $K_x = K_y = K_z$
Thus, the above equation reduces to

$$\frac{d^2h}{dx^2} + \frac{d^2h}{dy^2} + \frac{d^2h}{dz^2} = 0 \qquad (9.13)$$

which is the well-known Laplace's equation for groundwater flow.

9.3.5 Functional Form of Water-Table Position During Flow into Drain

Position of water table during flow into drain (steady or unsteady) can be described by the following functional form:

$$WT = \int (I, K, d, S, D, ..) \qquad (9.14)$$

where

- WT = water-table depth (or position)
- $\int$ = function
- I = recharge from rainfall or irrigation
- K = hydraulic conductivity of soil (vertical and horizontal direction, both upper and lower layer of drain)
- d = depth of drain from the soil surface
- S = drain spacing
- D = depth to impervious layer from the drain

9.3.6 Theory of Groundwater Flow Toward Drain

9.3.6.1 Steady State Problems

The steady-state drainage situation is one where a constant uniform accretion rate, I, is recharging the water table and the drain tubes are simultaneously draining the soil profile. After some time, a state of equilibrium will be established in which the water table does not change shape, and the drain discharge is constant (Fig. 9.11).

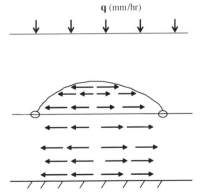

Fig. 9.11 Schematic of Dupuit-Forcheimer assumptions

9.3 Theory of Water Movement Through Soil and Toward Drain

Two types of approximate solutions of steady-state drainage condition have been proposed and are widely used:

(1) Based on horizontal flow assumption
(2) Based on radial flow assumption

Most techniques applied in analyzing the flow of water to subsurface drains are based on one-dimensional horizontal flow assumptions. Most solutions also assume that water percolates vertically from the soil surface through the unsaturated soil-water zone, so that the flux through the soil surface is also that through the water table. It is also commonly assumed that at steady state, the water table height at the top of the drain is negligible and that the drains do not flow completely full, a realistic assumption in all but extreme events and in systems with submerged outlets.

Horizontal Flow Assumption

Dupuit-Forcheimer's Assumption for Unconfined Flow

These assumptions regarding configuration of the flow and potential lines allow to derive solution of the flow of water toward drain without the use of Laplace's equation. These assumptions can be stated as follows:

(i) The flow lines (streamlines) are horizontal
(ii) The equipotential lines are vertical
(iii) The flow velocity in the plane at all depths is proportional to the slope of water table only and independent of the depth in the flow system.

The schematic of Dupuit-Forcheimer assumptions is shown in Fig. 9.11.

9.3.6.2 Non-steady State Drainage Situation

The non-steady state drainage situation is one in which the water table varies (falls or rises) with time. Although the non-steady state situation arises or exists in some cases and also in some parts of the operational period, steady-state assumptions are made for simplicity.

9.3.7 Sample Workout Problems

Example 9.2

A 1.2 m deep soil column consists of three layers, having 0.50, 0.4, and 0.3 m depth of the layers. The horizontal hydraulic conductivity of the layers are 0.20, 0.15, and 0.25 $m^3/m^2/h$, respectively. Determine the resultant horizontal hydraulic conductivity of the soil column.

Solution

We know

$$K_H = \frac{\sum K_i d_i}{\sum d_i}$$

Putting the values,

$$K_H = \frac{(0.20 \times 0.50) + (0.15 \times 0.4) + (0.25 \times 0.3)}{1.2}$$

$$= 0.235 \text{ m}^3/\text{m}^2/\text{h (Ans.)}$$

Example 9.3

A field soil has four distinct soil layers within 1.4 m depth from the surface. The depth and vertical hydraulic conductivity of the layers are 0.3, 0.4, 0.3, and 0.4 m; and 1.0, 0.5, 0.7, and 0.6 m³/m²/d, respectively. A constant water supply of 0.88 m/d is provided at the soil surface. What will be the resultant hydraulic conductivity of the soil column through the bottom of the soil layer? Comment on the pressure distribution at the up and bottom of the second layer.

Solution

The soil layers are schematically depicted in figure below:

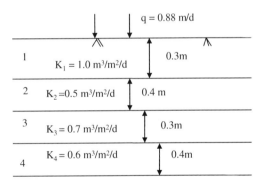

The second layer has the lowest conductivity and it is less than the supply (0.88 m/d), and hence, the vertical water movement will be restricted by this layer. Thus, the resultant hydraulic conductivity at the bottom of the layers will be 0.5 m³/m²/d.

At the top of the second layer (that is at the bottom of the first layer), the supply or the conductivity is higher than the conductivity of the second layer, thus positive pressure will exist. On the other hand, at the top of the third layer, the conductivity is higher than the conductivity (or the supply) of the second layer, thus negative pressure will exist at the bottom of the second layer.

9.4 Design of Surface Drainage System

9.4.1 Estimation of Design Surface Runoff

The runoff to be used in drain design is termed as "design runoff." Surface runoff to be generated from an area (design runoff) can be determined from the equations such as "Rational method" or "SCS method." Details of these methods are described in Chapter 6 (*Land and Watershed Management*), this volume. According to Rational method, peak surface runoff rate (Q) is

$$Q = CIA$$

where Q is the runoff rate (m³/h), A is the area from where runoff generates (drainage area) (m²), I is the peak rainfall intensity (m/h), and C is the runoff coefficient (dimensionless). Runoff coefficient is the fraction of rainfall which contributes to runoff. For agricultural field, its value ranges from 0.5 to 0.7 depending on initial soil moisture, rainfall intensity and duration, and soil condition/coverage. For design purpose, the value of I can be taken from long-term (20–50 years) peak rainfall records.

9.4.2 Design Considerations and Layout of Surface Drainage System

A network of surface drains is needed to remove the excess water (from rainfall and/or irrigation runoff) from the agricultural field. Drain layout should be based on the topography, shape of the farm/catchment, direction of natural slope, position of farm buildings and roads, and position/existence of natural depression, channel, or river. Drain layout should be done with consideration of minimum length of run and minimum crossing of roads. This will minimize wastage of land and minimize cost for culverts. Sometimes, land grading serves the purpose of surface drainage.

Sample typical layout of a farm/catchment is given in Fig. 9.12

9.4.3 Hydraulic Design of Surface Drain

Hydraulic design of surface drain (also termed as "drainage channel") is similar to the design of an open irrigation channel, as described in Chapter 1 (*Water Conveyance Loss and Designing Conveyance System*) of this volume. The capacity of the drainage channel should be based on the design peak surface runoff from the

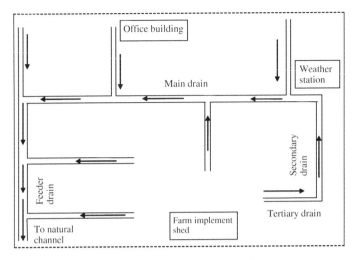

Fig. 9.12 Sample layout of surface drainage system in an agricultural farm

"drainage area" of the drain. "Drainage area" of a drain is the area from which the runoff falls to that drain.

For concrete channel, both rectangular and trapezoidal type can easily be constructed. For earthen channel (or earthen vegetative waterway), only trapezoidal type with smoothing of bottom corners is recommended.

9.4.4 Sample Work Out Problem

Example 9.4

Surface drainage should be planned for a new agricultural farm to drain out irrigation tail-water and seasonal rainfall runoff. Maximum rainfall intensity at the site in 20 years record is 35 mm/h. The tertiary drain would have to carry runoff from 4 ha land. The secondary drain would have to carry thrice of tertiary, and the main drain to carry discharge of four secondary drain (of similar flow). Determine the design discharge capacity of the (a) tertiary, (b) secondary, and (c) main drain.

Solution

We know, $Q = CIA$
Here,

Drainage area, $A = 4\,\text{ha} = 4 \times 10{,}000\,\text{m}^2 = 40{,}000\,\text{m}^2$
Design rainfall intensity, $I = 35\,\text{mm/h} = 9.72222\text{E} - 06\,\text{m/s}$
Runoff coefficient, $C = 0.6$ (as of agricultural land)

Putting the values, discharge capacity for the tertiary drain,

$$Q_t = 0.6 \times 9.72222\text{E-}06 \times 40{,}000 \text{ m}^3/\text{s}$$
$$= 0.233 \text{ m}^3/\text{s}$$

Discharge capacity for the secondary drain, $Q_s = Q_t \times 3 = 0.233 \times 3 = 0.7 \text{ m}^3/\text{s}$
Discharge capacity for the main drain, $Q_m = Q_s \times 4 = 0.7 \times 4 = 2.8 \text{ m}^3/\text{s}$(Ans.)

9.5 Equations/Models for Subsurface Drainage Design

9.5.1 Steady-State Formula for Parallel Drain Spacing

9.5.1.1 Hooghoudt's Equation

Definition Sketch

Hooghoudt's equation is widely used in subsurface drainage design. It can also be used for sloping land and to account for entrance resistance encountered by the water upon entering the drains. Hooghoudt's equation for drain spacing (Hooghoudt, 1940) can be described as

$$S^2 = \frac{4k_a h^2}{q} + \frac{8k_b d_e h}{q} \quad (9.15)$$

where

S = drain spacing (L)
q = drainage coefficient (L/T)
k_a = hydraulic conductivity of layer above the drain (L/T)
k_b = hydraulic conductivity of layer below the drain (L/T)
h = height of water at the midway between drains under stabilized condition (L)
d_e = equivalent depth (L)

The first term of the equation gives the spacing for the flow above the plane of the bottom of the drain, while the second term gives the spacing for the flow below the plane. The definition sketch of elements of Hooghoudt's drain spacing equation is given in Fig. 9.13

Here, "D" is the actual depth of impervious layer from the drain, and "d_e" is the equivalent depth of "D" to correct the resistance due to radial flow, which was assumed as horizontal flow. "Equivalent depth" represents the imaginary thinner soil layer through which the same amount of water will flow per unit time as in the actual situation. The resulting higher flow per unit area introduces an extra head loss, which accounts for (and thus resembles to) the head loss caused by converging flow lines.

Fig. 9.13 Schematic of definition sketch of Hooghoudt's drain spacing formula

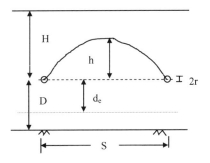

Hooghoudt's Equation Under Different Situation of Drain Placement

If the drain lies on the impervious layer, i.e., $D = 0$ (in Fig. 9.13), then consequently $d_e = 0$, and the Eq. (9.15) reduced to

$$S^2 = \frac{4k_a h^2}{q} \qquad (9.16)$$

If the head of water above the drainage base is too small and the flow from below the drain dominates, then the flow above the drain can be neglected and thus the Hooghoudt's formula reduces to

$$S^2 = \frac{8k_b d_e h}{q} \qquad (9.17)$$

Assumptions in Hooghoudt's Formula

Hooghoudt's formula is based on Dupuit-Forchheimer assumptions. The assumptions of Hooghoudt's formula can be summarized as follows:

- The soil is homogeneous
- Darcy's law is valid for flow of water through soil into the drain
- The drains are evenly spaced
- An impermeable layer underlies the drain
- There is a vertical recharge uniformly distributed in the horizontal plane
- Surface inflow rate, from irrigation or rainfall, is equal to the inflow to the drain
- The hydraulic gradient at any point in the flow regime is equal to the slope of water-table above the point
- The diameter of the drain is small compared to saturated thickness below the drain
- The drains have no entrance resistance
- The flow through drain is half-full

9.5 Equations/Models for Subsurface Drainage Design

Limitations of Hooghoudt Equation

The Hooghoudt's equation assumes an elliptical water table, which occurs below the soil surface. Sometimes excess precipitation may raise the water table to the soil surface, and ponded water remains on the surface for relatively long periods. For such conditions, the application of Hooghoudt's equation based on the D–F assumptions has limitations to calculate the subsurface drainage flux into the tile drains. In case of surface ponding, the D–F assumptions will not hold and the streamlines will be concentrated near the drains with most of water entering the soil surface in that vicinity (Kirkham, 1957).

Estimation of Equivalent Depth

The equivalent depth can be obtained as (Hooghoudt, 1940)

$$d_e = \frac{\pi S}{8 \ln \frac{S}{\pi r_0}} \tag{9.18}$$

where r_0 is the radius of drain.

Moody (1966) proposed the following approximation for d_e:

$$d_e = \frac{D}{1 + (D/S)[(8/\pi)\ln(D/r_0) - 3.4]}, \quad 0 < \frac{D}{S} \leq 0.3$$

$$= \frac{S}{(8/\pi)[\ln(S/r_0) - 1.15]}, \quad \frac{D}{S} > 0.3 \tag{9.19}$$

Closed-Form Solution for Equivalent Depth

Van der Molen and Wesseling (1991) developed closed-form expression for the equivalent depth (d_e) that can replace the Hooghoudt's tables:

$$d_e = \frac{\pi S}{8}\left[\ln \frac{S}{\pi r} + F(x)\right] \tag{9.20}$$

where

$$x = \frac{2\pi D}{S}$$

and

$$F(x) = \sum \frac{4e^{-2nx}}{n(1 - e^{-2nx})}, \text{ with } n = 1,3,5\ldots$$

The $F(x)$ converges rapidly for $x > 0.5$.

The above equation must be solved iteratively, as both "d_e" and "S" are unknown.

Van Beer Equation

Van Beers equation for equivalent depth is (ILRI, 1973):

$$d_e = \frac{D_S}{1 + \frac{8D}{\pi S} \times \frac{8D_S}{\pi^2 r}} \tag{9.21}$$

where, r is radius of drain pipe (m), D is drain depth (m), D_s is thickness of the aquifer below drain level.

Moustafa (1997) suggested from field investigation that a 5m depth instead of infinity for the impermeable layer in Nile Delta should be used in design purpose.

Important Parameters in Hooghout Drain Spacing Equation (and also in Others) and Their Interpretation

Form the equations of (9.15), (9.16) and (9.17), it is revealed that drain spacing is directly related to the hydraulic conductivity of the soil (K), and inversely related to the drainage discharge/outflow (q) from the field. So, these two parameters (input values) should be determined/estimated accurately.

K of Soil

If the estimated K value is less than the actual one, calculated drain spacing will be lower, and hence more financial involvement compared to actual need (i.e. financial loss). On the other hand, if the estimated K value is higher than the actual field value, the drain spacing will be higher than the actual need. Thus, there is a possibility of prolonged standing water in the root zone, and hence chance of crop loss.

q Value

The reverse will be true in case of q value.

9.5.1.2 Donnan's Formula

Donnan proposed the following formula for parallel drain spacing:

$$S^2 = \frac{4k}{q}\left[(D+h)^2 - D^2\right] \tag{9.22}$$

Solving for algebraic functions, the above equation reduces to

$$S^2 = \frac{4kh^2}{q} + \frac{8kDh}{q}$$

which is similar to Hooghoudt's equation.

9.5.2 Formula for Irregular Drain System

Most investigations, both theoretical and experimental, have focused on parallel drainage systems with equally spaced tiles. However, in many watersheds (e.g., as those in Illinois), parallel systems do not occur as frequently as irregular systems. Irregular tile systems predominate in areas where the majority of the tiles drain small depressional areas, and thus tile lines are placed at irregular angles and spacings. In these systems, a constant spacing parameter does not exist or cannot be easily defined.

Irregular and parallel tile systems are hydraulically different, as a single tile draws water from a semi-infinite distance on either side (Fig. 9.14), while the region of influence of each tile in a parallel system is constrained by the neighboring tiles.

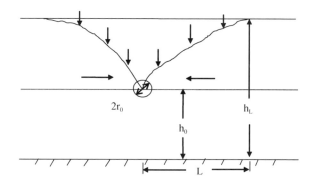

Fig. 9.14 Schematic of flow to single drain

Cook et al. (2001) derived equation for random and irregular tile drainage system. Their equation is

$$q_0 = \frac{K}{L}(h_L^2 - h_0^2) \qquad (9.23)$$

where

q_0 = the flow rate at the tile drain from each side (m³/s)
K = hydraulic conductivity (m³/m²/s)
L = the distance at which the water table becomes essentially horizontal (m)
h_L = height of water table above the impervious layer at distance L (m)
h_0 = height of drain above the impermeable layer (m)

The total drain outflow would be twice q_0, since water flows into the drain from both sides. It should be noted that this is identical to the result for parallel systems if L is replaced by the half spacing, $S/2$.

9.5.3 Determination of Drain Pipe Size

The pipe size should be 25–50% higher than the maximum design discharge, to compensate possible reduction in the net capacity of pipe due to deposition of silt. Discharge carrying capacity of the pipe can be calculated using Manning's equation (which is described in Chapter 1, this volume).

9.6 Design of Subsurface Drainage System

9.6.1 Factors Affecting Spacing and Depth of Subsurface Drain

Factors which influence drain spacing and depth include

- root zone depth of the proposed crop
- sensitivity of the crop to water logging or salinity
- soil texture (coarse or fine)
- salinity level of soil and/or groundwater
- depth of groundwater table
- root zone depth at saline sensitive growth stage
- depth of impervious soil layer
- hydraulic conductivity of the soil (horizontal and vertical direction)

9.6.1.1 Soil Salinity

Salinity distribution data of soil profile should be considered when selecting the drain depth.

9.6.1.2 Impact of Soil Texture on Drain Depth

Upflow *or* capillary rise of water through capillary tube (resembled to soil pore) can be expressed as

$$h_{cr} = \frac{2\tau}{r\rho} \cos\theta \tag{9.24}$$

where

h_{cr} = capillary rise of water (cm)
r = radius of tube (cm)
ρ = density of water (g/cm^3)
τ = surface tension (g/cm)
θ = angle of contact between meniscus and wall of tube (deg.)

Taking density of water, $\rho = 1$ gm/cm^3, $\tau = 0.074$ g/cm (for water of 20°C), and neglecting the angle (i.e., $\theta = 0$), we get

9.6 Design of Subsurface Drainage System

$$h_{cr} = \frac{0.15}{r} \qquad (9.25)$$

From the Eqs. (9.24) and (9.25), it is revealed that the h_{cr}, the capillary rise or height of upflow, is inversely proportional to the radius of the capillary tube. Thus for soil of clay type, i.e., soils having smaller diameter of pore, capillary rise will be higher. For the sandy soil, due to larger diameter of pore, the reverse will be true. Therefore, to minimize rise of groundwater within the root zone (or target soil depth), drain should be placed at greater depth in clayey or fine-textured soil than the coarse-textured soil (for similar crop and hydro-geologic condition).

9.6.2 Data Requirement for Subsurface Drainage Design

Data required for proper design of a subsurface drainage system include

- soil layering
- depth to layers restricting vertical flow
- soil hydraulic properties (for each layer)
- depth to water table
- salinity status of soil and ground water
- sources of drainage water other than deep percolation
- cropping pattern and crop root zone depth
- type of irrigation system
- irrigation schedule
- irrigation efficiency
- climate data

Soils data collection and analysis is common to all design procedures. Sampling and investigation of the soil must be done up to below the depth of potential drain placement (2–4 m) to determine the presence of a restricting layer. The soil salinity profile above the drain is useful to determine the need for remediation. The soil salinity profile below the drains is needed because this will be indicative of the potential salt load when the drains are in operation.

If the soil properties vary considerably within a farm, the area should be divided into "sub-areas" or "blocks." Then, drain spacing should be calculated for each sub-areas or blocks.

9.6.3 Layout of Subsurface Drainage

Layout of installing subsurface drain depends on shape of the farm/catchment, topographical feature, and drainage disposal facility. A typical layout is given in Fig. 9.15.

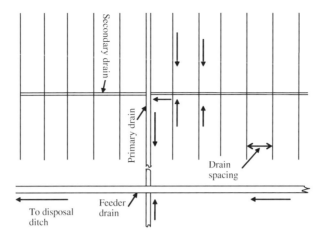

Fig. 9.15 Schematic of a typical layout of subsurface drainage system

9.6.4 Principles, Steps, and Considerations in Subsurface Drainage Design

9.6.4.1 Principles

The design of a subsurface drainage system requires developing criteria that

- specify the operation of the system and the physical configuration that fulfill the drainage objective(s).
- consider irrigation and drainage systems as an integrated water management system.
- minimize deep percolation losses through improved irrigation water management (source control)
- characterized by establishing the water-table depth at the mid-point between laterals and the drainage coefficient (that specifies the maximum volume, expressed as depth of water to be removed in a 24-h period)
- Specify option regarding reuse of drainage water for irrigation or stimulation of *in situ* use by crop through control of water table.

In arid areas an additional design criterion is the control of salt accumulation by capillary rise into the crop root zone, which is accomplished by managing the mid-point water-table depth to minimize upward flow of water and salt from the shallow ground water.

9.6.4.2 Steps

In designing subsurface drainage system in a command area or watershed, the following steps and procedures need to be followed:

9.6 Design of Subsurface Drainage System

(i) Investigate the soil profile and geo-hydrologic condition including groundwater quality
(ii) Measure the quality of proposed irrigation water
(iii) Estimate the sources of drainage water other than irrigation
(iv) Review and analyze the climatic data of the area
(v) Select appropriate crop(s)/cropping pattern
(vi) Measure the hydraulic conductivity of the root zone soil
(vii) Estimate drainage requirement or drainage coefficient
(viii) Optimize the depth of drain placement and lateral drain spacing (considering permissible mid-way water-table depth for the selected crop/cropping pattern, in situ crop water-use plan by capillary rise*, and cost of drain).

Different combinations of drain depth and spacing will result in same drainage coefficient. But the water quality of drainage may be different. For a particular drainage intensity, shallow depth of drains requires narrow spacing, thus drainage cost increases with decreasing drain depth.

First, a drain depth is specified and the spacing is calculated based on the recharge schedule and the mid-point water-table depth criteria. Subsequently, the drain depth will be varied to calculate a range of depths and spacing, and economic analysis should be performed for each case. The most economic drain depth and spacing is then selected from analyses of several drain system configurations.

(ix) Determine lateral pipe size and main pipe size (capable of carrying the maximum drainage rate)
(x) Design the drain envelop material
(xi) Design the drainage disposal system, or decide regarding reuse of drainage water for irrigation
(xii) Design the pump size to pump the maximum drainage discharge from the field (if need to be pumped)

* In situ use by the crop will affect the drainage design by reducing the irrigation requirement and the deep percolation losses that will be included in the drain system design procedure.

The USBR recommends installation of drains at a depth of 2.4 m, if possible, to provide a balance between the system cost and spacing.

9.6.4.3 Estimation of Drainage Requirement or Drainage Coefficient

For estimation of drainage requirement (or drainage intensity, or drainage coefficient), following steps may be followed:

- collect long-term rainfall and other weather data for the project area
- calculate daily average rainfall, evaporation, and evapotranspiration rate for the target crop season
- perform water-balance

Under natural rainfall condition, water balance can be expressed as

$$P = \text{ET} + R + D$$

or

$$D = P - \text{ET} - R \tag{9.26}$$

where

P = rainfall rate (mm/d)
ET = evapotranspiration rate (mm/d)
R = surface runoff amount, mm/d (if surface runoff is feasible)
D = deep percolation *or* subsurface drainage amount (mm/d)

The "D" value will indicate the drainage coefficient or drainage requirement. Instead of average rainfall, peak rainfall of 10–20 years recurrence interval may be used. Procedure for such determination has been described in Chapter 3 (*Weather*), Volume 1.

Drainage Requirement for Salinity Control

For salinity control purpose, leaching requirement (LR) should be calculated. Procedure for calculating LR has been described in Chapter 8 (*Management of Salt-Affected Soils*), this volume.

9.6.4.4 Criteria and Considerations

The USBR (United States Bureau of reclamation) recommends a minimum water-table depth from 1.1 to 1.5 m below land surface midway between lateral drains, depending on the crop rooting depth. From past records, this should result in achieving at least 90% of maximum crop production (US Department of Interior, 1993). The midpoint water-table recommendation ensures that the soil oxygen status is maintained in the root zone and reduces capillary transport of water and salt from shallow groundwater into the root zone and to the soil surface due to evaporative demand.

Deep placement of the drains generally results in a wide drain spacing that lowers the system cost relative to shallow one. However, in many cases deep placement has been shown to result in an excessive salt load being discharged with the drainage water (Christen and Skehan, 2001; Ayars et al., 1997). This is because the soil is normally more saline than the irrigation water salinity in most cases (Ali and Rahman, 2009; Ayars et al., 1997).

9.6.5 Controlled Drainage System and Interceptor Drain

9.6.5.1 Controlled Drainage

Introduction

In the past, subsurface drainage systems were typically designed to discharge water continuously, without regard to the environmental consequences and the effects on crop production. This philosophy has changed in humid areas of the world, as the environmental consequences and crop production impacts have been researched. Recently, controlled drainage has been identified as a potential management method in humid areas to reduce nitrate loading to surface water (Ayars et al., 2006; Hornbuckle et al., 2005; Lalonde et al., 1996; Doty and Parsons 1979).

Principle of the Method

In this drainage system, the water table is maintained at a shallower depth by a control structure which reduces deep percolation below the root zone by reducing hydraulic gradients and increases potential capillary upflow, as evapotranspiration depletes soil-water in the root zone (Fig. 9.16).

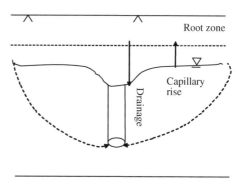

Fig. 9.16 Schematic of control drainage system

The flow lines are shallower than in the uncontrolled system and are more concentrated when close to the soil surface. In soil profiles with zones of lower soil salinity at the soil surface, this system results in decreased drain water salinity compared to the uncontrolled system. The reduced drain flows and lower salinity result in much reduced salt loads.

Management of Control Drainage System

Traditional drainage does not need much management aspects, but simply letting the system run continuously. In control drainage, active management measures are needed to regulate flow and reduce the impact of saline drainage water on the environment. The goal may be to reduce total drainage flow, reduce contaminant load, improve irrigation efficiency, or some combination of these outcomes. The effectiveness of drainage system management will depend on the crop, the ground water quality, and the water table position. Field research and model studies suggest that shallower placement of drainage laterals and reduced depth to mid-point water table will result in reductions in drainage volumes and salt loads.

9.6.5.2 Interceptor Drain

Where it is not practical to drain water out of a pocket point by other means (or if constraints exists), a collector system should be provided to drain water from the drainage layer. Collector systems may include plastic pipe slotted at the edge, or drain pipe installed in a longitudinal collector trench. This will limit the longitudinal seepage distance in the drainage layer, minimizing the drainage time and preventing the buildup of a hydrostatic head under the surface layer (Fig. 9.17).

Fig. 9.17 Schematic representation of interceptor drain

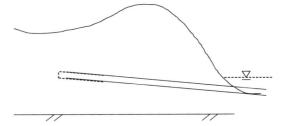

An interceptor drain may be defined as the drain that is constructed for the purpose of intercepting or cutting off groundwater which is moving down-slope. Its source may be of different origin. Interceptor drain may be either open or close construction.

Donnan (1959) raised several issues regarding interceptor drain: (1) How deep should the interceptor drain be placed? (2) How far upslope will the drawdown be effective? (3) Where does the post-installation water table become asymptotic coincide with the undisturbed (original) water table? (4) What is the shape of the drawdown curve on the down slope side? (5) How much of the total flow is intercepted? (6) What is the required capacity of the drain? (7) Should the drain be open or closed? He concluded in the way: drain should be placed as deep as it is practical (open ditch at the bottom of impervious layer), drawdown upslope will depend on the initial slope of the water table, down slope will be dependent on the water level in the drain device, the quantity of flow will vary directly with the depth of flow intercepted, and both open and tile drain are equally efficient.

9.6.6 Sample Workout Problems

Example 9.5

In an agricultural command area, the long-term average of daily maximum rainfall and evapotranspiration are 70 and 6 mm, respectively. The surface runoff on catchment basis can be considered as 40% of the rainfall. If the proposed crop does allow

9.6 Design of Subsurface Drainage System

ponding water more than 24 h, determine the subsurface drainage requirement for survival of the crop.

Solution

Considering water balance on daily basis,

$P = ET + R + D$
or, subsurface drainage, $D = P - ET - R$
Given,
$P = 70$ mm
$ET = 6$ mm
$R = 70 \times 0.4 = 28$ mm
Putting the values, $D = 70 - 6 - 28 = 36$ mm

That is, daily drainage requirement is 36 mm, or 1.5 mm/h (Ans.)

Example 9.6

In a saline agriculture, estimate the drainage requirement for successful crop production from the following information:

Crop: Wheat
$EC_e = EC_{dw} = 8.0$ dS/m
$EC_{iw} = 1.5$ dS/m
$ET = 5$ mm/d (peak rate)

Solution

$$LR = \frac{D_{dw}}{D_{iw}} = \frac{EC_{iw}}{5EC_e - EC_{iw}} = \frac{1.5}{5 \times 8 - 1.5} = 0.038$$

$$D_{iw} = \frac{ET}{1 - LR} = \frac{5}{1 - 0.0389} = 5.202$$

$$D_{dw} = 5.202 - 5.0 = 0.202 \text{ mm/d}$$

Considering irrigation application efficiency, $E_a = 0.8$, that is 20% is lost to deep percolation; the excess application is

$$= \frac{5}{0.8} - 5 = 1.25 \text{ mm/d, which is greater than } 0.202 \text{ mm/d}$$

Thus, the drainage coefficient, or drainage requirement $= 1.25$ mm/d (Ans.)

Example 9.7

A subsurface drainage should be provided through an agricultural farm of several kilometers long. Average expected recharge to the water table in the area is about 3 mm/d. It is required to maintain the water table not closer than 1.2 m from the soil surface. The value of hydraulic conductivity up to 2.0 m is 40 mm/d, and below that is 42 mm/d. Relatively, impervious layer exists 8 m below the soil surface. Determine the drain spacing.

Solution

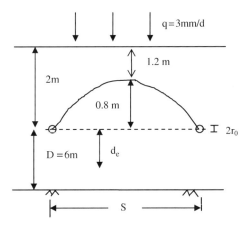

We know, Hooghoudt's drainage equation is

$$S^2 = \frac{4k_a h^2}{q} + \frac{8k_b d_e h}{q}$$

Given,

$q = 3$ mm/d $= 0.003$ m/d
$k_a = 40$ mm/d $= 0.04$ m/d
$k_b = 42$ mm/d $= 0.042$ m/d
$D = 6$ m
Assuming drain diameter $= 20$ cm. Thus, $r_0 = 20/2 = 10$ cm $= 0.1$ m
Placing the drain at 2.0 m below the surface, $h = 2.0 - 1.2 = 0.8$ m
Determination of equivalent depth (d_e):
We know,

$$d_e = \frac{\pi S}{8 \ln\left(\frac{S}{\pi r_0}\right)}$$

Writing the above formula and drain spacing formula in Excel worksheet, and solving for a series of values (trial values) for the variable "S" and "de," we can find that
when $S = 9.8$ m, then $d_e = 1.1$ m, and new value of $S = 9.98$, which is close to 9.8.
Thus, taking $d_e = 1.1$ m,
$S = 9.98$ m (Ans.)

9.7 Envelope Materials

9.7.1 Need of Using Envelop Material Around Subsurface Drain

Water moves to the drain due to hydraulic gradient. The hydraulic gradient that develops at the drain openings is often high enough to cause piping of soil material through the openings. For this reason an adequate amount of protective materials (termed as "envelope material") is needed around subsurface drain.

The functions of placing envelop material around subsurface drain conduits are the following:

(i) Prevent the movement of soil particles into the drains which may settle and close the drains (Barrier function)
(ii) Provide material in the immediate vicinity of the drain openings that is more permeable than the surrounding soils, thus increasing effective diameter of the drain (Hydraulic function)
(iii) Provide suitable bedding for the drain (Bed material function)

9.7.2 Need of Proper Designing of Envelop Material

The envelope material should be selected such that the material or any combination of materials gives adequate protection against siltation, yet providing relatively unrestricted water movement from the soil to the drain. To fulfill the objective of using envelope material, proper selection of envelope material and design is a must. Good design aims to assure an optimum combination of performance, long survive, and reasonable cost.

9.7.3 Materials for Envelope

Materials used for subsurface drain have included almost all permeable porous materials that are available economically in large quantities. Granular mineral materials such as coarse sand, fine gravel, and crushed stone have been used for decades in arid and semi-arid countries as a drainpipe envelope.

Envelope materials can be categorized into three groups:

(a) *Organic envelopes*: organic materials that are used as envelope materials. Frequently used materials are sawdust, chaff, cereal straw, flax straw, cedar, leaf, bamboo, corncobs, wood chips, reeds, heathers, bushes, grass sod, and coconut fiber.
(b) *In-organic/mineral envelopes*: the most common and most widely used envelope materials are naturally graded coarse sands and fine gravels.
(c) *Man-made/synthetic envelopes*: this type includes fiberglass, geo-textile sheets (<0.5 mm), successors of glass fiber sheets. Synthetic materials are relatively inexpensive and can be manufactured at large quantities of exact specification.

9.7.4 Design of Drain Envelope

9.7.4.1 Steps

The steps for designing drain envelope materials are the following:

- Make a mechanical analysis of both the soil and the proposed envelope material
- Prepare particle size distribution curve for each material
- Decide by some set of criteria whether the envelope material is satisfactory or not.

9.7.4.2 Criteria for Selecting Envelope Material

Many researchers have done work on specification of envelope materials, and different ratios were recommended. Terzaghi (1922), Karpoff (1955), Juusela (1958), Kruse (1962), Soil Conservation Service of USDA (SCS, 1971, 1988, 1994), the United States Army Corps of Engineers (US Army Corps of Engineers, 1941, 1978), the United States Bureau of Reclamation (USBR, 1973, 1978, 1993), Stuyt (1992) sequentially improved and suggested criteria for envelop materials. More recent work was carried out by Vlotman et al. (1994, 1995, 2000) and Stuyt et al. (2005).

Design Criteria for Synthetic Fiber Envelope

Dierickx (1993) and Vlotman et al. (2000) made detail review of various proposed retention criteria, primarily for geotextiles. The ratio of O_{90}/d_{90} is frequently used, with O_{90} and d_{90} the envelope pore size and the soil particle size, respectively, for which 90% of the pores or particles are smaller. Vlotman et al. (2000) propose that the retention criteria O_{90}/d_{90} will range from 2.5 to 5 for envelope thickness ranging from 1 to 5 mm.

Stuyt et al. (2005) suggested the following criteria for synthetic fiber envelope: for thickness ≥ 5 mm, $O_{90}/d_{90} \leq 5$; for thickness between 3 and 5 mm, $O_{90}/d_{90} \leq 4$; for thickness between 1 and 3 mm, $O_{90}/d_{90} \leq 3$; and for thickness ≤ 1 mm, $O_{90}/d_{90} \leq 2.5$. They noted that $O_{90}/d_{90} \geq 1$ minimized the risk of mineral clogging.

9.7 Envelope Materials

Design Criteria for Granular Mineral Envelopes

For gravel envelope, Vlotman et al. (2000) suggested the following criteria for the coarse boundary envelop material: for filter retention, $D_{15c} < 7d_{85f}$; for gradation curve guide, $D_{60c} = 5D_{15c}$; segregation criteria, $D_{100} < 9.5$ mm. For the fine boundary envelope materials, the criterions are the following: for hydraulic, $D_{15f} = >4d_{15c}$; for gradation curve guide (bandwidth), $D_{15f} = D_{15c}/5$ (based on $C_u^b \leq 6$ and bandwidth ratio ≤ 5); for hydraulic criterion, $D_5 > 0.074$ mm; for gradation curve guide (bandwidth), $D_{60f} = D_{60c}/5$; for retention criteria (bridging), $D_{85} = D_{60c}$.

In the above criterion, symbols "D" and "d" refer to particle size distribution of the gravel and base soil, respectively, where number gives the percentage with a smaller diameter. For example, D_{50} is the diameter of the sieve where 50% passes. The "C_u" means uniformity coefficient and the superscript "b" means base materials. The indices c and f in the subscripts refer to the upper and lower particle size diameters of the base-soil and the gravel.

Design Criteria by SCS (1971)

The SCS has combined other researcher's results into a specification. Their criterions are

$$\frac{D_{50f}}{D_{50b}} = 12 \text{ to } 50$$

$$\frac{D_{15f}}{D_{15b}} = 12 \text{ to } 40$$

Other criterions: filter materials should pass 1.5 in. sieve, 90% should pass the 0.75 in. sieve, and no more than 10% should pass the 0.01 in. sieve.

If the filter and base materials are nearly uniform, a filter stability ratio of less than 5 is generally safe, i.e., $\frac{D_{15f}}{D_{85b}} < 5$.

The uniformity co-efficient, C_u, (D_{60}/D_{10}) was also considered as a design factor. A low uniformity co-efficient indicates a uniform material. Kruse (1962) considered envelope material with a C_u of 1.78 to be uniform.

9.7.4.3 Envelope Thickness

As the amount of envelope material requirement increases with the square of the diameter, it is important to judicious selection of envelope thickness. The US Bureau of Reclamation recommended a minimum thickness of 10 cm around the pipe. The SCS (1971) recommends 8 cm minimum thickness.

9.7.5 Use of Particle Size Distribution Curve in Designing Envelop Material

Particle size distribution curve shows the relationship between grain size and percent larger (or finer) at that size. Particle size distribution curves of the soil and the proposed envelope material enable the selection of an envelope material, which best suits the soil. In designing, we have to compare the two particle size distribution curves and decide by the set of criteria (described earlier) whether the envelope materiel is satisfactory or not.

In determining maximum size of the perforations or joint openings of the drainage pipe, the 85% size of the envelope material is used from the particle size distribution curve.

9.7.6 Drain Excavation and Envelope Placement

Excavating and trenching machines, driven by steam engines, were introduced in 1890. Trenchers dig a trench at the required depth and grade and place the drain pipe at the bottom of the trench. Several types of trenchers are produced in various sizes and a wide range of capacities. They can install pipes to a depth of about 3 m in trenches up to 0.50–0.60 m in width. Trenchers have been developed in different modified models so that they can also be used to install drains in stony soils, in orchards, or in soils with high water tables.

Water enters the drain through the sides and bottom of the drain. The hydraulic gradient that develops at the drain openings is often high enough to cause piping of soil material at the openings. For this reason an adequate amount of envelope material is needed under the drain pipe as well as on top.

Where drains are laid manually, a layer of envelope material is placed and leveled to the design grade in the bottom of the trench (before the drain is laid). The drain pipe is then put into place and covered with envelope material to the required depth. The trench is then backfield with soil. Trenching machines can be fitted with two-hoppers for placing envelope material under and over a drain on a continuous basis.

Care should be taken to protect the drain-envelope system immediately following installation. No heave loads, mechanical or hydraulic, should be imposed until the loose back-fill material in the trench is consolidated naturally.

9.8 Models in Drainage Design and Management

Nowadays, models are useful and easy-to-way tool to design surface and subsurface drains, and management of drainage system. Some of the existing models are described below:

9.8.1 DRAINMOD

DRAINMOD is a deterministic hydrologic model developed to simulate a soil-water regime of surface and subsurface water management systems (Skaggs, 1978). It predicts surface runoff, infiltration, evapotranspiration, subsurface drainage, and seepage from tile drained landscapes. A basic relationship of the model is a water balance for a vertical soil column of unit surface area, which extends from the impermeable layer up to the soil surface, and located midway between adjacent drains.

9.8.2 CSUID Model

The Colorado State University Irrigation and Drainage (CSUID) model is a simulation model that can be used in the design and management of conjunctive irrigation and drainage systems (Manguerra and Garcia, 1995). The model is based on the numerical implementation of a quasi three-dimensional finite-difference model, which solves the Richards equation and the advective-dispersive transport equation for one-dimensional vertical flow and salt transport in the variably saturated zone above the water table; and it also solves the depth-averaged Boussinesq equation and two-dimensional advective-dispersive transport equation for areal flow and transport in the fully saturated zone below the water table. It is capable of drainage designing and management.

9.8.3 EnDrain

"EnDrain" is a software tool that calculates drainage discharge, hydraulic head, and spacing between parallel subsurface drains, pipes, or open ditches; with or without entrance resistance (Oosterbaan et al., 1996). It shows the curve of water table. The drain spacing calculations are based on the concept of the energy balance of groundwater flow. However, the traditional concepts based on the Darcy and water balance or mass conservation equations are also used. The program allows for the presence of three different soil layers with different hydraulic conductivity and permeability – one layer above and two below drain level. "EnDrain" can be used for the reclamation of saline soils. The software is free, and can be downloaded from the site: http://www.waterlog.info/

9.9 Drainage Discharge Management: Disposal and Treatment

9.9.1 Disposal Options

There are limited number of options available when trying to decide where and how to dispose of agricultural drainage water into the natural hydrological system.

The common option is to return the water either to natural depressions or lakes or rivers, or to salt sinks, such as ocean, or to return the water either to the land as part of the irrigation water supply. The options available to any single project may be limited because of water quality concerns. Downstream beneficial uses of any surface water body to which drainage water is added must be protected. For example, it may not be appropriate to discharge saline drainage water into a river of lake, when that surface water body is being used for domestic or agricultural water supplies. Disposal of runoff and drainage waters into natural depressions has been practiced for centuries. The impounded waters are dissipated by evaporation, seepage, and transpiration losses. The use of constructed disposal basins for saline agricultural drainage waters are also common worldwide where there are constraints on discharging into natural salt sinks such as the oceans and inland closed basins. In the Murray-Darling basin in Australia, some of the constructed evaporation basins are intended to hold saline water only temporarily. The stored waters are then released during high river flows.

9.9.2 Treatment of Drainage Water

Treatment approaches of drainage water can be divided into three general categories:

– physical
– chemical
– biological

Many processes exhibit both physical and chemical aspects and so are sometimes called physical/chemical or physiochemical treatment.

The first step in the selection of any treatment process for improving drainage water quality is to thoroughly define the problem and to determine what the treatment process has to achieve. In most cases, either regulatory requirements or the desire to re-use the water will be the driving force in defining the treatment issues to be selected for a particular drainage water. A thorough knowledge and understanding of these water quality criteria is required prior to select any particular treatment process.

An introductory description is given below for the most common treatment methods, which have application in treating agricultural drainage water. More detailed description of common treatment processes and design procedures of treatment plant (Waste water treatment) can be found in "Wastewater treatment" texts/books.

9.9.2.1 Physical/Chemical Treatment

Particle Removal

Several physical processes aim to remove suspended particulate matter. Subsurface drainage water itself is usually low in suspended particles. These processes might be used in an overall treatment process for the removal of particulates formed in another stages of the treatment, such as removal of bacteria from a biological system or removal of precipitates formed in a chemical treatment process. Particle removal unit processes include sedimentation, flotation, centrifugation, and filtration. Filtration further includes granular media beds, vacuum filters, belt presses, and filter presses.

Adsorption

Adsorption is the process of removing soluble contaminants by attachment to a solid. A common example is the removal of soluble organic compounds via adsorption onto granular activated carbon (GAC). GAC is useful for its ability to remove a wide range of contaminants. Certainly, if pesticides were a concern for the drainage water being examined, the use of GAC adsorption would be a leading candidate for treatment.

Another treatment for removing volatile compounds for water is air stripping. In a conventional counterpart of air stripping operation, the contaminated water is distributed at the top of a tall reactor vessel that is packed with materials or structures with a high surface area. As the water moves downward, clean air is introduced at the bottom of the reactor and moves upward. As the water and air make contact, volatile compounds are transferred from the liquid phase to the gas phase according to gas transfer theory.

Distillation

Distillation is a thermal process used for salt removal. Heat is used to vaporize the water, leaving the salts behind. The water vapor is condensed to a high quality water. The process is energy intensive. Reverse osmosis is used for desalination applications.

9.10 Economic Considerations in Drainage Selection and Installation

Economic analysis should be carried out before undertaking a drainage project. The result of economic analysis will help in deciding whether drainage system installation is feasible from economic point of view, or which type of drainage (e.g., surface (deep drain) or subsurface, or, random field lateral *or* uniform drainage) is more economical. Sometimes, the economic criterion is not only deciding criteria, but also depends on the national/regional food security issue, national priority, and government policy.

Detail procedure for performing economic analysis and decision criteria (along with sample example) has been discussed in Chapter 11 (*Economics in Irrigation Management*), Volume 1.

9.11 Performance Evaluation of Subsurface Drainage

9.11.1 Importance of Evaluation

Subsurface drainage is instrumental in the improvement of non-productive soils, and it can assist in avoiding unsuitable soil conditions during farming operations. Knowledge of subsurface drainage performance is important in order to use reclaimed land rationally, and to apply the scarce available financial means to repair improperly functioning drains.

9.11.2 Evaluation System

Discharge (q) and water-head midway between drains (h) are two important design parameters of subsurface drainage. To evaluate the performance of subsurface drainage, these two parameters, and water-head on top of the drains, are to be measured and compared with the design (and/or expected) value under the prevailing conditions. Measurement of water table should be carried out every 3–5 days depending on rainfall amount, ET demand, and hydraulic conductivity. The measurement should be continued for a reasonable length of time (at least one season or a part), and in several plots.

If the design parameters differed significantly from the designed/expected value, it indicates that original design criteria are not satisfied. The causes of drain malfunctioning may include the drains are too widely spaced, drain envelope materials are not properly designed, changes in hydraulic conductivity of the soil from the measured/gauged one (may be due to erroneous measurement of K, compaction of soil due to traffic or natural compaction/settlement). More elaborate measuring setup as well as long-term observation may be needed to ascertain the causes of failure or drain malfunctioning. Under natural rainfall condition, low rainfall can results in small discharge and low head.

In addition to the piezometer on top of the drain (which measures the water-head above the drain), another piezometer may be connected to the drain to ascertain whether the head above the drain is due to entrance resistance or to backpressure in the drain.

If we consider the drain center as reference level, the total head loss in the system (h_t) can be expressed as

$$h_t = h_m - r_0$$

where

h_m = head in the piezometer midway between drains (or the height of the groundwater table midway between drains above drain level) (m)
r_0 = outside radius of the drain (m)

The approach flow head loss or the head loss in the vicinity of the drainage system (drain pipe and envelope material), h_{ap}, is given by (Rimidis and Dierickx, 2003):

$$h_{ap} = h_v - r_0$$

where h_v is the head in the piezometer in the vicinity of the drain pipe (usually 40 cm away from the drain center). The approach flow resistance (W_{ap}) is

$$W_{ap} = h_{ap}/qL$$

where q is the specific discharge (m³/d/m) and L is the drain spacing (m).

Measured data can be expressed in equation similar to Hooghoudt equation. Rimidis and Dierickx (2003) expressed their experimental evaluation data in the form:

$$q = Ah_t^2 + Bh_t$$

The above equation is similar to the well-known equation of Hooghoudt (1940):

$$q = \frac{4Kh_t^2}{L^2} + \frac{8Kdh_t}{L^2}$$

where

$$A = 4K/L^2, \text{ and } B = 8Kd/L^2$$

9.12 Challenges and Needs in Drainage Design and Management

To meet the challenge of a sustainable irrigated agriculture, minimum impact on the environment should be ensured. Controlled drainage, a comparatively new approach of drainage management, is suffering from lack of design criteria for both humid and arid regions. Ayars et al. (2006) urged for development of new design criteria and management methods for controlled drainage system that should have minimum impact on environment.

Drain installation and maintenance still require a huge investment and skill operators. Low-cost drainage material with robustness should be sought in order to bring the technology to common farmers, who need it.

Relevant Journals

- Trans. ASAE
- Journal of Irrigation & Drainage Engg., ASCE
- ASAE Papers
- Land Management and Reclamation
- Irrigation and Drainage System
- Applied Engineering in Agriculture
- Irrigation Science

- Agricultural Water Management
- ICID Bulletin
- Egyptian Journal of Irrigation & Drainage
- Soil Science Society of America Journal
- Agronomy Journal
- Journal of Soil Water Conservation
- Journal of Environ. Quality

FAO/World Bank Papers

- FAO Irrigation and Drainage Paper 60 (Materials for Subsurface Land Drainage Systems, 2000)
- FAO Irrigation and Drainage Paper 61 (Agricultural drainage water management in arid and semi-arid areas, 2003)
- FAO Irrigation and Drainage Paper 62 (Guidelines and computer programs for the planning and design of land drainage systems, 2007)
- FAO Training Manual: Drainage of Irrigated Lands (Irri. Water Management Training Manual No.9, 1996)
- World Bank Technical Paper 195 (Drainage Guidelines, 1992)

Questions

(1) What are the benefits of drainage? Is drainage necessary in arid region?
(2) How you will assess the need of drainage in an agricultural farm?
(3) What factors will you consider during investigation and design for a drainage problem area?
(4) What suggestions will you made to lessen the drainage outflow from an area?
(5) What is drainable pore volume?
(6) Distinguish among homogeneous, isotropic, and homogeneous and isotropic media.
(7) Deduce the partial differential equation for flow through the homogeneous, anisotropic soil media. From this, derive the equation for homogeneous and isotropic soil.
(8) Derive an equation for calculating resultant/equivalent horizontal hydraulic conductivity for layered soil.
(9) Briefly describe the design steps and principles of surface drainage system.
(10) Discuss the conditions of resultant vertical conductivity for layer soil.
(11) Define with neat sketch the Hooghoudt's drain spacing equation for layered soil.
(12) What do you mean by "equivalent depth" in Hooghoudt's drain spacing equation?
(13) Draw a schematic layout for a surface drain design in a farm.

Questions

(14) What are the factors to be considered in designing subsurface drainage system?
(15) Briefly describe systematically the design procedure of subsurface drainage.
(16) Draw a schematic layout for a subsurface drainage system.
(17) How you will estimate the drainage intensity in an area?
(18) What is controlled drainage? Briefly discuss its principle and procedure.
(19) Do you think that envelop materials should be used around subsurface drain? Justify your answer.
(20) Why envelope materials should be designed and selected with proper care?
(21) Name some envelope materials from organic, mineral, and synthetic group.
(22) What are the design criteria for selecting envelope material? Briefly describe the procedure for selection and design of envelope material.
(23) Briefly describe the use of particle size analysis curve in designing envelope material.
(24) What are the methods of placing drain and envelop?
(25) What are the methods of disposing drainage outflow?
(26) How the drainage outflow can be treated?
(27) Narrate the economic considerations in drainage design and material selection.
(28) Briefly explain the concept, principle, and procedure of performance evaluation of drainage system.
(29) What are the challenges and needs in drainage design and management?
(30) A soil contains 42% moisture (by volume) at saturation. The moisture content becomes 35% when the capillary pressure is raised to 100 cm of water. Find the drainable porosity, void ratio, and the drainable water volume at 100 cm water tension from a (10 m × 10 m) plot of 1.0 m depth.
(31) A soil column consists of 3 layers, having 0.50 m depth of each. The horizontal hydraulic conductivity of the layers (from to bottom) are 0.25, 0.15, and 0.30 $m^3/m^2/h$. Determine the resultant horizontal hydraulic conductivity of the soil column.
(32) A surface drainage should be provided within an agricultural farm. Maximum rainfall intensity at the site in 20 years record is 40 mm/h. The tertiary drain would have to carry discharge from 5.0 ha land. The secondary drain would have to carry of 4 tertiary, and the main drain to carry discharge of 5 secondary drains (of similar discharge). Determine:

  (a) drainage outflow to be generated at tertiary, secondary, and main drain level
  (b) performance of hydraulic design of the secondary and main drainage canal

(33) Determine the drain spacing of parallel subsurface drains if the hydraulic conductivity of the soil is 15 cm/d, the recharge rate is 12 cm/d, drain diameter is 12 cm, drains are placed at 2.0 m deep, the water table is to be no closer than 1.2 m to ground surface, and the impervious layer is 3 m below the ground surface.
(34) A subsurface drainage should be provided through an agricultural farm. Maximum rainfall intensity at the site in 20 years period is 80 mm/d. The

secondary drain would have to carry 15 of tertiary, and the main drain to carry discharge of 4 secondary drain (of similar flow). It is required to maintain the water table not lower than 1.0 m from the soil surface. The soil is uniform in texture and other hydraulic properties. The value of K is 6 mm/d. A relatively impervious layer exists at 1.8 m depth from the surface. Determine:

(a) drain spacing at tertiary level
(b) size of secondary and main drain pipe to carry the generated outflow
Assume standard value of any missing data.

(35) It is proposed to design a subsurface drainage system to leach water through the root zone in an area where surface irrigation will be practiced with poor quality water. Parallel perforated GI pipe drains are to be laid to a slope of 3 in 1,000, so that the drains could discharge freely into the main collector ditch. Hydraulic conductivity of the soil is 10 mm/d. Calculate the drain spacing required to meet the requirements given below.

Irrigation water EC = 0.15 dS/m
Salt tolerance level = 1.8 dS/m
Maximum crop rooting depth = 0.8 m
Mean ET rate = 4.0 mm/d

Assume standard value of any missing data.

References

Ali MH, Rahman MA (2009) Effects of different levels of salinity on wheat yield in a saline area. Annual report for 2008–09, Bangladesh Institute of Nuclear Agriculture, Mymensingh, Bangladesh

Ayars JE, Christen EW, Hornbuckle JW (2006) Controlled drainage for improved water management in arid regions irrigated agriculture. Agric Water Manage 86:128–139

Ayars JE, Grismer ME, Guitjems JC (1997) Water quality as design criteria in drainage water management system. J Irrig Drain Eng 123:140–153

Christen EW, Skehan D (2001) Design and management of subsurface horizontal drainage to reduce salt loads. ASCE J Irrig Drain Eng 127:148–155

Cook RA, Badiger S, Garcia AM (2001) Drainage equations for random and irregular tile drainage systems. Agric Water Manage 48(3):207–224

Doty CW, Parsons JE (1979) Water requirements and water table variations for a controlled and reversible drainage system. Trans ASAE 22(3):532–539

Donnan WW (1959) Drainage of agricultural lands using interceptor lines. J Irrig Drain Div Proc ASCE 85:13–23

Dierickx W (1993) Research and developments in selecting subsurface drainage materials. Irrig Drain Syst 6:291–310

Hoogoudt SB (1940) Bijdragen tot de kennis van eenige natuurkundige grootheden van den grond. 7. Algemeene beschouwing van het problem van de detailontwatering en de infiltratie door middle van parallel loopende drains, greppels, slooten en kanalen. Versl Land Onderz 46(14):515–707

Hornbuckle JW, Christen EW, Ayars JE, Faulkner RD (2005) Controlled water table management as a strategy for reducing salt loads from subsurface drainage under perennial agriculture in a semi-arid Australia. Irrig Drain Syst 19:145–159

ILRI (1973) Drainage principles and applications. Publication 16, Vol. II, ILRI Wageningen, The Netherlands

References

Juusela T (1958) On the methods of protecting drain pipes and on the use of gravel as a protective material. Acta Agric Scand VIII 1:62–87

Karpoff KP (1955) The use of laboratory tests to develop design criteria for protective filters. Am Soc Test Mater 55:1183–1198

Kirkham D (1957) Theory of seepage of ponded water into drainage facilities. In: Luthin JN (ed) Drainage of agricultural lands Agron. Monogr. 7. ASA and SSSA, Madison, WI, pp 139–181

Kirkham D (1958) Seepage of steady rainfall through soil into drains. Trans Am Geophys Union 39:892–908

Kirkham D (1966) Steady-state theories for land drainage. J Irrig Drain Eng ASCE 92:19–39

Kruse EG (1962) Design of gravel packs for wells. Am. Soc Agric Eng Trans 5(2):179–199

Lalonde V, Madramootoo CA, Trenholm L, Broughton RS (1996) Effects of controlled drainage on nitrate concentrations in subsurface drain discharge. Agric Water Manage 29(2):187–199

Manguerra HB, Garcia LA (1995) Irrigation-drainage design and management model: validation and application. J Irrig Drain Eng 121(1):83–94

Moody WT (1966) Nonlinear differential equation of drain spacing. J Irrig Drain ASCE 92(2):1–9

Moustafa MM (1997) Verification of impermeable barrier depth and effective radius for drain spacing using exact solution of the steady state flow to drains. Irrig Drain Syst 11:283–298

Oosterbaan RJ, Boonstra J, Rao KVGK (1996) The energy balance of groundwater flow. In: Singh VP, Kumar B (eds) Subsurface-water hydrology, Proceedings of the international conference on hydrology and water resources, New Delhi, India, 1993, vol 2. Kluwer Academic Publishers, Dordrecht, The Netherlands, pp 153–160

Rimidis A, Dierickx W (2003) Evaluation of subsurface drainage performance in Lithuania. Agric Water Manage 59:15–31

SCS (1971) Drainage of agricultural land. USDA national engineering handbook, Section 16. USDA Soil Conservation Service (SCS), Washington, DC

SCS (1988) Standards and specifications for conservation practices. Washington, DC

SCS (1994) Gradation design of sand and gravel filters, NEH Part 63–26. National Engineering Publications from the US SCS. Water Resources Publications, Highlands Ranch, CO, p 37

Skaggs RW (1978) A water management model for shallow water table soils. Report 134, Water Resources Research Institute of the University of North Carolina, North Carolina State University, Raleigh, p 128

Stuyt LCPM (1992) The water acceptance of wrapped subsurface drains. Ph.D. Thesis, Agricultural University, Wageningen/DLO-Winand Staring Centre (SC-DLO), Wageningen, The Netherlands

Stuyt LCPM, Dierickx W, Martinez Beltran J (2005) Materials for subsurface land drainage systems. FAO Irrigation and Drainage Paper 60 Rev. 1, Rome, 183 pp

Terzaghi K (1922) Der grundbruch and stauwerken und seine verhutung. Forcheimer-Number der Wasserkraft 17:445–449

US Army Corps of Engineers (1941) Investigation of filter requirements for underdrains. Eng. Waterways Exp. Sta., Techn. Mem. 183–1

US Army Corps of Engineers (1978) Engineering and design; design and construction of levees. Engineer manual. Appendix E, Filter design. EM 1110–201913, 31 Mar 1978. Department of the Army, Office of the Chief of Engineers, Washington, DC, pp E1–E5

USBR (1973) Design of small dams, 2nd edn. United States Department of Interior, Bureau of Reclamation. Water Res. Tech. Publ. Denver, CO, pp 200–202

USBR (1978) Drainage manual. United States Department of Interior, Bureau of Reclamation, Denver, CO, pp 200–202

USBR (1993) Drainage manual, 2nd edn. United States Department of Interior, Bureau of Reclamation, Denver, CO

U.S. Department of Interior (1993) Drainage manual, 2nd edn. United States Department of Interior, Bureau of Reclamation, Denver, CO

Van der Molen WH, Wesseling J (1991) Solution in closed form and a series solution to replace the tables for the thickness of the equivalent layer in Hooghoudt's spacing formula. Agric Water Manage 19(1):1–16

Vlotman WF, Bhutta MN, Ali SR, Bhatti AK (1994) Design, monitoring and research fourth drainage project, Faisalabad, 1976–1994. (Draft Main Report). IWASRI Publication No 159. International Waterlogging and Salinity Research Institute, Lahore, Pakistan, p 90

Vlotman WF, Shafiq-ur-Rehman Y, Bhutta MN (1995) Gravel envelope test results and IWASRI design guidelines. IWASRI Publication No. 144. International Waterlogging and Salinity Institute, Lahore, Pakistan, p 49

Vlotman WF, Willardson LS, Dierickx W (2000) Envelope design for subsurface drains. Publication No. 57. ILRI, Wageningen, The Netherlands, p 358

Chapter 10
Models in Irrigation and Water Management

Contents

10.1	Background/Need of a Model	380
10.2	Basics of Model: General Concepts, Types, Formulation and Evaluation System	380
	10.2.1 General Concepts	380
	10.2.2 Different Types of Model	381
	10.2.3 Some related terminologies	386
	10.2.4 Basic Considerations in Model Development and Formulation of Model Structure	389
	10.2.5 Model Calibration, Validation and Evaluation	390
	10.2.6 Statistical Indicators for Model Performance Evaluation	391
10.3	Overview of Some of the Commonly Used Models	393
	10.3.1 Model for Reference Evapotranspiration (ET_0 Models)	393
	10.3.2 Model for Upward Flux Estimation	397
	10.3.3 Model for Flow Estimation in Cracking Clay Soil	397
	10.3.4 Model for Irrigation Planning and Decision Support System	402
	10.3.5 Decision Support Model	405
10.4	Crop Production Function/Yield Model	406
	10.4.1 Definition of Production Function	406
	10.4.2 Importance of Production Function	406
	10.4.3 Basic Considerations in Crop Production Function	407
	10.4.4 Pattern of Crop Production Function	407
	10.4.5 Development of Crop Production Function	408
	10.4.6 Some Existing Yield Functions/Models	408
	10.4.7 Limitations/Drawbacks of Crop Production Function	411
10.5	Regression-Based Empirical Models for Predicting Crop Yield from Weather Variables	411
	10.5.1 Need of Weather-Based Prediction Model	411
	10.5.2 Existing Models/Past Efforts	412
	10.5.3 Methods of Formulation of Weather-Based Prediction Model	413
	10.5.4 Discussion	415
	10.5.5 Sample Example of Formulating Weather-Based Yield-Prediction Model	415

Relevant Journals	419
Questions	419
References	420

In some cases, we need to know the knowledge of specific process within a complex system of interacting and interdependent phenomena, and then need to reintegrate such knowledge to obtain a comprehensive and accurate solution of the phenomena. Model is ideal to integrate the complex system and to obtain the answer if the condition is. It is merely a useful tool in obtaining answers in the choice of a decision or policy.

By definition, model is a simplified representation of reality. It may be a system of equations that shows how parameters and variables of interest are related to one another. A model can come in many shapes, sizes, and styles. Models of different types are discussed in this chapter. Different issues of modeling aspects such as calibration, sensitivity analysis, validation have been described. Different up-to-date statistical indicators for model performance evaluation have been discussed, and overview of some of the commonly used models is included. In addition, utility of crop production functions along with their development techniques/philosophy is described. And finally, development procedure of regressed-based empirical crop prediction model from weather variables is explained with sample data.

10.1 Background/Need of a Model

One of the main factors for the intensification of the agricultural production of plants is irrigation. In order to get the optimal results, it is necessary to calculate the required quantity of water in dependence of some parameters, for example of the environment, the subsurface geo-hydrological condition, type of crop, the stage of the growth, and growth rate. Some of the fundamental problems of agricultural and environmental research are how to obtain knowledge of specific process within a complex system of interacting and interdependent phenomena, and then how to reintegrate such knowledge so as to obtain a comprehensive and accurate solution of the phenomena. Model is ideal to integrate the complex system and to obtain the answer if the condition is.

Crop growth simulations are needed to help organize our knowledge of plant response to the environment for the purpose of assisting growers in management decisions, predicting impacts of land use decisions, and predicting the consequences of probable climatic variations.

10.2 Basics of Model: General Concepts, Types, Formulation and Evaluation System

10.2.1 General Concepts

A model is merely a useful tool in obtaining answers in the choice of a decision or policy. It may be a system of equations that show how parameters and variables of interest are related to one another. By definition, model is a simplified representation of reality. It serves as a means of predicting and examining an existing or proposed system's performance under a set of conditions specified by the users. A model can come in many shapes, sizes, and styles. It is important to emphasize that a model is not the real world but only an approximation of reality (human construct) to help us better understand real world systems. In general all models have an information input, an information processor, and an output of expected results (Fig. 10.1).

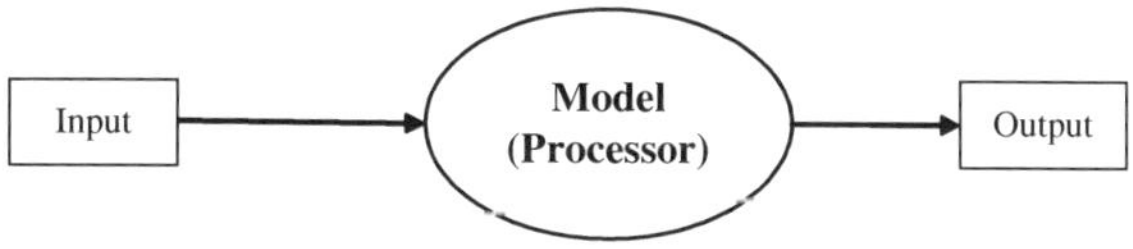

Fig. 10.1 Schematic representation of a model

Model is a tool to obtain promising field management strategies. With good models, realistic estimation of crop yield (or other expected outputs) can be simulated for various environmental conditions. The models are valuable for outscaling the experimental findings to new environments.

Models are not meant to exclusively represent all elements of a system. The trick is to balance the complexity of reality with the principle of parsimony (the simpler the better). The form of a model depends on the problem to be addressed, the client sponsor of model development, the state of knowledge about the prototype, and the use of the model by a potential decision maker.

10.2.2 Different Types of Model

Models can be broadly classified into two major groups:

- Physically based (or process based), and
- Empirical models

10.2.2.1 Physically Based or Process-Based Model

In this type of model, the processes are conceptually represented using mathematical equations. A process-based model normally represents (conceptualizes) all important physical processes occurring during an event as well as between events (if applicable). Such a model is derived from theoretical approaches to causal relationship.

For example, for modeling storm event pollution load from urban catchment, such models represent a description of the hydrological rainfall-runoff transformation process with associated erosion, pollution buildup and washoff, and other quality components.

Process-based models are accurate over a wide range of conditions.

10.2.2.2 Empirical Models (or Black Box Models)

This type of models is mathematical model, and inputs and outputs are related by empirical equation. The internal mechanism of the process (or causal relationship) is not explicitly described (or unknown, i.e., like a "black box") in this type of model. In this model, empirically derived formulas (relies upon experiments and observations) are used to predict the required output. The empirical data are derived from a limited sample of projects.

Various modelers or researchers categorized/classified models under different perspectives. Definitions of different category of models are summarized below:

10.2.2.3 Conceptual Model

Conceptual models (sometimes called abstract models) are theoretical models. A theoretical model often assumes away many complications while highlighting limited aspects of the object. A conceptual model is a representation of some phenomenon, data, or theory by logical and mathematical objects such as functions, relations, tables, stochastic processes, formulas, axiom systems, rules of inference.

The conceptual model is concerned with the real world view and understanding of data. Conceptual models are qualitative models that help highlight important connections in real world systems and processes. They are used as a first step in the development of more complex models. A conceptual model may include a few significant attributes to augment the definition and visualization of entities. No efforts need to be made to invent the full attribute population of such a model. A conceptual model may have some identifying concepts or candidate keys noted, but it explicitly does not include a complete scheme of identity, since identifiers are logical choices made from a deeper context. From the conceptual model, the mathematical model and validation experiment can be constructed.

For example, the development of computer models capable of realistic simulations of tornadic storms and tornadoes provide a means for careful quantitative evaluation of the physical processes which yield tornadoes.

Features of conceptual model:

– Conceptual models are a tool through which detailed technical concepts can be summarized in a non-technical way, and presented to end users.
– Conceptual models provide a physical background upon which the understanding derived from various scientific disciplines (e.g., ecology, chemistry and geology) can be integrated with the perspectives of other stakeholder groups for addressing management issues.
– Conceptual models help users to understand the often complex processes in a system (e.g., how things work, what drives these things, and major impacts) and demonstrate the links between them.
– Conceptual models can help users to identify any gaps in scientific understanding, monitoring, or natural resource management plans.

10.2.2.4 Mathematical and Statistical Models

Mathematical and statistical models involve solving relevant equation(s) of a system or characterizing a system based upon its statistical parameters such as mean, mode, variance, or regression coefficients. These types of models are obviously related, but there are also real differences between them.

Mathematical Models

When knowledge of a process is sufficiently concise that it can be represented with mathematical relationships, the relationship can be referred to as a mathematical model of the plant, crop, or process of concern. A mathematical model is defined by a series of equations, input factors, parameters, and variables aimed at characterizing the process being investigated. Statistically derived mathematical relationships are normally used to represent most processes. However, they do not contain information about mechanisms that cause events to happen; but only represent what happened.

Explicitly speaking, this type of model grow out of equations that determine how a system changes from one state to the next (differential equations) and/or how one variable depends on the value or state of other variables (state equations). These can also be divided into

– numerical models and
– analytical models.

Numerical Models

Numerical models are mathematical models that use some sort of numerical time-stepping procedure to obtain the models behavior over time. The mathematical solution is represented by a generated table and/or graph.

Analytical Models

Analytical models are mathematical models that have a closed form solution, i.e., the solution to the equations used to describe changes in a system can be expressed as a mathematical analytic function. For example, a model of personal savings that assumes a fixed yearly growth rate (r) in savings (S) implies that time rate of change in saving $d(S)/dt$ is given by,

$$d(S)/dt = r(S)$$

This example is also used to describe numerical models so that numerical and analytical models can be compared and contrasted more easily.

The analytical solution to this differential equation is

$$S = S_0 \times \text{EXP}(r \cdot t)$$

where S_0 is the initial savings, t is the time. This equation is the analytical model of personal savings with fixed growth rate.

Statistical Models

Statistical models include issues such as statistical characterization of numerical data, estimating the probabilistic future behavior of a system based on past behavior, extrapolation or interpolation of data based on some best-fit, error estimates of observations, or spectral analysis of data or model generated output. Statistical models are useful in helping identify patterns and underlying relationships between data sets.

10.2.2.5 Static and Dynamic Model

Static Model

A model that analyzes variables in the system in a single point in time. Time is not a variable.

Dynamic Model

A model (or system) that contains time as one of the variables. It captures important changes in, and inter-relationships between parameters and variables through time; i.e., output is projected as a function of time.

10.2.2.6 Mechanistic and Probabilistic Model

Mechanistic Model

When mathematical relationships based on physical, chemical, and/or biological principles (whatever may applicable) can be combined to represent a cause-effect process, the relationships can be referred to as a mechanistic model.

Mechanistic model is based on physical concepts and laws. Process-based mathematical model, which integrates the different processes (mechanisms) involve, is termed as mechanistic model. Mechanistic representations allow descriptions of why events happen to be integral parts of a mathematical model. Mechanistic representations can be combined in logical associations to provide a simulation of all, or portion of, large complex system.

Probabilistic Model

Model which is based on probability concepts and laws.

10.2.2.7 Deterministic and Stochastic Model

Deterministic Model

A model where all variables and parameters are known (non-random). In general, a deterministic model comprises of two components:

- the theoretical basis of the model, and
- the numerical implementation of model

Stochastic Model

Stochastic means *random*. A stochastic process is one whose behavior is non-deterministic, that is, the system's subsequent state is determined both by the process's predictable actions and by a random element. However, any kind of process, which is analyzable in terms of probability, deserved the name stochastic process. A model that relies on stochastic process (theory of probabilities) to predict the new/future values is termed as stochastic model.

For example, a stochastic model of stock prices includes in its ontology a sample space, random variables, the mean and variance of stock prices, various regression coefficients.

10.2.2.8 Time Series Model

A time series is a sequence of data points, measured typically at successive times spaced at uniform/equal time intervals. Time series analysis accounts for the fact that data points taken over time may have internal structure (such as trend or seasonal variation, autocorrelation) that should be accounted for. Time series model is that type of model which uses known past events to forecast future events. Time series analysis is used for many applications such as yield projections, resource depletion/degradation (e.g., groundwater table), demand forecasting, utility studies.

Time series model may be of two types: deterministic and stochastic time series model.

Deterministic Time Series Model

If the process is such that the future values are exactly determined by some mathematical function, and once have identified this function we can exactly predict future values of the time series, then the underlying process is said to be a deterministic time series model.

Stochastic Time Series Model

If the process is such that a future value of the process must be regarded as a random variable and identifying the underlying process only enables to identify the probability distribution of this random variable, then the underlying process is said to be a stochastic time series model.

10.2.2.9 Logical Model

The logical model is a generalized formal structure in the rules of information science.

10.2.3 Some related terminologies

10.2.3.1 Computer Model

Computer model is the numerical implementation of mathematical model, usually in the form of numerical discretization, solution algorithms, and convergence criteria.

10.2.3.2 Simulation

Simulation reproduces the essence of a system without reproducing the system itself. Mathematical simulation is a technique where by the behavior of a system is described using a combination of mathematical and logical relationships. A system is a part of reality that the engineers want to study.

10.2.3.3 Different Parameters and Variables in Modeling

Modeling: Modeling, especially scientific modeling, refers to the process of generating a model as a conceptual representation of some phenomenon.

Model parameter: An item that is constant in the model; a number.

Variable: An item of interest that can change value depending on the other components of the model.

State variable: A variable that describes the "status" of the system. There may be more than one state variable in a model.

Rate variable: A rate variable is a quantity that defines some process within the system at a given point in time. Rates always have dimensions of quantity per unit time; they cannot be measured instantaneously similar to "state variable."

Control variable: A variable that is controllable by human, reflecting one or more admissible policies. There may be more than one control variable in a model.

Driving variable: Driving variable is the variable which initiates or drives the main process of the system.

Constraint: Any condition that places bounds on the solution is a constraint.

Rate constant: Rate constant is the transfer rate of solute from one phase (e.g., solid, liquid, or gas) to another (in solute modeling).

Initial condition: Values of the state and control variables at the beginning of the time period of interest. The behaviors of all dynamic system are dependent upon their initial conditions. In classical mechanics, the initial conditions of systems are known.

Boundary condition: Boundary conditions are the set of conditions specified for the behavior of the solution to a set of differential equations at the boundary of its domain. Boundary conditions are important in determining the mathematical solutions to many physical problems.

Objective function: It is an equation that measures how well the system is attaining an objective defined by the modeler. The objective function is expressed in terms of the state variables, control variables, and parameters of the system.

Dynamic system: A system that contains time as one of the variables. The changes of parameters and variables and inter-relationship between them occur through time.

Empirical result/solution: A result (or solution) that is based on an empirical (data-driven) analysis.

Analytical solution: A mathematical solution to the objective that is based on an analytical analysis. Analytical solutions are more general than numerical results.

10.2.3.4 Sensitivity Analysis

Sensitivity analysis: Computer-based simulation models are used widely for the investigation of complex physical systems. These models typically contain parameters, and the numerical results can be highly sensitive to small changes in the parameter values.

Sensitivity analysis is a tool for characterizing the uncertainty associated with a model. It is the study of how variation (uncertainty) in the output of a mathematical model can be apportioned quantitatively, or qualitatively, to different sources of variation in the input of a model. In models involving many input variables, sensitivity analysis is an essential ingredient of model building and quality assurance. Sensitivity analysis is performed by changing a particular variable or parameter, while keeping all other variables or parameters constant, and observed how the output is changed. This is an important method for checking the quality of a given model, as well as a powerful tool for checking the robustness and reliability of its analysis.

Sensitivity information is useful in two ways. First, sensitivity indicates the simulation error that results if an error was made when assuming the original parameter values. Second, sensitivity indicates how changing a characteristic value can influence the simulation output or system comparisons. If a small change in a parameter results in relatively large change (in percentage form) in its outcomes, the outcomes are said to be sensitive to that parameter. The variable or parameter which is most sensitive, very accurate data for that variable or parameter has to be determined, or that the alternative has to be redesigned for low sensitivity.

Sensitivity analysis can be used to determine/ascertain:
1. The quality of model definition.
2. Factors that mostly contribute to the output variability.
3. The region in the space of input factors for which the model variation is maximum.
4. Optimal- or instability-region(s) within the space of factors for use in a subsequent calibration study.
5. Interactions between factors.
6. How a given model output depends upon the input parameters.

10.2.3.5 Model Validation

Verification/validation: The process of determining that a model implementation accurately represents the developer's conceptual description of the model and the solution to the model.

10.2.3.6 Confidence and Uncertainty

Confidence: Probability that a numerical estimate will lie within a specified range.

Uncertainty: A potential deficiency in any phase or activity of the modeling or experimentation process that is due to inherent variability or lack of knowledge.

10.2.3.7 Implicit Vs Explicit Scheme/Method, and Stability Issue

Numerical solution schemes are often referred to as being explicit or implicit. When a direct computation of the dependent variables can be made in terms of known quantities, the computation is said to be explicit. One can compute values on new level (time or space) by explicit formula.

When the dependent variables are defined by coupled sets of equations, and either a matrix or iterative technique is needed to obtain the solution, the numerical method is said to be implicit. To find a value on new level, one must solve the linear system of equations.

The limitation of an explicit scheme is that there is a certain stability criterion associated with it, so that the size of the time steps can not exceed a certain value. However, the use of an explicit scheme is justified by the fact that it saves a large amount of computer memory which would be required by a matrix solver used in an implicit scheme.

10.2.3.8 Finite Element and Finite Difference Scheme/Method

Finite Element Method

The finite element method is a method through which any continuous function can be approximated by a discrete model, which consists of a set of values of the given function at a finite number of pre-selected points in its domain, together with piecewise approximation of the function over a finite number of connected disjunct sub-domains. These sub-domains are called finite elements and are determined by the pre-selected points. The pre-selected points are called nodal points or nodes.

Finite Difference Method

Finite difference methods are numerical methods for approximating the solutions to differential equations using finite different scheme. It consists of replacing (transforming) the partial derivative expression with approximately equivalent difference quotients (equations) over a small interval. That is, it transforms the continuous domain of the state variables by a network or mesh of discrete points.

10.2.4 Basic Considerations in Model Development and Formulation of Model Structure

10.2.4.1 Basic Considerations

Successful model development always begins with a question or hypothesis, and the model complexity and scale should match those of the question. At that point, a decision must be made about the type of mathematical model the researcher will develop. For all cases, developing a model may not necessary. Simply developing a clear conceptual model can answer to a question. However, many complex problems (having complex interactions among different processes) cannot be solved with conceptual model alone. The process of developing a mathematical model can be extremely helpful in clarifying the relevant processes. It directs or forces the modeler to think clearly about the important processes and relationships that govern important features of the problem.

The key features that should be considered during development of any model are:

- simplifying assumptions must be made;
- boundary conditions or initial conditions must be identified;
- the range of applicability of the model should be understood.

10.2.4.2 Formulation of Model

Modeling process begins with a definition of goals. The goal may be find from the answer of the question – What will be the model be used for – to solve a specific problem, to learn more about the system, to evaluate alternative management strategies, etc? The answer will help in determining the structure of the model, the data required, and the expected output.

The next step is a conceptualization of the prototype, including selection of scales of time and space, dimensionality, and discretization. This is followed by formulation as a set of differential equations, statistical properties, empirical expressions, or combinations of these. Computational representation of model formulation implicitly requires selection of a method of solution. For illustration, a conceptualization of a simple irrigation scheduling model (conceptual model) is depicted in Fig. 10.2.

At this stage, the model is tested functionally under hypothetical but realistic boundary conditions using best estimates of uncertain parameters in the governing equations. Calibration of the model requires adjustment of parameters for a given set of boundary conditions and assumed values of key state variables. The modeler will decide whether agreement between model and prototype is acceptable or not.

Once the model is calibrated with a specific set of parameter values, it may be tested again against another set of prototype data in a process of *verification* to ascertain whether the degree of conformity with the prototype under the new conditions is consistent with that of calibration. If the model fails this test, recalibration may be necessary. Calibration/verification is sometimes considered a single process in model development, termed as *validation*.

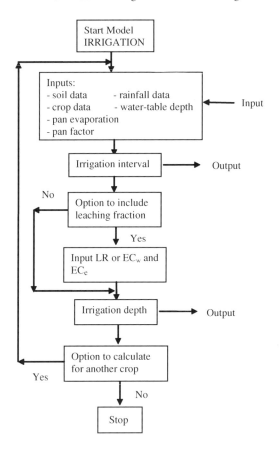

Fig. 10.2 Schematic of a conceptual model for irrigation

Sensitivity testing is an integral part of the model development process, usually accompanying the calibration/verification phases. It provides information on the relative importance of model parameters and their influence on model response in terms of state variables simulated. Within the period of model development, especially at the time of sensitivity analysis, it is appropriate to fix the character of the model, i.e., to provide documentation for the user.

The real test of model performance is in application. This may reveal deficiencies in model, requiring appropriate modification in structure, formulation, solution technique, parameter estimates, or basic data.

10.2.5 Model Calibration, Validation and Evaluation

The model evaluation procedure involves both calibration and validation of the model. Before simulation models are used, they need to be calibrated and validated for the conditions under which they will be used.

10.2.5.1 Model Calibration

The calibration process involves the adjustment of selected model parameters within an expected range so as to minimize the discrepancies between observed and predicted values. Calibration of the model is typically accomplished by adjusting input parameters and rate constants so that output values are as close as possible to measured values in the calibration data set.

This approach to calibrating a model has the disadvantage of requiring a calibration data set. In addition, the approach can be very time consuming, especially if the range of values explored for each parameter and rate constant is relatively large. Furthermore, since the process adjusts several input parameters and rate constants at the same time, inadequacies in the modeling of one process may be coincidentally compensated for when the input parameter or rate constant is changed for another process. Consequently, this approach may lead to good simulation for the calibration data set, but will not necessarily guarantee good simulations for independent data sets.

Another approach to calibration is to conduct laboratory studies to determine the rate constants and input parameters for each of the important processes in the model, and then use those rate constants and parameters to validate the model.

10.2.5.2 Model Verification/Validation

Validation is essentially an independent test of the model, where the model predictions are compared with data not used in the calibration testing. The goal of verification is to quantify confidence in the predictive capacity of the model by comparison with experimental data.

In contrast to traditional experiments, validation experiments are performed to generate high quality data for the purpose of assessing the accuracy of a model prediction. To qualify as a validation test, the specimen geometry, initial conditions, boundary conditions, and all other model input parameters must be prescribed accurately.

Evaluation of model performance should include both statistical criteria and graphical display. A model is a good representation of reality only if it can predict an observable phenomenon with acceptable accuracy and precision.

10.2.6 Statistical Indicators for Model Performance Evaluation

One of the most common issues raised concerning model usage is that of reliability. There is a pressing need to assess model validity in quantitative terms, e.g., statistical limits of model performance against the prototype system. The following statistics are generally used to indicate overall model performance (Fox, 1981; Willmott, 1982; Loague et al., 1989; Loague and Green, 1991; Retta et al., 1996; Ali, 2005a, b):

(i) Bias (i.e., error) or Mean Bias:

$$\text{Bias} = \frac{1}{N}\sum_{i=1}^{N}(S_i - M_i) \qquad (10.1)$$

where S and M are the simulated (or predicted) and measured (or observed) values for the ith observation and N is the number of observations.

(ii) Mean absolute bias:

$$\text{MAE} = \frac{1}{N}\sum_{i=1}^{i=N}|S_i - M_i| \qquad (10.2)$$

(iii) Root mean square error (RMSE), which quantifies the dispersion between simulated and measured (or observed) data:

$$\text{RMSE} = \sqrt{\frac{1}{N}\sum_{i=1}^{N}(S_i - M_i)^2} \qquad (10.3)$$

Ideally, the value of Bias, MAE, and RMS should be zero.

(iv) Relative error (RE):

$$\text{RE} = \frac{\text{RMSE}}{\bar{y}} 100 \qquad (10.4)$$

where $\bar{y}$ is the mean of observed values. RE indicates the percentage deviation with respect to mean observed value.

(v) Model efficiency (EF): Model efficiency is calculated as

$$\text{EF} = \frac{\sum(M - \bar{M})^2 - \sum(S - M)^2}{\sum(M - \bar{M})^2} \qquad (10.5)$$

where
$\bar{M}$ = measured mean
An ideal value of EF is unity.

(vi) Index of agreement (IA):

$$\text{IA} = 1 - \frac{\sum_{i=1}^{N}(o_i - s_i)^2}{\sum_{i=1}^{N}(O'_i + S'_i)^2}, \quad 0 \leq \text{IA} \leq 1 \qquad (10.6)$$

where $O'_i = |O_i - \bar{S}|$, $S'_i = |S_i - \bar{S}|$, O_i is the observed value, S_i is the simulated value, $\bar{S}$ is the simulated mean.

(vii) Coefficient of residual mass (CRM):

It is expressed as

$$\text{CRM} = \frac{(\sum O_i - \sum S_i)}{\sum O_i} \qquad (10.7)$$

(viii) Coefficient of determination (CD):
It is expressed as

$$\text{CD} = \frac{\sum (O_i - \overline{O})^2}{\sum (S_i - \overline{O})^2} \qquad (10.8)$$

10.3 Overview of Some of the Commonly Used Models

In recent years, mathematical modeling and simulation techniques, relying on the use of high speed computer, have been developed for the purpose of providing comprehensive quantitative predictions. The application of system analysis (modeling) techniques to agricultural cropping systems is now widespread. A number of models for scheduling irrigations under limited water supplies have been developed. Many models determine the optimal irrigation strategies using stochastic or probabilistic methods of weather variables (Mapp and Ediman, 1987; Rao et al., 1988; Rao et al., 1990). But in any season, the current weather variables can be significantly different from their probabilistic or stochastic estimates.

Some of the commonly used models are outlined here for general orientation.

10.3.1 Model for Reference Evapotranspiration (ET_0 Models)

Models for predicting reference evapotranspiration (ET_0) range from deterministically based combined energy balance-vapor transfer approaches to empirically relationships based on climatological variables or to evaporation from a standard evaporation pan. Updated procedures for calculating ET_0 were established by FAO (Smith et al., 1992).

10.3.1.1 CROPWAT Model

The FAO developed CROPWAT software (Clarke, 1998) to calculate ET_0 and irrigation planning. It is available both in DOS and Windows version. The ET_0 estimation is based on Penman-Monteith (P-M) equation. The inputs for ET_0 estimation are (i) maximum temperature, (ii) minimum temperature, (iii) average humidity, (iv) wind speed, (v) sunshine hour, (vi) latitude, and (vii) altitude of the location. The original form of the Penman-Monteith equation (Monteith, 1981) can be written as

$$\lambda ET_0 = \frac{\Delta(R_n - G) + c_p(e_a - e_d)\rho/r_a}{\Delta + \gamma(1 + r_c/r_a)} \quad (10.9)$$

where λET_0 is the latent heat of evaporation (kJm^2/s); γ the psychrometric constant (kPa/°C); R_n the net radiation flux at surface (kJm^2/s); G the soil heat flux (kJm^2/s); c_p the specific heat of moist air (kJ/kg/°C); ρ the atmospheric density (kg/m^3); Δ slope of vapor pressure curve (kPa/°C); r_a the aerodynamic resistance (s/m); r_c the bulk canopy resistance (s/m); and λ latent heat of vaporization (MJ/kg).

To facilitate the analysis of the combination equation, Smith et al. (1992) defined the aerodynamic and radiation term as

$$ET_0 = ET_{rad} + ET_{aero} \quad (10.10)$$

where

ET_0 = reference evapotranspiration of standard crop canopy [mm/d]
ET_{rad} = radiation term [mm/d]
ET_{aero} = aerodynamic term [mm/d]
For reference crop, crop canopy resistance, $r_c = 200/LAI$;

$$\text{and, } LAI = 24 h_c$$

where h_c = crop height = 0.12 m

From the original Penman-Monteith equation, the aerodynamic term can be written as

$$ET_{aero} = \frac{86.4}{\lambda} \frac{1}{\Delta + \gamma^*} \frac{\rho \cdot c_p}{r_a}(e_a - e_d) \quad (10.11)$$

where

ET_{aero} = aerodynamic term [mm/d]
86.4 = conversion factor to [mm/d]
γ^* = modified psychrometric constant

The radiation term is written as

$$ET_{rad} = \frac{1}{\lambda} \frac{\Delta}{\Delta + \gamma^*}(R_n - G) = \frac{0.408\Delta(R_n - G)}{\Delta + \gamma(1 + 0.34U2)} \quad (10.12)$$

where

ET_{rad} = radiation term [mm/d]
R_n = net radiation [MJ/m^2/d]
G = soil heat flux [MJ/m^2/d]
λ = latent heat of vaporization [MJ/kg] = 2.45

$$\text{Net radiation, } R_n = R_{ns} - R_{nl} \quad (10.13)$$

10.3 Overview of Some of the Commonly Used Models

where

R_{ns} = net incoming shortwave radiation [MJ/m²/d]
R_{nl} = net outgoing long-wave radiation [MJ/m²/d]

Net shortwave radiation is the radiation received effectively by the crop canopy taking into account losses due to reflection:

$$R_{ns} = (1 - \alpha)R_s \approx 0.77 R_s \qquad (10.14)$$

where

α = albedo or canopy reflection coefficient = 0.23 overall average for grass
R_s = incoming solar radiation [MJ/m²/d]

Shortwave radiation can be estimated from measured sunshine hours according to the following empirical relationship:

$$R_s = \left(a_s + b_s \frac{n}{N}\right) R_a \qquad (10.15)$$

where

a_s = fraction of extraterrestrial radiation (R_a) on overcast days $\approx$ 0.25 for average climate
$a_s + b_s$ = fraction of radiation on clear days $\approx$ 0.75
$b_s \approx$ 0.50 for average climate
n/N = relative sunshine fraction
n = bright sunshine hours per day (h)
R_a = extra-terrestrial radiation, MJ/m²/d

Available local radiation data can be used to carry out a regression analysis to determine the Angstrom coefficients a_s and b_s.

The extraterrestrial radiation can be calculated as

$$R_a = \frac{24 \times 60}{\pi} G_{sc} \cdot d_r(\omega_s \sin \psi \sin \delta + \cos \psi \cos \delta \sin \omega_s) \qquad (10.16)$$

where

R_a = extraterrestrial radiation (MJ/m²/d)
G_{sc} = solar constant (MJ/m²/d) = 0.0820
d_r = relative distance of earth and sun
ψ = latitude (rad)
δ = solar declination
ω_s = sunset hour angle (rad)

and

$$\omega_s = \arccos(-\tan\psi \tan\delta) \qquad (10.17)$$

$$d_r = 1 + 0.033 \cos\left(\frac{2\pi}{365}J\right) \qquad (10.18)$$

$$\delta = 0.409 \sin\left(\frac{2\pi}{365}J - 1.39\right) \qquad (10.19)$$

J = number of the day in the year (Julian day)

10.3.1.2 CropET$_0$ Model

The Excel template "CropET$_0$" was developed by Ali (2005a) following the FAO Penman-Monteith approach as described by Smith et al. (1992) and Allen et al. (1998). The basic inputs of the model are the monthly average values of the weather variables (i) maximum temperature, (ii) minimum temperature, (iii) humidity, (iv) wind speed, (v) sunshine hour, (vi) latitude, and (vii) altitude of the location. For north latitude, the value would be positive, and for south latitude, the value would be negative.

The user can adjust the parameter "Angstrom coefficients" (used in solar energy calculation) if the local calibrated values are available. Average values ($a_s = 0.25$, $b_s = 0.50$) as suggested by FAO can be used if local calibrated values are not available. Input data should be provided through "Model Interface," a data input platform.

Model Output

There is dynamic interaction between input and output in "CropET$_0$." The "CropET$_0$" produces the following:

(i) The monthly average reference evapotranspiration (mm/d) throughout the months of the year;
(ii) Display graph. As the model output ET$_0$ values are linked to a graphical display, the display graph respond instantaneously with the input value.

In addition, there are a "Print Output" and a "Print Input" window, from where the output values and input data can be printed. The graph can be printed from "Display Graph" window.

10.3.2 Model for Upward Flux Estimation

10.3.2.1 Model UPFLOW

UPFLOW is a software tool developed to estimate the expected upward flow from a shallow water table in a given soil profile and to evaluate the effects of environmental conditions on the upward flow (Raes, 2002; Raes and Deproost, 2003).

The input in the UPFLOW model are average evapotranspiration (ET) during the time period, expected mean soil water content, average depth of water table (WT) below the soil surface, crop type, the characteristics of the various layers of the soil profile and their thickness, and the salt content of the groundwater (if groundwater contains salts).

For the given environmental conditions, UPFLOW displays the expected steady upward flow [mm/day] from the water table to the topsoil, the soil water content [vol%] expected in the topsoil, the amount of salt transported upward during the given period [t/ha-year, if the water table contains salts,], the degree of water logging [%] in the root zone (if any), and a graphical display of the soil water profile above the water table.

In the model, the steady upward flow to the topsoil is estimated according to De Laat (1980, 1995):

$$z = \int_0^h \frac{K(h)}{q + K(h)} \, dh \tag{10.20}$$

where z (m) is the vertical co-ordinate, q the constant upward flux ($m^3/m^2/d$) of water, h the soil matric potential per unit weight of water (m), and $K(h)$ the hydraulic conductivity (m/d).

Given the K–h and θ–h relation for the various soil layers of the profile above the water table, UPFLOW is able to determine the maximum flux that can flow to the top soil by checking that the simulated soil water content (derived from the moisture profile) remains below the specified mean water content in the top soil. UPFLOW calculates the amount of water that the plant roots extract according to Feddes et al. (1978). Since the water flow inside the soil profile is assumed to be steady, the capillary rise from the water table to the topsoil can never exceeds the ET demand of the atmosphere. Mean soil evaporation or crop evapotranspiration for a given period from climate, soil, and crop parameters are calculated in the model according to Allen et al. (1998).

10.3.3 Model for Flow Estimation in Cracking Clay Soil

In clay soils swelling and shrinkage have important consequences for water transport. Several models have been developed to simulate water transport in clay soils.

10.3.3.1 FLOCR

FLOCR is a one-dimensional model that simulates water balance, cracking, and subsidence of clay soils (Bronswijk, 1988; Oostindie and Bronswijk, 1992). The main feature of "FLOCR" that distinguishes it from other soil-water simulation models is that this model computes a continuously changing system of soil matrix and shrinkage crack. The model considers only the vertical water flow through the soil matrix and through the shrinkage cracks.

Principle of FLOCR

- the soil is divided into soil matrix and cracks.
- The rainfall is divided into matrix infiltration and crack infiltration.
- The crack infiltration is added to the bottom of the cracks, and to the water content of the soil matrix at that depth.
- The matrix infiltration is added to the top compartment of the soil matrix.

Schematic conceptualization of simulation model "FLOCR" is given in Fig. 10.3.

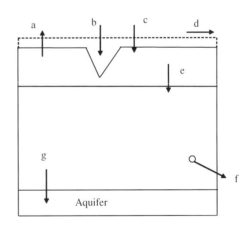

Fig. 10.3 Schematic conceptualization of simulation model 'FLOCR' (after Ali, 2000). **a** Evapotranspiration; **b** crack infiltrationl; **c** matrix infiltration; **d** runoff; **e** flux to groundwater; **f** flux to drain; **g** flux to aquifer

Governing Equations

Boundary Condition at the Top of the Soil Profile

The water balance of the soil surface within one time step is as follows (Oostindie and Bronswijk, 1992):

$$S = (P - E) - R - V \qquad (10.21)$$

where

$V =$ flux through the soil surface with $V = V_{\text{matrix}} + V_{\text{crack}}(m)$

10.3 Overview of Some of the Commonly Used Models

S = storage of water at soil surface (m)
P = precipitation (m)
E = actual evapotranspiration (m)
R = runoff (m)

Evapotranspiration

In FLOCR, water extraction from the soil profile by evapotranspiration is limited to the top compartment. When the top compartment dries, evapotranspiration is reduced in the following way:

$$\text{For } h > a, \quad E_{\text{r}} = E_{\text{p}}$$

$$\text{For } h < a, \quad E_{\text{r}} = E_{\text{p}} \left\{ 1 + \frac{h-a}{\frac{1}{b}+a} \right\} \quad (10.22)$$

In which

E_{p} = potential evapotranspiration (cm/d)
E_{r} = reduced evapotranspiration (cm/d)
h = pressure head of the top compartment (cm)
a = pressure head above which $E_{\text{r}} = E_{\text{p}}$ (cm)
b = soil parameter. The value $-1/b$ describes the pressure head which below no water is extracted from the soil profile.

Surface Runoff

Surface runoff is only possible when cracks are closed. When cracks are closed runoff occurs after a certain threshold value (i.e., maximum surface storage) is exceeded.

$$R = S - S_{\text{max}} \quad (10.23)$$

where

R = surface runoff (cm)
S = computed storage of water at soil surface (cm)
S_{max} = maximum storage of water at soil surface before runoff begins (cm)

Infiltration into Matrix and Cracks

When rain water reaches the surface of a cracked clay soil, part of the water infiltrates into the soil matrix and part of it flows into the cracks. In addition, a certain amount of rain falls directly in the cracks. Surface runoff only occurs when cracks are closed. Matrix infiltration and crack infiltration at a given rainfall intensity is calculated as follows (Oostindie and Branswijk, 1992):

$$\begin{aligned} P<I_{max} : I &= A_m \times P \\ I_c &= A_c^* P \\ P>I_{max} : I &= A_m \times I_{max} \\ I_{c.1} &= A_m^*(P - I_{max}) \\ I_{c.2} &= A_c^* P \\ I_c &= I_{c.1} + I_{c.2} \end{aligned} \qquad (10.24)$$

In which

P = Rainfall intensity (m/s)
I_{max} = maximum infiltration rate (m/s)
I = Infiltration rate in soil matrix (m/s)
I_c = Infiltration rate in crack (m/s)
A_m, A_c = Relative areas of soil matrix and cracks respectively
$I_{c.1}$ = The part of total crack infiltration caused by rainfall intensity exceeding maximum infiltration rate (m/s)
$I_{c.2}$ = The part of total crack infiltration caused by rain falling directly into the crack (m/s)

Defined in this way, all infiltration rates are based on total surface area.

Conductivity

To enable the analytical solution of the Darcy equation, the $K(h)$ relation is expressed by an exponential relation. One $k(h)$ relation is described by three different exponential $k(h)$ sections, as for many soils an exponential $k(h)$ relation is valid only for a narrow range of pressure heads. Each line segment is described by

$$k(h) = k_0 e^{ah} \qquad (10.25)$$

In which,

k = calculated hydraulic conductivity (cm/d)
α = gradient
h = pressure head (cm)
k_0 = saturated hydraulic conductivity (cm/d)

The pressure heads that each line covers are specified. That is, the model input is the intersection of the first and second line segment and the intersection of the second and third line segment.

10.3 Overview of Some of the Commonly Used Models

Boundary Condition at the Bottom of the Soil Profile

The boundary condition at the bottom of the soil profile is defined by a flux to the open water system via drains and a flux to the underlying aquifer:

$$V_b = V_d + V_a \tag{10.26}$$

In which

V_b = bottom boundary flux (cm/d)
V_d = flux to the open water system via drains (cm/d)
V_a = flux to the aquifer (cm/d)

Subsidence and Cracking

The shrinkage characteristic is defined in the model as the relationship between moisture ratio (volume of moisture/volume of solids) and void ratio (volume of voids/volume of solids) of the soil. Introducing the shrinkage characteristic of a soil into model calculations allows each soil compartment to have its own relationship with water content, pressure head, hydraulic conductivity, and matrix volume. Each water content change (in comparison with saturation) in the model can be converted into a volume change of the soil matrix ΔV using the shrinkage characteristic. ΔV can then be converted into a change in compartment thickness and a change in crack volume by Oostindie and Branswijk (1992):

$$\Delta z = z_1 - \left[\left(\frac{V_2}{V_1}\right)\frac{1}{r_s}\right] z_1 \tag{10.27}$$

$$\Delta V_{cr} = \Delta V - z_1^2 \cdot \Delta z \tag{10.28}$$

where

ΔV = change in soil matrix volume (m^3)
Δz = change in compartment thickness due to shrinkage (m)
ΔV_{cr} = change in crack volume (m^3)
V_1, V_2 = volume of cube of soil matrix at saturation and after shrinkage (m^3)
z_1 = compartment thickness at saturation (m)
r_s = geometry factor (–)

Time Step

The model computes the maximum allowable time step. The time step depends on the wet range of the hydraulic conductivity curve. The critical value for the time step is calculated as

$$\Delta t < \frac{\partial \theta}{\partial k}\left[\frac{e^{\alpha \Delta z} - 1}{e^{\alpha \Delta z} + 1}\right]\Delta z$$

where

Δz = change in compartment thickness
α = soil constant/gradient; different for different $k(h)$ line segment

10.3.3.2 MACRO

The MACRO model is a comprehensive model of both the field water balance and solute transport and transformation processes in the soil/crop system (Jarvis, 1994). An important feature of the model is that it may be run in either one or two flow domains. In two domains, the total porosity is partitioned into macropores and micropores. Each domain is characterized by a degree of saturation, conductivity, and a flux, while interaction terms account for convective and diffuse exchange between flow domains. In one domain, the interaction terms are redundant and the model simply reduces to standard numerical solutions of the Richard's and convection-dispersion equations.

Soil-Water Balance

Water movement in the micropores is calculated with the Richard's equation including a sink term to account for root water uptake (Jarvis, 1994). The soil hydraulic properties are described by the functions of Brooks and Corey and Mualem. Water flow in macropores is calculated with Darcy's law assuming a unit hydraulic gradient and a simple power law function to represent the unsaturated hydraulic conductivity. Driving variables in the model consists of measured rainfall data at a given solute concentration, daily potential evapotranspiration and, if pesticide transport is being considered, daily maximum and minimum air temperatures.

The surface boundary condition in MACRO partitions the net rainfall into an amount taken up by micropores and an excess amount of water flowing into macropores. This partitioning is determined by a simple description of the infiltration capacity of the micropores. Alternative options for the bottom boundary condition are available in the model.

Root water uptake is predicted as a function of the evaporative demand, root distribution and soil water content using a simple empirical model. The calculated water uptake is assumed to be preferentially satisfied from the water stored in the macropores.

10.3.4 Model for Irrigation Planning and Decision Support System

10.3.4.1 CROPWAT

CropWat for Windows is a program that uses the FAO Penman-Monteith method (Smith et al., 1992) for calculating reference crop evapotranspiration. These estimates are used in crop water requirements and irrigation scheduling calculations

10.3 Overview of Some of the Commonly Used Models

(Clarke, 1998). The program uses a flexible menu system and file handling, and extensive use of graphics. Graphs of the input data (climate, cropping pattern) and results (crop water requirements, soil moisture deficit) can be drawn and printed with ease. Complex cropping pattern can be designed with several crops with staggered planting dates.

The input data manus are the following:

CLIMATE – to enter monthly climatic data to calculate ETo
ETO – to enter your own monthly ETo
RAINFALL – to enter monthly rainfall data
CROPS – to enter cropping pattern and crop coefficient data
SOIL – to enter data for your soil type

Once the data has been entered, it can be seen in a table with the *Tables* main menu option or plotted in a graph with the *Graphs* main menu option. The *Schedule* menu lets one to define how irrigations are calculated and to management groups of data files (climate, rain, crop, soil). CROPWAT for Windows give the users a wide range of methods to calculate an irrigation schedule. After the criterion of scheduling is defined, the program will calculate the irrigation dates and amounts. A useful feature of CROPWAT is the ability to enter the user's own irrigation amounts and timing with the *User Defined* irrigation options.

10.3.4.2 CropSyst

Overview of the Model

CropSyst (Cropping Systems Simulation Model) is a multi-year, multi-crop, daily timestep crop growth simulation model (Stockle et al., 1994; Stockle and Nelson, 1994). The model simulates the soil water budget, soil-plant nitrogen budget, crop phenology, crop canopy and root growth, bio-mass production, crop yield, residue production and decomposition, soil erosion by water, and pesticide fate. The management options include cultivar selection, crop rotation (including fallow years), irrigation, nitrogen fertilization, tillage operations, and residue management.

The water budget in the model includes rainfall, irrigation, runoff, interception, water infiltration and redistribution in the soil profile, crop transpiration, and evaporation. The evapotranspiration model is the predominate component of CropSyst crop growth model. Daily crop growth is expressed as bio-mass increase per unit ground area. The model accounts for four limiting factors to crop growth: water, nitrogen, light, and temperature.

Main Features of CropSyst

- Link of CropSyst simulations with GIS software is possible
- Optional runtime graphic display
- Report data graphics viewer

- Input parameter file editor
- Report format editor

Technical Description

Growth is described at the level of whole plant and organs. Integration is performed with daily time steps using the Euler's method. Users may select different methods to calculate water redistribution in the soil profile and reference evapotranspiration. Water redistribution in the soil profile is handled by a simple cascading approach or by a finite difference approach to determine soil water fluxes. CropSyst offers three options to calculate reference ET. In decreasing order of required weather data inputs, these options are Pemman-Monteith model, the Preiestley-Tayler model, and a simpler implementation of Preiestley-Tayler model which requires only temperature. Crop ET is determined from a crop coefficient at full canopy and ground coverage determined by canopy leaf area index. Irrigation can be set for deficit irrigation.

Crop development is simulated based on thermal time required to reach specific growth stages. The thermal time (degree days) are accumulated from planting. The accumulation of thermal time may be accelerated by water stress. Thermal time may be also modulated by photo-period and vernalization requirement whenever pertinent. Given the common pathway for carbon and vapor exchange of leaves, there is a conservative relationship between crop transpiration and biomass production. The increase of leaf area during the vegetative period, expressed as leaf area per unit of soil area (leaf area index, LAI), is calculated as a function of biomass accumulation, specific leaf area, and a partitioning coefficient. Leaf area duration, specified in terms of thermal time and modulated by water stress, determines canopy senescence. Root growth is synchronized with canopy growth, and root density (by soil layer, a function of root depth penetration). The prediction of yield is based on the determination of a harvest index. Although an approach based on the prediction of yield components could be used. The harvest index is determined using as base the unstressed harvest index, a required crop input parameter, modified according to crop stress (water and nitrogen) intensity and sensitivity during flowering and grain filling.

Models Inputs

Four input data files are required to run CropSyst: Location, Soil, Crop, and Management files. A "Simulation Control file" combines the input files as desired to produce specific simulation runs. In addition, the "Control file" determines the start and ending day for the simulation, and set the values of all parameters required for initialization.

Model Output

The model outputs are yield (grain yield and biomass), water balance, nitrogen balance, residue production and decomposition, and soil erosion. There are three types of output variables:

10.3 Overview of Some of the Commonly Used Models

- Daily variables computed each day.
- Yearly or annual variables provided an annual summary of values accumulated throughout the calendar year.
- Harvest variables provide harvest yield and relevant crop and soil conditions at harvest time accumulated throughout the growing season.

Output Time Step

The user may restrict the amount of information printed in the daily report by specifying a time step interval.

Limitations/Shortcomings

No specific limitations are documented. CropSyst does not handle leap year. It is not tested in case of vertisol.

10.3.5 Decision Support Model

In developing or developed countries, detailed and long-term field experiments are often difficult to conduct due to financial or personnel limitations. Decision support models can provide effective information by extrapolating field experimental results to a range of production scenarios than is possible with field trials, reducing the amount of repetitive, laborious and time-consuming experimentation. For example, the effects of soil types and seasons, alternative production with direct reference to the farmers' resource base, sustainability, and long-term suitability of production system, can all be studied. By doing so, it can help to identify knowledge gaps, gain insight into situations where experimental results are lacking or are incomplete. This in turn may help to develop new or refine existing fertilizer recommendations for a wide range of production conditions.

10.3.5.1 DSSAT

The Decision Support System for Agrotechnology Transfer (DSSAT) (Tsuji et al., 1994) is a comprehensive decision support system for assessing agricultural management options. It has been widely used in both developed and developing countries. The model handles management strategies that involve crop rotations, irrigation, fertilization, and organic applications. The DSSAT contains various submodels. The CERES-Wheat within DSSAT can simulate the main process of crop growth and development such as timing of phenological events, the development of canopy to intercept photosynthetically active radiation, and dry matter accumulation. It allows the inclusion of cultivar-specific information that makes it possible to predict the cultivar variations in plant ontogeny, yield component characteristics and their interactions with weather. The biomass calculated is partitioned between leaves, stems, roots, ears and grains. The proportion partitioned to each organ is determined by the stage of development and growing conditions, modified when deficiencies of water and nutrient supplies occur. Crop yields are determined as a product of the grain numbers per plant times the average kernel weight at physiological maturity. The grain weight is calculated as a function of cultivar-specific optimum growth rate multiplied by the duration of grain filling, which is reduced below optimum value when there is an insufficient supply of assimilate from either the biomass produced or stored biomass in the stem.

10.3.5.2 APSIM

APSIM stands for "Agricultural Production Systems sIMulator." It is a modeling framework that provides capabilities to simulate cropping systems over variable time periods using available meteorological data (McCown et al., 1996; APSIM Initiative, 2010).

The framework provides a "plug-in-pull-out" facility, allowing users to select modules for modeling crops and their environments, under a constraint conditions. Once the required modules have been selected, the behavior of the simulation is controlled through user-defined management criteria. The user can define and run simple simulations using options on the menu-bar. For more complex simulations, APSFront includes the capability to use template files to generate the APSIM parameter files needed for a simulation run.

10.4 Crop Production Function/Yield Model

10.4.1 Definition of Production Function

The relationship between crop growth or crop yield and water use is termed as crop production function. On the other hand, a mathematical model is an equation or set of equations which represents the behavior of a system. The behavior of crop production function can be explained with an example, shown in Fig. 10.4. The figure shows the yield curve of wheat that might result from an experiment in which irrigation water is supplied at different rates. Response of diminishing return type, as in Fig. 10.4, are quite common in biology and elsewhere, such as response of crops to fertilizer and photosynthetic response to light.

10.4.2 Importance of Production Function

The basic information needed to solve problems of optimum water management on farms consists in a precise knowledge of the water consumption of each crop and its response to irrigation. In other words, we must know the production functions in relation to water. Profits and risks inherent in irrigation management decisions depend directly on the underlying crop-water production functions. Irrigation scheme managers are often confronted with the issue of intensive (full irrigation to meet evapotranspiration demand) versus extensive (to cover more area with deficit

Fig. 10.4 Wheat yield versus applied water

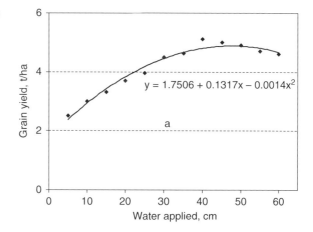

irrigation) to satisfy individual versus community demand for more production. Production functions are used to optimize on-farm irrigation and economic evaluation of irrigation water application. The crop production functions are also essential for improving water productivity (also termed as "water use efficiency," WUE) and effective allocation of water resources among crops/cropping patterns and soil types in a region under conditions of water shortage.

10.4.3 Basic Considerations in Crop Production Function

The yield of dry-land crops can be considered in terms of water use, dry matter production per unit of water use, and dry matter distribution to the grain or harvest index (or harvest ratio). In case of crop production, photosynthetically active radiation (PAR) is used by the plant as the energy in the photosynthesis process to convert CO_2 into biomass using the water, transpired by plant. Obviously, soil evaporation is not a function of crop production. Thus, actually yield is a function of transpiration. Since independent measurement of evaporation and transpiration is hard and requires specific equipment and expertise, evapotranspiration (ET) is normally used instead of transpiration.

10.4.4 Pattern of Crop Production Function

Crop yield relationship with seasonal evapotranspiration (ET) has been reported as linear by many investigators (Singh et al., 1979; Mogensen et al., 1985; Musick et al., 1994). The functional relationship between water applied and yield can be quadratic, polynomial, or exponential (Hexem and Heady, 1978; Martin et al., 1984). Zhang and Oweis (1999) developed quadratic crop production function with the total applied water for wheat in the Mediterranean region. For the same amount of water use, the impact on yield may be differed due to differential response of plant to irrigation timing and stress history (Ali et al., 2007). Hence, a singular production function may not appropriate for all circumstances.

10.4.5 Development of Crop Production Function

Generally two approaches for estimation of crop-water production function are available in the literature. One approach synthesizes production functions from theoretical and empirical models of individual components of the crop-water process. The second approach estimates production functions by statistical inference from observations of the effect of different water applications and salinity levels on crop yield.

The first set of approach is quite useful but due to their implicit assumptions and complication, their applications are restricted; whereas, the second set of approach estimates production by direct relationship. By using the second set of approach, many production functions have been estimated both for saline and non-saline water. Such empirically derived water production functions are usually valid only for a single crop at a single location under conditions of optimal deficit sequence. These functions are usually highly empirical and difficult to generalize. Economic solutions derived from such empirical functions are only useful for specific situations. These functions are based on the assumption that, considering all the other factors of production at their optimum level, it is the water scarcity that limits the final yield.

The effect of water stress on crop yield during individual growth stages has been investigated using an additive model (Minhas et al., 1974; Howell and Hiler, 1975) and a multiplicative model (Jensen, 1968; Hanks, 1974). An additive production function implies that the total absence of water at any stage would only result in some discrete reduction in yield, whereas the multiplicative model implies that the plant would die if water input falls to zero at any growth stage. Singh and Aggarwal (1986) reported that three-degree polynomial crop yield model without the consideration of the time of water stress gave the best results. However, the quadratic crop yield model was found to be more suitable for practical use.

10.4.6 Some Existing Yield Functions/Models

Model of de Wit

According to the theory of de Wit (1958), crop yield (Y) is a linear function of its transpiration (T):

$$Y = mT/E_0 \tag{10.29}$$

10.4 Crop Production Function/Yield Model

In which E_0 represents the evaporation of a free water surface and "m" a crop coefficient. This relationship was the basis for several models to predict yield from evapotranspiration.

Stewart and Hagan Model

The model proposed by Stewart and Hagan (1973) is

$$Y = Y_m - Y_m K_y \text{ETD}/\text{ET}_m \quad (10.30)$$

where

Y = crop yield
Y_m = maximum crop yield under the same condition of soil texture, fertility, etc.
K_y = crop coefficient
ETD = cumulative evapotranspiration deficit during the growth period, calculated as

$$\text{ETD} = \text{ET}_m - \text{ET}_a \quad (10.31)$$

where ET_m = maximum evapotranspiration; ET_a = actual evapotranspiration.

The model is simple and practical and can be used when the sensitivity to moisture stress is the same during the whole growing period.

Stewart et al. Model

For the case that the sensitivity differs significantly among growth periods, Stewart et al. (1977) proposed a model that takes into account the effect of moisture stress during successive phenological stages. They used a different coefficient for each stage, according to

$$\frac{y}{y_m} = \prod_{n=1}^{m} \left[1 - k_{y(n)} \left(1 - \frac{ET}{ET_m} \right)_n \right] \quad (10.32)$$

where n is generic growth stage, and m is the number of growth stage considered, and k_y is the crop or yield response factor.

Stewart's formula is based on the theory that, considering all other factors of production at their optimum level, it is the water scarcity factor (estimated as the ratio of actual to maximum evapotranspiration, ET/ET_m) that limit the final yield. This is an attempt to generalize the production function.

Determination of Response Factor of Stewart Model

To determine yield response factor k_y, the following procedure may be followed (Doorenbos and Kassem, 1979):

- Determine maximum yield (Y_m) of the crop as dictated by climate, under conditions of full water requirement.
- Calculate the maximum evapotranspiration (ET_m) that prevails when crop water requirements are fully met by the available supply.
- Quantify the effect of water deficit on the crop yield by the relationship between relative yield decrease and relative ET deficit:

$$1 - Y_a/Y_m = k_y(1 - ET_a/ET_m) \tag{10.33}$$

or,

$$k_y(i) = \frac{1 - \dfrac{Y_a(i)}{Y_m(i)}}{1 - \dfrac{ET_a(i)}{ET_m(i)}} \tag{10.34}$$

Jensen Model

According to Jensen (1968), the effect of water deficit during certain growth stages on grain yield is

$$\frac{Y}{Y_m} = \prod_{i=1}^{n} \left(\frac{ET_i}{ET_m}\right)^{\lambda_i} \tag{10.35}$$

where

Y = grain yield (t/ha)
Y_m = maximum yield from the plot without water stress during the growing season (t/ha)
ET_i = actual evapotranspiration during the growing season stage i (mm)
ET_m = maximum evapotranspiration corresponding to Y_m (mm)
λ_i = sensitivity index of crop to water stress (–)
i = growth stage (i=1 to n)
n = total number of growth stages

Determination of Sensitivity Index (λ_i)

The sensitivity index (λ_i) can be determined as follows:

The above form of production function, expressing the effect of water deficiency on yield at different stages of growth, in the form of power function, can be expressed as (Tsakiris, 1982):

$$\frac{y}{y_0} = \prod_{i=1}^{m} w_i^{\lambda_i} \tag{10.36}$$

10.5 Regression-Based Empirical Models

where λ_i usually refers to as sensitivity index, express the response of the crop to water deficiency during the ith period within the growth cycle; w_i is the relative water content of the root zone.

If the evapotranspiration is suppressed only during a certain stage, say the lth (i.e., $I = l$), then $w_i = 1$ for all except the lth stage. Therefore Eq. (10.36) yields

$$\frac{y}{y_0} = W_l^{\lambda_l} \qquad (10.37)$$

or,

$$\log(y/y_0) = \lambda_l \log w_l \qquad (10.38)$$

Thus by taking the logarithms of the data i.e., y/y_0 and w_l, the sensitivity index for the lth period λ_l, can be determined. Similarly, the sensitivity index can be obtained for all the m stages.

10.4.7 Limitations/Drawbacks of Crop Production Function

Although yield-seasonal ET relations have been widely used for management purposes in water-deficient areas as a guideline for irrigation, they do not account for the effect of timing of water application. Water stress during certain growth stages may have more effect on grain yield than similar stress at other stages. In addition, effect of alternate stress and previous stress history have an impact on subsequent stress (or stress hardening), and hence on yield.

10.5 Regression-Based Empirical Models for Predicting Crop Yield from Weather Variables

10.5.1 Need of Weather-Based Prediction Model

Extensive experiments and research have provided the basic information regarding the culture and methods of treatment needed for the successful production of crops under field conditions. Such information includes the selection of appropriate locations, optimum fertilization, irrigation requirement and effective methods of plant protection. Even so, crop yields fluctuate considerably from year to year, a fact attributable to the variation in environmental conditions.

Crop yield depends on certain uncontrollable events (weather elements) such as timing and amount of rainfall, exposure to sunlight, humidity, temperature, wind speed. Variation of rainfall is an inter-seasonal as well as intra-seasonal problem. Crop yield also depends on many controllable events such as timing and density of planting, the timing of harvesting, the timing and amount of maintenance inputs. Maintenance inputs include water for irrigation and salt leaching, fertilizers, pesticides, and herbicides.

Multiple linear regression approach is extensively used for modeling situations where more than one factor influences response variable. The underlying model can be expressed as

$$Y = a_0 + a_1 X_1 + a_2 X_2 + \cdots + \varepsilon \qquad (10.39)$$

where Y is response variable; $x_1, x_2, \ldots$ are explanatory variables; and ε is error term. Using this approach, influence of climatic variables on yield of various crops has been investigated by various researchers. Some researchers use non-parametric regression method to model the effect of weather elements on crop yield (Chandran and Prajneshu, 2004).

A time series of crop yield may be divided into three components; the mean yield, the trend in yield with time, and the residual variation. The mean yield is determined by the interacting effects of climate, soil, management, technological and economic factors. The trend is probably mainly due to long-term economic and/or technological changes. The third component is the variation between years and it is a prime objective of agricultural meteorologist to understand the role of weather in this variation.

Uncertainty in weather creates a risky environment for agricultural production. Crop models that use weather data in simulating crop yields have the potential for being used to assess the risk of producing a given crop in a particular environment and assisting in management decisions that anticipate appropriate measures.

10.5.2 Existing Models/Past Efforts

Granger (1980) investigated the effects of variations in agro-climatically significant variables on yield of four crops (oats, barley, almonds and walnuts). They obtained statistical relationships between agro-climatic variables and yield through stepwise multiple regression techniques and polynomial curve fitting. Quadir et al. (2003) tried to develop a functional relation of Aman rice yield anomaly with the rainfall. But to eliminate the trend term, they plotted the temporal yield data corresponding to year (absolute) value; which seems inappropriate (should be relative year value). They found a quadratic relationship of national Aman rice yield anomaly with monsoon rainfall of Dhaka. Parthasarathy et al. (1992) examined the relationship between all-India monsoon rainfall and rainy-season food grain production. They expressed the rainfall as percentage of mean, and food grain as percentage of technological trend. They developed a simple linear regression equation between the two indices.

However, because yield anomalies can vary considerably over scales of hundreds of kilometers, national or international yield fluctuations are generally not simple extrapolations of local or single-state values. The spatial complexities inherent in the analysis of large-scale yield anomalies are due largely to fluctuations in meteorological variables over scales of hundreds to thousands kilometers. So, location specific model is useful.

10.5.3 Methods of Formulation of Weather-Based Prediction Model

Crop-production statistics are composed of two main components:

– the area under the crop (e.g., hectare, ha)
– the production rate (t/ha) and/or total production (t)

The following steps may be followed to construct a regression based prediction model (Ali and Amin, 2006):

(i) *Zoning of crop area*
The total land area of the location under consideration is to be divided into a number of zones considering agro-climatic regions and/or soil resource based regions. But for all these zones, the yield and weather data may not be readily available. In that case, for model development, the total land area is to be divided into a number of crop zones keeping in mind the agro-climatic or soil resource based zones and available crop-reporting districts.

(ii) *Removing trend component*
The technological trend component is resulted from the use of high yielding cultivars, improved fertilization, management, etc. It is depended only on time. The trend curve of yield (of the crop under consideration) is to be examined for technological trend/advancement through time series analysis of the data. For this purpose, the yield rate of the crop (t/ha) is plotted against the relative year value (e.g., for 1980–1997 years data, relative year values or year ranks are 1–18), yield being the dependent variable. The slope of the plot represents the technological trend (t/ha/yr). The trend component is subtracted from the original yield series.

(iii) *Choice of weather variables*
The agro-climatic variables are selected based on the cause and effect relationship. The principal weather elements affecting yield of crops are rainfall, temperature (day-time and/or night temperature), light (photo period and/or bright sunshine hour), humidity, solar radiation and wind speed. For a particular crop variety, all of the factors or elements may not contribute to yield anomaly. To test a weather element whether it is attributable to yield, perform regression analysis of the yield series with the corresponding weather series. If the regression coefficient is significant at 5% level, consider this element for

final regression analysis. Discard elements which are not relevant for a particular crop species. Weather elements may be considered as single or composite index. Different composite indices are described in Chapter 3 (Weather), volume 1. A most commonly used composite thermal index is degree-days, also called growing-degree-days (GDD), as considered for whole growing period.

(iv) *Regression approach*

For a particular zone, multiple regression analysis is to be performed with the yield series (after subtracting the trend component from original yield series) and the selected weather elements (single or composite) where yield being the dependent or explained variable. Different coefficients are obtained for different weather elements, and also an intercept or constant value. These values are used for yield prediction. The predicted yield rate is the sum of the trend component and the component for weather elements. The yield (i.e., total yield) is obtained by multiplying the yield rate (t/ha) with the cultivated area (ha) for that zone.

The same procedure is to be applied for each of the zones. Finally, the simulation sub-models may be compacted in one platform to predict the regional or national yield.

10.5.3.1 The Typical Form of a Model

The typical form of a weather-based multiple regression model is

$$Y_i = C_1(i) \times t + C_2(i) + C_3(i) \times \text{GDD}(i) + C_4(i) \times R(i) + \cdots$$

where

Y_i = predicted yield rate of zone "i"
$C_1(i)$ = coefficient of trend for zone "i"
$C_2(i)$ = intercept of regression curve for zone "i"
$\text{GDD}(i)$ = growing degree days for zone "i"
$C_3(i)$ = Coefficient of GDD for zone "i"
$C_4(i)$ = Coefficient of rainfall for zone "i"
$R(i)$ = rainfall (cm) for the zone "i"
t = (simulation year − base year)

The simulation year is the year for which yield is to be predicted.

10.5.3.2 Basic Assumptions in Regression-Based Prediction Model

The basic assumptions in the model are (i) the effect of weather variables on yield is linear, and (ii) the technological trend over time is similar.

10.5.4 Discussion

Output of a model depends on the accuracy of the input data and the framework of the model itself. The yield and weather data should be taken from reliable source (or it should be primary source data). The model should be tested on independent weather and crop data. If the model predicts yield with reasonable accuracy, it may be used with confidence.

Better climatic data, particularly measurements of solar radiation, sunshine hour, and a greater number of rainfall stations, and yield data for homogeneous sub-areas or micro-zones might improve the yield prediction. The sub-area could be constructed on the basis of soil-climate zones or using spatial analysis and interpolation tool available with GIS software. Forecasted yields from sub-areas could then be aggregated to obtain the forecast for the whole country. In order for the concept of the sub-areas to succeed for improving the yield prediction, the availability of yield data for such small arial units will also be required.

The trend component of simulation may be updated from time to time to incorporate the non-linear development in the future.

Since the crop yield depends on many types of variables viz. weather factors, performance of plant during crop growth stages, agricultural inputs, it will be better if all these aspects can be considered while forecasting crop yield. For this, composite index (weather, input, management; and weighting them) for forecasting may be tried.

In case of large area (say, for a country), the same date of crop transplanting or swing and consequently the same date of physiological maturity may not appropriate. This factor can be more realistically included if prediction can be done for micro-zones.

10.5.5 Sample Example of Formulating Weather-Based Yield-Prediction Model

For example, for a particular region (or crop district), the yield rate and weather variables for the effective crop growing period are given as follows:

Year	Yield rate (t/ha)	Degree-day	Rainfall (cm)
1980	0.8970	1,270	37.5
1981	0.9725	1,260	28.2
1982	0.8069	1,292	8
1983	0.9285	1,239	15.4
1984	0.8512	1,276	40.8
1985	0.8797	1,330	16.5
1986	1.0856	1,217	45.2
1987	1.0334	1,320	32

(continued)

Year	Yield rate (t/ha)	Degree-day	Rainfall (cm)
1988	1.0736	1,305	73.5
1989	0.9963	1,354	47.6
1990	1.2589	1,175	68.6
1991	1.2876	1,355	47.4
1992	1.2484	1,364	25.6
1993	1.2101	1,350	61.7
1994	1.4541	1,314	39.5
1995	1.4134	1,367	44.1
1996	1.4197	1,325	43.3
1997	1.3648	1,289	46.7

Now, we have to formulate regression based yield prediction model based on weather variables.

10.5.5.1 Formulation

Step 1

Yield rate data are arranged corresponding to relative year. For this purpose, the starting year of the data series is taken as year 1 (one).

Year	Relative year	Yield rate (t/ha)
1980	1	0.8970
1981	2	0.9725
1982	3	0.8069
1983	4	0.9285
1984	5	0.8512
1985	6	0.8797
1986	7	1.0856
1987	8	1.0334
1988	9	1.0736
1989	10	0.9963
1990	11	1.2589
1991	12	1.2876
1992	13	1.2484
1993	14	1.2101
1994	15	1.4541
1995	16	1.4134
1996	17	1.4197
1997	18	1.3648

Step 2

Plot the yield data against the relative year value, yield being the dependent variable.

The slope of the regression line represents the "trend." From the below figure, the trend (slope) is 0.0369 t/ha/yr.

10.5 Regression-Based Empirical Models

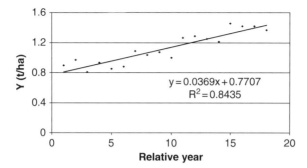

Step 3

Deduct the trend component from the original yield series. The yield data becomes as follows:

Year	Yield rate after deducting trend (t/ha)
1980	0.8601
1981	0.8987
1982	0.6962
1983	0.7809
1984	0.6667
1985	0.6583
1986	0.8273
1987	0.7382
1988	0.7415
1989	0.6273
1990	0.8530
1991	0.8448
1992	0.7687
1993	0.6935
1994	0.9006
1995	0.8230
1996	0.7924
1997	0.7006

Step 4

Test the dependency of the yield on weather elements one by one. Perform simple linear regression analysis of yield and a weather element, say, rainfall.

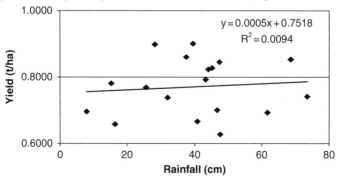

Step 5

Arrange the yield data and weather elements (single or composite) for multiple linear regression analysis as follows:

Yield rate (t/ha)	Degree-day	Rainfall (cm)
0.8601	1,270	37.5
0.8987	1,260	28.2
0.6962	1,292	8
0.7809	1,239	15.4
0.6667	1,276	40.8
0.6583	1,330	16.5
0.8273	1,217	45.2
0.7382	1,320	32
0.7415	1,305	73.5
0.6273	1,354	47.6
0.8530	1,175	68.6
0.8448	1,355	47.4
0.7687	1,364	25.6
0.6935	1,350	61.7
0.9006	1,314	39.5
0.8230	1,367	44.1
0.7924	1,325	43.3
0.7006	1,289	46.7

Step 6

Perform multiple regression analysis. The results (obtained through Microsoft Excel tool) are as follows:

Summary output	
Regression statistics	
Multiple R	0.33849
R square	0.11458
Adjusted R square	−0.00348
Standard error	0.08500
Observations	18

	Coefficients	Standard error	t stat	P-value	Lower 95%	Upper 95%	Lower 95.0%	Upper 95%
Intercept	1.42676	0.508	2.807	0.013	0.343	2.510	0.343	2.510
X Variable 1	−0.00052	0.000	−1.335	0.202	−0.001	0.000	−0.001	0.000
X Variable 2	0.00035	0.001	0.301	0.767	−0.002	0.003	−0.002	0.003

ANOVA	df	SS	MS	F	Significance F
Regression	2	0.0140	0.0070	0.9705	0.4014
Residual	15	0.1084	0.0072		
Total	17	0.1224			

Summary results are obtained as

$$Y = 0.0369(N - B) + (1.42676 - 0.00052 \times \text{GDD} + 0.00035 \times R)$$

where

Y = predicted yield rate for the year N (say, 2012)
B = Base year (the year from which the simulation data is based, e.g., for the above case it is 1980)
and GDD, R, corresponds to year N.

The same procedure should be applied for each zone. The total yield of a zone is obtained by multiplying the yield rate by the cultivated area.

Relevant Journals

- Agricultural Systems
- Precision Agriculture
- Computers and Electronics in Agriculture
- Computing in Science and Engineering
- Trans. ASAE
- Trans. ASCE
- Water Resources Management (Springer)
- Advances in Water Resources (Elsevier)
- Agricultural Water Management
- Irrigation Science
- Journal of Hydrology
- Irrigation and Drainage systems
- J. of Agrometeorology
- J. of Agricultural and Forest Meteorology
- J. of Agricultural Science

Questions

(1) What is a model? Explain the need of model.

(2) Classify the models
(3) Write short note on the following:
   Conceptual model, mathematical and statistical model, mechanistic and probabilistic model, static and dynamic model, deterministic and stochastic model, time series model
(4) What are the considerations during development of a model?
(5) Describe the process of model calibration and validation
(6) Write short note on
   Sensitivity analysis, implicit vs explicit scheme, finite element vs finite difference method
(7) Explain different statistical parameters for model evaluation.
(8) Briefly describe the following models (with respect to capabilities, input, output, process/mechanisms involved, governing equations, and limitations): FLOCR, UPFLOW, CROPWAT, CROPET0, MACRO, DSSAT
(9) Explain the need of weather-based prediction model.
(10) Describe the procedure for formulating weather-based prediction model.

References

Addiscott TM, Whitemore AP (1987) Computer simulation of changes in soil mineral nitrogen and crop nitrogen during autumn, winter and spring. J Agric Sci Camb, 109:141–157

Ali MH (2000) Behavior study of cracking clay soil with special emphasis on water relations. M. Engg. Thesis, Submitted to the Department of Civil & Environmental Engineering, The University of Melbourne, Australia

Ali MH (2005a) CropET$_0$: a computer model to estimate reference evapotranspiration from climatic data. Bangladesh J Agric Eng 16(1 & 2):25–37

Ali MH (2005b) E-STAT: a computer program to perform statistical analysis of experimental data. J Bangladesh Agric Univ 3(1):133–138

Ali MH, Amin MGM (2006) 'AmanGrow': a simulation model based on weather parameters for predicting transplanted 'Aman rice' production in Bangladesh. Ind J Agril Sci 76(1): 50–51

Ali MH, Hoque MR, Hassan AA, Khair MA (2007) Effects of deficit irrigation on wheat yield, water productivity and economic return. Agric Water Manage 92:151–161

Allen R, Pereira L, Raes D, Smith M (1998) Crop evapotranspiration (guidelines for computing crop water requirements). FAO Irrigation and Drainage Paper No 56. Rome, Italy, p 300

APSIM Initiative (2010) APSIM – agricultural production systems simulator. www.apsim.info. Accessed on Mar 10, 2010

Bronswijk JJB (1988) Modeling of water balance, cracking and subsidence of clay soils. J Hydrol 97:199–212

Chandran KP, Prajneshu (2004) Modeling the effect of sunshine and temperature on rice tiller production using non-parametric regression. Ind J Agril Sci 74(10):563–565

Clarke D (1998) CropWat for windows: user guide. FAO, Rome, p 23

de Wit CT (1958) Transpiration and crop yields. Agric Res Rep. 64.6, Pudoc, Wegeningen, 88 pp

De Laat PJM (1980) Model for unsaturated flow above a shallow water-table. Applied to a regional sub-surface flow problem. PUDOC, Doctoral thesis, Wageningen, The Netherlands, p 126

De Laat PJM (1995) Design and operation of a subsurface irrigation scheme with must. In Pereira, LS, van den Broek, BJ, Kabat, P, Allen RG (eds) Cropwater-simulation models in practice. Wageningen Pers, The Netherlands, pp 123–140

Donnigan AS (1983) Model predictions vs. field observations: the model validation/testing process. In: Swannand RL, Eschenroder A (eds) Fate of chemicals in the environment. American Chemical Society Symposium Series 225. ACS, Washington, DC, pp 151–171

Doorenbos J, Kassam AH (1979) Yield response to water. FAO Irrigation and Drainage Paper No. 33, FAO, Rome

Feddes RA, Kowalik PJ, Zaradny H (1978) Simulation of field water use and crop yield. Simulation Monographs. PUDOC, Wageningen, The Netherlands, p 189

Fox MS (1981) An organizational view of distributed systems. IEEE Trans Syst Man Cybernet 11:70–80

Gabrielle B, Kengni L (1996) Analysis and field-evaluation of the CERES model's soil components: nitrogen transfer and transformations. Soil Sci Soc Am J 60:142–149

Granger OE (1980) The impact of climatic variation on the yield of selected crops in three California countries. Agril Meteor 22:367–386

Hanks RJ (1974) Model for predicting plant yield as influenced by water use. Agron J 66:660–665

Hexem RW, Heady EO (1978) Water production functions for irrigated agriculture. Iowa State University Press, Ames, IA, p 215.

Howell TA, Hiler EA (1975) Optimization of water use efficiency under high frequency irrigation. I. Evapotranspiration and yield relationship. Trans ASAE 18(5):873–878

Jarvis N (1994) The MACRO model (Version 3.1): technical description and sample simulations. Monograph, Department of Soil Sciences, Reports and Dissertations – 19, Swedish University of Agricultural Sciences, Uppsala

Jensen ME (1968) Water consumption by agricultural plants. In: Kozlowski TT (ed) Water deficit and plant growth, vol 2, Academic press, New York, pp 1–22

Lecina S, Martinez-Cob A, Perez PJ, Villalobos FJ, Baselga JJ (2003) Fixed versus variable bulk canopy resistance for reference ET estimation using the Penman-Monteith equation under semi-arid conditions. Agril Water Manage 60:181–198

Loague K, Green RE (1991) Statistical and graphical methods for evaluating solute transport models: overview and application. J Contam Hydrol 7:51–73

Loague KM, Yost RS, Green RE, Liang TC (1989) Uncertainty in a pesticide leaching assessment for Hawai. J Contamin Hydrol 4:139–161

Mapp HP, Eidman VR (1978) Simulation of soil-water-crop yield systems: the potential for economic analyses. South J Agric Econ 7(1):47–53

Martin DL, Watts DG, Gulley JR (1984) Model and production function for irrigation management. J Irrig Drain Eng, ASCE 110(2):149–164

McCown RL, Hammer GL, Hargreaves JNG, Hozworth DP, Freebairn DM (1996) APSIM: a novel software system for model development, model testing and simulation in agricultural systems research. Agric Syst 50:255–271

Minhas BS, Parikh KS, Srinivasan TN (1974) Towards the structure of a production function for wheat yields with dated inputs of irrigation water. Water Resour Res 10:383–393

Mogensen VO, Jensen HE, Rab MA (1985) Grain yield, yield components, drought sensitivity and water use efficiency of spring wheat subjected to water stress at various growth stages. Irrig Sci 6:131–140

Monteith JL (1981) Evaporation and surface temperature. Quart J R Meteor Soc 107:1–27

Musick JT, Jones OR, Stewart BA, Dusek DA (1994) Water-yield relationships for irrigated and dryland wheat in the U.S. Southern Plains. Agron J 86:980–986

Oostindie K, Bronswijk JJB (1992) FLOCR – a simulation model for the calculation of water balance, cracking and surface subsidence of clay soils. Report 47, Agricultural Research Department, The Winand Staring Centre for Integrated Land, Soil and Water Research, Wageningen, The Netherlands

Parthasarathy BK, Kumar BR, Munot AA (1992) Forecast of rainy-season food grain production based on monsoon rainfall. Ind J Agril Sci 62(1):1–8

Quadir DA, Khan TMA, Hossain MA, Iqbal A (2003) Study of climate variability and its impact of rice yield in Bangladesh SAARC. J Agric 1:69–83

Raes D (2002) UPFLOW – water movement in a soil profile from a shallow water table to the top soil (capillary rise). Reference manual version 2.1, Department of Land Management, K.U. Leuven University, Belgium

Raes D, Deproost P (2003) Model to assess water movement from a shallow water table to the root zone. Agric Water Manage 62:79–91

Rao NH, Sarma PBS, Chander S (1988) Irrigation scheduling under a limited water supply. Agric Water Manage 15:165–175

Rao NH, Sarma PBS, Chandar S (1990) Optimal multicrop allocation of seasonal and intra-seasonal irrigation water. Water Reour Res 26:551–559

Retta A, Vanderlip RL, Higgin RA, Moshier LJ (1996) Application of SORKAM to simulate shattercane growth using forage sorghum. Agron J 88:596–601

Singh P, Aggarwal MC (1986) Improving irrigation water use efficiency and yield of cotton by different agrotechniques. In: Proceedings of the National Seminar, water management the key to development of agriculture. Indian National Science Academy, New Delhi (1986)

Singh NT, Singh R, Mahajan PS, Vig AC (1979) Influence of supplemental irrigation and pre-sowing soil water storage on wheat. Agron J 71:401–404

Smith M, Allen R, Monteith JL, Perrier A, Pereira LS, Segeren A (1992) Expert consultation on revision of FAO methodologies for crop water requirements. Food and Agriculture Organization of the United Nation (Land and Water Development Division), Rome, p 60

Stewart JI, Hagan EM (1973) Functions to predict effects of crop after deficit. J Irrig Drain Div, ASCE, 99(IR4):429–439

Stewart JI, Hagen RM, Pruitt WO, Hanks RJ, Denilson RE, Franklin WT, Jackson EB (1977) Optimizing crop production through control of water, salinity levels. Utah Water Research Laboratory PRWG 151, Logan, Utah, 191 pp

Stockle CO, Martin SA, Campbell GS (1994) CropSyst, acropping systems simulation model: water/nitrogen budgets and crop yield. Agric Syst 46(3):335–359

Stockle CO, Nelson RL (1994) CropSyst user's manual (version 1.0). Biological Systems Engineering Department, Washington State University, Pullman, WA

Stockle CO, Nelson RL (2000) CropSyst user's manual (version 3.0). Biological Systems Engineering Department, Washington State University, Pullman, WA

Tsakiris GP (1982) A method for applying crop sensitivity factors in irrigation scheduling. Agric Water Manage 5(4):335–343

Tsuji GY, Uehara G, Balas S (eds) (1994) DSSAT v.3. University of Hawaii, Honolulu

Willmott CJ (1982) Some comments on the evaluation of model performance. Am Meteorol Soc Bull 63:1309–1313

Chapter 11
GIS in Irrigation and Water Management

Contents

11.1	Introduction	424
11.2	Definition of GIS	424
11.3	Benefits of GIS Over Other Information Systems	424
11.4	Major Tasks in GIS	425
11.5	Applications of GIS	425
11.6	Techniques Used in GIS	427
11.7	Implementation of GIS	427
11.8	Data and Databases for GIS	428
11.9	Sources of Spatial Data	428
11.10	Data Input	429
11.11	GIS-Based Modeling or Spatial Modeling	429
11.12	Remote Sensing Techniques	430
	Relevant Journals	431
	Questions	431
	References	431

In most part of the world, water resources are finite and most of the economically viable development has already been implemented. In addition, population growth and the effects of cyclic droughts on irrigated agriculture have put pressure on the available water resources. Such prevailing conditions have the effect of creating an imbalance between the increasing water demand and limited available water supply. Under this perspective, effective planning and management can only be obtained on the basis of reliable information on spatial and temporal patterns of farmer's water demand, on farming irrigation practices, and on physical and operational features of large-scale irrigation systems. The timely and reliable assessment and monitoring of water resources and systematic exploration and developing new ones is of paramount importance. For this, it is necessary to employ modern methods of surveying, investigations, design, and implementation. Remote sensing and GIS are viewed as an efficient tool for irrigation water management.

Geographic Information System (GIS) and remote sensing techniques can provide managers and planners with the visualizing effects resulting from various management strategies, under different climatic and operational conditions. They can be used as analytical tools and can significantly enhance the ability of researchers and practitioners responsible for investigating water-management alternatives.

11.1 Introduction

Where resources are scarce, proper planning and decision-making at different levels is essential. In today's high-tech world, information technology provides easy solutions for decision making, where the key is the collection and collation of different information at usable format. A geographic information system (GIS) allows users to bring all types of information based on the geographic and locational component of the data. GIS provides the power to create maps, integrate information, visualize scenarios, solve complicated problems, present powerful ideas, and develop effective solutions like never before. More than that, GIS lets one model scenarios to test various hypotheses and see outcomes visually to find/identify the outcome that meets the needs of the stakeholders. Now a days, GIS and related technologies are increasingly being recognized as useful tools for natural resources inventorying studies and management because of their capability to bring together geographically referenced data from a variety of subject matters to aid in processing, interpretation, and analysis of such data.

11.2 Definition of GIS

Geographic Information System (GIS) is computer-based system used to store and manipulate geographic information. A widely used definition of GIS is "an organized collection of database, application, hardware, software, and trained manpower capable of capturing, manipulating, managing, and analyzing the spatially reference database and production of output both in tabular and map form." In a more generic sense, GIS is a tool that allows users to create interactive queries, analyze the spatial information, and edit data.

11.3 Benefits of GIS Over Other Information Systems

GIS is fundamentally used to answer questions and make decisions. A GIS like other information systems provides the following four sets of capabilities to handle geographic data:

(i) Input
(ii) Data management (data storage and retrieval)

(iii) Manipulation and analysis
(iv) Output

In addition, GIS is designed for the collection, storage, and analysis of objects and phenomena, where geographic location is an important characteristic or critical to analysis. The spatial searching and overlay of (map) layers are the unique functions of GIS. For example, maps of crop potential and ground/surface water situation can be combined in a GIS to produce map of the crop/land suitability on a temporal and spatial basis.

Since in a real world situation, complexity is large (e.g., in agriculture for decision making, data on land, soil, crop, climate, hydrology, forestry, livestock, fisheries, social and economic parameters are required), and the physical computing capacity to manipulate data is low and time consuming; GIS is an advanced and excellent planning tool for the resource managers and decision makers. GIS thus is the central element in the configuration of modern information technology (IT).

11.4 Major Tasks in GIS

GIS performs major six types of jobs:

- *Input*: Digitalization from paper map, scanning and vector processing, image classification.
- *Manipulation*: Before all the information are integrated, they must be transformed into same scale of resolution.
- *Management*: The spatial and attribute database management.
- *Query and viewing*: Once the data base is prepared, user can do any query on the data through GIS, e.g., where is the soil having land type MHL and clay-textured soils.
- *Analysis*: GIS has many powerful tools to generate "what-if" scenario. For example, "Does drought exists in an area? How intensive is it? What is the extent and what will be the crop yield loss?"
- *Visualization and printing*: Preparation of maps, legends, symbology, and other related elements, and providing facility to print from printers.

11.5 Applications of GIS

The GIS can be used in the following fields/disciplines:

(a) Agriculture

- Climatic constraints of growing crops
- Soil resources availability, assessment, and planning
- Potential suitability of crops/cropping pattern

- Crop hazard mapping and yield loss estimation
- Agro-ecosystem characterization
- Determining potential area suitable for crops
- Extrapolation area delineation of agricultural technologies
- Agricultural extension
- Agricultural research planning
- Agricultural development planning
- Crop production planning

(b) Natural resources management

- Water resources management
  - spatial map of soil hydraulic properties
  - map of groundwater table depth
  - estimation of crop water stress and developing water demand map
  - irrigation scheduling
  - estimating water logging condition
  - spatially distributed data sets that can be utilized for better management of large-scale irrigation systems and for supporting decision-making processes
- Environmental impact assessment
- Forest resources management
- Flood and other natural disaster mitigation planning

(c) Urban and rural planning

- Planning and zoning
- Infrastructure planning
- Land information system
- Percale mapping
- Tax assessment based on present land use

(d) Others

- Business development/marketing
- Crime analysis
- Utilities mapping and management

GIS has a wide range of potential applications in agricultural research. One aspect of agricultural research/planning for which the integration and analytical capabilities of GIS can be effectively utilized is characterization of crop-growing environments using agro-ecological approach. A GIS can convert existing digital information, which may not be in map form, into forms it can recognize and use. For example, digital satellite images generated through remote sensing can be analyzed to produce a map-like layer of digital information about vegetative covers.

Thus, the changes in vegetation vigor through a growing season can be animated to determine when drought is most extensive in a particular region.

GIS and related technology will help greatly in the management and analysis of large volumes of data, allowing for better understanding of terrestrial processes and better management of human activities to maintain world economic vitality and environmental quality.

11.6 Techniques Used in GIS

GIS software is the main method through which geographic data is accessed, transferred, transformed, overlaid, processed, and displayed. GIS-based modeling deals with different layers or information concerns. The model needs some rules or criterion based on which the analysis is done. The relevant scientists, who know the interactions of the layers and can apply their knowledge for the analysis, can develop the rules or criterion. The overlay technique of GIS provide the users flexibility in dealing with the interactions between the parameters concerned spatially and thereby demonstrate the results spatially as map form. The rules and the relevant database development are crucial for GIS modeling.

11.7 Implementation of GIS

It is generally understood that a system means a combination of workable hardware and software, but a GIS system is different, which includes the data base, trained personnel, and a methodology to operate the data for the application. Requirements of a GIS should be defined in terms of the applications the system is expected to support. A data base is the foundation needed to perform any application. GIS software is the central to the professional analysis and presentation of GIS data.

The following different steps have to follow to carry out a GIS application:

(e) Outline and elaborate the objective(s)
(f) Outline the analysis need
(g) Formulate the rule for analysis
(h) Outline the data need and their format
(i) Look the data availability at different sources
(j) Identify the data need to capture, collect, and collate
(k) Coding the data, assess its accuracy
(l) Capture the data, edit and clean it with proper documentation
(m) Do necessary analysis
(n) Produce output and validate (if applicable)
(o) Prepare a report

11.8 Data and Databases for GIS

GIS data are handled in a database or databases, which will have special functional requirements as well as the general characteristics of any standard database. Geographical data are inherently a form of spatial data. Spatial data that pertain to a location on the earth's surface are often termed geo-referenced data. Geographic data are commonly characterized as having two fundamental components:

i. *Attribute data*: The phenomenon being reported such as a physical dimension or class
ii. *Spatial data*: The spatial location of the phenomenon

Examples of a physical dimension might be the height of a forest canopy, the population of a city, or width of a river. The class could be a land type, a vegetation type. The location is usually specified with reference to a common coordinate system such as latitude and longitude.

A third fundamental component to geographic information is time. The time component often is not stated explicitly, but it often is critical. Geographic information describes a phenomenon at a location as it existed at a specific point of time. A land cover map describes the location of different classes of land cover as they existed at the time of data collection.

An agro-ecological zone (AEZ) is a geographical area delineated through a unique combination of physiography, soil, and hydrological and agro-climatic characteristics. Overlay of the agro-climatic inventory on the land resources inventory (physiography and soil map) produces agro-climatic zone/region. AEZ is an effective planning tool for agricultural development purposes. Thirty agro-ecological zones and 88 sub-zones in Bangladesh have been created which are relevant for land use and for assessing agricultural potential.

11.9 Sources of Spatial Data

The primary data sources are remote sensing and global positioning system (GPS).

Remote Sensing: Captures digital data by means of sensors on satellite or aircraft that provide measurement of reflectances or images of portions of the earth. Remotely sensed data are usually in raster structure.

Global Positioning System: Allows capturing terrestrial position and vehicle tracking, using a network of navigation satellites. Data are captured as a set of point position readings and may be combined with other attributes of the object by means of textual/numerical devices such as data loggers. The data are structured as sequence of points, that is, vector format.

The spatial data is stored as X, Y coordinate pairs. Since the world is spherical (3D), to store the geographic features in 2D system, one must project the geographic data to get the real world coordinate system. One of the key issues for the GIS database development is to specify a common coordinate system for the database.

11.10 Data Input

There are five types of data entry systems commonly used in GIS: keyboard entry, coordinate geometry, manual digitizing, scanning, and the input of existing digital files.

> *Keyboard entry and coordinate geometry procedure*: Attributes data are commonly input by keyboard, where spatial data are rarely entered in this way. Coordinate geometry procedures are used to enter land record information.
> *Scanning*: Scanning, also termed scan digitizing, is a more automated method for entering map data. It converts an analog data source (usually a printed map) into a digital data set.
> *Digitizers*: Manual digitizing is the most widely used method for entering spatial data from maps. It provides a means of converting an analog spatial data source to a digital data set with a vector structure.

11.11 GIS-Based Modeling or Spatial Modeling

Geographic information systems (GIS) are quite common and generally accepted in surveying and mapping, cartography, urban and regional planning, land resources assessment, and natural hazard mapping, e.g., crop loss estimation; for environmental applications such as hydrological modeling, climate change modeling, and land management. Current trends in GIS in interrogation of vector and raster data sets resulting a hybrid GIS make easy to accommodate database layers, specifically the remote-sensing-based satellite images and subsequent classification of the images for spatial modeling.

Modern techniques such as remote sensing (RS) and geographic information systems (GIS) have the capabilities of water resource management and conservation tool. RS/GIS analysis can show where water enters a system and how it leaves through evapotranspiration and runoff. Using this information, planners can identify areas where there is potential for development of new water resources; where water can be reallocated from one use or one basin to another; and identify potential areas of water scarcity before water shortages occur. The GIS technique helps in integration of satellite and ground information to evaluate the system performance and to diagnose the inequality in the performance to aid in improving the water management.

Approaches for modeling under GIS could be of two types: a decision support system based on rule formation for individual layers or integrating the layers by a series of equations, which expresses the relationship among the parameters and finally providing an output. The second type of model also called lumped parameter model. If under GIS environment all the basic functions (data input, management, analysis, and output generations) are integrated with the rule-based or other necessary tools for the modeling under one graphical users interface (GUI), then it is called the GIS-based modeling or decision-support system.

GIS have been used in many areas such as agricultural watershed management (Sarker, 2002), hydrology (Chandrapala and Wimalasuriya, 2003), water productivity estimation (Ines et al., 2002), monitoring of irrigation delivery (Rowshown et al., 2003), groundwater assessment (Chowdary et al., 2003), estimating water demand (Satti and Jacobs, 2004), hydrologic impacts of land-use change (Bhaduri et al., 2000), and water resources management (McKinney and Cai, 2002; Knox and Weatherfield, 1999).

11.12 Remote Sensing Techniques

Remote sensing is the collection of data through imaging sensor technologies, usually on an aircraft or satellite. The remote sensing techniques involve close examination of enlarged landsat imageries and aerial photographs followed by identification, interpretation, and mapping of various water and agricultural resource data connected with water bodies, forests, soil and land types, crop areas, roads, villages, etc. In addition to collection and mapping of various resources data, correlation studies between two or three types of data can easily be made from the imageries and air-photos. Remote sensing technique can be used in water resources management (Chandrapala and Wimalasuriya, 2003).

Crop yield estimation can be improved by using agrometeorological and remote sensing data which involve the development of crop-specific and area-specific crop growth and yield models. Soil moisture also plays an important role under both irrigated and rain-fed conditions. Thus monitoring of soil moisture status in a region using remote sensing data would be useful for assessing the crop condition as well as for advising the farming community for providing life saving irrigation, thinning operation, post-ponding fertilizer, and pesticide application.

Remote sensing techniques are useful for irrigation water management in the following areas:

- Identifying, inventory and assessment of irrigated crops
- Determination of irrigation water demand over space and time
- Assessment of water availability in reservoir for optimal management of water to meet the irrigation demand
- Distinguishing lands irrigated by surface water bodies
- Estimating crop yield
- Water logging and salinity problems in irrigated land
- Irrigation system performance evaluation

The information on the crop and water availability derived from remote sensing methods form a reliable database for further investigation and analysis across space and time. Integration of remote sensing and ground inputs ca be very effectively organized and analyzed in GIS environment.

Relevant Journals

- Agricultural Water Management
- Irrigation Science
- Precision Agriculture
- Computers and Electronics in Agriculture
- Computing in Science & Engineering
- Agricultural Systems
- Trans. ASAE
- Trans. ASCE
- Water Resources Management (Springer)
- Advances in Water Resources (Elsevier)
- Journal of Hydrology
- Irrigation and Drainage systems
- Natural Resources Research
- The Geographical journal
- Environmental Management

Questions

(1) What is GIS? What are the capabilities of GIS software?
(2) Narrate the special characteristics of GIS software compared to other software tools?
(3) What tasks GIS can perform? Explain the application of GIS in different fields.
(4) What are the steps to follow to carry out a GIS application?
(5) What types of data can GIS handle?
(6) How you can obtain spatial data? How they can be entered into computer?
(7) What are the approaches for modeling under GIS?
(8) Explain the principles of GIS and Remote sensing (RS).
(9) Discuss the links between RS and GIS.
(10) Explain the use of GIS in natural resources management.
(11) Write short note on (a) GPS, (b) Remote sensing technique.

References

Bhaduri B, Harbor J, Engel B, Grove M (2000) Assessing watershed-scale, long-term hydrologic impacts of land-use change using a GIS-NPS model. Environ Manage 26(6):643–658

Chandrapala L, Wimalasuriya M (2003) Satellite measurement supplemented with meteorological data to operationally estimate evaporation in Sri Lanka. Agric Water Manage 58:89–107

Chowdary VM, Rao NH, Sarma PBS (2003) GIS-based decision support system for groundwater assessment in large irrigation project areas. Agric Water Manage 62:229–252

Ines AVM, Gupta AD, Loof R (2002) Application of GIS and crop growth models in estimating water productivity. Agric Water Manage 54:205–225

Knox JW, Weatherfield EK (1999) The application of GIS to irrigation water resource management in England and Wales. Geograph J 165:90–98

McKinney DC, Cai X (2002) Linking GIS and water resources management models: an object-oriented method. Environ Model Softw 17:413–415

Rowshown MK, Kwok CY, Lee TS (2003) GIS-based scheduling and monitoring irrigation delivery for rice irrigation system. Agric Water Manage 62:105–126

Sarker MMH (2002) Application of GIS in agricultural watershed management through hydrological modeling for winter (vegetable) crops. Proceedings of the 2nd annual paper meet, Agricultural Engineering Division, The Institution of Engineers, Bangladesh, 20 November 2002, pp 100–106

Satti SR, Jacobs JM (2004) A GIS-based model to estimate the regionally distributed drought water demand. Agric Water Manage 66:1–13

Chapter 12
Water-Lifting Devices – Pumps

Contents

12.1	Classification of Water-Lifting Devices	435
	12.1.1 Human-Powered Devices	435
	12.1.2 Animal-Powered Devices	436
	12.1.3 Kinetic Energy Powered Device	436
	12.1.4 Mechanically Powered Water-Lifting Devices	437
12.2	Definition, Purpose, and Classification of Pumps	437
	12.2.1 Definition of Pump	437
	12.2.2 Pumping Purpose	437
	12.2.3 Principles in Water Pumping	438
	12.2.4 Classification of Pumps	438
12.3	Factors Affecting the Practical Suction Lift of Suction-Mode Pump	442
12.4	Centrifugal Pumps	442
	12.4.1 Features and Principles of Centrifugal Pumps	442
	12.4.2 Some Relevant Terminologies to Centrifugal Pump	443
	12.4.3 Pump Efficiency	445
	12.4.4 Specific Speed	446
	12.4.5 Affinity Laws	446
	12.4.6 Priming of Centrifugal Pumps	448
	12.4.7 Cavitation	449
12.5	Description of Different Types of Centrifugal Pumps	449
	12.5.1 Turbine Pump	449
	12.5.2 Submersible Pump	451
	12.5.3 Mono-Block Pump	454
	12.5.4 Radial-Flow Pump	455
	12.5.5 Volute Pump	456
	12.5.6 Axial-Flow Pump	456
	12.5.7 Mixed-Flow Pump	456
	12.5.8 Advantage and Disadvantage of Different Centrifugal Pumps	457
	12.5.9 Some Common Problems of Centrifugal Pumps, Their Probable Causes, and Remedial Measures	457

12.6	Other Types of Pumps		458
	12.6.1	Air-Lift Pump	458
	12.6.2	Jet Pump	459
	12.6.3	Reciprocating Pump/Bucket Pump	461
	12.6.4	Displacement Pump	462
	12.6.5	Hydraulic Ram Pump	462
	12.6.6	Booster Pump	462
	12.6.7	Variable Speed Pump	462
12.7	Cavitation in Pump		463
	12.7.1	Cavitation in Radial Flow and Mixed Flow Pumps	463
	12.7.2	Cavitation in Axial-Flow Pumps	463
12.8	Power Requirement in Pumping		464
12.9	Pump Installation, Operation, and Control		465
	12.9.1	Pump Installation	465
	12.9.2	Pump Operation	466
	12.9.3	Pump Control	467
12.10	Hydraulics in Pumping System		468
	12.10.1	Pressure Vs Flow Rate	468
	12.10.2	Pressure and Head	468
	12.10.3	Elevation Difference	469
12.11	Pumps Connected in Series and Parallel		469
12.12	Pump Performance and Pump Selection		469
	12.12.1	Pump Performance	469
	12.12.2	Factors Affecting Pump Performance	469
	12.12.3	Selecting a Pump	470
	12.12.4	Procedure for Selecting a Pump	470
12.13	Sample Workout Problems on Pump		473
Questions			476

Every day, millions of water pumps deliver water from wells to homes, farms, and businesses. A pump is the link to the water resource. Without pumps, we could not access ground water. As energy costs continue to increase, developing more efficient equipment will contribute to saving energy. Significant opportunities exist to reduce pumping system energy consumption through smart design, retrofitting, and operating practices. Cost of water supply depends on the appropriate selection (type and size for the prevailing condition) and efficiency of the pumping equipments. Each type of pump has its merits and demerits. Proper selection of pumps, motors, and controls to meet the requirement is essential to ensure that a pumping system operates effectively, reliably, and efficiently. The basics of pump control systems and hydraulic system are useful in pump operation. A proper discussion of pumping considers not just the pump, but the entire pumping "system" and how the system components interact. The recommended systems approach to evaluation and analysis includes both the supply and demand sides of the system. Greater details regarding the characteristics of each pump, factors influencing optimum pump selection, and procedure for selecting a pump have been described in this chapter.

12.1 Classification of Water-Lifting Devices

According to power sources, water-lifting devices can be classified as

(a) Human-powered devices
(b) Animal-powered devices
(c) Kinetic energy powered devices
(d) Mechanically powered devices

Mechanically powered devices are usually termed as pumps.

12.1.1 Human-Powered Devices

Man has a limited physical power output, which may be in the range of 0.08–0.1 hp. This power can be used to lift water from shallow depths for irrigation. The common man-powered devices are

(i) Swing basket
(ii) Don
(iii) Archimedean screw
(iv) Paddle wheel
(v) Counterpoise lift

12.1.1.1 Swing Basket

The device consists of a basket made from the cheap materials like bamboo strips, leather, or iron sheet to which four ropes are attached. Two persons hold the basket facing toward each others, dip the basket in water source, and by swinging, the basket is lifted and filled in water course from where the water flows to the fields. The device is useful up to a depth of 0.5 m and discharge may vary from 3,500 to 5,000 l/h.

12.1.1.2 Don

The don consists of a trough made from wooden log or iron sheet, closed at one end and open at the other. The open end of the trough is connected to a hinged pole with a counter weight through rope. For operation, the trough is lowered by exerting pressure on it by pulling the rope and also by foot of the operator till the closed end is submerged in water. Upon releasing pressure the trough comes to its original position due to action of counter weight along with water. Water can be lifted from this device from a depth of 0.8–1.2 m.

12.1.1.3 Archimedean Screw

It consists of a helical screw mounted on spindle, which is rotated inside a wooden or metallic cylinder. One end of the cylinder remains submerged in water and is placed at an inclined position at an angle of 30 degrees. It is used for lifting of water from a depth of 0.6–1.2 m and may discharge 1,600 l/h.

12.1.1.4 Paddle Wheel

It is also known as Chakram and is mostly used in coastal regions for irrigating paddy fields. It consists of small paddles mounted radially to a horizontal shaft, which moves in close fitting concave trough, thereby pushing water ahead of them. The number of blades depends on the size of wheel, which may be 8 for 1.2 m and up to 24 for 3–3.6 m diameters. The wheel having 12 blades may lift about 18,000 l/h from a depth of 0.45–0.6 m.

12.1.2 Animal-Powered Devices

Animal power is available in Indian subcontinent, but decreasing day by day. They are used for lifting of water, besides other field operations and processing works. A pair of bullocks may develop approximately 0.80 hp. They can lift water from the depth of 30 m or more. Of course the rate of discharge will go down with increase in lift. Some of the devices used for irrigation operated by animal power are Persian wheel, rope-and-bucket lift, self-emptying bucket, two bucket lift, chain pump, etc.

12.1.3 Kinetic Energy Powered Device

12.1.3.1 Hydraulic Ram

Hydraulic ram is a device to lift the water without any prime mover by utilizing the kinetic energy of flowing water. In this system the impact of water is converted into shock waves, which is called water hammer. This energy is utilized for lifting of water. The essential components of the system are check dam, supply pipe, hydraulic ram, storage tank, and discharge pipe. Except for changing of washers in the valves, there is no repair and maintenance required and the ram can operate 365 days in a year without any trouble. For fixing a hydraulic ram, a check dam is constructed on flowing water of a river or streams to create low head. Due to velocity and pressure of the water, the valve of the ram closes suddenly which creates a water hammer in the system. This causes building up of high pressure, which opens the tank valve and water rushes to the tank. The tank is enclosed from all sides and the air present in it creates further pressure on the water, which enters the tank and closes the valve of tank, thus discharging water from it. This discharged water is lifted by the hydraulic ram to higher head than the supply head. During this action, part of the water in the supply pipe also starts flowing in reverse direction and the water valve is opened due to its own weight and the water again starts running in supply pipe. This action continues unless the action of waste valve is stopped. The magnification factor of head and efficiency of hydraulic ram can be known by the following formula:

$$q \times h = Q \times H \times e$$

where

q = amount of water lifted by the ram,
h = head to which water is lifted,
Q = amount of water supplied to supply pipe,
H = head due to which water enters the supply pipe, and
e = efficiency of the ram.

12.1.4 Mechanically Powered Water-Lifting Devices

Mechanically powered water-lifting devices are usually termed as pumps, which are operated with the help of auxiliary power sources such as engine or electric motor. These pumps are capable of lifting large quantity of water to higher heads and are usually employed for the irrigation of field and horticultural crops, city water supply, etc.

12.2 Definition, Purpose, and Classification of Pumps

12.2.1 Definition of Pump

There are different definitions of pump. Simply speaking, it is a device or machine which is used for transferring fluids and/or gases from one place to another, or to increase the pressure of a fluid, or to create a vacuum in an enclosed space. Common definition of pump in the literature is that pump is a device which converts mechanical energy into pressure energy.

12.2.2 Pumping Purpose

The primary function of a pump is to transfer energy from a power source to a fluid, and as a result to create flow, lift, or greater pressure on the fluid. A pump can impart three types of hydraulic energy to a fluid: lift, pressure, and velocity. In irrigation and drainage systems, pumps are commonly used to lift water from a lower elevation to a higher elevation and/or add pressure to the water. Most pumps are used to take water from a standing (or non-pressurized) source and move it to another location. For example, a pump might take water from a lake and move it to a sprinkler system.

A booster pump, on the other hand, is used to increase the water pressure of water that is already on its way somewhere.

12.2.3 Principles in Water Pumping

Basically there are four principles involved in pumping water:

(1) Atmospheric pressure
(2) Centrifugal force
(3) Positive displacement
(4) Movement of column of fluid caused by difference in specific gravity.

Pumps are usually classified on the basis of operation, which may employ one or more of the above principles.

12.2.4 Classification of Pumps

Whether you are selecting an irrigation pump to a new-install system or replace the one in your existing system, you will first need to understand the different types available. There are pumps designed to fit each system.

The pumps can be classified based on different perspectives, such as

(a) mode of intake of fluid to pumps
(b) position of motor or prime mover
(c) the type of use (field of use)
(d) the principle by which energy is added to the fluid
(e) specific geometries commonly used
(f) design of the pump

12.2.4.1 Classification Based on Mode of Intake of Fluid to Pumps

According to the mode of intake of fluid to pumps, pumps can be classified into two groups:

(i) Suction mode pump
(ii) Force mode pump

Suction mode pumps are the pumps which draw water into the pump casing by applying suction force. The pump is located above the water level (at the soil surface or at specified location of the surface). The theoretical limit of lifting water from the soil surface is equal to the atmospheric pressure of the location concerned. The reciprocating pump falls into this category.

In case of force mode pump, the pump is installed below the water level and the lifting capacity is not limited by atmospheric pressure, rather on the force of the pump (or prime mover). Turbine pumps and other submersible pumps fall into this category.

12.2.4.2 Classification Based on the Position of Prime Mover

Based on the position of motor (power), the pumps can be grouped into

- Surface mounted
- Submersible pump

Surface-mounted pumps have motors which are above ground – although deep well types may have pump parts hundreds of feet below the surface. Submersible pumps are designed to spend most of there life underwater, only being pulled out every several years for routine maintenance.

12.2.4.3 Classification Based on the Type of Water Use (or Field of Use)

Based on the type of water use, the pumps may be categorized into

- Water pumps
- Wastewater pumps
- Well pumps
- Sump pumps
- Sampling pumps
- Drum pumps

> *Water pumps* are designed to move water that does not contain suspended solids or particulates. Applications include water supply, irrigation, land and mine drainage, sea water desalination.
> *Wastewater pumps* are used in the collection of sewage, effluent, drainage, and seepage water.
> *Well pumps* are most commonly used to bring water from wells and springs to the surface.
> *Sump pumps* are used in applications where excess water must be pumped away from a particular area. Sump pumps generally sit in a basin or sump that collects this excess water.
> *Sampling pumps* are used to monitor liquids, air, and gases. They are usually portable and developed for specific tasks.
> *Drum pumps* are used to transfer materials from a container into a process or other container. They may be electrically, hydraulically, or pneumatically powered depending on the working environment or application.

12.2.4.4 Classification Based on the Principle by Which Energy Is Added to the Fluid

Under this system of classification, all pumps may be divided into two major categories:

(1) *Dynamic pumps* – where continuously added energy increases velocity of the fluid and later this velocity is changed to pressure, and
(2) *Displacement pumps* – where periodically added energy directly increases pressure.

Dynamic Pumps

Dynamic pumps can be classified as one of several types of centrifugal pumps and a group of special effect pumps. In centrifugal pumps, energy is imparted to a fluid by centrifugal action often combined with propeller or lifting action. Centrifugal pumps can be classified by impeller shape and characteristics.

With respect to type of impeller, all centrifugal pumps can be classified into three following groups:

- *Radial-flow pumps* – which develop head mainly by the action of centrifugal force.
- *Axial-flow pumps* – produce flow by the lifting action of the propellers or vanes.
- *Mixed-flow pumps* – use both centrifugal force and some lifting action.

In addition, a centrifugal pump can be classified in one of four major groups depending on its design and application:

- Volute pumps
- Diffuser pumps
- Turbine pumps
- Propeller pumps

Turbine pump can be classified into

- Deep-well turbine pump
- Submersible turbine pump

Deep-well turbine pump can be further divided into

- Single stage
- Multistage

Single-stage pump have one impeller in one case. In multistage pump, two or more impellers enclosed in one casing to increase pressure (like pumps connected in

12.2 Definition, Purpose, and Classification of Pumps

series). Each separate section is called a stage, and the greater the number of stages, the greater the pressure or lift that is created.

Displacement Pumps

Displacement pumps have limited capacities and are not suitable for pumping large amounts of water required for irrigation or drainage. They are used mainly for chemical injection in agricultural irrigation systems.

12.2.4.5 Classification Based on the Means by Which the Energy Is Added

According to this approach, all pumps fall into two general categories:

- rotodynamic (centrifugal, mixed flow and axial flow), and
- positive displacement.

A rotodynamic pump converts kinetic energy to potential or pressure energy. The pumping units have three major parts: (i) the driver (that turns the rotating element), (ii) the impeller and shaft (the rotating element), and (iii) the stationary diffusing element.

Positive displacement (PD) pumps use gears, pistons, or helical rotors with tight tolerance to the casing so that pressure can build up beyond normal rating.

PD pumps can be classified into two main groups: rotary and reciprocating. Rotary pumps transfer liquid from suction to discharge through the action of rotating screws, gears, rollers, etc., which operate within a rigid casing. It typically works at low pressure (up to 25 bars or 360 lb per square inch (psi)). In case of reciprocating pumps, the rotary motion of the driver (such as diesel engine or electric motor) is converted to reciprocating motion by a crankshaft, camshaft, or swash-plate. It typically works at high pressure (up to 500 bar).

Other Types of Pumps

Other pump types less common in irrigation include:

> *Non-positive displacement pumps* have an impeller which spins to create pressure but does not have close tolerance to the casing. This type of pump will not build pressure beyond its normal rating.
> *Helical rotor pumps* force water through with an auger type action.
> *Booster pumps* are in-line pumps (series connection) used to increase the operating pressure of the system.
> *DC powered pumps* use direct current from motor, battery, or solar power to move liquids such as acids, chemicals, lubricants and oil, as well as water, wastewater, and potable water.
> *Hydraulic pumps* deliver high-pressure fluid flow to the pump outlet. Hydraulic pumps are powered by mechanical energy sources to pressurize fluid.

Submersible pumps can be mounted into a tank with the liquid media. The pump's motor is normally sealed in an oil-filled cavity that is protected from contact with the liquid.

12.3 Factors Affecting the Practical Suction Lift of Suction-Mode Pump

The following factors affect or limit the practical suction lift of pump:

(i) Elevation above the mean sea level, or actual atmospheric pressure at specified location
(ii) Density and viscosity of the fluid
(iii) Temperature of the fluid
(iv) Friction loss in suction pipe and well loss (entrance and formation loss, if applicable)
(v) Air-bubbling point of the liquid

Maximum theoretical suction lift in a location is equal to the atmospheric pressure at that location. Temperature affects the density and viscosity of the fluid, which consequently affect the friction loss. Air-bubbling point depends on the temperature of water and air pressure above it.

12.4 Centrifugal Pumps

12.4.1 Features and Principles of Centrifugal Pumps

Almost all irrigation pumps fall into this category. These types of pumps use centrifugal force (hence the name, centrifugal pump) to push water out. Water entering the pump hits an impeller (sort of like a propeller, but a little different) that imparts circular motion to the water, causing it to spin. This spinning action moves the water through the pump by means of centrifugal force, and forces it outward to the pump wall. As this happens, the water picks up speed, which becomes pressure as the water exits the pump. Centrifugal pumps may be "multistage," which means they have more than one impeller and casing, and the water is passed from one impeller to another with an increase in pressure occurring each time. Each impeller/casing combination is referred to as a "stage." Almost all turf-irrigation pumps are centrifugal pumps.

All centrifugal pumps must have a "wet inlet," that is, there must be water in both the intake (inlet) pipe and the casing when the pump is started. They can't suck water up into the intake pipe like you can suck soda up into an empty straw. They must be "primed" before the first use. To prime them, you simply fill the intake pipe with water and then quickly turn "on" the pump. To put it simply, this type of pump can not suck air, only water, so if there is no water already in the pump it would not pull any water up into it. Once it gets water in it the first time, it will hold the water with a small valve so the pump does not need to be primed again every time you turn it on.

The force created by centrifugal action of the pump depends on the density of the fluid within the pump. As the density of air is too low (~1.2 kg/m^3, in contrast 1,000 kg/m^3 for water), the force created is not enough to suck water from the water source.

With respect to type of impeller, all centrifugal pumps can be classified into the three following groups:

– Radial-flow pumps
– Axial-flow pumps
– Mixed-flow pumps

Specific speed range of different pumps is given in Table 12.1:

Table 12.1 Specific speed of different centrifugal pumps

Specific speed range	Pump type
Below 5,000	Radial flow pumps
4,000–10,000	Mixed flow pumps
9,000–15,000	Axial flow pumps

12.4.2 Some Relevant Terminologies to Centrifugal Pump

Important concepts associated with the operation of centrifugal pumps include pump efficiency, net positive suction head, specific speed, affinity laws, cavitation, and priming. Good design, efficient operation, and proper maintenance require understanding of these concepts.

12.4.2.1 Suction Lift

The absolute pressure on the water at the water source is the driving force for the water moving into the eye of the impeller. Theoretically, if a pump could create a perfect vacuum at the eye of the impeller, and if it were operating at sea level, the atmospheric pressure of approximately 14.5 psi would be the driving force pushing water into the eye of the impeller. This pressure could lift water a distance of about 8 m (1 psi = 2.31 ft of water). In practice, this lift is much smaller due to lack of perfect vacuum in the impeller and friction losses in the intake pipe. The practical value of maximum lift differs between pumps, but it is usually no greater than 24 ft.

12.4.2.2 Net Positive Suction Head Available

Net positive suction head available (NPSHa) is the absolute pressure of the water at the eye of the impeller. It is atmospheric pressure minus the sum of vapor pressure of the water, friction losses in the intake pipe, and suction head or lift. Since any variation of these four factors will change the NPSHa, NPSHa should be calculated using

Eq. (12.1). If the water source is located above the eye of the pump impeller (submerged pump), suction head (SH) must be added instead of subtracted. An accurate determination of NPSHa is critical for any centrifugal pump application.

$$\text{NPSHa} = \text{BP} - \text{SH} - \text{FL} - \text{VP} \qquad (12.1)$$

where

    BP = barometric pressure at pump level (m)
    SH = suction head or lift (m)
    FL = friction losses in the suction pipe (m)
    VP = vapor pressure of the liquid at given temperature (m)

12.4.2.3 Net Positive Suction Head required (NPSHr)

It is a measure of the head necessary to transfer water into the impeller vanes efficiently and without cavitation. The NPSHr required by a specific centrifugal pump depends on the pump design and flow rate. It is constant for a given head, flow, rotational speed, and impeller diameter. However, it changes with wear and different liquids, since it depends on the impeller geometry and on the density and viscosity of the fluid. For a given pump, NPSHr increases with increases in pump speed, flow rate, and water temperature.

The value of NPSHr is provided by the manufacturer for each specific pump model, and it is normally shown as a separate curve on a set of pump characteristic curves. To avoid cavitation, NPSHa must be always equal to or greater than NPSHr.

12.4.2.4 Static and Dynamic Head

Static Head

Static Suction Head

Static suction head (or static suction lift) is the vertical distance from the static water level in the suction pipe to the centerline of the impeller.

Static Discharge Head

Static discharge head is the vertical distance from the center of the impeller to the discharge outlet, or liquid level when discharging into the bottom of a water tank.

Dynamic Head

Dynamic Suction Head

Dynamic suction head is the sum of static suction lift plus friction loss in the suction pipe, plus the loss of head in the formation (when the pipe is installed in a aquifer). During pumping, actual water level lowers than the static level due to lifting of water column. This is observed outside the suction pipe by observation well.

12.4 Centrifugal Pumps

In addition, water entering the screen of the suction pipe through the formation, encounters resistance hence requires energy. This is called formation loss

Dynamic Discharge Head

It is the sum of static discharge head, friction head for discharge pipe, and the velocity head of the discharging fluid.

12.4.2.5 Total Head

Total dynamic head, or simply '*total head*' is the sum of dynamic suction head and dynamic discharge head.

This is the total pressure, in meter, that the pump must overcome to perform its work as designed. Numerically, it is the sum of the suction head, delivery head (if any), velocity head, friction head for the suction and discharge pipe, and formation loss (in terms of head) (if applicable). That is,

$$H_T = DH + SH + VH + FH + FL \tag{12.2}$$

DH = delivery head or discharge head (m)
SH = suction head or lift (m)
VH = velocity head (due to velocity of discharging water)
FH = friction losses in the suction pipe and delivery pipe (m)
FL = formation loss (m)

12.4.3 Pump Efficiency

The efficiency of a pump is a measure of its hydraulic and mechanical performance. It is defined as the ratio of the useful power delivered by the pump (water horsepower) to the power supplied to the pump shaft (brake horsepower). The efficiency of the pump is expressed in percent and can be calculated using the following equation:

$$E = \frac{WHP}{BHP} \times 100\% \tag{12.3}$$

where

E = pump efficiency
WHP = Water horsepower
BHP = Brake horsepower

The efficiency range to be expected varies with the pump size, type, and design. However, it is normally between 70 and 80%. A pump should be selected for a given application so that it will operate close to its point of maximum efficiency.

12.4.4 Specific Speed

Two pumps are geometrically similar when the ratios of corresponding dimensions in one pump are equal to the same ratios of the other pump. Specific speed is a constant for any geometrically similar pump. It is an index number correlating pump flow, head, and speed at the optimum efficiency point which classifies pump impellers with respect to their geometric similarity. Specific speed is usually expressed as

$$N_s = \frac{N\sqrt{Q}}{H^{3/4}} \quad (12.4)$$

where

N_s = specific speed of the pump (rpm)
N = rotational speed of pump at optimum efficiency (rpm)
Q = flow of pump at optimum efficiency (m³/s)
H = head at optimum efficiency (m).

The specific speed is an index which is used when selecting impellers to meet different conditions of head, capacity, and speed. Knowing this index is very helpful in the determination of the maximum permissible suction lift, or minimum suction head, which is necessary to avoid cavitation under different capacities, heads, and pump speeds. For a given head and capacity, suction lift is greater for a pump with lower specific speed.

The calculation of specific speed allows for determination of the pump type required for a given set of conditions to be determined. Usually high-head impellers have low specific speeds and low-head impellers have high specific speeds.

There is often an advantage in using pumps with high specific speeds since, for a given set of conditions, their operating speed is higher, and the pump is therefore smaller and less expensive. However, there is also some trade-off since pumps operating at higher speeds will wear faster.

12.4.5 Affinity Laws

Effect of Change of Speed

Affinity laws state that for a given pump, the capacity will vary directly with a change in speed, the head will vary as the square of speed, and the required horsepower will vary as the cube of speed. Mathematically, affinity laws can be expressed as

$$\begin{array}{l}\text{For flow: } Q \infty N \\ \text{For head: } H \infty N^2 \\ \text{For BHP: } BHP \infty N^3\end{array} \quad (12.5)$$

12.4 Centrifugal Pumps

where

Q = pump capacity (gpm)
H = pump head (ft)
N = rotational speed of the pump (rpm)
BHP = required brake horse power

That is:

$$\frac{Q_1}{Q_2} = \frac{N_1}{N_2}$$

$$\frac{H_1}{H_2} = \left(\frac{N_1}{N_2}\right)^2$$

$$\frac{BHP_1}{BHP_2} = \left(\frac{N_1}{N_2}\right)^3$$

The above equations assume that the diameter of the pump impeller is constant. In some cases the size of the impeller can be changed. Often a pump is very precisely matched to a specific application by trimming the impeller. It is not feasible to increase impeller diameter.

Basically, the above relationships mean that an increase in pump speed will produce more water at a higher head but will require considerably more power to drive the pump. These calculated values are very close to actual test results, provided pump efficiency does not change significantly. However, when conditions are changed by speed adjustment, usually there is no appreciable change in efficiency within the range of normal pump operation speeds.

For increase in pump speed, the NPSHr increases but it cannot be determined from the affinity laws. Also the laws do not say anything about how the efficiency of the pump will change with speed, but generally this is not a significant change. NPSHr and efficiency changes must be obtained from the pump manufacturer's data (pump characteristic curves).

Effect of Change of Diameter

There is a second set of affinity laws (Eq. 12.6), which describes the relationships between the same variables when the impeller size is changed under constant speed conditions. These laws relate the impact of impeller diameter changes to changes in pump performance. Since change of impeller diameter changes other design relationships in a pump, therefore, this second set of affinity laws does not yield the accurate results of the first three laws discussed above and must be applied with caution.

$$\text{Law 1 (for discharge):} \frac{Q_1}{Q_2} = \frac{D_1}{D_2}$$

$$\text{Law 2 (for head):} \frac{H_1}{H_2} = \left[\frac{D_1}{D_2}\right]^2 \qquad (12.6)$$

$$\text{Law 3 (for impeller diameter):} \frac{\text{BHP}_1}{\text{BHP}_2} = \left[\frac{D_1}{D_2}\right]^3$$

where

D_1 = initial diameter of the impeller
D_2 = diameter of the impeller after changes

This second set of affinity laws strictly applies only to radial-flow pumps. They are only approximate for mixed-flow impellers. In addition, these equations only hold for small changes in impeller diameter. Calculations for a trim of more than 10% of the original diameter can be significantly in error.

12.4.6 Priming of Centrifugal Pumps

All centrifugal pumps must be primed by filling them with water before they can operate. The objective of priming is to remove a sufficient amount of air from the pump and suction line to permit atmospheric pressure and submergence pressure to cause water to flow into the pump, when pressure at the eye of the impeller is reduced below atmospheric as the impeller rotates.

When axial-flow and mixed-flow pumps are mounted with the propellers submerged, there is normally no problem with re-priming of these pumps because the submergence pressure causes water to refill the pumps as long as air can readily be displaced. On the other hand, radial-flow pumps are often located above the water source, and they can lose prime. Often, loss of prime occurs due to an air leak on the suction side of the pump. Volute or diffuser pumps may lose prime when water contains even small amounts of air or vapor. Prime will not be lost in a radial-flow pump if the water source is above the eye of the impeller and flow of water into the pump is unrestricted.

In some cases pumps are primed by manually displacing the air in them with water every time the pump is restarted. Often, by using a foot valve or a check valve at the entrance to the suction pipe, pumps can be kept full of water and primed when not operating. If prime is lost, the water must be replaced manually, or a vacuum pump can be used to remove air and draw water into the pump.

A self-priming pump is one that will clear its passages of air and resume delivery of liquid without outside attention. Centrifugal pumps are not truly self-priming. So called self-priming centrifugal pumps are provided with an air separator in the form of a large chamber or reservoir on the discharge side of the pump. This separator allows the air to escape from the pump discharge and entraps the residual liquid necessary during re-priming. Automatic priming of a pump is achieved by the use of a recirculation chamber which recycles water through the impeller until the pump is primed, or by the use of a small positive displacement pump which supplies water to the impeller.

12.4.7 Cavitation

Pump cavitation is defined as the formation of cavities on the back surface of an impeller and the resulting loss of contact between the impeller and the water being pumped. These cavities are zones of partial vacuum which fill with water vapor as the surrounding water vaporizes due to the reduced pressure in the cavities. The cavities are displaced with the flowing water along the pump impeller surfaces toward the outer circumference of the impellers. As they move toward the circumference, the pressure in the surrounding water increases, and the cavities collapse against the impellers with considerable force. The force created by the collapse of the cavities often causes erosion and rapid wear of the pump impellers as well as a characteristic noise during pump operation.

The process of cavitation is caused by the reduction in pressure behind the impellers to the point that the water vaporizes. Thus, it can be caused by any combination of factors which allow pressure to drop to that point, including inadequate submergence or excessive suction lift, so that little pressure is available to move water into the pump, high impeller speeds which cause extremely low pressures to be generated behind the impellers, restricted pump intake lines which prevent water from moving readily into the pump, and high water temperatures which decrease the pressure at which water vaporizes.

Cavitation can occur in all types of pumps and can create a serious problem. In some cases of mild cavitation, the only problem may be a slight drop in efficiency. On the other hand, severe cavitation may be quite destructive to the pump and result in pitting of impeller vanes. Since any pump can be made to cavitate, care should be taken in selecting the pump for a given system and planning its installation.

Pump manufacturers specify the Net Positive Suction Head required (NPSHr) for the operation of a pump without cavitation. Pump cavitation can be avoided by assuring that the net positive suction head available (NPSHa) is always greater than that required (NPSHr) by the pump.

12.5 Description of Different Types of Centrifugal Pumps

12.5.1 Turbine Pump

A turbine pump is basically a centrifugal pump mounted underwater and attached by a shaft to a motor mounted above the water (Fig. 12.1). The shaft usually extends down the center of a large pipe. The water is pumped up this pipe. Turbine pumps are

Fig. 12.1 Schematic of a submersible turbine pump

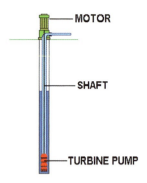

very efficient and are used primarily for larger pump applications. They are typically the type of pumps used on municipal water system wells, irrigation systems.

12.5.1.1 Deep-Well Turbine Pumps

Under this category are grouped all types of pumps that are suspended by the discharge column within which the drive shaft is located. The name, deep-well turbine pump, is applied only to pumps operating on the centrifugal principle and having diffuser vanes within the bowl or case. They can be single-stage or multi-staged for higher pressure applications. Pump bowls which contain impellers and diffusers are located below the water surface, and they should be submerged under pumping conditions. The drive shaft is located in the center of a discharge pipe and it can be either oil or water lubricated.

Features of Turbine Pump

These pumps are used in tube wells or in open wells where the water level is below the practical suction limit of centrifugal pump. The pump unit remains submerged in water where as the prime mover (motor or engine) is kept above the ground level. The pump unit is connected to the motor with the help of long vertical shaft supported on bearings, which may be water or oil lubricated. Although the turbine pump also operates on centrifugal principle, it differs from volute type as stationary guide vanes guide upward the water thrown by the impeller to the periphery. The gradual enlarging vanes guide the water to the casing, thus converting kinetic energy into potential energy. Therefore, the turbine pump generates heads several times that of a volute pump. Turbine pumps are most effective for tube wells, and applications requiring high heads and discharge. A well-maintained pump provides trouble free service for several years. These pumps are available in wide range of models with various flow capabilities and head levels.

A generalized specification of this type of pump is given below:

Diameter of tube well (mm)	Discharge (l/min)	Head per stage (m)	Speed (rpm)
150–900	150–62,000	1.2–49.0	960–2,880

Uses

It is used for lifting of water from tube wells, open wells for irrigation, domestic and industrial applications.

12.5.2 Submersible Pump

A submersible pump is a pump which has a hermetically sealed motor close-coupled to the pump body. The whole assembly is submerged in the fluid to be pumped (Fig. 12.2). The advantage of this type of pump is that it can provide a significant lifting force as it does not rely on external air pressure to lift the fluid.

A system of mechanical seals is used to prevent the fluid being pumped entering the motor and causing a short circuit. The pump can either be connected to a pipe, flexible hose, or lowered down guide rails or wires so that the pump sits on a "ducks foot" coupling, thereby connecting it to the delivery pipe-work.

Submersible pumps are found in many applications – small irrigation farm, drinking, household, office, commercial places, etc. Single stage pumps are used

Fig. 12.2 Schematic of submersible pump

for drainage, sewage pumping, general industrial pumping, and slurry pumping. Multiple stage submersible pumps are typically lowered down a borehole and used for water abstraction, or in water wells.

Submersible pumps are also used in oil wells. By increasing the pressure at the bottom of the well significantly, more oil can be produced from the well compared to natural production. Submersible pumps are used in many varying applications:

(i) It is very useful where the water level has gone below the suction limit of the pump.
(ii) Vertical deep-well pumps can be used where shaft elongation problems are created by very deep settings or when the well is not straight.
(iii) Vertical booster pumps can be installed in open sumps for many applications where flooding can be a threat to surface-mounted motors, where space is at a minimum, and where appearance prohibits surface installations.
(iv) Horizontal in-line booster pumps work efficiently to provide constant added pumping pressure that may be required in the line. They may be installed at any angle.

12.5.2.1 Suitability of Submersible Pump

– Their compact and streamlined design makes them ideal for wells and other jobs where space is limited.
– Submersibles have the advantage of being able to work in the water source being pumped. As a result the submersible is not a subject to the suction lift limitations of other typical pumps. No suction hose is required helping to save money and time while eliminating a potential source of problems. The pump is limited only by the discharge head it is capable of producing.
– The pump motors use a vertical drive shaft to turn an impeller and generate the velocity needed to create the discharge pressure. Water flows in through the bottom and is discharged out of the top of the pump casing.
– A high-quality pump will have its motor housed in a watertight compartment and equip it with thermal overload sensors that shut down the motor to prevent damage from over heating.
– Maintenance is minimal and generally consists of periodically inspecting the electrical cord and the mechanical seal lubricant. There are none of the concerns common with engine driven pumps such as noise, fuel, or emissions.
– Control boxes and float switches are available for unattended operation of submersible pumps. The boxes provide protection against voltage fluctuations and incorrect phasing, while float switches turn the pump on and off according to fluctuating water levels. A number of different accessories are available but care should be taken that they meet the electrical requirements of the pump.

Combining electricity and water obviously brings a certain element of risk. Further, it is difficult and often impossible to know if there is a problem once the

12.5 Description of Different Types of Centrifugal Pumps

pump is submerged. As a result the pump should provide some built-in protections to ensure safety and guard against damage to the equipment.

12.5.2.2 Features

– No suction limitations
– Easy to install without pump house
– Easy to handle as compared to vertical turbine pumps
– No possibility of theft

12.5.2.3 Applications

Agriculture, industry, water supply schemes, pressure boosting, mining, sprinklers, lift irrigation, fountains, etc.

12.5.2.4 Requirements for Submersible Pump

Submersible Pump Wire

Submersible water well pumps need wire capable of use in under-water situations. The maximum distance from the power source to the submersible pump motor includes the UF burial cable from the power source to the well head, and the distance down into the well with the submersible pump wire.

12.5.2.5 Wide Ranges of Submersible Pumps

A range of submersible pumps are available in the market – four inch (4″), three inch (3″) submersible pump. The 3″ submersible pump can deliver twice as much water as a jet pump with less horsepower. The pressure is always better with a submersible pump. This 3″ submersible pump comes with 1/3, 1/2, 3/4, 1, and 1.5 hp. Pumping ranges from 5 to 30 gal/min.

12.5.2.6 Open Well Submersible Pump

Features and Benefits

- Pump set works under the water and hence suction and priming problems do not arise
- Easy installation requiring no foundation and pump house arrangements
- Noiseless operation
- Maintenance-free operation due to water-lubricated bearings
- Single shaft for motor and pump, permanent correct alignment
- High operating efficiency resulting in lower power consumption

Applications

- Open well pumping for irrigation
- River and canal lift-irrigation
- Pumping from in-well bores
- Sprinkler and drip irrigation systems
- Pumping from sumps for high-rise buildings and industry

12.5.2.7 Open Well Submersible Pumps

Single phase open submersible pumps are extensively used for domestic water supply, irrigation-open, drip, sprinklers, high-pressure washing garages and poultry farms (Fig. 12.3).

Fig. 12.3 Views of single phase open-well submersible pump

Applications

- Domestic water supply
- High-raise buildings
- Gardening and sprinkler
- Industries, for clean water handling
- Storage tank filling

12.5.3 Mono-Block Pump

12.5.3.1 Features

It is one of the most common types of centrifugal pump employed for irrigation. It consists of impeller or rotor and progressively widening spiral or volute casing. The pump is directly connected to the prime mover, which may be electric motor or engine. The direct coupling feature reduces transmission losses. Upon rotation of the impeller, the water enters at the eye, which is thrown radially outward to the periphery. Such an action causes vacuum at the eye and thus more water enters the suction pipe to maintain the continuous flow. The impeller accelerates the water to a high velocity and the casing converts this velocity head into pressure head due to volute design. Volute pumps usually employ closed type of impeller for irrigation, which has curved vanes for smooth flow of liquid. There are other kinds of impellers like open, semi-open, and nonclogging impeller, but these are used for purposes other than irrigation.

A generalized specification of the reciprocating pump is given below

Suction size (mm)	Delivery size (mm)	Impeller diameter (mm)	Total head (m)	Discharge (l/s)	Revolution per min	Motor rating (kW)
50–125	40–125	185–300	4–45	1.4–73	1,380–1,480	1.5–7.5

12.5.3.2 Uses

Monoblock pumps are used for agricultural purposes. These are also suitable for domestic and industrial applications of pumping water.

12.5.4 Radial-Flow Pump

Basically, a centrifugal radial-flow pump has two main parts: (1) a rotating element (impeller and shaft) and (2) a stationary element (casing, stuffing box, and bearings). Water enters the pump near the axis of the high-speed impeller, and by centrifugal force is thrown radially outward into the pump casing. The velocity head imparted to the fluid by the impeller is converted into pressure head by means of a volute (Fig. 12.4) or by a set of stationary diffusion vanes surrounding the impeller.

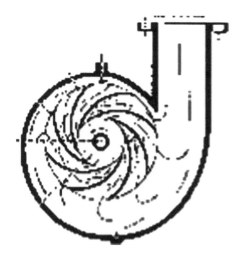

Fig. 12.4 Schematic of a radial flow pump

12.5.5 Volute Pump

The centrifugal volute pump is the most common type of radial-flow centrifugal pump. It has an impeller housed in a progressively widening spiral casing. Water enters the eye of the impeller and is thrown radially outward. This type of pump does not have diffuser vanes to reduce the velocity of the water. Instead, velocity is reduced by the shape of the volute itself. This design creates an unequal pressure distribution along the volute which may result in a heavy thrust load on the impeller, creating deflection of the shaft, and increasing the probability of its failure.

Volute pumps can be single-suction or double-suction pumps. A single-suction pump impeller is exposed to a large axial hydraulic thrust resulting from the unbalanced hydraulic pressures on the impeller. In a double-suction pump, water is fed from both sides of the impeller, significantly improving its hydraulic balance. As a result, double-suction volute pumps can produce higher pressures than single-suction pumps.

Volute pumps are commercially available as single-stage (single impeller) or multistage (multiple impellers) pumps. The main reason for the multistage configuration is to increase the head produced by the pump. If a multistage pump has single suction impellers, the impellers are usually arranged with equal numbers discharging in opposite directions to counteract the hydraulic imbalance on each of the impellers.

Volute pumps are used where irrigation water is obtained from depths generally less than 20 ft.

12.5.6 Axial-Flow Pump

Axial-flow pumps, also called propeller pumps, produce flow by the lifting action of the propellers. Axial-flow pumps are designed for conditions where the capacity is relatively high and the head developed by the pump is low.

An axial pump does not produce high pressure or lift, but can have significant flow capacity if the pump is large enough. Most axial-flow pumps operate on installations where suction lift is not required. Generally, these pumps are mounted vertically or on an incline from vertical, since it is necessary to submerge the impeller of an axial-flow pump. In some applications where high volumes of water are required and ample submergence above the pump is available, it is possible to mount an axial-flow pump in a horizontal position.

12.5.7 Mixed-Flow Pump

Mixed-flow centrifugal pumps use both centrifugal force and some lifting action to move water. Water is discharged both radially and axially into a volute-type casing. The process is a combination of processes occurring in volute and axial-flow types

12.5 Description of Different Types of Centrifugal Pumps

of pumps. Mixed-flow impellers are often used in deep-well turbine and submersible turbine pumps.

12.5.8 Advantage and Disadvantage of Different Centrifugal Pumps

Table 12.2 presents advantages and disadvantages of various centrifugal pumps. This comparison may be helpful in selecting a centrifugal pump for a particular application.

Table 12.2 Advantages and disadvantages of various centrifugal pumps

	Volute pumps	Diffuser pumps	Turbine pumps	Propeller pumps
Advantages				
Available in a wide range of sizes	✓	✓	✓	✓
Simple construction	✓			✓
Relatively quiet operation	✓	✓	✓	✓
Robust with a long life	✓	✓		✓
Available in a wide variety of materials	✓	✓		✓
Can handle liquids containing solids	✓	✓		✓
Can handle liquids with a high proportion of vapor			✓	
Self-priming			✓	
Variable speed drive units not required to adjust the capacity	✓	✓	✓	✓
Pressure and power developed are limited at shutoff	✓	✓	✓	✓
Disadvantages				
Unsuitable for pumping high viscosity liquids	✓	✓	✓	✓
Heads developed are limited	✓	✓	✓	✓
Close clearances			✓	

12.5.9 Some Common Problems of Centrifugal Pumps, Their Probable Causes, and Remedial Measures

Problem	Possible causes	Remedial measures
No liquid delivered	– Pump not primed – Insufficient available NPSH – Suction line strainer clogged – End of suction line not in water – System total head higher than pump total head at zero capacity	Prime the pump Increase the NPSHa Check the depth to water level

Problem	Possible causes	Remedial measures
Pump delivers less than rated capacity	– Air leak in suction line or pump seal	Close the air leak
	– Insufficient available NPSH	Increase available NPSH
	– Suction line strainer partially clogged or of insufficient area	Wash out the strainer
	– System total head higher than calculated	
	– Partially clogged impeller	Clean the impeller
	– Impeller rotates in wrong direction	Stop the pump, check the rotation of the impeller, install in correct direction
	– Suction or discharge valves partially closed	Check the valves and clean if necessary
	– Impeller speed too low	
Loss of prime while pump is operating	– Water level falls below the suction line intake	
	– Air leak develops in pump or seal	
	– Air leak develops in suction line	
	– Water vaporizes in suction line	
Pump is noisy	– Cavitation	Increase the NPSHa
	– Misalignment	
	– Foreign material inside pump	Align the pump correctly
	– Bent shaft	
	– Impeller touching casing	
Pump takes too much power	– Impeller speed too high	Decrease the speed
	– Shaft packing too light	
	– Misalignment	Correct the alignment
	– Impeller touching casing	
	– System total head too low causing the pump to deliver too much liquid	
	– Impeller rotates	
	– Impeller installed in wrong direction	

12.6 Other Types of Pumps

12.6.1 Air-Lift Pump

Air-lift pumps are used where the water level is beyond the suction limit. Air-lift pumps operate on the principle that a mixture of air and water will rise in a pipe surrounded by water. An air-lift pump basically consists of a vertical pipe partially submerged in water and an air supply tube allowing compressed air to be fed into the pipe at a considerable distance below the static water surface. The mixture of water and air is lighter than the water outside the pipe and it rises in the pipe (Fig. 12.5). The head which can be produced depends on the depth of submergence of the air tube.

Fig. 12.5 Schematic of air lift pump

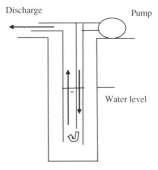

Air-lift pumps are relatively inefficient. Typical efficiencies range between 30 and 50%. Generally, air-lift pumping is most efficient when the static water level is high, the casing diameter is relatively small, and the well depth is not excessive in relation to the pressure capability of the compressor.

The volume of air needed to lift the water depends on the total pumping lift, the submergence, the length of air line, and the casing length and diameter. A useful rule of thumb for determining the proper compressor capacity for air-lift pumping is to provide about 3/4 cfm (cubic feet per minute) of air for each 1 gpm of water at the anticipated pumping rate.

Air-lift pumps have some advantages over the other pumps discussed above. They do not have any moving parts, can be used in a corrosive environment, and are easy to use in irregularly shaped wells where other deep-well pumps cannot fit. Air-lift pumps are not available from suppliers, but they are very simple to build. The main disadvantages of air-lift pumps are their low efficiencies and requirement.

12.6.2 Jet Pump

A jet pump is a combination of a volute centrifugal pump and a nozzle-venturi arrangement. The driving force lifting the water in this type of pump is provided by a high-pressure nozzle, which creates a low pressure region in a mixing chamber (Fig. 12.6). This low pressure causes water to flow into the pump. A diffuser following the mixing chamber slows down the water and converts velocity head into pressure head. The jet nozzle is installed in the pipe conveying the water. For a shallow well the nozzle is frequently located outside the well next to the centrifugal pump. However, for a deep well, the nozzle can be placed inside the well in the intake line. This location increases the jet pump lift capability considerably beyond that which is practical for the volute centrifugal pump. The role of the centrifugal

Fig. 12.6 Schematic of a jet pump

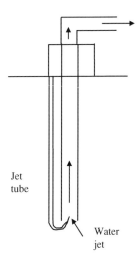

pump in a jet pump is to produce the flow to the nozzle and maintain the combined flow through the intake pipe beyond this point.

Jet pumps are self-priming, have no moving parts, and do not require lubrication. Their efficiency is typically low (on average about 40%), and they provide low flows at high pressure. Because of this characteristic, they are not suitable for large-scale irrigation. However, they are frequently used for home-water supplies and irrigation of lawns and gardens.

12.6.2.1 Features

A jet pump is a diffuser pump that is used to lift water from both shallow and deep wells. During working, the output of the diffuser is split, and half to three-fourths of the water is sent back down the well through the pressure pipe. At the end of the pressure pipe, water is accelerated through a cone-shaped nozzle. The water goes through a venturi in the suction pipe. The venturi has two parts: the venturi throat, which is the pinched section of the suction tube; and above that is the venturi itself which is the part where the tube widens and connects to the suction pipe. The venturi speeds up the water causing a pressure drop which sucks in more water through the intake at the very base of the unit. The water goes up the suction pipe and through the impeller, most of it for recirculation around to the venturi.

General Specifications

Bore well size (mm)	Suction lift (m)	Discharge head (m)	Discharge (l/min)	Power (hp)
75–115	9–30	12–18	14–45	0.5–1.0

12.6 Other Types of Pumps

12.6.2.2 Uses

The jet pump is used for lifting water from shallow and deep wells for irrigation and domestic applications.

12.6.3 Reciprocating Pump/Bucket Pump

12.6.3.1 Features

Reciprocating pumps are normally used for drinking water supply in addition to irrigation. The main parts of the reciprocating pumps are the pump cylinder in which an airtight piston or plunger moves up and down with the help of pump rod, handle for operation of pump, valves, pipe, and strainer (Fig. 12.7). As the plunger rises, water is drawn through a non-return valve at the bottom of cylinder into the cylinder, and on the downward stroke the water is released to the upper side of plunger. On the next upward movement of plunger, water is raised to pump head and discharged through the spout. By changing either the frequency of reciprocation or stroke length of the piston, the discharge rate can be varied. The reciprocating pumps are available in various designs and models, which can be operated manually, with animal power and auxiliary power sources.

A generalized specification of the reciprocating pump is given below

Height (mm)	Diameter (mm)	Weight (kg)	Strokes per minute (no.)	Water depth optimum (m)	Discharge (l/min)
715–1,150	80–200	30–40	40–45	28–34	12–14

12.6.3.2 Uses

Reciprocating pumps are used to lift water from underground sources; therefore, if the water level is deep, the pump cylinder has to be lowered close to water surface

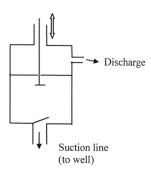

Fig. 12.7 Schematic of reciprocating or bucket pump

to reduce the suction head. The number of cylinders can be increased according to power sources.

12.6.4 Displacement Pump

Displacement pumps force the water to move by displacement. This means pumps such as piston pumps, diaphragm pumps, roller-tubes, and rotary pumps. These pumps are used for moving very thick liquids, or creating very high pressures. They are used in fertilizer injectors, spray pumps, air compressors, and hydraulic systems for machinery. With the exception of fertilizer injectors (used for mixing fertilizer into irrigation water), it is not typically used for irrigation systems.

12.6.5 Hydraulic Ram Pump

A hydraulic ram pump is a motor-less low flow rate pump. It uses the energy of flowing water to operate. It is suitable for use where a large flow rate is not required. The flow rate of typical commercially available units is limited to approximately 14 gal/min or 20,000 gal/day. The head produced by the pump depends on the quantity and velocity of water flow at the pumping source. Water can be lifted up to 400 ft depending upon the quantity and velocity of water flow in the delivery pipe. Hydraulic ram pumps can be used for domestic water supply or livestock watering. Usually, their flow rates are too small to consider them for other applications, such as irrigation.

12.6.6 Booster Pump

Booster pumps are used to increase the water pressure. Therefore the required booster pump pressure is simply the desired pressure minus the existing pressure. For most pump brands the pressure must be expressed in feet of head, not PSI.

$$PSI \times 2.31 = \text{feet head}, \text{Feet head} \times 0.433 = PSI$$

12.6.7 Variable Speed Pump

Many pumping systems require a variation of flow or pressure. Either the system curve or the pump curve must be changed to get a different operating point. Where a single pump has been installed for a range of duties, it will have been sized to meet the greatest output demand. It will therefore usually be oversized, and will be operating inefficiently for other duties. Consequently, there is an opportunity to achieve an energy cost savings by using control methods, such as variable speed, which reduce the power to drive the pump during the periods of reduced demand.

12.7 Cavitation in Pump

12.7.1 Cavitation in Radial Flow and Mixed Flow Pumps

In radial-flow and mixed-flow types of centrifugal pump, when the water enters the eye of the impeller, an increase in velocity takes place. As a result of this velocity increase, water pressure is reduced (as the water flows from the inlet of the pump to the entrance to the impeller vane) resulting in cavitation.

A concentrated transfer of energy during cavitation creates local forces capable of destroying metal surfaces. The more brittle the material which the impeller is constructed of, the greater is the damage. In addition to causing severe mechanical damage, cavitation causes a loss of head, reduces pump efficiency, and results in noisy pump operation.

If cavitation is to be prevented, volute or diffuser pumps must be provided with water under absolute pressure which exceeds the NPSHr. The following conditions should be avoided in volute and diffuser pump installations:

(a) Heads much lower than head at peak efficiency of pump.
(b) Capacity much higher than capacity at peak efficiency of pump.
(c) Suction lift higher or submergence head lower than recommended by manufacturer of the pump.
(d) Water temperature higher than that for which the system was originally designed.
(e) Speeds higher than manufacturer's recommendation.

12.7.2 Cavitation in Axial-Flow Pumps

In axial-flow pumps cavitation cannot be explained in the same way as for radial-flow and mixed-flow pumps. The water enters an axial-flow pump in a large bell-mouth inlet and is guided to the smallest section, called the throat, immediately ahead of the propeller. The capacity at this point should be sufficient to fill the ports between the propeller blades. When the head is increased beyond a safe limit, the capacity is reduced to a quantity insufficient to fill up the space between the propeller vanes, creating cavities of almost a perfect vacuum. When these cavities collapse, the water hits the propeller vane with a force sufficient to pit the surface of the vane. The first two cavitation prevention rules listed for volute and diffuser pump are different for an axial-flow pump. Avoid:

– Heads much higher than head at peak efficiency of pump.
– Capacity much lower than capacity at peak efficiency of pump.

The last three rules are the same for all centrifugal pumps.

12.8 Power Requirement in Pumping

Power requirement in pumping can be expressed as follows:

$$P = \frac{mgh}{t} = \frac{(Q_T \times \rho) \times g \times H}{t} = \frac{Q_T}{t} \times \rho \times g \times H$$

i.e.,

$$P = Q \times \rho \times g \times H \qquad (12.8)$$

where

P = power requirement, in watt (W)
m = mass of fluid delivered, in kg
Q_T = total discharge for the time t, in m^3
ρ = density of fluid, in kg/m^3 ($\sim$1,000 kg/m^3 for normal water)
g = acceleration due to gravity ($\sim$9.81 m/s^2)
H = total head of water, in m
t = pumping period, in seconds
$\frac{Q_T}{t} = Q$ = discharge rate, m^3/s

Let the efficiency of the electric motor to be used is E_m, then the motor size *or* capacity would be

$$P_m = (Q \times \rho \times g \times H)\frac{1}{E_m} \qquad (12.9)$$

where P_m = motor capacity, in W

If the density of water, $\rho = 1{,}000$ kg/m^3, $g = 9.81$ m/s, then the above equation can be written as

$$P_m = (Q \times 9.81 \times H) \times \frac{1}{E_m} [\text{kW}]$$

Units of other elements will be same as mentioned earlier.
1 kilowatt (kW) = 1,000 W

Kilowatt can be converted into horse power by the relation:

$$1\,\text{kW} = 1.341\,\text{hp}$$

Using $\rho = 1{,}000$ kg/m^3, $g = 9.81$ m/s, 1 m^3 = 1,000 l, and the relation of 1 kW = 1.341 hp; energy of discharging water, or "water horse power" can be expressed as

$$\text{WHP} = (\text{discharge in l/s} \times H)/76$$

where H = total head, in m

12.9 Pump Installation, Operation, and Control

12.9.1 Pump Installation

12.9.1.1 Installation

The pump should be located as close as practical to the liquid source, this will minimize inlet losses. Improper location of the pump may result in decreased pump performance. Other points that should be considered are the following:

- The foundation should be designed to hold the pump assembly rigid and to absorb any vibration or external strain that may be encountered. A concrete foundation on a solid base should be adequate.
- Pump and driver should be accessible for inspection and maintenance.
- On permanent installations, it is recommended that the pumping assembly be secured to the foundation by anchor bolts.

12.9.1.2 Alignment

Some pumps are shipped on baseplates without drivers. For these units, install and tighten each coupling half on driver and pump shafts. Place driver on baseplate and set proper distance between shafts and coupling hubs.

Final alignment of pump and driver should take place after unit is secured to foundation, and after the suction and discharge piping is connected to the pump. The objective of any aligning procedure is to align shafts (not align coupling hubs) by using methods that cancel out any surface irregularities, shaft-end float, and eccentricity. Grouting is recommended to prevent lateral shifting of baseplate, not to take up irregularities in the foundation. For installations requiring grouting, a baseplate designed specifically for this purpose is needed. For pumps driven through a separate gearbox or other device, first align device relative to pump, and then align driver relative to device.

Alignment of pump and its driver should be checked and corrected, if necessary, at least every 6 months. If system experiences an unusual amount of vibrations or large variations in operating temperatures, this should be done often. Well-maintained alignment will help insure maximum equipment life.

12.9.1.3 Installation of Submersible Pump

The following steps should be followed in installing a submersible pump:

- Lift the motor vertically and lower into well, resting the motor on the "U" plate.
- Make temporary electric connection to motor, bump starter to determine correct rotation (counter clockwise).
- With the motor suspended in the well, on the "U" plate attach an elevator or clamps to pump end, lifting it vertically over the motor.

- Align the motor coupling with the motor shaft, while lowering the pump end onto the motor.
- Replace all capscrews and lockwashers furnished with the pump and bolt up tightly.

12.9.2 Pump Operation

12.9.2.1 Suction Line

The suction line should be designed so pump inlet pressure, measured at pump inlet flange, is greater than or equal to the minimum required pump inlet pressure (also referred to as Net Positive Inlet Pressure Required or NPIPR). Suction line length should be as short as possible with piping diameters being equal to or larger than pump's inlet size. All joints in suction line must be leak free.

12.9.2.2 Field Test

When a field test of the pump's performance is required, the following reading should be taken:

- Discharge rate measurement
- Total head and horsepower measurement
- Rotating speed
- Liquid temperature

Compare the results of the field test with the performance curve for your pump.

Capacity (Discharge Rate) Measurement

Measure the rate of flow (volume per unit time) from the pump discharge, preferably in liter per second (l/s) or m^3/s. The volume and/or flow rate measurement may be done using any of the following equipment: accurately measured reservoir, calibrated venturi meter, orifice meter, flume, etc.

Total Head Measurement

The total pumping head consists of distance from the water level in the sump (when pumping) to the center of the discharge pressure gauge, plus the discharge gauge reading, and the velocity at the discharge. Convert gauge reading using the relation $P = \omega h$, where ω is the density of water.

Horse Power Requirement

Measure the horsepower consumption of the pump by a direct reading of a wattmeter and applying the reading to the following formula:

$$BHP = (kW\ Input \times Efficiency)/0.746$$

12.9.2.3 Operational Check

- Check to ensure that the pump is delivering liquid; if not, stop the driver immediately, determine the cause.
- Check the pump for excessive vibration, localized heating of components, noise, and leakage. If any of the conditions exists, stop the pump and investigate and take corrective action.
- Check the pump discharge pressure and intake vacuum to manufacturer's specifications. If not to specification, investigate and take corrective action.

12.9.3 Pump Control

The basic of pump control systems is that the pump is started and stopped by turning "ON" or "OFF" the power supply. All pump controls are no more than a variation on this same procedure. When we need water, we can simply plug in the extension cord that ran to the pump, and to turn the water off we can simply unplug the cord.

The next step up in pump control is to get rid of the plug and outlet and use a permanent switch to turn "ON" and "OFF" the power flow. Some small pumps are stopped and started in exactly this way. But with the higher voltage and amperes commonly used for irrigation pumps, a simple switch will not be practical. The switch would have to be huge to handle the load. Most pump control circuits are designed using a relay circuit that isolates the user from the pump voltage. The relay circuit is like a messenger. You tell the relay to start the pump and the relay starts it for you. Most relays use 12 or 24 V, a few use 120 V (Fig. 12.8).

12.9.3.1 Automating the Control Circuit

To automate the control of the pump, you simply replace the manual switch in the circuit with an automatic switch. There are several types commonly used:

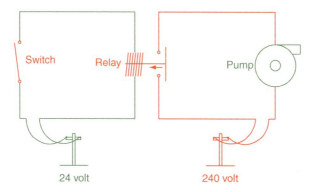

Fig. 12.8 Typical relay control circuit

timer, pressure switch, irrigation controller, flow switch, and any combination of the above.

12.10 Hydraulics in Pumping System

If you randomly select one pump, it will not automatically fit the needs of your system; unless you have calculated the amount of flow your system needs and determined how much pressure it will take to achieve the desired flow.

Velocity is used to increase pressure. Velocity of water being discharged is determined by the size of the casing and impeller as well as the speed of the impeller spinning inside the case. Since water is essentially a noncompressible liquid, it exhibits the unique trait of transferring pressure horizontally when in a confined space. What this means is that water in a pipe exhibits the same pressure as it would if the pipe were perfectly vertical, even if the pipe is not.

12.10.1 Pressure Vs Flow Rate

It is to remember here that there is always an inverse relationship between pressure and flow (for a fixed conduit or pipe). Higher pressures mean lower flows. Lower pressures result in higher flows.

12.10.2 Pressure and Head

The pumping head is normally expressed in feet of water; it is simply height of elevation. Pressure is normally expressed as "pounds per square inch" (PSI). It is the *weight* of the water in pounds per square inch. Pressure and head are related by the equation:

$$P = \omega h$$

where ω is the density of the fluid, and h is the height of the fluid column.

In most cases we measure water pressure in the static state when designing irrigation systems (or any other water piping system for that matter). Then we use calculations to figure out the friction loss that will occur in the pipes and subtract it from the static pressure to arrive at the dynamic pressure.

The following is oriented toward wells.

12.10.2.1 Dynamic Water Depth

When the pump is running, the water level in the well drops below the water table. It may drop a few inches or more than several meters depending on the type of soil (or rock) the well is drilled into and the discharge rate. Dynamic water depth is the depth of the water below the top of the well, when the pump is running. When a well company drills a new well they insert a temporary pump to "break in" and test the well. They refer to this as "developing" the well. As part of this process they also measure the dynamic water depth of the new well at various pumping rates.

12.10.3 Elevation Difference

It is the difference between the top of your well and the highest point in the area to be irrigated. That is, how much higher (or lower) is the highest point in the irrigated area than the top of the well. This may be a negative number if the well is higher than the irrigated area.

12.11 Pumps Connected in Series and Parallel

Multiple pumps can operate in series or parallel. Pumps placed in parallel provide additional flexibility in the range of flow rates. When pumps are connected together in series (or multi-staged), the total flow (GPM) will stay the same, while the pressures generated by the pumps will be additive. When pumps are connected in parallel, the pressure stays the same, while the flow volume is additive.

Pumps should not be operated in series or parallel unless specifically procured for this purpose, since serious equipment damage may occur. For parallel operation, the pumps must have approximately matching head characteristics, otherwise the system operating head may exceed the shut-off head of one or more pumps and the result in zero output flow.

12.12 Pump Performance and Pump Selection

12.12.1 Pump Performance

When selecting an irrigation pump, no relationship is more important than that of pressure and flow. The performance of a pump depends on it. How much water the pump is moving directly affects the pressure it is creating, and is the determining factor as to whether the pump is suitable for your irrigation system. It is important to understand that pressure and flow have an inverse relationship: as flow *increases*, pressure in the pump *decreases*.

12.12.2 Factors Affecting Pump Performance

For each model there are two variables which affect the pump performance. The first is the horsepower of the motor attached to the pump. Remember, what we commonly refer to as a pump is actually a pump and motor. The pump is the part that moves the water, the motor is the part that moves the pump. Most pumps can be attached to several different sizes of motor. Bigger motors mean more volume and pressure.

The second variable is the size of the impeller. The impeller spins inside the case and this is what moves the water. Larger impellers fit tighter in the case leaving less room for slippage. This results in higher pressures. But we don't always want higher pressures, as pressures higher than what you need just waste energy.

12.12.3 Selecting a Pump

Significant opportunities exist to reduce pumping system energy consumption through smart design, retrofitting, and operating practices. Operating characteristics as supplied by the manufacturer, include relationships between flow rate, pumping head, rotational speed, and power required. Optimum pump selection for agriculture depends on consideration of the following:

  (i) discharge rate required
 (ii) source of water
(iii) well characteristics (if applicable)
(iv) quality of water to be pumped
 (v) type, grade, and characteristics of the pump

Special attention must be given to the pumping plant when the irrigation system changes, as may occur when groundwater levels decline or operating pressure changes due to changing the type of irrigation system. Such changes may result in temptation to add a "booster" pump or connect two or more pumps together into a common distribution pipeline. These changes require a careful assessment if the desired result is to be achieved. Economic operation of the entire irrigation system can depend upon proper selection of the pump, power unit, and fuel type as well as proper routine maintenance, testing, and adjustment. The performance of a pump varies depending on how much water the pump is moving and the pressure it is creating. This is an important relationship because (i) it determines whether the pump is suitable for your irrigation system, and (ii) it is the pump characteristics which allow you to control the operation of your pump.

12.12.4 Procedure for Selecting a Pump

All pumps have oddities and special and unique requirements. Before installing a new pump, it is better to read the manufacturer's manual.

The basic procedure for selecting a pump for a new irrigation system is summarized below:

12.12 Pump Performance and Pump Selection

(a) Estimate your flow (GPM) and pressure (feet of head) requirements and select a preliminary pump model to use.
(b) Using your preliminary pump information, create a first draft irrigation design.
(c) Once you have a first draft of your irrigation, you may be able to fine tune your pump selection based on that design. Would a different pump lower your irrigation costs or better fit your irrigation system design? Return to the pump selection process and re-evaluate your pump selection. Make your final pump selection.
(d) Return once again to your irrigation design. Can it be fine tuned to better match your final pump selection? Make any necessary adjustments.

Although this method requires considerable effort, it will give one an excellent balance between pump and irrigation system, leaving one with a very efficient irrigation system.

12.12.4.1 Pump Selection from Pump Curves

A pump curve (or pump performance curve) is a simple graph which shows the performance characteristics of a particular pump. Pump curves are created by the pump manufacturer based on test results of the various pump models the manufacturer produces.

The pump manufacturer should be able to provide you with performance curves for the pumps you are considering. Here is a sample pump curve for study (Fig. 12.9):

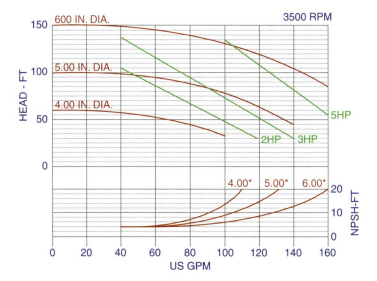

Fig. 12.9 Schematic of a pump curve

Each pump curve typically reflects a single model of pump made by the manufacturer. At the top right of the chart it gives the pump speed. Finding the proper pump is just a matter of selecting a model and size that will produce both the head and GPM that you need. To do this, you select the horsepower and impeller size that will give you the desired performance.

The red color curved lines (the top one in the pump curve above is labeled "6.00 in. dia.") represent the various impeller sizes. The green color straight lines represent the motor horsepower ratings available for this pump. Together the impeller curves and horsepower lines represent the best performance the pump is capable of, if that horsepower or impeller size is selected. Some pump curves do not have horsepower lines, and some pump curves combine the horsepower and impeller lines into one single line. This is usually because the pump only is available with one motor, so you don't get to select the horsepower. The pump may also only come with one size of impeller, so you will only see a single line on the entire pump curve.

To use the curves, you select the pressure you want on the left and then move horizontally across the chart to the vertical line that corresponds with the flow (GPM) that you want. You then select an impeller size curve and horsepower line that are above this point to determine the impeller size and horsepower you will need for your pump.

For example, a farmer wants a pump that produces 120 ft of head while pumping 100 gal/min (GPM). For that, start at 120 ft head on the left of the pump curve. Now move straight across the curve to the right until you reach the line that goes down to 100 GPM on the bottom of the curve. From the point where the two lines intersect move up the chart to see what horsepower pump will be needed. In this case a 5 hp will be needed, as the next horsepower line above the intersection point is the 5 hp line. Similarly, the impeller curve must also be higher in the chart than our line intersection, so a 6.00 in. diameter impeller will work.

For the case the curve is midway between the 5.00 in. curve and the 6.00 in. curve, then you would need a 5.50 in. impeller.

On many pump curves, additional set of ellipses labeled "efficiency" or simply with percentages labeled on them exists. These ellipses indicate the efficiency of the pump. To use them, simply look for the smallest ellipse that your line intersection point is inside. This is the efficiency at which the pump will operate. A high-efficiency pump uses less energy to operate than a low-efficiency pump. It is better to avoid any pump that has an efficiency of less than 60%.

12.12.4.2 Pump Selection Based on Pump Characteristics and Well Characteristics

Pump can be selected based on the well characteristics, which best suits for the particular well. In this approach, the well-characteristic curve is drawn with the same scale as that of pump characteristic curve, made a copy on tracing paper. Then the well-characteristic curve is placed over the pump curve (supplied by the manufacturer). From the matching (intersection), a pump is selected.

12.13 Sample Workout Problems on Pump

Example 12.1

Determine the net positive suction head available at the pump inlet from the following data:

suction head = 5 m
friction loss = 1 m
vapor pressure of the liquid at water temperature = 0.5 m
barometric (or atmospheric) pressure at pump level = 10 m

Solution

We know, net positive suction head available,
 NPSHa = BP − SH − FL − VP
Given,
 Barometric pressure at pump level, BP = 10 m
 Suction head, SH = 5 m
 Friction loss, FL = 1 m
 Vapor pressure of the liquid, VP = 0.5 m
Putting the values,
 NPSHa = 10 − 5 − 1 − 0.5 = 3.5 m (Ans.)

Example 12.2

A centrifugal pump has been installed to a depth of 35 m. The pump is discharging 2.5 cusec (ft³/s) water. Determine the capacity of the motor to operate the pump. Assume motor efficiency of 82%.

Solution

Given,
 $Q = 2.5$ ft³/s = 0.0708 m³/s [1 m³ = 35.3 ft³]
 $H_d = 35$ m
 $E_m = 82\% = 0.82$
 Taking discharge velocity of water = 2.0 m/s, velocity head = $V^2/2g$
   = $(2.0)^2/(2 \times 9.81)$
   = 0.2038 m
 Assuming friction loss = 5% of discharge head = $5 \times 35/100$ = 1.75 m
 Total head, H = discharge head + friction head + velocity head
   = 25 + 1.75 + 0.204
   = 36.95 m
 We get, power of the motor, $P_m = Q \times 9.81 \times H/E_m = 0.0708 \times 9.81 \times 36.95$
   = 31.31 kW (Ans.)

Example 12.3

A submersible pump lifts 70,500 l of water/h against a total head of 25 m. Determine the power requirement to lift the water in (i) kilowatt, and (ii) horse power.

Solution

Given,
$Q = 70{,}500 \text{ l/h} = 0.0195833 \text{ m}^3/\text{s}$
$H = 25 \text{ m}$

(i) We know, power, $P = Q \times 9.81 \times H = 0.0195833 \times 9.81 \times 25$

$= 4.833 \text{ kW (Ans.)}$

(ii) We know, 1 kW = 1.341 hp
Thus, $P = 4.803 \times 1.341 \text{ hp} = 6.441 \text{ hp (Ans.)}$

Example 12.4

In a wheat growing area, the cultivable land is 80 ha and wheat will be cultivated to all of the lands. The permissible interval between two irrigations at peak period is 15 days and the depth of irrigation required for that particular soil & agro-climatic region at peak period is 6.0 cm. If the total head for pumping is 25 m, pump efficiency is 85%, motor efficiency is 80%, and the maximum allowable operating period of the pump is 16 h/day, determine:

(a) The pump capacity required for that command area,
(b) Capacity of the motor

Solution

Given,
$A = 80 \text{ ha} = 800{,}000 \text{ m}^2$
Irri. interval = 15 days
Depth of irri., $d = 6 \text{ cm} = 0.06 \text{ m}$
Daily pump operating period = 16 h
Total head, $H = 25 \text{ m}$
Pump efficiency, $E_p = 85\% = 0.85$
Motor efficiency, $E_m = 80\%$
We get, total volume of water required for irrigation in 15 days,
$V = (800{,}000 \text{ m}^2 \times 0.06 \text{ m})$
$= 48{,}000 \text{ m}^3$
Total pumping period in 15 days, $t = 15 \times 16 \times 3{,}600 = 864{,}000 \text{ s}$

12.13 Sample Workout Problems on Pump

(a) Pump capacity, $Q = V/t = 48{,}000/864{,}000 = 0.05556$ m³/s (Ans.)
(b) Considering pump efficiency, capacity of the motor (output rated capacity),

$$P_m = (Q \times 9.81 \times H)/E_p = (0.05556 \times 9.81 \times 25)/0.85$$

$$= 16.03 \text{ kW}$$

(c) Rated capacity of the motor $= P_m/E_m = 16.03/0.8 = 20.04$ kW (Ans.)

Example 12.5

Maize crop of 20 ha is to be irrigated from a submergible pump. The maximum permissible interval between two irrigations at peak period is 12 days and the depth of each irrigation is 60 mm. If the maximum allowable operating period of the pump is 10 h/day, determine the pump capacity to meet the water demand of the farm.

Solution

Area of the field, $A = 20$ ha $= 200{,}000$ m²
Depth of irri. req., $d = 60$ mm $= 0.06$ m
Max. interval period $= 12$ days
Daily max. pumping period $= 10$ h
We get, total volume of water to be discharged in 12 days, $V = A \times d$
$= 200{,}000 \times 0.06$ m³
$= 12{,}000$ m³
Total pumping period in 12 days, $t = (12 \times 10) \times 60 \times 60 = 332{,}000$ s
Pumping rate or pump capacity required, $Q = V/t = 12{,}000/432{,}000$
$= 0.027778$ m³/s (Ans.)
Or, 0.98154 ft³/s (or cfs or cusec)
[Since 1 ft³/s $= 0.0283$ m³/s]

Example 12.6

In a residential area having population of 2,000 and expected population growth rate of 5%, the average daily water demand per capita is 100 l/day. Projecting for a time period of 30 years, determine the required capacity of the pump to satisfy the water demand of that area. Assume that the pump can be operated 8 h/day at its maximum.

If the pump is installed at 25 m below the ground surface, the velocity head of the flowing water is 1.5 m/s, friction loss in the discharge pipe and within pump casing is 10% of the discharge head, determine the optimum size of the motor to operate the pump.

Solution

Given,
Present population, $P = 2,000$ nos
Growth rate, $r = 5\% = 0.05$
Projection time period in yr, $t = 30$
Daily per capita water $= 100$ l
We get, population after 30 yrs $= P(1 + r)^t = 2,000 (1 + 0.05)^{30} = 8,643.9$
Assuming no change in per capita demand, daily total demand (V) will be
$= 8,643.9 \times 100 = 864,390$ l $= 864.39$ m^3
Daily pump operating period, $t = 8$ h $= 8 \times 3600 = 28,800$ s
Pump discharge rate req., $Q = V/t = 0.030$ m^3/s (Ans.)
Given, delivery head, $H_d = 25$ m
Velocity of flowing water, $v = 1.5$ m/s
Thus, velocity head, Hv $= v^2 = (0.5)^2/(2 \times 9.81) = 0.115$ m
Friction loss (or friction head), Hf $= 25 \times 10/100 = 2.5$ m
Thus, total head, $H = H_d + $ Hv $+$ Hf $= 25 + 0.115 + 2.5 = 27.625$ m
Power required to deliver the water, $P = Q \times 9.81 \times H$
$= 0.030 \times 9.81 \times 27.625 = 8.13$ kW
Considering efficiency of the pump $= 80\%$
Capacity of the motor, $P_m = P/E_p = 8.13/0.80 = 10.16$ kW (Ans.)

Example 12.7

Four pumps are connected in series, each one pumping 30 GPM at 25 PSI. What is the total output in flow volume and pressure?

Solution

Total flow $= 30$ GPM,
Total pressure $= 25 + 25 + 25 + 25 = 100$ PSI

Example 12.8

Four pumps are connected in parallel, each one pumping 30 GPM at 25 PSI. What is the total output in flow volume and pressure?

Total pressure $= 25$ PSI,
Total flow $= 25 + 25 + 25 + 25 = 100$ GPM

Questions

(1) What is a pump? Describe its purpose.
(2) What are the principles of water pumping?
(3) Classify the pumps under different perspectives.

(4) Explain the factors affecting the practical suction lift of pump.
(5) What are the features of centrifugal pump? Describe the working principles of centrifugal pump.
(6) Write short note on the following (in connection to centrifugal pump): (a) Priming, (b) Cavitation, (c) Specific speed, (d) Affinity laws.
(7) Describe the features, working principle, merits and limitations of the following pumps: (a) Turbine pump, (b) Submergible pump, (c) Volute pump, (d) Air-lift pump, (e) Bucket pump, (f) Jet pump.
(8) Deduce the formula for calculating power requirement of pump.
(9) Briefly describe the pump installation, operation and control principles and procedures.
(10) What are the hydraulics of water pumping?
(11) Explain the discharge and pressure calculation for pump connection in series and parallel.
(12) What is pump curve? Briefly explain how you will select a pump for its best performance.
(13) Six pumps are connected in series, each one pumping 40 GPM at 35 PSI. What is the total output in flow volume and pressure?
(14) Five pumps are connected in parallel, each one pumping 15 GPM at 22 PSI. What is the total output in flow volume and pressure?
(15) The discharge capacity of a pump is 0.068 m^3/s. The pump is lifting water from a depth of 33 m. Calculate the water power.
(16) A centrifugal pump has been installed to a depth of 27 m. The pump is discharging 2.8 cusec (ft^3/s) water. Determine the capacity of the motor to operate the pump. Assume motor efficiency of 88%.
(17) A submersible pump lifts 60,000 l of water per h against a total head of 30 m Determine the power requirement to lift the water in (i) kilowatt, and (ii) horse power.
(18) In a maize growing area, the cultivable land is 90 ha and wheat will be cultivated to all of the lands. The permissible interval between two irrigations at peak period is 12 days and the depth of irrigation required for that particular soil and agro-climatic region at peak period is 5.0 cm. If the total head for pumping is 35 m, pump efficiency is 87%, motor efficiency is 85%, and the maximum allowable operating period of the pump is 12 h/day, determine:

(i) the pump capacity required for that command area,
(ii) capacity of the motor, and
(iii) electric power required.

Chapter 13
Renewable Energy Resources for Irrigation

Contents

13.1	Concepts and Status of Renewable Energy Resources		480
	13.1.1	General Overview	480
	13.1.2	Concept and Definition of Renewable Energy	481
	13.1.3	Present Status of Uses of Renewable Energy	482
13.2	Need of Renewable Energy		482
13.3	Mode of Use of Renewable Energy		483
13.4	Application of Solar Energy for Pumping Irrigation Water		483
	13.4.1	General Overview	483
	13.4.2	Assessment of Potential Solar Energy Resource	484
	13.4.3	Solar or Photovoltaic Cells – Theoretical Perspectives	485
	13.4.4	Solar Photovoltaic Pump	485
	13.4.5	Uses of Solar System Other than Irrigation Pumping	489
	13.4.6	Solar Photovoltaic Systems to Generate Electricity Around the Globe	490
13.5	Wind Energy		491
	13.5.1	Wind as a Renewable and Environmentally Friendly Source of Energy	491
	13.5.2	Historical Overview of Wind Energy	491
	13.5.3	Causes of Wind Flow	492
	13.5.4	Energy from Wind	493
	13.5.5	Advantages of Wind Energy	493
	13.5.6	Assessing Wind Energy Potential	494
	13.5.7	Types of Wind Machines	495
	13.5.8	Suitable Site for Windmill	495
	13.5.9	Application of Wind Energy	496
	13.5.10	Working Principle of Wind Machines	497
	13.5.11	Wind Power Plants or Wind Farms	498
	13.5.12	Calculation of Wind Power	498
	13.5.13	Intermittency Problem with Wind Energy	500
	13.5.14	Wind and the Environment	501
	13.5.15	Sample Work Out Problems	501

13.6	Water Energy		502
	13.6.1	Forms of Water Energy	503
	13.6.2	Wave Energy	503
	13.6.3	Watermill	504
	13.6.4	Tide Mill	505
	13.6.5	Exploring the Potentials of Water Power	505
13.7	Bio-energy		506
	13.7.1	Liquid Biofuel	507
	13.7.2	Biogas	508
13.8	Geothermal Energy		508
13.9	Modeling the Energy Requirement		509
13.10	Factors Affecting Potential Use of Renewable Energy in Irrigation		509
	13.10.1	Groundwater Requirement and Its Availability	510
	13.10.2	Affordability of the User	510
	13.10.3	Willingness of the User to Invest in a Renewable Energy Based Pump	510
	13.10.4	Availability of Alternate Energy for Irrigation and Its Cost	511
	13.10.5	Alternate Use of Renewable Energy	511
13.11	Renewable Energy Commercialization: Problems and Prospects		511
	13.11.1	Problems	512
	13.11.2	Prospects/Future Potentials	514
	13.11.3	Challenges and Needs	516
Relevant Journals			516
Questions			517
References			518

The growing energy use all over the world and the increase in energy cost result in an increase in the prices of energy-dependent products, such as crops, meats, etc. Environmental pollution caused by the use of traditional energy sources, such as fossil fuel, makes it necessary to find new solutions for this problem. In addition, sources of fossil fuel are being rapidly depleted and energy consumption is increasing at an exponential rate. Under such situation, renewable energies seem to be the alternative. Renewable energy effectively uses natural resources such as sunlight, wind, rain, tides, and geothermal heat, which may be naturally replenished. Using renewable energy resources to power an irrigation system is a means of decreasing the dependency of food products on the prices of fuel and minimizes the impact of the irrigation system on the environment.

13.1 Concepts and Status of Renewable Energy Resources

13.1.1 General Overview

We use energy to do work and make all movements. Energy is defined as the ability or the capacity to do work. When we eat, our bodies transform the food into energy to do work. When we run or walk or do some work, we "burn" energy in our bodies.

Cars, planes, trolleys, boats, and machinery also transform energy into work. Work means moving or lifting something, warming or lighting something.

There are many sources of energy that help to run the various machines invented by man. The discovery of fire by man led to the possibility of burning wood for cooking and heating thereby using energy. For several thousand years human energy demands were met only by renewable energy sources – sun, biomass (wood, leaves, straw), hydel (water), and wind power. As early as 4,000–3,500 BC, the first sailing ships and windmills were developed harnessing wind energy. With the use of hydropower through water mills, things began to move faster. Solar energy is being used for drying, heating, and generating electricity.

13.1.2 Concept and Definition of Renewable Energy

Renewable energy effectively uses natural resources such as sunlight, wind, rain, tides, and geothermal heat, which may be naturally replenished. Renewable energy technologies include solar power, wind power, hydroelectricity, biomass, and biofuels for transportation. According to the International Energy Agency, "Renewable energy is derived from natural processes that are replenished constantly." In its various forms, it derives directly from the sun, or from heat generated deep within the earth. Included in the definition is electricity and heat generated from solar, wind, ocean, hydropower, biomass, geothermal resources, and biofuels and hydrogen derived from renewable resources.

Most renewable energy comes either directly or indirectly from the sun. Sunlight, or solar energy, can be used directly for heating and lighting homes and other buildings, for generating electricity, and for hot water heating, solar cooling, and a variety of commercial and industrial uses.

The sun's heat also drives the winds, whose energy is captured with wind turbines. Then, the winds and the sun's heat cause water to evaporate. When this water vapor turns into rain or snow and flows downhill into rivers or streams, its energy can be captured using hydropower.

Along with the rain and snow, sunlight causes plants to grow. The organic matter that makes up those plants is known as biomass. Biomass can be used to produce electricity, transportation fuels, or chemicals. The use of biomass for any of these purposes is called biomass energy.

Hydrogen also can be found in many organic compounds, as well as water. It's the most abundant element on the earth. But it doesn't occur naturally as a gas. It is always combined with other elements, such as with oxygen to make water. Once separated from another element, hydrogen can be burned as a fuel or converted into electricity.

Not all renewable energy resources come from the sun. Geothermal energy taps the earth's internal heat for a variety of uses, including electric power production, and the heating and cooling of buildings. The energy of the ocean's tides comes from the gravitational pull of the moon and the sun upon the earth.

The ocean can produce thermal energy from the sun's heat and mechanical energy from the tides and waves. Flowing water creates energy that can be captured and turned into electricity. This is called hydroelectric power or hydropower.

Each of these sources has unique characteristics which influence how and where they are used. Every country in the world is seeking new energy sources as alternatives to fossil fuels. For the best use of such resources, we need to know how much is available, what the limitations are, and how to make efficient devices to convert them into suitable energy form, or directly use them for various purposes.

13.1.3 Present Status of Uses of Renewable Energy

In 2006, about 18% of global final energy consumption came from renewables, with 13% coming from traditional biomass, like wood-burning. Hydropower was the next largest renewable source, providing 3%, followed by hot water/heating which contributed 1.3%. Modern technologies, such as geothermal, wind, solar, and ocean energy together provided some 0.8% of final energy consumption. The technical potential for their use is very large, exceeding all other readily available sources (WEA, 2001).

13.2 Need of Renewable Energy

The growing energy use all over the world and the increase in energy cost results in an increase in the prices of energy-dependent products, such as crops and meats. Environmental pollution caused by the use of traditional energy sources, such as fossil fuel, makes it necessary to find new solutions for this problem. In addition, sources of fossil fuel are being rapidly depleted and energy consumption is increasing at an exponential rate. The International Energy Outlook 2009 (IEO, 2009) predicts strong growth for worldwide energy demand over the period from 2006 to 2030. Using renewable energy resources to power an irrigation system is a means of decreasing the dependency of food products on the prices of fuel and minimizing the impact of the irrigation system on the environment.

Renewable energy is an essential part of many countries low emissions energy mix, and is important to today's energy security. It plays a strong role in reducing greenhouse gas emissions and helping nations stay on track to meet its Kyoto target and beyond. Governments should support for renewable energy, assist industry development, reduce barriers to the national electricity market, and provide community access to renewable energy.

Renewable energy can be particularly suitable for developing countries. In rural and remote areas, transmission and distribution of energy generated from fossil fuels can be difficult and expensive. Producing renewable energy locally can offer a viable alternative. Renewable energy projects in many developing countries have demonstrated that renewable energy can directly contribute to poverty alleviation by providing the energy needed for creating businesses, employment, and safe water supply. Renewable energy technologies can also make indirect contributions to alleviating poverty by providing energy for cooking, space heating, and lighting. Renewable energy can also contribute to education, by providing electricity to schools.

13.3 Mode of Use of Renewable Energy

The main renewable energy technologies for irrigation water pumping are:

- solar photovoltaic pump
- windmill pump
- producer gas based dual fuel engine pump
- biogas-based dual fuel pump
- water energy through hydroelectric power plant

Renewable energy available at a specific site may be used at that site through energy converting tools. For example, wind energy may be used to operate pumps through wind mills, solar energy may be used through solar or photovoltaic cells (in isolation from the national grid). Alternatively, generating electricity or biodiesel with the renewable energy where it is available, integrating the generated electricity through national grid, and then using the electricity (or biodiesel) to operate irrigation pumps, and also for other purposes. For water energy (sea wave or waterfall), geothermal, and bio-energy, integration is the only way to effective use of the energies in irrigation pumping.

13.4 Application of Solar Energy for Pumping Irrigation Water

13.4.1 General Overview

Power sources (fuel and electricity) are one of the basic components of mechanized irrigation equipment. The growing energy demand for expansion of irrigated agriculture has created the need to search and develop alternative energy sources, i.e., renewable energy sources. One of the renewable energy sources is solar energy. It is nondepletable, site dependent, nonpolluting and its benefits are numerous in terms of socio-economic, energy security, and environmental friendliness. Throughout the world, there are pilot projects for irrigating crops by photovoltaic pumps and generating electricity to add national grid.

Solar system does not require any kind of conventional fuels. This is a unique advantage of this new technology because of a great relief in the situation of energy crisis. Secondly, there is no moving part in this system (Fig. 13.1), thus resulting into quiet functioning leading to durability and soundless environment. With the growing use of power-based equipment in irrigated agriculture, the demand of stable supply of power has become more and more important for attainment of crop production. Solar power can be a supplement of power supply in irrigation purpose.

Fig. 13.1 Solar photovoltaic cells using for generating electricity

13.4.2 Assessment of Potential Solar Energy Resource

There is no shortage of solar-derived energy on Earth. Indeed the storages and flows of energy on the planet are very large relative to human needs. The amount of solar energy intercepted by the Earth every minute is greater than the amount of energy the world uses in fossil fuels each year. Tropical oceans absorb 560 trillion gigajoules (GJ) of solar energy each year, equivalent to 1,600 times the world's annual energy use. Annual photosynthesis by the vegetation in the United States is about 50 billion GJ, equivalent to nearly 60% of the nation's annual fossil fuel use.

There are many different ways to assess potentials:

13.4.2.1 Theoretical Potential

The theoretical potential indicates the amount of energy theoretically available for energy purposes, such as, in the case of solar energy, the amount of incoming radiation at the earth's surface.

13.4.2.2 Technical Potential

The technical potential is a more practical estimate of how much could be put to human use by considering conversion efficiencies of the available technology and available land area.

The technical potentials generally do not include economic or other environmental constraints, and the potentials that could be realized at an economically competitive level under current conditions and in a short time-frame is still lower.

Ideally, detailed solar data for each location should be used in evaluating the potential of using a solar photovoltaic (SPV) pump. However, to make an initial macro-level assessment, broad (nation or state wise) solar radiation availability characteristics are readily available in the literature. It is recommended by UNESCAP that for installing SPV pump, the average daily solar radiation in the least sunny month should be greater than 3.5 kW/m² on a horizontal surface (UNESCAP, 1991).

If measured data of solar irradiance is not available, it can be estimated from other easily measurable weather data such as temperature. Procedure for such indirect estimation and direct measurement by instrument has been described in detail in Chapter 3 (*Weather*) of Volume 1.

13.4.3 Solar or Photovoltaic Cells – Theoretical Perspectives

Photovoltaic system uses solar energy for generation of electricity. Photovoltaic is a technology that converts radiant light energy (photo) to electricity (voltaic) due to energy transfer that occurs at the subatomic level. Silicon is one of the elements used as base material for the production of photovoltaic cells. A silicon atom has four valences of electrons that are shared with adjacent silicon atoms in covalent bonding. The boron atoms occupy a lattice position within the silicon structure, and a positive-charged hole forms in place of the missing fourth electron. Silicon material with boron impurities is called a p-type semiconductor. The phosphorous atoms occupy a lattice position within the silicon structure and form a negative or n-type semiconductor. Putting a layer of p-type and a layer of n-type semiconductor material together makes photovoltaic cell (solar cells). Solar energy comes in small packages called photons. These photons hit the outer-level electrons in the photovoltaic cells. When the photons in solar radiation strike a photovoltaic cell, the kinetic energy of the photons is transferred to the valence level of electrons. The freed electrons and positive-charged holes attract each other and create positive–negative pairs. The formation of these pairs creates electricity (Garg, 1987). The power produced by photovoltaic cell is the product of the current and voltage being produced at any given time. Photovoltaic power is directly proportional to the level of irradiance available at any given time.

13.4.4 Solar Photovoltaic Pump

Photovoltaic array is composed of two or more photovoltaic modules. Photovoltaic module is a combination of photovoltaic cells wired together in series and parallel for the purpose of generating specific current and voltage at a given level of irradiance. To increase voltage output, photovoltaic cells are wired in series, and to increase amperage output photovoltaic cells are wired in parallel.

13.4.4.1 Photovoltaic System Types

Photovoltaic system types in pumping water for irrigation may be grouped into three categories:

i. Stand-alone photovoltaic system
ii. Stand-alone hybrid system
iii. Grid connected system

Stand-Alone Photovoltaic System

Stand-alone photovoltaic system produces power independently of other energy sources. It may be in three configurations: (a) direct coupled, (b) DC loads with battery, (c) DC loads with battery and charge controller.

Batteries require strong energy for use at night time or on cloudy days. These (with battery) stand-alone systems comprise the majority of photovoltaic installation in remote region because they are often the best, most cost-effective choice for applications far from the utility grid, which requires reliable power supply and low maintenance. Sun tracker may be used to increase the output of the photovoltaic array.

Stand-Alone Hybrid System

In stand-alone hybrid system, the photovoltaic array is coupled with one or more additional power sources. Hybrid system may be the cost-effective choice to provide power for large-scale irrigation, especially when another power source is already available or greater system is desired.

Grid System

In a grid connected system, the photovoltaic system acts as a miniature power plant, feeding excess power direct to utility grid during the day, and drawing power from the grid at night, in periods of low sunshine or load shedding time.

13.4.4.2 Photovoltaic System Design for Irrigation Water Pumping

In designing photovoltaic pumping system, possible supply and demand of energy have to be matched. Energy demand can be calculated from the peak power (water) demand during minimum irrigation interval required, and pumping plant and irrigation efficiencies.

Steps/Procedures for the Designing of PV Pumping System

The steps in designing PV systems are:

- estimate the peak water demand of the irrigation scheme (considering minimum irrigation interval)
- calculate power demand for the peak water demand
- calculate the requirement of PV cells for power generation (considering minimum expected solar irradiance during the peak period).

Mathematical Formulation

The peak water demand per day (Q_p) is estimated by taking into account the irrigation efficiency, i.e.,

$$Q_p = Q_c/E_{ir} \qquad (13.1)$$

13.4 Application of Solar Energy for Pumping Irrigation Water

where

Q_p = The peak water demand per day (m³/day)
Q_c = Peak crop water demand (for minimum irrigation interval period) (m³/day)
E_{ir} = Irrigation efficiency

The procedure for estimating daily crop water demand for an irrigation scheme is described in Chapter 10 (*Crop Water Requirement*) of Volume 1. The highest value is the peak demand.

The pumping (i.e., discharge) rate is estimated taking into account the minimum possible solar hour (i.e., here pumping hour) from the relation:

$$Q_P = (Q \times H_S) \times 3600$$

or,

$$Q = Q_P/(H_S \times 3600) \qquad (13.2)$$

where

Q_p = The peak water demand per day (m³/day)
Q = the discharge rate of the pump (m³/s)
H_s = minimum possible solar (bright sunshine) hour (i.e., pumping hour) per day for the peak period
$3,600$ = conversion factor, to convert hour into second

The peak energy demand (kW), P_E, is calculated by taking into account the total head requirement and pumping plant efficiency (E_p):

$$P_E = (Q \times 9.81 \times H) / E_P \qquad (13.3)$$

where

P_E = The peak energy demand (kW)
Q = the discharge rate of the pump (m³/s)
H = total pumping head required (m)
E_p = pumping plant efficiency

Here, H includes suction head, delivery head, velocity head, and head loss due to friction in pipe and pump casing.

The number of PV array (N_{PV}) required to meet peak energy demand is calculated as:

$$N_{PV} = P_E/(R_S \times A_{PV}) \qquad (13.4)$$

where

P_E = Peak energy demand (kW)
R_S = Solar irradiation per unit surface under the location in question (kW/m²)
A_{PV} = Area of each PV array (m²)

Considering the efficiency of photovoltaic cells, and an overall design safety factor, the equation for calculating N_{PV} takes the form:

$$N_{PV} = (P_E \times S_F)/(R_S \times A_{PV} \times E_{PV}) \quad (13.5)$$

where

S_F = Overall design safety factor (1.1–1.2)
E_{PV} = efficiency of photovoltaic cells to convert solar irradiance to electricity (35–42%)

13.4.4.3 Economics of Photovoltaic Irrigation Pumping

The cost of photovoltaic-powered water pumping system is progressively decreasing. The cost of photovoltaic modules has already fallen 400% in the last 30 years (due to technological improvement and commercialization system), and this trend continues. On the other hand, the cost of power derived from fossil fuel is increasing.

In a case study in Bangladesh, Bhuiyan et al. (2000) evaluated the economics of stand-alone photovoltaic power to test the feasibility in remote and rural areas and to compare renewable generators using method of net present value analysis. They found that the life-cycle cost of photovoltaic energy is lower than the life-cycle cost of conventional energy where there is no grid, and the photovoltaic generator is economically feasible in remote and rural areas.

13.4.4.4 Sample Work Out Problems

Example 13.1

How much energy is required to lift a discharge of 0.05 m³/s from a depth of 10 m?

Solution

We know, required power,
$P = Q \times 9.81 \times H$
Here,
$Q = 0.05$ m³/s
$H = 10$ m
Putting the values,
$P = 4.905$ kilowatt (kW)(Ans.)

Example 13.2

In a small agricultural farm, the peak water demand rate is 250 m³/d and total head requirement is 10 m. The minimum and maximum solar irradiance through the cropping period at the site are 6 MJ/m²/d and 8 MJ/m²/d. Determine the number of PV array required to meet the power demand, if the area of each array is 3 m².

13.4 Application of Solar Energy for Pumping Irrigation Water

Solution

We know, Peak energy demand, $P_e = Q \times 9.81 \times H/E_P$

Given,

Discharge rate = 250 m³/d = 0.009921 m³/s [Assuming daily bright sunshine (solar) hour, i.e., pumping hour = 7]

Head, $H = 10$ m

Assuming efficiency of the pump = 70% = 0.07

Putting the values, $P_e = 1.3903$ kW

Now, the number of PV array, $N_{PV} = P_e/(R_S \times A_{PV} \times E_{PV})$

Here, Assured solar irradiance, $R_S = 6 \text{MJ/m}^2/\text{d} = 0.23809 \text{ kJ/m}^2/\text{s}$

= 0.23809 kW/m² [considering one (1) solar day = 7 h, as mentioned above]

Area of single PV unit, $A_{PV} = 3$ m²

Assuming efficiency of the PV cells, $E_{PV} = 30\% = 0.30$

Putting the values, $N_{PV} = 6.488 \approx 7$ nos (Ans.)

Example 13.3

Determine the increase in power from a solar panel if the efficiency of PV cells increase from 30 to 50%.

Solution

Consider a PV cell of 1 m², and solar irradiance = R_s MJ/m²/d

Power to be generated by PV cell, $P_{PV} = R_S \times 1 \times E_{PV}$

As the power is directly related to the efficiency of the PV cell, the increase in power due to increase in efficiency will be (50–30%) = 20%.

Relative power increase = (20/30) ×100 = 66.66% (Ans.)

13.4.5 Uses of Solar System Other than Irrigation Pumping

In this context, "solar energy" refers to energy that is collected from sunlight. Solar energy can be applied in many ways, including, to

- generate electricity using photovoltaic solar cells.
- generate electricity using concentrated solar power.
- generate electricity by heating trapped air which rotates turbines in a solar updraft tower.
- heat buildings, directly, through passive solar building design.
- heat foodstuffs, through solar ovens.
- heat water or air for domestic hot water and space heating needs using solar-thermal panels.
- heat and cool air through use of solar chimneys.

- generate electricity in geosynchronous orbit using solar power satellites.
- dry various food stuffs.
- solar air conditioning.

13.4.6 Solar Photovoltaic Systems to Generate Electricity Around the Globe

Since 2004 there has been renewed interest in solar thermal power stations, and two plants were completed during 2006/2007: the 64 MW Nevada Solar One and the 11 MW PS10 solar power tower in Spain. Three 50 MW trough plants were under construction in Spain at the end of 2007, with 10 additional 50 MW plants planned. In the United States, utilities in California and Florida have announced plans (or contracted for) of at least eight new projects totaling more than 2,000 MW.

In developing countries, three World Bank projects for integrated CSP/combined-cycle gas-turbine power plants in Egypt, Mexico, and Morocco were approved during 2006/2007. There are several solar thermal power plants in the Mojave Desert which supply power to the electricity grid. Solar Energy Generating Systems (SEGS) is the name given to nine solar power plants in the Mojave Desert which were built in the 1980s. These plants have a combined capacity of 354 megawatts (MW) making them the largest solar power installation in the world.

The Moura photovoltaic power station, located in the municipality of Moura, Portugal, is presently under construction and will have an installed capacity of 62 MW. The first stage of construction has finished in 2008, and the second and final stage is scheduled for 2010, making it one of the largest photovoltaic projects ever constructed. Construction of a 40 MW solar generation power plant is underway in the Saxon region of Germany. The Waldpolenz Solar Park will consist of some 550,000 thin-film solar modules. The direct current produced in the modules will be converted into alternating current and fed completely into the power grid. Completion of the project is expected in 2009. Three large photovoltaic power plants have recently been completed in Spain: the Parque Solar Hoya de Los Vincentes (23 MW), the Solarpark Calveron (21 MW), and the Planta Solar La Magascona (20 MW).

Another photovoltaic power project has been completed in Portugal. The Serpa solar power plant is located at one of Europe's sunniest areas (Fig. 13.2). The 11 MW plant covers 150 acres and comprises 52,000 PV panels. The panels are raised 2 m off the ground and the area will remain productive grazing land. The project will provide enough energy for 8,000 homes and will save an estimated 30,000 tonnes of carbon dioxide emissions per year.

A $420 million large-scale solar power station in Victoria is to be the biggest and most efficient solar photovoltaic power station in the world. Australian company Solar Systems will demonstrate its unique design incorporating space technology in a 154 MW solar power station connected to the national grid. The power station

13.5 Wind Energy

Fig. 13.2 Solar power plant near Serpa, Portugal (38°1'51"N, 7°37'22"W)

will have the capability to concentrate the sun by 500 times onto the solar cells for ultra high power output. The Victorian power station will generate clean electricity directly from the sun to meet the annual needs of over 45,000 homes with zero greenhouse gas emissions.

13.5 Wind Energy

13.5.1 Wind as a Renewable and Environmentally Friendly Source of Energy

Wind power is renewable and produces no greenhouse gases during operation, such as carbon dioxide and methane. Wind is called a renewable energy source because the wind will blow as long as the sun shines. Wind power is one of the most environmentally friendly sources of renewable energy. At one time, wind was the major source of power for pumping water, grinding grain, and transporting goods by sailing ships. Present day applications of wind power include water pumping and the generation of electricity.

13.5.2 Historical Overview of Wind Energy

Wind power has been harnessed by mankind since ancient times. Over 5,000 years ago, the ancient Egyptians used wind to sail ships on the Nile River. Later, people built windmills to grind wheat and other grains. Wind filled the sails of explorers and fueled trade throughout the world. The Dutch used wind to claim land from the ocean, and early European settlers in the American West used wind to pump water for farms. The earliest known windmills were in Persia (Iran). These early windmills looked like large paddle wheels. Centuries later, the people of Holland improved the basic design of the windmill. They gave it propeller-type blades, still made with sails. Holland is famous for its windmills. During the first half of the twentieth century, wind was used to provide electricity for many rural areas before the expansion of the electrical grid and offered a cheaper and more reliable form of power.

American colonists used windmills to grind wheat and corn, to pump water, and to cut wood at sawmills. As late as the 1920s, Americans used small windmills to generate electricity in rural areas without electric service. When power lines began to transport electricity to rural areas in the 1930s, local windmills were used less and less, though they can still be seen on some Western ranches.

The oil shortages of the 1970s changed the energy picture for the country and the world. It created an interest in alternative energy sources, paving the way for the re-entry of the windmill to generate electricity. In the early 1980s wind energy really took off in California, partly because of state policies that encouraged renewable energy sources. Support for wind development has since spread to other states, but California still produces more than twice as much wind energy as any other state. The first offshore wind park in the United States is planned for an area off the coast of Cape Cod, Massachusetts.

13.5.3 Causes of Wind Flow

Wind is simple air in motion. It is caused by the uneven heating of the earth's surface by the sun (Fig. 13.3). Since the earth's surface is made of very different types of

Fig. 13.3 Schematic of mechanism of air-flow

land and water, it absorbs the sun's heat at different rates. During the day, the air above the land heats up more quickly than the air over water. The warm air over the land expands and rises, and the heavier, cooler air rushes in to take its place, creating winds. At night, the winds are reversed because the air cools more rapidly over land than over water.

In the same way, the large atmospheric winds that circle the earth are created because the land near the earth's equator is heated more by the sun than the land near the North and South Poles.

13.5.4 Energy from Wind

Airflows can be used to run wind turbines. Modern wind turbines range from around 600 kW to 5 MW of rated power, although turbines, with rated output of 1.5–3 MW, have become the most common for commercial use. The power output of a turbine is a function of the cube of the wind speed, so as wind speed increases, power output increases dramatically. Areas where winds are stronger and more constant, such as offshore and high altitude sites, are preferred locations for wind farms.

Since wind speed is not constant, a wind farm's annual energy production is never as much as the sum of the generator nameplate ratings multiplied by the total hours in a year. The ratio of actual productivity in a year to this theoretical maximum is called the *capacity factor*. Typical capacity factors are 20–40%, with values at the upper end of the range in particularly favorable sites. For example, a 1 MW turbine with a capacity factor of 35% will not produce 8,760 MWh in a year, but only $1 \times 0.35 \times 24 \times 365 = 3,066$ MWh, averaging to 0.35 MW.

Globally, the long-term technical potential of wind energy is believed to be five times of total current global energy production, or 40 times current electricity demand. This could require large amounts of land to be used for wind turbines, particularly in areas of higher wind resources. Offshore resources experience mean wind speeds of ~90% greater than that of land, so offshore resources could contribute substantially more energy. This number could also increase with higher altitude ground-based or airborne wind turbines.

13.5.5 Advantages of Wind Energy

Wind energy has many advantages over conventional energies. The significant advantages are the following:

- A wind farm, when installed on agricultural land, has one of the lowest environmental impacts of all energy sources.
- It occupies less land area per kilowatt-hour (kWh) of electricity generated than any other energy conversion system, apart from rooftop solar energy, and is compatible with grazing and crops.

- It generates the energy used in its construction in just 3 months of operation, yet its operational lifetime is about 20–25 years.
- Greenhouse gas emissions and air pollution produced during its construction are tiny and declining due to advancement of technology.
- There are no emissions or pollution produced by its operation.
- In substituting for base-load coal power, wind power produces a net decrease in greenhouse gas emissions and air pollution, and a net increase in biodiversity.
- Modern wind turbines are almost silent and rotate so slowly (in terms of revolutions per minute) that they are rarely a hazard to birds.

There is still the problem of what to do when the wind isn't blowing. At those times, other types of power plants must be used to make electricity.

13.5.6 Assessing Wind Energy Potential

The water output of a windmill pump is very sensitive to any variation in wind speed. Wind speed at any location depends upon a variety of site specific factors. Within a country or state, large variation in annual monthly mean wind speeds exists. Therefore, analysis for the estimation of potential of windmill pumps for irrigation water pumping should be carried out using data of small area, such as district or other local unit. Other data such as surface water irrigation facility and depth to groundwater table should be taken into account to estimate potential uses.

The wind speed at any location increases with an increase in height from the ground level. In most of the prevailing windmill pump designs, rotor is fixed on a tower at about 10 m height from the ground. From the available data on measured values of annual monthly mean wind speed at different locations, the annual monthly mean wind speeds at the 10 m height from the ground level should be estimated by using standard relationship. One of the forms of such relationship is (Kumar and Kandpal, 2007)

$$V_{10} = V_h \left(\frac{10}{h}\right)^\alpha$$

where

V_h = annual monthly mean wind speed measured at a reference height of h meters from the ground level
α = ground surface friction coefficient

The value of α varies from 0.1 to 0.5 depending upon the terrain of the location. For crops and shrubs type terrain, a value of 0.2 is generally recommended. For water or smooth flat ground, α is taken as 0.1.

Water output of commercially available windmills at wind speeds below 10 km/h is very low. The areas with annual monthly mean wind speeds greater than 10 km/h should be considered for windmill-pump-based irrigation.

13.5.7 Types of Wind Machines

There are two types of wind machines (turbines) used today based on the direction of the rotating shaft (axis):

- horizontal-axis wind machines, and
- vertical-axis wind machines

The size of wind machines varies widely. Small turbines used to power a single home or business may have a capacity of less than 100 kW. Some large commercial-sized turbines may have a capacity of 5 million watts, or 5 MW. Larger turbines are often grouped together into wind farms that provide power to the electrical grid.

13.5.7.1 Horizontal-Axis

Most wind machines being used today are the horizontal-axis type. Horizontal-axis wind machines have blades like airplane propellers. A typical horizontal wind machine stands as tall as a 20-story building and has three blades that span 200 ft across. Wind machines stand tall and wide to capture more wind.

13.5.7.2 Vertical-Axis

Vertical-axis wind machines have blades that go from top to bottom and the most common type looks like a giant two-bladed egg beaters. The type of vertical wind machine typically stands 100 ft tall and 50 ft wide. Vertical-axis wind machines make up only a very small percent of the wind machines used today.

The Wind Amplified Rotor Platform (WARP) is a different kind of wind system that is designed to be more efficient and use less land than wind machines in use today. The WARP does not use large blades; instead, it looks like a stack of wheel rims. Each module has a pair of small, high capacity turbines mounted to both of its concave wind amplifier module channel surfaces. The concave surfaces channel wind toward the turbines, amplifying wind speeds by 50% or more.

13.5.8 Suitable Site for Windmill

Wind plant owners must carefully plan where to locate their machines. One important thing to consider is how fast and how much (time period) the wind blows. Wind speed varies throughout the country. It also varies from season to season. As a rule, wind speed increases with altitude and over open areas with no windbreaks. Good sites for wind plants are the tops of smooth, rounded hills; open plains or shorelines; and mountain gaps that produce wind funneling. For using plain land, its value for alternate use should be taken into consideration.

13.5.8.1 Characterization of the Resource

Vast areas with high wind power potential exist in many parts of the world. Wind power resource is categorized according to wind power class. Wind class 1 denotes very light winds; higher numbers indicate stronger winds.

13.5.9 Application of Wind Energy

13.5.9.1 Mode of Use of Wind Power

Wind turbines can be used as stand-alone applications, or they can be connected to a utility power grid or even combined with a photovoltaic (solar cell) system. For utility-scale sources of wind energy, a large number of wind turbines are usually built close together to form a wind plant. Several electricity providers today use wind plants to supply power to their customers.

Stand-alone wind turbines are typically used for water pumping. However, homeowners, farmers, and ranchers in windy areas can also use wind turbines as a way to cut their electric bills.

Small wind systems also have potential as distributed energy resources. Distributed energy resources refer to a variety of small, modular power-generating technologies that can be combined to improve the operation of the electricity delivery system.

13.5.9.2 Water Pumping Windmill

A windmill could be installed on an open well, bore well, pond, etc., and at a site which is free from any obstacles such as high-rise buildings and tall trees that could restrict the availability of wind to the rotor of the windmill.

A typical windmill comprises of several bladed rotor of 2–3 m diameter, and is installed on a tower of 10 m height. The rotor through the gear mechanism drives the connecting rod and the pump, which pump water from the well. A 18-bladed rotor of 3 m diameter can pump water from a maximum depth of 30 m, at an average wind speed of 8–10 km/h. The approximate rate of pumping under ideal conditions ranges from 1,000 to 1,200 l/h, which could cater to the irrigation needs of about half to one hectare area depending upon the cropping pattern and its water requirement.

13.5.9.3 Wind in Production of Electricity

The technology used to convert wind power to electricity is fairly simple. The blades of the turbine are similar to airplane blades. There are some other designs, but the basic two or three bladed turbine is widely used. The rotation of the blades spins a turbine. Turbines are used in almost all electrical generating technologies. Within the turbine, there are magnets whose rotating motion causes an electrical field, and we get a current flow. The generated electricity can be used at any area for irrigation water pumping.

A wide range of generating capacities allow wind turbines to fill a wide range of applications and sites. Individual producers, generating electricity for their own home needs may use a turbine as small as 10 kW (this would power one hundred 100 W light bulbs at peak power). Larger utility-based wind farms may use turbines from 50 kW all the way up to 2 MW (1 MW = 1,000 kW). The typical sizes for modern utility applications are between 200 and 300 kW.

13.5.10 Working Principle of Wind Machines

Like old-fashioned windmills, today's wind machines use blades to collect the wind's kinetic energy (Fig. 13.4). The wind flows over the airfoil-shaped blade, the blade acts much like an airplane wing. When the wind blows, a pocket of low-pressure air forms on the downwind side of the blade. The low-pressure air pocket then pulls the blade toward it, causing the rotor to turn. This is called lift. The force of the lift is actually much stronger than the wind's force against the front side of the blade, which is called drag. The combination of lift and drag causes the rotor to spin like a propeller, and the turning shaft spins a generator to make electricity.

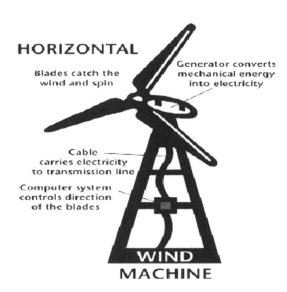

Fig. 13.4 Schematic view of a wind machine

13.5.11 Wind Power Plants or Wind Farms

Wind power can be tapped in a variety of places. Wind can also be tapped on water, or on farms. Wind turbines, like windmills, are mounted on a tower to capture the most energy. At 30 m or more aboveground, they can take advantage of the faster and less turbulent wind. Turbines catch the wind's energy with their propeller-like blades. Usually, two or three blades are mounted on a shaft to form a rotor.

Wind-power plants or wind farms are clusters of wind machines used to produce electricity (Fig. 13.5). A wind farm usually has dozens of wind machines scattered over a large area. The world's largest wind farm, the Horse Hollow Wind Energy Center in Texas, has 421 wind turbines that generate enough electricity to power 230,000 homes per year.

13.5.11.1 Seasonal Variation of Wind Energy

Operating a wind-power plant is not as simple as just building a windmill in a windy place. In Tehachapi, California, the wind blows more from April through October than it does in the winter. This is because of the extreme heating of the Mojave Desert during the summer months. The hot air over the desert rises, and the cooler, denser air above the Pacific Ocean rushes through the Tehachapi mountain pass to take its place. In a state like Montana, on the other hand, the wind blows more during the winter. Fortunately, these seasonal variations are a good match for the electricity demands of the regions. In California, people use more electricity during the summer for air conditioners. In Montana, people use more electricity during the winter months for heating.

13.5.12 Calculation of Wind Power

Wind turbines convert the kinetic energy in the wind into mechanical power. This mechanical power can be used for pumping water, or a generator can convert this mechanical power into electricity.

Consider a segment of air shaped like a horizontal cylinder. The energy in it depends on its volume, density, and speed. The kinetic energy (E_k) of the wind segment is

$$E_k = \frac{1}{2}MV^2 \tag{13.6}$$

The mass per unit time for the air volume, $M = \rho AV$
where

ρ = density of air (kg/m^3)
A = swept area (m^2)
V = wind speed (m/s)

13.5 Wind Energy

Fig. 13.5 Wind turbine (**a**) typical modern wind turbine at field (**b**) typical wind farm in a hilly area (modern wind turbines tower above an old windmill) (**c**) Wind turbine at sea

Substituting the mass of the air, Eq. (13.6) yields

$$E_k = \frac{1}{2}\rho A V^3 \qquad (13.7)$$

The density of air varies with temperature and pressure (and hence on altitude), and can be estimated from the relation:

$$\rho = 3.48\ P/T\ (\text{kg/m}^3)$$

where P is the pressure in kPa and T is the temperature in Kelvin. At 25°C temperature and 1 atm (or 101 kPa) pressure, the value of air density is close to 1.2 kg/m³.

The swept area is the area that the blades cover when they rotate, and is the total area in which the kinetic energy of the wind is captured by the turbine. The swept area of rotor, $A = \pi r^2$, where r is the rotor radius or the distance from the hub to blade tip.

Thus, the theoretical model for power of wind turbine (P_w) becomes

$$P_w = \frac{1}{2}\rho n \pi r^2 V^3 \qquad (13.8)$$

where n is the number of blades.

Introducing the efficiency term for turbine and generator, the energy model becomes

$$P_w = \frac{1}{2}\rho n \pi r^2 V^3 E_T E_G \qquad (13.9)$$

where

E_T = efficiency of turbine (or windmill) (40–50%)
E_G = efficiency of generator (80–85%)

Because of the cubed factor governing the wind speed, a small increase in wind speed leads to a much greater increase in power. Thus using the average wind speed does not reflect the actual power over time. Let us consider the wind speed of 5, 10, and 15 m/s. The average velocity here is 10 m/s. Due to increase in wind speed to double and triple, the resultant energy will be increased by 8 and 27 times. The average power is about 150% compared to that of average velocity.

Considering the above facts, for practical estimation, a general rule of thumb is to double the power found using average wind speed.

13.5.13 Intermittency Problem with Wind Energy

One particular problem with wind energy is the intermittent nature of the resource. The wind does not always blow. This is perhaps the leading criticism of wind power

13.5 Wind Energy

today. For the home generator, batteries can be used to store the power; however, this increases costs and reduces the environmental benefits. A solution is to use the electric grid as a "storage" mechanism. During times when the turbine is producing excess capacity, the electric company purchases your power. During times when you need more power than you can produce, you can "buy back" the power you sold.

13.5.14 Wind and the Environment

In the 1970s, oil shortages pushed the development of alternative energy sources. In the 1990s, the push came from a renewed concern for the environment in response to scientific studies indicating potential changes to the global climate, if the use of fossil fuels continues to increase.

Wind energy is an economical power resource in many areas of the world. Wind is a clean fuel; wind farms produce no air or water pollution because no fuel is burned. Growing concern about emissions from fossil fuel generation, increased government support, and higher costs for fossil fuels (especially natural gas and coal) have helped wind power capacity in the United States grow substantially over the last 10 years.

The most serious environmental drawbacks to wind machines may be their negative effect on wild bird populations and the visual impact on the landscape. To some, the glistening blades of windmills on the horizon are an eyesore; to others, they're a beautiful alternative to conventional power plants.

13.5.15 Sample Work Out Problems

Example 13.4

In an area, the average wind speed at the turbine level is 20 km/h. The area swept by the rotating blades of the wind turbine is 0.5 m². Determine the wind power to be generated from the turbine.

Solution

We know, mechanical power from wind turbine,
$$P_w = 1/2 \rho A V^3 E_T$$
Given,
$V = 20$ km/h $= 5.56$ m/s
$A = 0.5$ m²
Assuming, air density, $\rho = 1.2$ kg/m³
Assuming efficiency of the turbine, $E_T = 55\% = 0.55$
Putting the values in above equation, $P_w = 28.29$ W (Ans.)

Example 13.5

Determine the potentials of electricity generation from wind energy per wind turbine having average wind speed 15 km/h, 3 blades, radius of the blades 2 m, turbine efficiency of 60%, and generator efficiency of 85%.

Solution

We get,
$$P_w = 1/2 \rho n \pi r^2 V^3 E_T E_G$$
Given,
$V = 15$ km/h $= 4.1667$ m/s
$n = 3$
$r = 2$ m
$E_T = 60\%; = 0.60$
$E_G = 85\% = 0.85$
Assuming air density, $\rho = 1.2$ kg/m^3
Putting the values, $P_w = 834.4855$ W (Ans.)

Example 13.6

Determine the increase in electricity production from wind energy in an area if the wind speed increases from 10 to 20 km/h.

Solution

We get, $P_w = 1/2 \, \rho A V^3 E_T E_G$
Here, all but 'wind speed' are constant, thus we can write,
$$P_w = CV^3$$
Given,
$V_1 = 10$ km/h $= 2.778$ m/s
$V_2 = 20$ km/h $= 5.556$ m/s
For 1st speed, power, $P_{w-1} = C(2.778)^3 = C \times 21.43$ W
For 2nd speed, power, $P_{w-2} = C(5.556)^3 = C \times 171.47$ W
$P_{w-2} - P_{w-1} = C(171.47 - 21.43) = C \times 150.03$ W
%increase in power $= (C \times 150.03 \times 100/(C \times 21.43) = 700.0$ (Ans.)

13.6 Water Energy

Energy in water (in the form of motive energy or temperature differences) can be harnessed and used. Since water is about 800 times denser than air, even a slow

13.6 Water Energy

flowing stream of water, or moderate sea wave, can yield considerable amounts of energy.

13.6.1 Forms of Water Energy

There are many forms of water energy:

- Hydroelectric energy is a term usually reserved for large-scale hydroelectric dams. Examples are the Grand Coulee Dam in Washington State and the Akosombo Dam in Ghana.
- Micro-hydro systems are hydroelectric power installations that typically produce up to 100 kW of power. They are often used in water-rich areas as a remote area power supply (RAPS). There are many of these installations around the world, including several delivering around 50 kW in the Solomon Islands.
- Damless hydro systems derive kinetic energy from rivers and oceans without using a dam.
- Wave power uses the energy in waves. The waves usually make large pontoons go up and down in the water, leaving an area with reduced wave height in the "shadow." Wave power has now reached commercialization.
- Tidal power captures energy from the tides in a vertical direction. Tides come in, raise water levels in a basin, and tides roll out. Around low tide, the water in the basin is discharged through a turbine.
- Tidal stream power captures energy from the flow of tides, usually using underwater plant resembling a small wind turbine. Tidal stream power demonstration projects exist, and the first commercial prototype is installed in Strangford Lough in September 2007.
- Ocean thermal energy conversion (OTEC) uses the temperature difference between the warmer surface of the ocean and the colder lower recesses. To this end, it employs a cyclic heat engine. OTEC has not been field-tested on a large scale.
- Blue energy is the reverse of desalination. This form of energy is in research.

13.6.2 Wave Energy

Waves are a free and sustainable energy resource created as wind blows over the ocean surface. Waves are caused by wind blowing over water, and winds are generated by the sun heating the earth. Wave energy technology is very young compared to generating electricity from wind turbines.

Water covers around three-quarters of the earth's surface. The World Energy Council estimates that the energy that can be harvested from the world's oceans is equal to twice the amount of electricity that the world produces now. The UK has by far the highest potential in the European Union for converting wave energy into electricity. The amount of energy in waves around the UK is more than the energy

Fig. 13.6 Ocean Wave Power engines in the harbour of Peniche/Portugal

currently used to meet people's demand for electricity. Scotland has particularly high potential: much of the western coastline is exposed to the Atlantic Ocean.

Portugal now has the world's first commercial wave farm, the *Aguçadora Wave Park*, established in 2006 (Fig. 13.6). The farm will initially use three Pelamis P-750 machines generating 2.25 MW. Initial costs are put at €8.5 million. Subject to successful operation, a further €70 million is likely to be invested in 2009 on a further 28 machines to generate 525 MW.

13.6.3 Watermill

A watermill is a structure that uses a water wheel or turbine to drive a mechanical process such as flour or lumber production, or metal shaping (rolling, grinding or wire drawing). The ancient Greeks and Romans were the first to have used water for powering mills. A watermill that generates electricity is usually called a hydroelectric plant.

13.6.3.1 Working Principle of Watermills

Typically, water is diverted from a river or impoundment or mill pond to a turbine or water wheel, along a channel or pipe (variously known as a flume, head race, mill race, leat, leet, lade (Scots), or penstock). The force of the water's movement drives the blades of a wheel or turbine, which in turn rotates an axle that drives the mill's other machinery. Water leaving the wheel or turbine is drained through a tail race, but this channel may also be the head race of yet another wheel, turbine, or mill. The passage of water is controlled by sluice gates that allow maintenance and some

13.6 Water Energy

measure of flood control; large mill complexes may have dozens of sluices controlling complicated interconnected races that feed multiple buildings and industrial processes.

13.6.3.2 Types

Watermills can be divided into two kinds:

(a) one with a horizontal waterwheel on a vertical axle
(b) the other with a vertical wheel on a horizontal axle.

The oldest of these were horizontal mills in which the force of the water, striking a simple paddle wheel set horizontally in line with the flow, turned a runner stone balanced on the rynd which is atop a shaft leading directly up from the wheel. The bedstone does not turn. The problem with this type of mill arose from the lack of gearing; the speed of the water directly set the maximum speed of the runner stone which, in turn, set the rate of milling.

Most watermills in Britain and the United States had a vertical waterwheel, one of three kinds: undershot, overshot, and breast-shot. The horizontal rotation is converted into the vertical rotation by means of gearing, which also enabled the runner stones to turn faster than the waterwheel.

13.6.4 Tide Mill

A different type of water mill is the tide mill. This mill might be of any kind, undershot, overshot, or horizontal but it does not employ a river for its power source. Instead a mole or causeway is built across the mouth of a small bay. At low tide, gates in the mole are opened allowing the bay to fill with the incoming tide. At high tide the gates are closed, trapping the water inside. At a certain point, a sluice gate in the mole can be opened allowing the draining water to drive a mill wheel or wheels. This is particularly effective in places where the tidal differential is very great, such as the Bay of Fundy in Canada where the tides can rise fifty feet, or the now derelict village of Tide Mills in the UK.

Other water mills can be set beneath large bridges where the flow of water between the stanchions is faster. At one point, London Bridge had so many water wheels beneath it that bargemen complained that passage through the bridge was impaired.

13.6.5 Exploring the Potentials of Water Power

Hydropower provides a developing economy with opportunities to develop appropriate technologies. One of the first things a country can do is to assess its opportunities for developing alternative energy sources. Hydropower in developing countries has the following distinguishing characteristics: sustainability, dependence on local resources, cost effectiveness, durability, flexibility, simplicity, ability to fit into existing systems, accessibility to isolated rural communities, and ability to meet multiple purposes.

The sustainability of hydropower arises from the fact that it uses a renewable source of energy – water. It is essentially nonpolluting. It is environmentally sound and acceptable. Hydropower makes maximum use of local resources and, thus, compared with thermal power, is usually much more appropriate for conditions in many developing countries, which face shortages of the foreign exchange required to import fuel oil. Hydropower is largely cost effective and is, to some extent, insulated from inflation. No fuel is required and heat is not involved, so operating costs are low. Approximately 650 kWh production by a hydropower plant will reduce the requirement for oil (or its fuel equivalent) by 1 bbl (1 bbl = about 0.16 m^3). Because of this and the durability of the facilities, a hydropower installation is to some extent inflation proof. Because no heat is involved, the equipment has a long life, and malfunctioning is uncommon. Dams and control works can perform for decades, and limited maintenance is required. Hydropower's reliability and flexibility of operation, including fast start-up and shutdown times in response to rapid changes in demand, makes it an especially valuable part of a large power system of a developing country. The relative simplicity of a small-scale, hydro-based enterprise makes energy instantly available. Small-scale hydropower fits nicely into the energy balance of a country. It can contribute to inter-regional equity by meeting the needs of isolated rural communities. It can be made available in small installations and with relative ease in remote areas of developing countries.

The strengths of water mills are that they make use of locally available materials and are accessible to poor households in remote and inaccessible areas. Water mills provide a striking but a rare case of a foreign technology that has been almost fully "indigenized" in rural area. The technology can be fitted nicely into the local farming system. Appropriate policy instruments should be designed to encourage the expansion of watermills in areas where water is available. Some of the measures that could be taken are (1) removing the taxes imposed on water mills, (2) establishing a water-mill-promotion project within the Rural Technology Promotion Department of the Ministry of Agriculture or Power, and (3) commissioning feasibility studies.

There is an opportunity of using irrigation pump discharge (water energy) to generate electricity. This micro-hydropower can be of the most valuable resources for electricity generation. Several researchers pointed out that application of pump instead of turbine could be an important alternative solution to turbines, presenting reasonable efficiencies (Ramos and Borga, 1999; Garey, 1990). Thus, there is an ample scope to generate electricity from irrigation pumps on a sustainable basis.

13.7 Bio-energy

Biomass is plant or animal matter. Using biomass (or fuels or wastes derived from biomass) as a source of energy entails burning it to yield heat that can then drive engines or generate electricity. The energy in biomass is chemical in nature; it

13.7 Bio-energy

does not suffer from the problem of intermittence that is inherent to wind and solar resources. In this respect, biomass more nearly resembles fossil fuels than it does other renewables. Indeed, geologists tell that fossil fuels are simply fossilized biomass.

For most of recorded history, biomass was mankind's principal energy source, mainly in the form of wood used for cooking and heating, and as foods to "fuel" human labor and beasts of burden. With the industrial revolution, fossil fuels captured this dominant role. Today, biomass still accounts for 15% of worldwide primary energy consumption, but, significantly, the fraction is much higher in developing nations than in developed ones.

Wastes generated by the forest products industry of East Texas include logging residues left behind after harvest as well as bark, wood chips, and sawdust generated at mills. In general, the wood wastes generated by modern mills are highly utilized; indeed, forest mills are the largest biomass energy users in the nation today, generating more than half of their large energy requirement on-site. Many mills, including currently five in Texas, generate electricity for local use or occasionally for resale to the grid.

Urban sources of biomass may represent some of the best opportunities for increasing biomass near-term presence in the energy mix. Wastes that would otherwise be landfilled are a particularly good potential fuel source, since the producer is charged a tipping fee for their disposal. Methane gas generated and captured at existing landfills or at municipal sewage treatment facilities is another important form of urban bio-energy. A final advantage of these wastes is that their supply is surprisingly reliable, much more than agricultural commodities that fluctuates annually with the vagaries of markets, weather, and government policy.

Plants use photosynthesis to grow and produce biomass. Also known as biomatter, biomass can be used directly as fuel or to produce liquid biofuel. Agriculturally produced biomass fuels, such as biodiesel, ethanol, and bagasse (often a by-product of sugar cane cultivation) can be burned in internal combustion engines or boilers. Typically, biofuel is burned to release its stored chemical energy. Research into more efficient methods of converting biofuels and other fuels into electricity, utilizing fuel cells is an area of very active work.

13.7.1 Liquid Biofuel

Liquid biofuel is usually either a bioalcohol such as ethanol fuel or a bio-oil such as biodiesel and straight vegetable oil. Biodiesel can be used in modern diesel vehicles with little or no modification to the engine and can be made from waste and virgin vegetable and animal oil and fats (lipids). Virgin vegetable oils can be used in modified diesel engines. In fact diesel engine was originally designed to run on vegetable oil rather than fossil fuel. A major benefit of biodiesel is lower emissions. The use of biodiesel reduces emission of carbon monoxide and other hydrocarbons by 20–40%.

In some areas corn, cornstalks, sugarbeets, sugar cane, and switchgrasses are grown specifically to produce ethanol (also known as grain alcohol), a liquid which can be used in internal combustion engines and fuel cells. Ethanol is being phased into the current energy infrastructure. E85 is a fuel composed of 85% ethanol and 15% gasoline that is sold to consumers. Biobutanol is being developed as an alternative to bioethanol. There is growing international criticism about biofuels from food crops with respect to issues such as food security, environmental impacts (deforestation), and energy balance.

According to the International Energy Agency, new biofuels technologies being developed today, notably cellulosic ethanol, could allow biofuels to play a much bigger role in the future than previously thought. Cellulosic ethanol can be made from plant matter composed primarily of inedible cellulose fibers that form the stems and branches of most plants. Crop residues (such as corn stalks, wheat straw and rice straw), wood waste, and municipal solid waste are potential sources of cellulosic biomass. Dedicated energy crops, such as switchgrass, are also promising cellulose sources that can be sustainably produced in many regions of the United States.

13.7.2 Biogas

Biogas can easily be produced from current waste streams, such as paper production, sugar production, sewage, animal waste, and so forth. These various waste streams have to be slurried together and allowed to naturally ferment, producing methane gas. This can be done by converting current sewage plants into biogas plants. When a biogas plant has extracted all the methane it can, the remains are sometimes better suitable as fertilizer than the original biomass.

13.8 Geothermal Energy

Geothermal energy is an energy obtained by tapping the heat of the earth itself, usually from kilometers deep into the Earth's crust. Ultimately, this energy derives from heat in the Earth's core. It is expensive to build a power station to harness this energy, but operating costs are low resulting in low energy costs for suitable sites. The International Energy Agency classifies geothermal power as renewable.

Three types of power plants are used to generate power from geothermal energy:

- dry steam
- flash
- binary

Dry steam plants take steam out of fractures in the ground and use it to directly drive a turbine that spins a generator. Flash plants take hot water (usually at temperatures over 200°C) out of the ground, and allows it to boil as it rises to the surface then separates the steam phase in steam/water separators and then runs the steam through a turbine. In binary plants, the hot water flows through heat exchangers, boiling an organic fluid that spins the turbine. The condensed steam and remaining geothermal fluid from all three types of plants are injected back into the hot rock to pick up more heat.

The geothermal energy from the core of the Earth is closer to the surface in some areas than in others. Where hot underground steam or water can be tapped and brought to the surface it may be used to generate electricity. Such geothermal power sources exist in certain geologically unstable parts of the world such as Chile, Iceland, New Zealand, United States, the Philippines, and Italy. The two most prominent areas for this in the United States are in the Yellowstone basin and in northern California. Iceland produced 170 MW geothermal power and heated 86% of all houses in the year 2000 through geothermal energy. Some 8,000 MW of capacity is operational in total.

There is also the potential to generate geothermal energy from hot dry rocks. Holes at least 3 km deep are drilled into the earth. Some of these holes pump water into the earth, while other holes pump hot water out. The heat resource consists of hot underground radiogenic granite rocks, which heat up when there is enough sediment between the rock and the earth's surface. Several companies in Australia are exploring this technology.

13.9 Modeling the Energy Requirement

Model can be used to design, simulate, and predict the performance of an irrigation system powered by a renewable energy source. The model predicts the energy demand based on the water demand for the irrigation system.

Different approaches may be used in predicting water demand. The water demand may be predicted using metrological data, and a statistical distribution may be used to predict the expected upper limits of water demand. The lower limits of available energy are predicted using the metrological data and also by stochastic methods. The amount of energy to be required is calculated based on the minimum irrigation interval demanded, and then the renewable energy system is designed to assure the desired reliability.

13.10 Factors Affecting Potential Use of Renewable Energy in Irrigation

The use of renewable energy-based pumps in irrigation depends on several factors:

- availability of the renewable energy
- cost of the system (pump and the energy harnessing system)
- groundwater requirement for irrigation and its availability
- alternate energy availability and its cost
- reliability of the renewable energy

- affordability of the farmers
- alternate use of renewable energy
- propensity of the user to invest in renewable energy
- promotional activity
- government subsidy

Availability of the renewable energy may vary considerably from one technological option to another, while the issues concerning the remaining factors are expected to be more or less common.

13.10.1 Groundwater Requirement and Its Availability

As surface water irrigation is usually the cheapest option for irrigation, farmers having access to sufficient surface water for irrigation may not choose any other options. Therefore, the areas in a country with surface water availability should not be included in estimating potential area for solar pumping. On the other hand, for groundwater use, the area can be considered where groundwater table is within specific depth. For example, a centrifugal pump coupled with a solar photovoltaic (SPV) array is considered for shallow well water pumping. The maximum suction head for such a pump is about 7 m. The SPV system with a submersible pump is capable of deep-well irrigation pumping up to about 65 m. But such deep-well irrigation pumping may not always be economical. Thus, the SPV water pumping for irrigation is limited in areas with groundwater table less than 10 m. Similarly, the water output of commonly available windmill pumps is low for a head of more than 20 m. Water table at the end of dry season (pre-monsoon period) should be used for comparison and interpretation.

13.10.2 Affordability of the User

The average annual cost associated with the renewable energy pumping system (installation, repair, and maintenance) should be comparable with the user's annual income from agriculture and should be within the affordable limit. As the capital cost is required at the initial time, a fraction of the capital cost may be provided as capital subsidy by the government.

13.10.3 Willingness of the User to Invest in a Renewable Energy Based Pump

The propensity of the user to invest in a renewable energy based pump depends on the awareness and knowledge of the user about the technology for water pumping and also on the availability, reliability, and easiness of conventional options to him.

13.10.4 Availability of Alternate Energy for Irrigation and Its Cost

Availability of alternate energy for irrigation and its cost is a major factor influencing the use of renewable energy.

13.10.5 Alternate Use of Renewable Energy

If there is alternate use of renewable energy from where the farmers may achieve higher return, will be easily motivated to that direction.

13.11 Renewable Energy Commercialization: Problems and Prospects

The need for community support for renewable energy is clear. Several of the technologies, especially wind energy, but also small-scale hydro power, energy from biomass, and solar thermal applications, are economically viable and competitive. The others, especially photovoltaic (silicon module panels directly generating electricity from the sun's light rather than heat), depend only on (how rapidly) increasing demand and thus production volume to achieve the economy of scale necessary for competitiveness with central generation. In fact, looking at the various sector markets in early 2003, it is probably not over-optimistic to conclude that the lion's share of remaining market resistance to renewables penetration relates to factors other than economic viability. This should be seen against the rapidly improving fiscal and economic environment being created in the EU both by European legislation itself swinging into full implementation and the Member States' own programs and support measures, which despite the short-term macro-economic background, are accelerating rapidly.

The problems (constraints) associated with the renewable energy commercialization and popularization are

- cost
- market size
- fluctuating energy supply
- aesthetics
- environmental and social considerations
- land area required
- incapability of generating large amount of energy
- reliability
- longevity issues
- sustainability
- transmission

13.11.1 Problems

13.11.1.1 Costs

Still now, cost of renewable energy is higher than the other competing energy sources. Renewable energy systems encompass a broad, diverse array of technologies, and the current status of these can vary considerably. Some technologies are already mature and economically competitive (e.g., geothermal and hydropower), others need additional development to become competitive without subsidies.

The developed world can make more research investments to find better cost-efficient technologies, and manufacturing could be transferred to developing countries in order to use low labor costs. The renewable energy market could increase fast enough to replace and initiate the decline of fossil fuel dominance and the world could then avert the looming climate and peak oil crises. Fund from the Government is needed for research in renewable technology, to make the production cheaper and generation more efficient. Another mechanism is imposition of fossil fuel consumption and carbon taxes, and channel the revenue earned toward renewable energy development.

13.11.1.2 Availability and Reliability

Availability and reliability of the renewable energy is a great problem for wide acceptance to the customers. The challenge of variable power supply may be readily alleviated by energy storage. Available storage options include pumped-storage hydro systems, batteries, hydrogen fuel cells, and thermal mass. Initial investments in such energy storage systems may be high, although the costs can be recovered over the life of the system.

Wave energy and some other renewables are continuously available. A wave energy scheme installed in Australia generates electricity with an 80% availability factor.

13.11.1.3 Aesthetics

Both solar and wind generating stations have been criticized from an aesthetic point of view. However, methods and opportunities exist to deploy these renewable technologies efficiently and unobtrusively: fixed solar collectors can double as noise barriers along highways, and extensive roadway, parking lot, and roof-top area is currently available; amorphous photovoltaic cells can also be used to tint windows and produce energy. Advocates of renewable energy also argue that current infrastructure is less aesthetically pleasing than alternatives, but cited further from the view of most critics.

13.11.1.4 Environmental and Social Considerations

While most renewable energy sources do not produce pollution directly, the materials, industrial processes, and construction equipment used to create them may generate waste and pollution. Some renewable energy systems actually create environmental problems. For instance, older wind turbines can be hazardous to flying birds.

13.11.1.5 Land Area Required

Another environmental issue, particularly with biomass and biofuels, is the large amount of land required to harvest energy, which otherwise could be used for other purposes or left as undeveloped land. However, it should be pointed out that these fuels may reduce the need for harvesting nonrenewable energy sources, such as vast strip-mined areas and slag mountains for coal, safety zones around nuclear plants, and hundreds of square miles being strip-mined for oil sands. These responses, however, do not account for the extremely high biodiversity and endemism of land used for ethanol crops, particularly sugar cane.

In the United States, crops grown for biofuels are the most land- and water-intensive of the renewable energy sources. In 2005, about 12% of the nation's corn crop (covering 11 million acres (45,000 km^2) of farmland) was used to produce four billion gallons of ethanol – which equates to about 2% of annual US gasoline consumption. For biofuels to make a much larger contribution to the energy economy, the industry will have to accelerate the development of new feedstocks, agricultural practices, and technologies that are more land and water efficient. Already, the efficiency of biofuels production has increased significantly and there are new methods to boost biofuel production.

13.11.1.6 Incapability of Generating Large Amount of Energy

In most cases of renewable energies (except hydroelectric system), the system is not capable of producing large amount of energy. For this reason, it is a problem to transmit the generated energy, and the cost becomes high.

13.11.1.7 Longevity Issues

Though a source of renewable energy may last for billions of years, renewable energy infrastructure, like hydroelectric dams, will not last forever, and must be removed and replaced at some point. Events like the shifting of riverbeds, or changing weather patterns could potentially alter or even halt the function of hydroelectric dams; lowering the amount of time they are available to generate electricity.

Although geothermal sites are capable of providing heat for many decades, eventually specific locations may cool down. It is likely that in these locations, the system was designed too large for the site, since there is only so much energy that can be stored and replenished in a given volume of earth.

13.11.1.8 Sustainability

Renewable energy sources are generally sustainable in the sense that they cannot "run out" as well as in the sense that their environmental and social impacts are generally more benign than those of fossil. However, both biomass and geothermal energy require wise management if they are to be used in a sustainable manner. For all of the other renewables, almost any realistic rate of use would be likely to approach their rate of replenishment by nature.

13.11.1.9 Transmission

If renewable and distributed generation were to become widespread, electric power transmission and electricity distribution systems might no longer be the main distributors of electrical energy but would operate to balance the electricity needs of local communities. Those with surplus energy would sell to areas needing "top ups." That is, network operation would require a shift from "passive management" – where generators are hooked up and the system is operated to get electricity "downstream" to the consumer – to "active management," wherein generators are spread across a network and inputs and outputs need to be constantly monitored to ensure proper balancing occurs within the system. Some governments and regulators are moving to address this, though much remains to be done. One potential solution is the increased use of active management of electricity transmission and distribution networks. This will require significant changes in the way that such networks are operated.

However, on a smaller scale, use of renewable energy produced on site reduces burdens on electricity distribution systems. Current systems, while rarely economically efficient, have shown that an average household with an appropriately sized solar panel array and energy storage system needs electricity from outside sources for only a few hours per week. By matching electricity supply to end-use needs, advocates of renewable energy and the soft energy path believe electricity systems will become smaller and easier to manage, rather than the opposite.

13.11.2 Prospects/Future Potentials

In view of the increasing global climate concerns, interests in the development and dissemination of renewable energy technologies have been revived/renewed. Climate change concerns coupled with high oil prices, peak oil, and increasing government support are driving increasing renewable energy legislation, incentives, and commercialization. European Union leaders reached an agreement in principle in March 2007 that 20% of their nations' energy should be produced from renewable fuels by 2020, as part of its drive to cut emissions of carbon dioxide, blamed in part for global warming. Investment capital flowing into renewable energy climbed from $80 billion in 2005 to a record $100 billion in 2006. This level of investment combined with continuing increases each year has moved what once was considered alternative energy to mainstream. Wind was the first to provide 1% of electricity, but solar is not far behind. Some very large corporations such as BP, General Electric, Sharp, and Royal Dutch Shell are investing in the renewable energy sector.

13.11 Renewable Energy Commercialization: Problems and Prospects

Table 13.1 Typical renewable energy cost (WEC, 2004)

	2001 energy cost	Potential future cost
Electricity		
Wind	4–8 ¢/kWh	3–10 ¢/kWh
Solar photovoltaic	25–160 ¢/kWh	5–25 ¢/kWh
Solar thermal	12–34 ¢/kWh	4–20 ¢/kWh
Large hydropower	2–10 ¢/kWh	2–10 ¢/kWh
Small hydropower	2–12 ¢/kWh	2–10 ¢/kWh
Geothermal	2–10 ¢/kWh	1–8 ¢/kWh
Biomass	3–12 ¢/kWh	4–10 ¢/kWh
Coal (comparison)	4 ¢/kWh	
Heat		
Geothermal heat	0.5–5 ¢/kWh	0.5–5 ¢/kWh
Biomass – heat	1–6 ¢/kWh	1–5 ¢/kWh
Low temp solar heat	2–25 ¢/kWh	2–10 ¢/kWh

Note: All costs are in 2001 US$-cent per kilowatt-hour.

Sustainable development and global warming groups propose a 100% renewable energy source supply, without fossil fuels and nuclear power.

Renewable energy sources are becoming cheap (Table 13.1) and convenient enough to place it, in many cases, within reach of the average consumer. By contrast, the market for renewable heat is mostly inaccessible to domestic consumers due to inconvenience of supply, and high capital costs. Solutions such as geothermal heat pumps may be more widely applicable, but may not be economical in all cases. Proponents advocate the use of "appropriate renewables," also known as soft energy technologies, as these have many advantages.

Renewable energy technologies are sometimes criticized for being intermittent or unsightly, yet the market is growing for many forms of renewable energy. Wind power is growing at the rate of 30% per annum, has a worldwide installed capacity of over 100 GW, and is widely used in several European countries and the USA. The manufacturing output of the photovoltaic industries reached more than 2,000 MW in 2006, and PV power plants are particularly popular in Germany. Solar thermal power stations operate in the USA and Spain, and the largest of these is the 354 MW SEGS power plant in the Mojave Desert. The world's largest geothermal power installation is The Geysers in California, with a rated capacity of 750 MW. Brazil has one of the largest renewable energy programs in the world, involving production of ethanol fuel from sugar cane, and ethanol now provides 18% of the country's automotive fuel. Ethanol fuel is also widely available in the USA.

While there are many large-scale renewable energy projects and production, renewable technologies are also suited to small off-grid applications, sometimes in rural and remote areas, where energy is often crucial in human development. Kenya has the world's highest household solar ownership rate with roughly 30,000 small (20–100 W) solar power systems sold per year.

As of April 2008, worldwide wind farm capacity was 100,000 MW, and wind power produced some 1.3% of global electricity consumption, accounting for approximately 18% of electricity use in Denmark, 9% in Spain, and 7% in Germany. The United States is an important growth area and latest American Wind Energy Association figures show that installed US wind power capacity has reached 18,302 MW, which is enough to serve 5 million average households.

The renewable market will boom when cost efficiency attains parity with other competing energy sources. Other than market forces, renewable industry often needs government sponsorship to help generate enough momentum in the market. Many countries and states have implemented incentives – like government tax subsidies, partial co-payment schemes, and various rebates over purchase of renewables – to encourage consumers to shift to renewable energy sources.

Initiative is needed such as development of loan programs that stimulate renewable favoring market forces with attractive return rates, buffer initial deployment costs, and entice consumers to consider and purchase renewable technology. An example is the solar loan program sponsored by UNEP helping 100,000 people finance solar power systems in India. Success in India's solar program has led to similar projects in other parts of developing world like Tunisia, Morocco, Indonesia, and Mexico.

Also oil peak and world petroleum crisis and inflation are helping to promote renewables. Most importantly, renewables is gaining credence among private investors as having the potential to grow into the next big industry. Many companies and venture capitalists are investing in photovoltaic development and manufacturing.

13.11.3 Challenges and Needs

At present, the renewable energy technologies are not reachable to the doorstep of general people; neither in terms of economic competitiveness nor in technical easiness. A breakthrough of technology is required in manufacturing renewable technologies (i.e., solar photovoltaic cells, wind turbine) at low lost to compete with other nonrenewable energy sources. Development of such a technology is a challenge to the researchers. Real breakthrough of creative ideas/technologies is needed to develop batteries at low cost/affordable prices to store electricity from renewable sources, as these energies are intermittent in nature. Without breakthrough of such technology, it is a challenge to go ahead and achieve a success with the renewable energy technologies.

Relevant Journals

– Energy
– Solar Energy
– Renewable Energy

- Renewable & Sustainable Energy Reviews
- Wind Engineering
- Journal of Applied Meteorology
- Agricultural & Forest Meteorology
- Biomass Bioenergy
- Resource and Energy Economics (Elsevier)
- Energy Sources (Taylor & Francis)
- Applied Energy
- Energy Policy
- Energy Conversion and Management
- Power System Research
- IEEE Trans. On Power Systems
- Trans. ASAE on Energy Conversion
- Journal of Hydrometeorology
- Journal of Climate & Applied Meteorology
- Bioresource Technology
- Journal of Wind Engineering and Industrial Aerodynamics
- Irrigation Science
- Natural Resources Research
- Sustainable Development

Questions

(1) What do you mean by renewable energy? Why renewable energy is important at present day?
(2) How renewable energy can be used in irrigation water pumping?
(3) How you can assess the potential solar energy in your area/locality?
(4) Briefly explain how the electricity is produced from photovoltaic (PV) cells in the presence of solar irradiance.
(5) What are the different types of photovoltaic system?
(6) Write down the steps (along with formula) to calculate the requirement of PV cell for an irrigation scheme.
(7) Analyze the economic PV technology in your locality.
(8) Analyze the potentiality of wind energy as an energy resource in your area.
(9) What are the different types of wind machines?
(10) Narrate the criteria for selecting site for windmill.
(11) Briefly explain the mechanism of electricity production from wind energy.
(12) What is the mechanism of operating wind turbine/machine?
(13) Write down the equation for calculating wind power, and define the parameters.
(14) Write short note on the following:
   Wave energy, tide mill, bio-energy, liquid biofuel, geothermal energy.
(15) Describe the working principle of watermill.

(16) What are the advantages of hydropower-based electricity production?
(17) Discuss the factors affecting potential use of renewable energy in irrigation.
(18) What are the problems and prospects of widespread use (Commercialization) of renewable energy in your province/state/country?
(19) What are the challenges the renewable energy facing now?
(20) How much energy is required to lift one ha-m of water 40 m up a hill?
(21) The peak period water demand in a crop field is 0.02 m^3/s and total head requirement is 4 m. The minimum solar irradiance at the site is 5 MJ/m^2/day. Find out the required number of PV array to meet the power for irrigation. Given, the area of each array is 2 m^2, efficiency of the solar cells is 20%, efficiency of the generator is 60, and average bright sunshine hour during the period is 7 h.
(22) Determine the potentials of electricity generation from wind energy per wind turbine having average wind speed 25 km/h, area swept by the rotating blades is 5 m^2, and turbine efficiency of 55%.
(23) Determine the percentage increase in electricity production from wind energy in an area if the efficiency of the wind turbine increased from 35 to 55%.
(24) In an area, the average wind speed is 30 km/h. The area swept by the rotating blades of the wind turbine is 10 m^2. Determine the wind power to be generated at that site.

References

Bhuiyan MMH, Asgar MA, Mazumder RK, Hussain M (2000) Economic evaluation of a stand-alone residential photovoltaic power system in Bangladesh. Renewable Energy 21:403–410

Garey PN (1990) Using pumps as hydro-turbines. Hydro Review 52–61

Garg HP (1987) Advances in solar energy technology, vol 3. Reidel Publishing, Boston, MA

IEO (2009) International energy outlook 2009. Highlights section. http://www.eia.doe.gov/oiaf/ieo/highlights.html. Accessed on 15 Apr 2010.

Kumar A, Kandpal TC (2007) Renewable energy technologies for irrigation water pumping in India: a preliminary attempt towards potential estimation. Energy 32:861–870

Ramos H, Borga A (1999) Pumps as turbine: an unconventional solution to energy production. Urban Water 1:261–263

UNDP (2000). World energy assessment: Energy and the challenges of sustainability. United Nations development program, 508p.

UNESCAP (1991) Solar powered pumping in Asia and the Pacific. UNESCAP, Bangkok, Thailand, The United nations Building, Rajadamnern Nok Aveneue

WEA (World Energy Assessment) (2001). Renewable energy technologies, Chapter 7

WEC (World Energy Council) (2004) World energy assessment – overview: 2004 update. http:www.undp.org/energy/weaover2004.htm

Subject Index

Note: The letters 'f' and 't' following the locators refer to figures and tables respectively.

A
Abstract models, 382
Adaptation alternatives to climate change, 186–188
Adaptive management, 232
Adsorption, 371
Advance function, 90
Advance rate, 89
Advance ratio, 69
Advection coefficient, 258
Affinity laws, 446–447
    effect of change of diameter, 447–448
    effect of change of speed, 446–447
Agricultural pollution, extent of, 243–250
    agricultural pollutants, impact of, 248–250
        eutrophication impact, 248–249
        pesticides and chemicals, impact of, 249–250
        reduction in biodiversity, 249
    factors affecting solute contamination, 244–247
        crop root length and density, 245
        crop rotation, 246
        crop type and growth rate, 245
        drier condition and fertilizer residue, 245
        excess use of fertilizer, 245
        inconsistence with irrigation water, 245
        irrigation or rainfall, 246
        irrigation schedule and fertilizer type, 245
        management factors, 246
        nutrient retention in soil, 246f
        soil type and organic matter content, 245–246
    hazard of nitrate (NO_3–N) pollution, 248
    major pollutant ions, 243–244
    mode of pollution by nitrate and pesticides, 247–248
        fate of pesticide residues in environment, 247f
Agro-ecological zone (AEZ), 428
Aguçadora Wave Park, 504
AGWA (Automated Geospatial Watershed Assessment Tool), 227
Air density, 500, 502
Airfoil-shaped blade, 497
Air-lift pumps, 458, 459f
    advantages and disadvantages, 458
    principle, 458
Alazba's empirical model, 72
Alkalinity, 283
Alternate furrow irrigation, 153, 265, 311
Alternate wetting and drying, 153
Amelioration of saline soil
    principles/approaches of salinity management, 302
    salinity management options, 302–316
        biological reduction of salts, 311
        chemical practices, 308–309
        control of saline water, 302–303
        developing/cultivating salt tolerant crops, 313
        engineering practices, 302–308
        increasing water use of annual crops and pastures, 314
        irrigation and water management practices, 309–311
        other management options, 311–313

Amelioration of saline soil (*cont.*)
    policy formulation, 314
    removing surface salts/scraping, 302
American Water Works Association, 153
American Wind Energy Association, 516
Amorphous photovoltaic cells, 512
Analysis of breakthrough curve, 260
Analytical models, 383
Analytical solution, 387
Angstrom coefficients, 395
Animal-powered devices, 436
Anti-transpirants, 157
Apiculture/honeybee, hazards to, 249
Apparent diffusion coefficient or diffusion-dispersion coefficient, 253
Application efficiency, 69, 100, 112–113, 113t, 123, 130, 133
Application rate, 100, 106
Applications of GIS, 425–427
    agriculture, 425–426
    natural resources management, 426
    urban and rural planning, 426
APSIM (Agricultural Production Systems sIMulator), 406
Aquifer property, 258
Arsenic contamination, 167, 188
    See also Groundwater
Artificial drainage, 303
    subsurface drainage
        buried pipe drain, 332, 333f
        deep open drain, 331, 332f
        evaluation system, 372–373
        importance of evaluation, 372
        typical length of different pipe drain, 332
    surface drainage
        land forming, 331
        open field drains, 331
        surface drainage, 331
Artificial recharge, 162, 167, 302–303, 307
ASCE, 112, 174
Aspect ratio, 80
Attribute data, 428
Average application depth, 101
Average velocity, 251, 258, 341, 500
Axial-flow pumps, 440, 456, 463

B
BASCAD, 84
Basin irrigation, 37, 40–43, 41f
    concept and characteristics, 40–41
    evaluation, 130
    suitabilities and limitations, 41–43
        advantage and disadvantage, 43
        attainable efficiencies, 42
        cost and economic factor, 43
        crop, 41
        labor and energy requirement, 42–43
        required depth of irrigation application, 42
        soil and topography, 42
        water quantity, 42
Basin irrigation design, 79–84
    existing models, 84
        BASCAD, 84
        COBASIM, 84
    factors affecting, 79–81
        aspect ratio, 80
        basin longitudinal slope, 80
        elevation difference between adjacent basins, 81
        flow rate, 80
        irrigation depth, 81
        local surface micro-topography, 80–81
        number of check bank outlets, 81
        soil type, 80
    hydraulics, 81–82
    simulation modeling, 82–84
        hydrodynamic model, 82–83
        other approaches, 83–84
        zero-inertia model, 83
    water flow pattern, 82f
Basin longitudinal slope, 80
Basin sustainable yield, 163–164
BC2C model, 320
Benefits from drainage, 330
Bernoulli's equation/energy equation, 23
Bias (error) or mean bias, 392
Binary plants, 509
Bioalcohol, 507
Biobutanol, 508
Biodiesel, 483, 507
Bio-energy, 506–507
Biofuel, 481, 507–508, 513
Biogas, 508
Biological reduction of salts, 311
Biomass energy, 481
Biomatter, *see* Biomass energy
Bio-oil, 507
Black alkali soil, 281
Black box models, *see* Empirical models (or black box models)
18-Bladed rotor, 496
Blending, 161, 310
Blue baby syndrome, 248
Blue energy, 503

Subject Index

Booster pumps, 441, 452, 462
Border irrigation, 37–40, 38f
    concept and features, 38–39
    evaluation, 128–129
    suitabilities/capabilities/limitations, 39–40
        advantages and disadvantages, 40
        attainable irrigation efficiency, 39
        crop suitability, 39
        economy and financial involvement, 39
        labor requirement, 40
        soil and land suitability, 39
Border irrigation system design, 68–79
    design approaches and procedures, 71–72
        empirical models, 71–72
    design parameters, 70
    examples, 73–76
    factors affecting, 70
        border inflows, 70
        irrigation depth, 70
        longitudinal slope, 70
        soil type and infiltration characteristics, 70
    general guidelines, 79
    simulation modeling, 76–77
    software tools/models, 77–78
        BORDEV, 77
        SADREG, 78
        SIRMOD, 77–78
        WinSRFR, 78
    typical border parameters, 79t
Border strip, 68
Boundary condition, 386, 398, 401
Boussinesq equation, 369
Brake horse power (BHP), 445–447
Breakthroughs, 516
Breakthrough curve (BTC), 259, 259f
Breakthrough experiment, 259–260
Breeding program, 296, 311, 313
Broad bed and furrow (BBF) system, 235
Brundtland Commission, 173–174
Buckingham-Darcy law, 257
Building block method (BBM), 149–150

C

Calcareous soil, 317
Calibration, 389–391
Canal lining
    benefits of, 8–9
    decision on, 9
    definition, 10
    materials, 9–10
Capacity factor, 493
Capillary irrigation, 55

Cation exchange capacity (CEC), 281, 283, 287–288
Cation exchange reaction
    reclamation by acid formers or acids, 317–318
        elemental S and sulfuric acid, 317
        iron sulfate, 317–318
    reclamation by gypsum, 317
Cavitation
    in axial-flow pumps, 463
    NPSHa, 449
    NPSHr, 449
    process of, 449
    pump cavitation, definition, 449
    in radial flow and mixed flow pumps, 463
Cellulosic ethanol, 508
Center pivot sprinkler, 49–50, 50f
Centrifugal force, 438, 440, 442, 455–456
Centrifugal pumps
    affinity laws, 446–448
    cavitation, 449
    features and principles, 442–443
    priming of, 448–449
    pump efficiency, 445
    specific speed, 443t, 446
    terminologies, 443–445
Challenges in water resources management
    quality degradation due to continuous groundwater pumping, 188
    risk and uncertainties, 188
    WT lowering and increase in pumping cost, 189
Channel bed, 10
Channel density, 121, 135t
Channel design, 11–12, 18–21
Channel geometry, 11
Check-basin irrigation, 41
Chemical (and physical) treatment of drainage, 370–371
    adsorption, 371
    distillation, 371
    particle removal, 370
Chemigation, 51, 54
Chezy's equation, 12–13
Christiansen's uniformity coefficient, 114–115
Classification of pumps, 437–442
    based on means by which energy is added, 441–442
        positive displacement (PD) pumps, 441
        rotodynamic pump, 441
    based on mode of intake of fluid to pumps, 438–439
        force mode pump, 439

Classification of pumps (*cont.*)
    suction mode pump, 438
    based on position of prime mover, 439
        submersible pumps, 439
        surface-mounted pumps, 439
    based on principle by which energy is added to fluid, 440–441
        displacement pumps, 441
        dynamic pumps, 440
    based on type of water use (or field of use), 439
        drum pumps, 439
        sampling pumps, 439
        sump pumps, 439
        wastewater pumps, 439
        water pumps, 439
        well pumps, 439
Climate change
    addressing, 238
        groundwater focus, 238
    concerns, 514
    impact on water resource management, 185–188
Clip-on meter, 128
Cloud seeding, 168
Coastal flooding, 186
COBASIM (contour basin simulation model), 84
Coefficient of determination (CD), 393
Coefficient of residual mass (CRM), 392–393
Cold spell, 186
Colebrook-White equation, 28
Collector drain, 330, 332
Colorado State University Irrigation and Drainage (CSUID) model, 369
Community-based approach to watershed management, 231–234
    challenges associated, 233
    characteristics, 231–232
Compartmentalized ponds, 155
Competition in water resource, 141
Components of head loss, 26–27
Components of watershed management, 228
Computer models
    definition, 386
    LEACHC, 322
    SALTMOD, 322
    SGMP, 322
Conceptual model, 382
    abstract models, 382
    features, 382
Confidence, 388
Conflicts in water resources management
    analysis and causes of, 184–185
    in integrated water resources management process
        economic conflicts, 179–180
        legal conflict, 180
        perspective, 180
        social conflicts, 179
    scales of
        global scale, 181–182
        regional scale, 182–184
        upstream and downstream relationship, 184
Conservation of energy, 21
Conservation of matter, 21
Conservation of momentum, 22
Constraint, 386
Contaminant/contamination, 244
Continuity equation/conservation of mass, 22–23
Contour area method (CAM), 211–212, 212f
Controlled drainage system, 361–362
    management of, 361
    principle of method, 361
Control variable, 386
Convection, 251
Convection-dispersion equation (CDE), 254, 260, 268, 402
Conventional fuel, 483, 488, 493, 501, 510
Conversion of wind power to electricity, 496
Conveyance efficiency, 2, 120–122
Conveyance loss, 2–10
Conveyance system, *see* Water conveyance loss and conveyance system designing
Cracking clay soil, model for flow estimation in, 397–402
    FLOCR, 398–402
    MACRO, 402
Criteria for selecting envelope material
    design criteria by SCS (1971), 366
    design criteria for granular mineral envelopes, 367
    design criteria for synthetic fiber envelope, 367
Crop and water productivity/acreage
    description, 124–125
    performance, 119
CropET$_0$ model, 396
Crop production function/yield model
    basic considerations in, 407
    definition of, 406
        wheat yield *vs.* applied water, 407f
    development of, 405–406
    existing yield functions/models, 408–411

Jensen model, 410–411
Stewart and Hagan model, 409–410
Stewart model, 409–410
importance of, 406–407
limitations/drawbacks of, 411
pattern of, 407
Crop productivity, 125
Crop rotation, 158, 246
CropSyst (Cropping Systems Simulation Model), 321, 403–405
Crop tolerance to soil salinity and effect of salinity on yield
examples, 300
factors influencing tolerance to crop, 295–297
environmental factors, 297
growth stage, 296
rootstocks and salinity tolerance, 297
soil moisture, 297
soil type, 297
varietal differences in salt tolerance, 296
relative salt tolerance of crops, 298
crop selection, 298
sensitive crops, 298
use of saline water for crop production, 298–299
yield reduction due to salinity, 299–300
CROPWAT, 393–396, 402–403
CLIMATE, 403
CROPS, 403
ETO, 403
RAINFALL, 403
SOIL, 403
Crop yield estimation, 430
Cross slope, 204
2CSalt model, 320
CSUID model, *see* Colorado State University Irrigation and Drainage (CSUID) model
Curve-fitting method, 260
Cutback, 68
Cutback ratio, 69
Cutoff ratio or advance ratio, 89
Cutoff time, 68–69
Cutting and filling, 207
Cyclone, 186

D

Damless hydro systems, 503
Darcy's equation, 342, 345
Darcy-Weisbach formula, 27–28
Databases for GIS, 428

Data processing, 131
DC powered pumps, 441
Decision support model, 405–406
APSIM, 406
DSSAT, 405–406
Deep percolation fraction, 124
Deep percolation ratio, 124
Deep-well turbine pump, 440, 450–451, 450f
multistage, 440
single stage, 440
Deficit irrigation, 151, 154, 266, 404
Delivery head (DH), 445
Delivery ratio, 221
Demand and supply of water, 144–150
Demand estimation
for domestic/industrial/commercial uses, 145
in-stream demand, 146
irrigation demand, 145–146
non-irrigated evaporative demand, 146
Demand management
approaches of, 150–154
in Bangladesh urban/agricultural sector, 161
concept, 150
obstacles in implementing, 159–161
reducing evaporation, 154–159
Department of Environment and Resource Management, The State of Queensland (Australia), 320
Depth area method (DAM), 209
Design approaches, 71–72
Design parameters, 70
Design surface runoff, estimation of, 349
rational method or SCS method, 349
runoff coefficient, 349
Design variables, 91–92
Deterministic model, 384–385
Deterministic time series model, 385
Diffuser pumps, 440, 448, 457, 460
Diffuse source, 242–243
Diffusion, 251–252
Digitizers, 429
Dimensionless time, 259
Discharge area, 276
Discharge pipe, 436, 450, 475
Discharge velocity
definition, 341
See also Velocity of flow in porous media
Dispersion, 251, 253, 293, 295
coefficient, 253–258

Dispersivity, 258
  length, 259
Displacement pumps, 440–441, 462
Distillation, 371
Distribution efficiency or uniformity, 115, 119
Distribution uniformity (DU), 69, 71, 88, 96, 101, 130, 132–134
Donnan's formula, 354
Drainable pore space, 336
  in different textured soils, 337, 337t
Drainable pore volume
  under negative pressure (or suction), 338
Drainable porosity, *see* Drainable pore space
Drainable water, 337–338
  water-filled pores with depth to water table, 338f
Drainage, 307–307
  artificial recharge of rainwater to aquifer through recharge well, 307
  channel, *see* Hydraulic design of surface design
  harvesting rainwater at farm pond or canal, 308
  mole drainage, 307
  requirement, 341, 359
  selection/installation, 371
  subsurface drainage, 307
  surface drainage, 307
  vertical drainage, 307
Drainage design and management, models in
  CSUID model, 369
  DRAINMOD, 368
  EnDrain, 369
Drainage discharge management
  disposal options, 369
  treatment of drainage water, 370–371
Drainage of agricultural lands
  concepts and benefits of drainage
    benefits from drainage, 329–330
    concepts, 329
    effects of poor drainage on soils and plants, 329
    goal and purpose of drainage, 329
    irrigation/drainage channel, difference between, 334
    merits/demerits of deep open/buried pipe drains, 333
    types of drainage, 330–332
  design of subsurface drainage system
    controlled drainage system and interceptor drain, 361–362
    data requirement for, 357
    examples, 362–365
    factors affecting spacing/depth of, 356–357
    layout of subsurface drainage, 357
    principles/steps/considerations in, 358–360
  design of surface drainage system
    design considerations/layout of, 349
    estimation of design surface runoff, 349
    examples, 350
    hydraulic design of surface drain, 349–350
  drainage discharge management
    disposal options, 369
    treatment of drainage water, 370–371
  economic considerations, 371
  envelope materials
    around subsurface drain, 365
    design of drain envelope, 366–367
    drain excavation and envelope placement, 368
    materials for envelope, designing, 365
    use of particle size distribution curve, 367
  equations/models for subsurface drainage design
    formula for irregular drain system, 355
    steady-state formula for parallel drain spacing, 351–354
  models in drainage design and management
    CSUID model, 369
    DRAINMOD, 368
    EnDrain, 369
  performance evaluation of subsurface drainage
    evaluation system, 372–373
    importance of evaluation, 371
  physics of land drainage
    examples, 340
    soil pore space and soil-water retention behavior, 334
    terminologies, 335–338
    water balance in drained soil, 338–340
  water movement theory
    examples, 347–349
    functional form of water-table position, 346
    Laplace's equation for groundwater flow, 345
    resultant or equivalent hydraulic conductivity of layered soil, 342–345
    terminologies, 341–342

Subject Index

theory of groundwater flow toward drain, 346–347
velocity of flow in porous media, 341
Drain envelope, design of, 366–367
   envelope material selection, 365–366
      design criteria by SCS (1971), 366–367
      design criteria for granular mineral envelopes, 367
      design criteria for synthetic fiber envelope, 366
   envelope thickness, 367
   steps, 366
DRAINMOD, 368
Drip irrigation, 38, 52–54, 52f, 309
   concept and features, 52
   suitabilities/capabilities/limitations, 53–54
      advantages and disadvantages, 53–54
      attainable efficiency, 53
      crop suitability, 53
      fertilizer application, 53
      utilities of buried drip system, 53
      water supply, 53
Drip/micro-irrigation evaluation, 134
Driving variable, 386
Drought, 186–187
Drum pumps, 439
Dryland salinity, 282
Dry steam plants, 508
DSSAT (Decision Support System for Agrotechnology Transfer), 405
D_{10}-uniformity co-efficient (Cu), 367
D_{60}-uniformity co-efficient (Cu), 367
Dupuit-Forcheimer's assumption for unconfined flow, 346f, 347
Duration of irrigation, 100–101
Duty of discharge, 124–125
Dynamic model, 384
Dynamic pressure, 24
Dynamic pumps
   centrifugal pumps, 440
      axial-flow pumps, 440
      design and application, 440
      mixed-flow pumps, 440
      radial-flow pumps, 440
Dynamic system, 387

E

E85, 508
Earth Summit, 175
Earth work volume estimation, 208–212
   contour area method (CAM), 211–212, 212f
   cut and fill on site, 208f
   depth area method (DAM), 209
   end area method (EAM), 210–211, 211f
   grid method, 209–210, 210f
   prismoidal formula (PF), 211
Economic conflicts, 179–180
Economic cross-section, 2
Economic good, water as, 142–143
Economic value of water, 182
Effect of change of diameter, 447–448
Effect of change of speed, 446
Efficiency, irrigation, 112–116
Efficiency of generator, 500
Electrical conductivity (EC), 279, 282
   value interpretation based on, 290t
Electric meter, 128
Elements of watershed, 197
Emission uniformity, *see* Low-quarter distribution uniformity
Empirical models (or black box models), 381–382
Empirical result/solution, 387
End area method (EAM), 210–211, 211f
EnDrain, 369
Energy, *see* Renewable energy resources
Energy consumption, 480, 482
Energy demand, 481–483
Energy-dependent products, 480
Energy equation, 23
Energy grade line, 25
Energy use efficiency, 122
Engineering indicators, 118–122
Envelope materials
   around subsurface drain, 365
   design of drain envelope, 366–367
      envelope thickness, 367
      material, selection, 366–367
      steps, 366
   drain excavation and envelope placement, 368
   materials for envelope, 365
      in-organic/mineral envelopes, 366
      man-made/synthetic envelopes, 366
      organic envelopes, 366
   need of proper designing of, 365
   use of particle size distribution curve, 367
Envelope thickness, 367
ENVIRO-GRO, 323
Environment, 244
Environmental flow (EF), 146–147, 149
   concept of, 148–149
   methods of
      building block method, 149
      drift method, 149

Environmental pollution, 244, 480, 482
Environmental protection, 178
Environment and water, 141
Equations/models for subsurface drainage design
　formula for irregular drain system, 355
　　schematic of flow to single drain, 355f
　steady-state formula for parallel drain spacing, 351–354
　　Donnan's formula, 354
　　Hooghoudt's equation, 351–354
Equipotential line, 342
Equity of water delivery (EWD), 122
Equivalent (or resultant) hydraulic conductivity of layered soil, 342–345, 344f
　of horizontal and vertical direction, 345
　horizontal hydraulic conductivity, 343–344, 343f
　vertical hydraulic conductivity, 344
Estimating water demand, 430
Estimation of solute load, 261–264
　determination of solute concentration, 262–264
　　examples, 263–264
　　solute transport study by tracer, 263
　　sampling from controlled lysimeter box, 261
　　sampling from crop field, 261–262
　　water sampling equipments, 262f
Ethanol fuel, 507–508, 515
European Union (EU), 503, 511, 514
Eutrophication, 248–249
Evaluation, see Performance evaluation, irrigation projects
Evaporation, 151, 154–156
Evaporative demand, 145–146, 157, 275, 360, 402
Evapotranspiration (ET), 157, 266, 393–396
　CropET$_0$ model, 396
　CROPWAT model, 393–396
Excessive pumping, 189
Exchangeable sodium percentage (ESP), 283
　estimation of, 289
Explicit scheme, 388

F

Factors affecting pump performance, 469–470
Factors affecting watershed, 198
Fanning's friction factor, 28–29
Farmer's income ratio, 126
Farm irrigation efficiency, 69
Fertigation, 51, 54, 56, 265–266
Fick's law, 251–254

Field water indicators
　description, 120–126
　performance, 119
Finite difference method, 388
Finite element method, 388
First-order kinetic diffusion process, 257
First order rate constant, 258
Fisheries, hazards to, 249
Fixed solar collectors, 512
Flash flood, 185
Flash plants, 508
Floating objects, 154
FLOCR, 398–402, 398f
　boundary condition at bottom of soil profile, 398
　conductivity, 400
　equations, 398
　evapotranspiration, 399
　infiltration into matrix and cracks, 399–400
　principle, 398
　subsidence and cracking, 401
　surface runoff, 399
　time step, 401
Flood irrigation, 37–38, 161, 200
Flood mitigation strategies, 187
Flour/lumber production, see Watermill
Flow rate, basin irrigation, 79t, 80
Flume, 504
Force mode pump, 439
Formula for irregular drain system, 355
Formulation of model
　irrigation/water management models, 389
　　conceptual model for irrigation, 390f
　weather-based prediction model, 411–412
　　basic assumptions in, 414
　　choice of weather variables, 413
　　crop-production statistics, 413
　　methods of formulation of, 413–414
　　regression approach, 414
　　removing trend component, 413
　　typical form of model, 414
　　zoning of crop area, 413
　weather-based yield-prediction model, 415–416
Fossil fuel, 480, 482, 484, 488, 501, 507, 512, 510
Freeboard, 10, 12, 14f, 16f, 19f, 20
Frictional head loss, 27
Friction factor, 28–29
Friction losses (FL), 134, 443–445
Fundamental aspects of water resources management, 144
FURDEV, 96–97

Furrow design considerations, 92
Furrow grades, 204
Furrow irrigation, 38, 43–45, 44f
　concepts and features, 43–44
　evaluation, 130–131
　suitability and limitations, 44–45
　　advantages and disadvantages, 45
　　attainable efficiencies, 45
　　cost and economic factor, 45
　　crop suitability, 44
　　labor and energy requirement, 45
　　level of technology, 45
　　required depth of irrigation application, 45
　　soil and topography, 44
　　water quantity, 44
Furrow irrigation system design, 84–98, 85f
　average depth of flow estimation, 95–96
　controllable variables and design variables, management, 91–92
　examples, 97–98
　factors affecting, 90–91
　　cutoff ratio, 91
　　length of run, 91
　　soil characteristics, 90
　　stream size, 90
　　tailwater reuse, 91
　　wetted perimeter, 91
　furrow design considerations, 92
　hydraulics, 85
　irrigation models, 96–97
　　FURDEV, 96–97
　mathematical description of water flow, 86
　　unsteady gradually varied surface water flow, 86–87
　modeling, 92–94
　　furrow design variables, simulation, 93–94
　　thumb rule for furrow design, 94–95
Furrow length, 77, 85f, 88, 91–94
Furrow shape, 89, 91, 93, 95

G

Geographic Information System (GIS), *see* GIS in irrigation and water management
Geometric similarity, 446
Geo-referenced data, 428
Geothermal energy, 508–509
GIS based modeling, 429–430
GIS environment, 430
GIS in irrigation and water management, 423–431
　applications, 425–427
　benefits, 424
　data and databases
　　attribute data, 428
　　spatial data, 428
　data input
　　coordinate geometry procedure, 429
　　digitizers, 429
　　keyboard entry, 429
　　scanning, 429
　definition, 424
　GIS-based modeling/spatial modeling, 429–430
　implementation, 427
　major tasks
　　analysis, 425
　　input, 425
　　management, 425
　　manipulation, 425
　　query and viewing, 425
　　visualization and printing, 425
　remote sensing techniques, 430–431
　sources of spatial data, 428–429
　　global positioning system, 428
　　remote sensing, 428
　techniques, 427
GIS software, 415, 427
Global positioning system (GPS), 428
Global scale of water conflicts
　economic value of water, 182
　increasing environmental concern, 181–182
　lack of accessible water, 181
Global warming, 185–186, 514–515
Goal and purpose of drainage, 329
Grain alcohol, *see* Ethanol fuel
Granular activated carbon (GAC), 371
Graphical users interface (GUI), 430
Gravity irrigation, 36–37, 204
Greenhouse gas emissions, 482, 491, 494
Groundwater
　aquifers, 141, 174, 182
　assessment, 147, 430
　contamination, 243, 245, 263
　depletion, 189, 196
　development in saline/coastal areas, 148
　discharge, seepage salting/salty, 276–277
　fed salinity or seasonal salinity, 291
　mining, 165
　overabstraction/groundwater mining, adverse effect of, 165
　resource, 147–148, 163–164
　shallow saline, 274
　storage through recharge, 167

Groundwater availability
    for pumping in terms of potential
        recharge, 147
    in terms of safe yield, 147–148
Groundwater flow toward drain, theory of,
        346–347
    examples, 347–349
    non-steady state drainage situation, 347
    steady state problems, 346–347
        horizontal flow assumption, 347
        radial flow assumption, 347
Growing-degree-days (GDD), 414
Gypsum, 274t, 280, 280t, 282–283, 288, 289t,
        308–309, 317, 319
    cation exchange reaction, 317
        reclamation by acid formers or acids,
            318–319
        reclamation by gypsum, 317
    pure and rate of, 319
    requirement, 288, 289t, 317

H
Hagan and Stewart model, 409
Hagen-Poiseuille equation, 29–30
Half-life, 244
Halophytes, 282
Hand pump, 170
Hand watering, 54–55
Harvesting rainwater, *see* Rainwater harvesting
Hazard of nitrate, 248
Head loss, causes/components, 26–27
Head race, 504
Headworks efficiency, 120
Heat wave, 186
Helical rotor pumps, 441
Homogeneous and isotropic media, 342, 342f
Hooghoudt's equation, 351–354
    assumptions in Hooghoudt's formula, 352,
        352f
    definition, 351–352
    of drain placement, 352
    estimation of equivalent depth, 353
        closed-form solution for equivalent
            depth, 353
        limitations of, 353
Horizontal-axis wind machines, 495
Horizontal flow assumption, 347
Horizontal in-line booster pumps, 452
Horizontal waterwheel, 505
Human health hazards, 249
Human-powered devices, 435–436
    Archimedean screw, 436
    don, 435

    paddle wheel, 436
    swing basket, 435
Hydraulic design of surface drain, 349–350
    drainage area, 350
    drainage channel, 349
Hydraulic grade line or hydraulic gradient, 25
Hydraulic pumps, 441
Hydraulic radius, 11
Hydraulic ram pumps, 462
Hydraulics and fluid flow theories, 22–23
Hydraulics of furrow, 85
    infiltration advance
        cross-section of furrow, 85f
        throughout furrow length, 85f
Hydrodynamic dispersion, 253–254, 258
Hydrodynamic model, 82–83
Hydroelectric dams, 503, 513
Hydroelectric energy, 503
Hydroelectric plant, 504
Hydroelectric power, *see* Hydropower
Hydrograph, 77, 129, 215
Hydrologic cycle, 185, 188
Hydrologic Unit Code (HUC), 198–199
Hydropower, 481–482, 505–506, 512, 515t

I
ICARDA, 154
Illegal fishing, 182
Impact of global climate change on water
        resources management
    adaptation alternatives
        conservation measure, 187
        drought protection strategies, 187
        educate people, 188
        flood mitigation measures, 187
        new technology, 187
    issues on
        agricultural water demand, 185–186
        municipal/industrial use, 186
        water supply, 185
Impact of salinity on soil and crop production
    clay swelling and dispersion, 293
    impacts on plant nutrition, 294
        osmotic effect, 294
        plant growth inhibition, 294
        slow mineralization, 294
        specific ion effect, 294, 294t
        uptake of competitive nutrients, 294
    plant available water content (PAWC), 293
Impact of sodicity on soil and plant growth
    dispersion, 295
    problems, 295
    scalding, 295

Subject Index

stunting, 295
Impacts of water policy research, 178
Impaired watersheds, 236
Implementation of GIS, 427
Implicit scheme, 388
Importance of watershed management, 198
Improving irrigation efficiency, 156
Improving irrigation performances, 134–136
Index of agreement (IA), 392
Indicators, engineering, 120–122
Infiltration equations, 214–215
Infiltration opportunity time, 85, 88, 131, 214
Inflow/irrigation discharge, 129
Inflow–outflow method, seepage measurement, 5–6
Initial condition, 386
In-organic/mineral envelopes, 366
In-stream demand, 146
In-stream flow, 146, 150, 197, 238
Intake rate, 87
Integrated approach, 171, 175, 234
Integrated CSP/combined-cycle gas-turbine power plants, 490
Integrated water resources management (IWRM)
  in Bangladesh, 172
  capacity building for, 178
  conflicts of, 179–180
  coordinated decision making across sectors, 171f
  definition of, 170–171
  strategies for, 170–173
  sustainability, 177
Interceptor drain, 361–362
Intergovernmental Panel on Climate Change (IPCC), 186
International Energy Agency, 481, 508
International Energy Outlook 2009, 482
International Rice Research Institute (IRRI), 313
International rivers, sharing of, 162, 167–168
International Water Management Institute (IWMI), 145–146
Intrinsic velocity, 258
Ions, definition, 282
Ion specific effects, 282
Ions to EC, 285
Irrigated area, 46, 49, 114, 116, 121, 124, 126, 469
Irrigation and water management practices, 309–311

alternate/cyclic irrigation with saline and fresh water, 310
at alternate furrows, 311
frequency management
  drip irrigation, 309
  subsurface irrigation, 310
methods
  drip irrigation, 277
  sprinkler irrigation, 277
mixing/blending of salinity and fresh water, 310
pre-sowing (or pre-plant) irrigation, 309
with saline water at less sensitive growth stages, 310
Irrigation benefit–cost ratio, 126
Irrigation cost per unit area, 126
Irrigation demand, 145–146
Irrigation efficiency, 69, 112–116
Irrigation intensity, 124
Irrigation method/technologies, adopting efficient
  improving irrigation efficiency, 156
  irrigation scheduling, 156
Irrigation planning and decision support system, 402–406
  CropSyst, 403
  CROPWAT, 402–403
Irrigation projects performance evaluation
  irrigation efficiencies
    application efficiency, 112–113, 113t, 119
    low-quarter distribution uniformity, 115–116
    storage efficiency/water requirement efficiency, 114
    uniformity, 114–115
  performance evaluation
    concept/objective/purpose of, 116–117
    factors affecting, 117–118
    improving, 134–136
    indicators, description of, 120–126
    performance indices/indicators, 118–119
    procedure, 126–128
    under specific irrigation system, 128–134
Irrigation set time, 89
Irrigation system designing, 65–109
  basin irrigation, 79–84
  border irrigation, 68–79
  furrow irrigation, 84–98
  sprinkler system, 98–106

Irrigation system designing (cont.)
  surface irrigation, 66–68
Irrigation system performance, problems/faults/measures, 134–136
Irrigation uniformity, 114–115
Irrigation water productivity (IWP), 125
ISO 1400, 181

J
Jensen model, 410
  determination of sensitivity index, 410
Jet pumps, 459–461, 460f
  features, 460
  general specifications, 460
  uses, 461

K
Kinematic-wave model, 93–94
KINEROS (Kinematic Runoff and Erosion Model), 227
Kinetic energy, 21, 436, 441, 450, 485, 497–498, 500, 503
Kinetic energy powered device, 436–437
  hydraulic ram, 436–437
Kostiakov infiltration equation, 72, 129
Kyoto target, 482

L
Laminar flow, 25–26, 29–30
Land and watershed management, 193–239
  background and issues, 195–197
    floods/landslides/torrents, 196
    population pressure and land shrinkage, 196–197
    water pollution, 196
    water scarcity, 196
  climate change addressing, 238
  fundamental aspects, 197–199
    elements of watershed, 197
    factors affecting, 198
    importance, 198
    naming watershed, 198–199
    watershed functions, 198
  land grading, 199–212
    activities and design considerations, 203–205
    concept/purpose/applicability, 199–200
    earth work volume estimation, 208–212
    factors affecting, 201–203
    methods, 205–212
    precision grading, 200–201
  runoff and sediment yield, 212–227
    erosion and sedimentation control, 225–226
    factors affecting runoff, 213–214
    factors affecting soil erosion, 219–221
    modeling runoff and sediment yield, 226–227
    runoff and erosion processes, 212–213
    runoff volume estimation, 214–219
    sediment yield and its estimation, 221–223
  watershed management, 227–236
    community-based approach, 230–233
    components, 228
    land use planning and practices, 230, 234–235
    pond management, 231
    problem identification, 227–228
    structural management, 230
    sustainable watershed management, 235–236
    tools for watershed protection, 230
    watershed planning and management, 229
  watershed restoration and wetland management, 236–238
    drinking water systems using surface water, 236–237
Land drainage, physics of
  examples, 340
  soil pore space and soil-water retention behavior, 334
    different types of water in soil, 335f
    solid and pore space in soil bulk volume, 335f
  terminologies, 335–338
    drainable pore space or drainable porosity, 337
    drainable pore volume under negative pressure (or suction), 338
    drainable water, 337–338
    moisture concentration, 336
    pore water, 336
    porosity, 336
    relation between porosity and void ratio, 336
    void ratio, 335
  water balance in drained soil, 338–340
    artificial drained and natural drained soil, 339f
Land grading, 199–212
  activities and design considerations, 203–205
    cross slope, 204
    furrow grades, 204

Subject Index 531

maximum length of runs for irrigation, 204–205
other considerations, 205
concept, 199
earth work volume estimation, 208–212
essential conditions and applicability, 200
factors affecting, 201–203
  climate and microclimate, 201
  cost and benefit factors, 202–203
  existing regulatory context, 202
  field size, 202
  indigenous vegetation and wildlife habitats, 202
  proposed plan of use, 202
  soil/hydrologic/geologic conditions, 201
  sustainability issues, 202
  topography, 201
methods, 205–212
  construction guidelines, 207–208
  cutting and filling, 207
  earth work volume estimation, 208–212
  maintenance considerations, 208
  planning and early surveying, 205–207
precision grading, 200–201
purpose, 199–200
Land improvement, 199
Land leveling, 39, 42–43, 45, 51, 57–58, 152, 199, 204–205, 207–208, 312, 331
Landscaped buffer, 234
Landslides, 196
Land use planning, 230, 314
Laplace's equation for groundwater flow, 345
Large-scale groundwater schemes, 170
Leachate, 263, 265, 268
LEACHC, 322
Leaching, 283
  equation for total irrigation depth, derivation, 304–305
  method, considerations in, 306
  other aspects/options in leaching calculation, 306
  perspectives on calculating LF, 306
    case 1: no drainage limitation, 306–307
    case 2: internal drainage, limitation, 307
  recharge well, 308f
  requirement, calculation of, 303
Leaching fraction (LF), 283, 303, 315
Leaching requirement (LR), 303, 306, 360
Legal conflict, 180
Length of run, 67f, 89, 91, 117, 134, 204, 225, 349

Level basin irrigation, 40, 83
Level instrument, 129
Liquid biofuel, 507–508
Localized irrigation, 55
Logical model, 385
Lot benching, 206–207
Low energy precision application (LEPA), 50, 309
Low-quarter distribution uniformity, 115–116
Lumped parameter model, 430

M
MACRO model, 402
Management, *see* Water resources management
Management controllable variables, 91
Man-made/synthetic envelopes, 366
Manning's equation, 12–13, 18, 20, 71
Manning's N values, 18, 77, 118, 129
Manning's roughness coefficient, 13, 13t, 87
Mathematical models, 145, 260, 381–384, 386–387, 389, 393, 406
Maximum allowable fluctuation of water table (MAWT), 148
Maximum non-erosive flow rate, 95
Maximum permissible limit, 244
Mean absolute bias, 392
Measuring EC, 289
Mechanical energy, 437, 441, 481
Mechanically powered water-lifting devices, 437
Mechanistic model, 384
Metal shaping, *see* Watermill
Metering and pricing, 152
Methemoglobinemia, 248
Micro-hydro systems, 503
Micro-irrigation, 54, 134
Micro-topography, 80–81, 83, 199
Mineral envelopes, 366–367
Minor loss, 27, 30, 30t
Mixed-flow pumps, 440, 443, 448, 456–457, 463
Model calibration, 390–391
Model efficiency (EF), 392
Model evaluation, 390
Modeling
  definition, 386
  process, 389
Model parameter, 386, 390–391
Model performance, 380, 390–393
Model performance evaluation, statistical indicators for, 391–393
  bias (error) or mean bias, 392
  coefficient of determination (CD), 393

Model performance (cont.)
  coefficient of residual mass (CRM), 392–393
  index of agreement (IA), 392
  mean absolute bias, 392
  model efficiency (EF), 392
  relative error (RE), 392
  root mean square error (RMSE), 392
Models in irrigation/water management, 379–419
  commonly used models
    decision support model, 405–406
    for flow estimation in cracking clay soil, 397–402
    for irrigation planning and decision support system, 402–405
    for reference evapotranspiration (ET_0 models), 393–396
    for upward flux estimation, 397
  considerations, 389–390
  crop production function/yield model
    basic considerations in, 407
    definition of, 406
    development of, 408
    existing yield functions/models, 408–411
    importance of, 406–407
    limitations/drawbacks of, 411
    pattern of, 407–408
  formulations, 389–390
    conceptual model for irrigation, 390f
  general concepts, 380–381
    schematic representation of model, 381f
  model calibration, 390–391
  regression-based empirical models
    existing models/past efforts, 412–413
    weather-based prediction model, 411–412
    weather-based yield-prediction model, 415–419
  statistical indicators for model performance evaluation, 391–393
    bias (error) or mean bias, 392
    coefficient of determination (CD), 393
    coefficient of residual mass (CRM), 392–393
    index of agreement (IA), 392
    mean absolute bias, 392
    model efficiency (EF), 392
    relative error (RE), 392
    root mean square error (RMSE), 392
  terminologies, 386–388
    computer model, 386
    finite element and finite difference scheme/method, 388
    implicit *vs.* explicit scheme/method and stability issue, 388
    simulation, 386
  types of model, 381–385
    conceptual model, 382
    deterministic and stochastic model, 384–385
    empirical models (or black box models), 381–382
    logical model, 385
    mathematical and statistical models, 382–384
    mechanistic and probabilistic model, 384
    physically based or process-based model, 381
    static and dynamic model, 384
    time series model, 385
  verification/validation, 391
Models in solute transport, 266–267
  DRAINMOD-N, 267
  HYDRUS-2D, 267
  IRRSCH, 267
  LEACHN, 267
  MACRO, 267
  NLEAP, 267
  Opus, 267
  SOIL, 267
  SOILN, 267
  SWMS_2D, 267
Models/tools in salinity management
  BC2C model, 320
  CropSyst, 321
  2CSalt model, 320
  ENVIRO-GRO, 323
  LEACHC, 322
  SALT, 320
  SALTMOD computer model, 322
  SGMP computer model, 322
  SIMPACT model, 322
  SIWATRE, 321
  SIWM model, 322
  SWAM, 322
  SWASALT/SWAP, 323
  SWMS-3D, 321
  TARGET model, 320
  UNSATCHEM, 321
  WATSUIT, 321
Model validation, 388
Modern wind turbines, 493–494, 499f
Moisture concentration, 336

Molecular diffusion, 247, 252–253, 258
Mole drainage, 307
Mono-block pumps
  features, 454–455
  specification of reciprocating pump, 455
  uses, 455
Monomolecular films, 154
Moody Chart, 28, 29f
Motor capacity, 464
Moura photovoltaic power station, 490
Mulching, 151, 154–155, 205, 225, 246, 302
  and crop residue management, 311
Multi-meter, 128
Multiple basins, 79, 81, 83
Multiple linear regression analysis, 412, 418–419
Multi-stage pump, 440, 456
64 MW Nevada Solar One, 490
11 MW PS10 solar power tower, Spain, 490

N
Naming watershed, 198–199
Natural buffer, 205, 234
Net positive suction head available (NPSHa), 443–444, 449, 457t, 473
Net positive suction head required (NPSHr), 444, 447, 449, 463
Network rehabilitation, 162, 168
Nitrogen management, 265–266
  deficit irrigation, 266
  method of application, 266
  nitrate accumulation in soil, minimization, 266
  nitrogen application correction, 265
  slow-release fertilizer, uses, 266
  soil nitrate test, 266
Nodal points or nodes, 388
Nonconventional water measures, 161–162
  saline water
    and fresh water, irrigation of, 166
    irrigation at less sensitive growth stages, 166
    mixing with fresh water, 165–166
Nonpoint source, 172, 198, 226, 228, 231, 234, 242–243, 263
Non-positive displacement pumps, 441
Non-reactive solute, 243, 253
n-type semiconductor, 485
Numerical models, 260, 383, 388

O
Objective function, 386
Obstacles in demand management implementation
  financial/economic, 159–160
  public perception, 160–161
  technical/institutional, 160
Ocean thermal energy conversion (OTEC), 503
Ocean Wave Power engines, 504f
On-farm water loss, 119, 123–124, 135–136, 155
Open channel, flow velocity calculation of
  Chezy's equation, 12–13
  Manning's equation/N values, 13, 13t
Open well submersible pump, 453
  applications, 453
  features and benefits, 453
Organic envelopes, 366
Organic manuring, 302, 312
Organic matter (OM), 54, 158, 218, 223–225, 245, 250, 262, 281, 287, 295, 319, 329, 335, 481
Osmotic potential, 284–285
  alkalinity, 283
  calcareous, 283
  dryland salinity, 282
  gypsum, 283
  halophytes, 282
  ion specific effects, 282
  leaching, 283
  leaching fraction, 283
  saline groundwater, 282
  salinity control regions, 282
  salinity mitigation, 282
  salt tolerant, 282
  sodicity, 283
Overall efficiency (OE), 112, 119, 122
Overhead irrigation, 37
Overland flow, 68, 76, 81–82, 84, 212–213, 215, 329
Overlapping factor, 103, 106

P
Panel on Climate Change (IPCC), 186
Parque Solar Hoya de Los Vincentes, 490
Partition coefficient, 258
Pascal's law, 22
Pattern efficiency, *see* Low-quarter distribution uniformity
Peak energy demand, 487–489
Peak power (water) demand, 486, 497
Penman-Monteith equation, 393–394
Percolation, 2, 8, 55, 66, 123, 136, 145, 246, 265–266, 274, 308
Performance, *see* Irrigation projects performance evaluation
Performance evaluation, irrigation projects, 116–136

Performance evaluation (cont.)
  ideal/normal field condition, 128
  queries, 127
  steps and techniques, 126–127
Performance indices, basin irrigation, 130
Permanent salinity or saline soil, 292
Permissible velocity, 11, 31
Persistent pollutants, 244, 249
Photons, 485
Photosynthetically active radiation (PAR), 405, 407
Photovoltaic array, 485–486
Photovoltaic cells, 483, 484f, 485, 488, 512, 516
Photovoltaic module, 485
Photovoltaic power, 485, 488, 490
Photovoltaic pump, 485–489
Photovoltaic system, 485–488, 490–491
  design
    mathematical formulation, 486–487
    steps/procedures for, 486
  types
    grid system, 486
    stand-alone hybrid system, 486
    stand-alone photovoltaic system, 486
Photovoltaic technology, 485
Physically based model, 226, 381
Physical treatment, see Chemical (and physical) treatment of drainage
Physicochemical treatment, see Chemical (and physical) treatment of drainage
Physiography, 428
Planta Solar La Magascona, 490
Plant available water content (PAWC), 293, 295
Plumbing fixture, 153
  and devices, 153
Point source, 113t, 242
Pollutant/pollution, 228, 243–244, 246
  sources, 242–243
Pollution from agricultural fields and its control, 241–268
  extent of agricultural pollution, 243–250
    agricultural pollutants, impact of, 248–250
    factors affecting solute contamination, 244–247
    hazard of nitrate (NO_3–N) pollution, 248
    mode of pollution by nitrate and pesticides, 247–248
  models in solute transport, 266–267
  pollution sources, 242–243

nonpoint, 242–243
point, 242
solute leaching control, 265–266
  cultural management, 266
  irrigation management, 265
  nitrogen management, 265–266
solute load estimation, 261–264
  sampling from controlled lysimeter box, 261
  sampling from crop field, 261–262
  solute concentration determination, 262–264
  water sampling equipments, 262f
solute transport parameters, measurement of, 258–260
  breakthrough curve and breakthrough experiment, 259–260
  different parameters, 258–259
solute transport processes in soil, 250–257
  basic solute transport processes, 251–253
  governing equation through homogeneous media, 254–256
  one-dimensional solute transport with nitrification chain, 256–257
  transport of solute through soil, 250–251
  water flow/solute transport in heterogeneous media, 257
types of pollutants/solutes
  nonreactive solute, 243
  reactive solute, 243
Ponding method, 4–5, 5f
Population pressure and land shrinkage, 196–197
Pore water or drainable water, 336
Pore water velocity, 254–258
Porosity, definition, 337
Positive displacement (PD) pumps, 438, 441, 449
Potential evapotranspiration (PET), 147, 399, 402
Potential recharge, 147–148
Potential water supply estimation
  groundwater resource, 147–148
  surface water resource, 146–147
Power requirement, 103
  in pumping, 464
Power sources, 435, 437, 453, 461–462, 483, 486, 505, 509
Precipitation rate, 99–100
Pre-sowing (or pre-plant) irrigation, 157, 309
Pressure distribution

Subject Index 535

in closed vessel, 22f
in vertical water tank, 24f
in water column/tank, 24
Pressure *vs.* flow rate, 24
Pressurized irrigation, 36–37
Pricing subsidies removal, 151, 153
Primary salinity, 292
Prime mover, 439
    submersible pumps, 439
    surface-mounted pumps, 439
Priming, 152, 157, 443, 448–449, 453, 457t, 460
Principles in water pumping, 438
    atmospheric pressure, 438
    centrifugal force, 438
    movement of column of fluid, 438
    positive displacement, 438
Prismoidal formula (PF), 211
Probabilistic model, *see* Chemical (and physical) treatment of drainage
Process-based model, 226, 381
Project application efficiency, 122
Propeller pumps, *see* Axial-flow pumps
Propeller-type blades, 492
PSI (pounds per square inch), 23, 103, 462, 468, 476
p-type semiconductor, 485
Public awareness approaches, 151, 158
Pump characteristics, 470, 472
Pump control, 467–468
    automating control circuit, 467–468
    typical relay control circuit, 467f
Pump curves, 471–472
Pump efficiency, 128, 443–445, 463, 474–475
Pumping plant efficiency, 118, 120, 127–128, 487
Pump installation, 465–468
    alignment, 465
    submersible pump, 465–466
Pump operation, 466–467
    field test, 466
    check, 467
    suction line, 466
Pump performance
    factors affecting, 469–470
Pumps connected in series and parallel, 469

Q

Quality degradation
    due to continuous pumping of groundwater, 188
    reduction of, 168
Quasi-analytical method, 260

R

Radial flow assumption, 347
Radial-flow pumps, 440, 443, 448, 455, 455f
Rainwater harvesting, 162, 166, 308
Rain-water ponding at crop field by high levees, 158
Rate constant, 255, 258, 386, 391
Rate variable, 386
Reactive solute, 243, 254
Reallocation, 168
Recession flow, 69
Recharge well, 303, 307–308, 308f
Reciprocating pump/bucket pump, 461–462, 461f
    features, 461
    specification, 461
    uses, 461–462
Recordable information during sprinkler evaluation, 132
Recreational areas, 197
Rectangular basin, 83
Rectangular channel, 14–15, 14f
Reducing evaporation, 151, 154–156
Reduction of seepage, 8–10
Regional scale water conflicts, 182–184
    Indian Subcontinent, 183
    Middle East, 183–184
Regression-based empirical models
    existing models/past efforts, 412
    weather-based prediction model
        methods of formulation of, 413–414
        need of, 411–412
    weather-based yield-prediction model
        sample example of formulating, 415–419
Rehabilitation, network, 168
Relative error (RE), 392
Relative yield, 299–300, 410
    vs. EC, 300f
Remote area power supply (RAPS), 503
Remote sensing (RS), 423, 426, 428–431
Renewable energy resources
    bio-energy, 506–508
        biogas, 508
        liquid biofuel, 507–508
    commercialization, problems and prospects, 511–516
        aesthetic criticism, 512
        availability and reliability, 512
        challenges and needs, 516
        cost, 512
        environmental and social considerations, 512–513

Renewable energy resources (*cont.*)
    incapability of generating large amount of energy, 513
    longevity issues, 513
    prospects/future potentials, 514–516
    required land area, 513
    sustainability, 513–514
    transmission, 514
concept and definition of, 481–482
factors affecting potential use of, 509–511
    alternate energy, 511
    groundwater requirement and availability, 510
    user affordability and willingness, 510
geothermal energy, 508–509
modeling energy requirement, 509
mode of using, 483
need for, 482–483
    IEO's prediction, 482
    poverty alleviation in developing countries, 483
    reducing greenhouse gas emissions, 482
renewable energy cost, 515t
solar energy, 483–491
    assessment, 484–485
    other uses, 489
    solar/photovoltaic cells/systems, 485, 488
    solar photovoltaic pump, 485–489
status of, 480
water energy, 502–506
    forms of, 503
    potentials of, 505–506
    tide mill, 505
    watermill, 504–505
    wave energy, 503–504
wind energy, 491–502
    advantages, 493–494
    application of, 496–497
    causes of wind flow, 492–493
    energy from wind, 493
    and environment, 501
    potential assessment, 494–495
    problem with, 500–501
    as renewable and environmentally friendly, 491
    suitable site for windmill, 495–496
    types of wind machines, 495
    wind power calculation, 498–500
    wind power plants or wind farms, 498
    working principle of wind machines, 497
Renewable energy source, 483
Renewables penetration, 511
Retardation factor, 254–255, 258
Return flow, 39, 69, 92
Reynolds number, 25–26
Richards' equation, 214, 250, 369
Ridge bed system, 312
Ring infiltrometer, 129
Root mean square error (RMSE), 392
Rotodynamic pump, 441
Row/seed bed management, 312, 312f
Runoff and sediment yield, 212–227
    erosion and sedimentation control, 225
    factors affecting runoff, 213–214
        land use, 214
        mulches and crop residues, 214
        other management practices, 214
        rainfall, 213
        soil cover management, 214
        soil type/infiltration rate, 213
        surface depressions/ponds/natural water storage, 214
        topography, 213–214
        vegetation, 213
    factors affecting soil erosion, 219–221
        climate, 219
        land use, 220
        other management practices, 220–221
        soil cover management/cultural practices, 220
        soils, 219–220
        topography, 220
        vegetative cover, 220
    modeling runoff and sediment yield, 226–227
    runoff and erosion processes, 212–213
    runoff volume estimation
        examples, 218–219
        rational method, 214–219
        SCS method, 216–218
    sediment yield and its estimation, 221–223
        concept, 221
        cover and management factor, 223
        examples, 223–225
        mechanism, 221
        rainfall and runoff factor, 222
        soil erodibility factor, 222
        support practice factor, 223
Runoff fraction/tailwater ratio, 124
Runoff volume estimation, 214–219
    rational method, 215–216
        average recurrence interval (ARI), 216
        design peak flow, 215–216

Subject Index 537

peak runoff from single storm event, 215
runoff coefficient, 216
time of concentration, 216
SCS method, 216–218
Rural Technology Promotion Department of the Ministry of Agriculture or Power, 506
RUSLE2, 227

S

Safe yield, 147–148, 162–163, 165
Saint-Venant equation, 77, 82, 86
Saline and sodic soils
    classification and characteristics of salt-affected soils, 279–282
        saline-sodic soil, 282
        saline soil, 280–281
        sodic soil, 281
    diagnosis of salinity and sodicity, 285–290
        determination of intensity, 289–290
        by field observation, 286
        by laboratory determination, 286–289
    management of, 319
    salinity mapping and classification, 290–293
    terminologies and conversion factors, 282–285
        EC, 283
        ESP, 283
        ions to EC, 285
        salinity, 284
        salinity and osmotic potential, 284–285
        salinization, 284
        SAR, 284
        TSS or TDS, 283
        unit of EC, 285
Saline groundwater, 165, 273–274, 284, 292, 311
Saline-sodic soil, 282
    characteristics, 282
Saline soil, 280–281
    characteristics, 280
    or permanent salinity, 292
    salts and their solubility in water, 280t
Saline water/poor quality water, 161
Salinity, 284
    amelioration, *see* Amelioration of saline soil
    avoidance, 302, 311
    estimation of, 289
    factors affecting, 277–278
        climate, 278

        depth to water table, 277
        method of irrigation, 277
        soil factor, 277–278
        soil moisture, 277
    hazard, mechanism of, 278
    management, principles and approaches of, 301
    mapping, 290–291
    mitigation, 282
    and osmotic potential, 284–285
    problems, 167, 206, 273–275, 277, 282, 289, 292–293, 295, 302–303, 314, 430
    on soil and crop production, 293–294
    threshold level, 299
Salinity and sodicity
    determination of intensity of hazards, 289–290
        estimation of ESP, 289–290
        estimation of salinity, 289
        estimation of SAR, 290
        interpretation based on EC value, 290t
        interpretation based on Na/SAR/ESP/Ca:Na ratio, 290t
        measuring EC, 289
    diagnosis by field observation, 286
        field symptoms, 286
    diagnosis by laboratory determination, 286–288
        test required for different problem identification, 289t
    and high soil pH, symptoms of, 287t
    impact of salinity on soil and crop production, 293–294
    impact of sodicity on soil and plant growth, 294–295
    problem, extent of, 272
    salinity affected crop fields, 287f
Salinity classification, 291–293
    based on causes/sources of salinity, 291–292
        groundwater fed salinity or seasonal salinity, 291–292
        saline soil or permanent salinity, 292
    based on development mode, 292
        primary salinity, 292
        secondary salinity, 292
    based on salt content or EC, 292–293
Salinity development, causes of, 273–277
    climatic condition, 275
    from irrigation water, 273–274
    landscape feature, 275

Salinity development (*cont.*)
  salinity development in discharge area, 275f
  man-made activity, 275–276
  salts in soil, 273
    salts in saline area and their equivalent weight, 274t
    seepage salting/salty groundwater discharge, 276–277
  shallow saline groundwater, 274
  from urban area, 276
Salinity management options, 302–316
  biological reduction of salts, 311
  chemical practices (reclamation/treatment of saline soil), 308–309
  control of saline water, 302–303
  developing/cultivating salt tolerant crops, 313
  engineering practices, 303–308
    drainage, 307
    leaching, 303–307
  increasing water use of annual crops and pastures, 314
  irrigation and water management practices, 309–311
    alternate/cyclic irrigation with saline and fresh water, 310
    alternate furrows, 311
    frequency management, 309
    mixing/blending of salinity and fresh water, 310
    pre-sowing (or pre-plant) irrigation, 309
    saline water at less sensitive growth stages, 310
  other management options, 311–313
    application of sand (sanding), 312
    appropriate/well adjusted fertilization, 313
    crop choice/growing salt tolerant crop species or cultivar, 313
    mulching/crop residue management, 311
    organic manuring, 312
    physical management, 311–312
    row/seed bed management, 312, 312f
    salinity avoidance, 311
  policy formulation, 314
  removing surface salts/scraping, 302
Salinization, 284
SALT, 320
Salt-affected soils, management of, 271–325
  crop tolerance to soil salinity
    examples, 300
    factors influencing tolerance to crop, 295–297
    relative salt tolerance of crops, 297–298
    use of saline water for crop production, 298–299
    yield reduction due to salinity, 299–300
  development of salinity and sodicity
    causes, 273–277
    factors, 277–278
    mechanism, hazard, 278
    salt balance at farm level, 278–279
  diagnosis and characteristics
    classification, 279–282
    salinity and sodicity, 285–290
    salinity mapping/classification, 290–293
    terminologies/conversion factors, 282–285
  extent of problem, 272
  impact of salinity and sodicity
    impact on crop production, 293–294
    impact on soil and plant growth, 294–295
  management/amelioration of saline soil
    models/tools, 320–323
    salinity management, 302–316
  management of sodic and saline-sodic soils
    saline-sodic soil, 319
    sodic soil, 317–319
Salt balance at farm level, 278–279
Salt loads, 282, 320, 357, 360–361
SALTMOD computer model, 322
Salt scalds, 282
Salt tolerance of crops, relative, 297–298
  of annual field crops and forages, 298t
Salt tolerant crops, 272, 282, 286, 290, 292, 295–296, 299, 302, 313, 319, 323
Sampling pumps, 439
Sanding (application of sand), 312
Scalding, 295, 330
Scanning, 425, 429
Scarce resource, 143
  See also Water resources management
Scarcity and misuse, *see* Sustainability
Seasonal salinity or groundwater fed salinity, 291
Seasonal variation of wind energy, 498
Secondary climate impacts, 186
Secondary salinity, 292
Sedimentation, 8, 143t, 200, 205, 208, 225–226, 272, 293, 370
Sediment yield, 221–227
Seepage

definition of, 2
expression of, 3
loss, 2–4, 4f, 8–10, 155
measurement, 4
    inflow–outflow method, 5–6
    ponding method, 4–5
    seepage meter method, 6–7
meters, 4, 6–7, 6f
rate, 2–7, 9t
reduction methods
    biological, 8
    chemical, 8
    physical, 8
Selecting pump
    based on pump/well characteristics, 472
    procedure for, 470–472
    from pump curves, 471–472
Selection of irrigation method, 56–63
    factors affecting, 56–62
        attainable irrigation efficiency, 59
        climate, 57
        crop to be irrigated/type of crop, 59
        depth and frequency of irrigation application, 59–60
        economic factors, 58
        energy requirement, 58
        existing farm equipments compatibility, 59
        farm machinery and equipment requirement, 60
        field shape/geometry and topography, 57
        labor requirement, 58
        level of technology at locality, 60
        personal preference/cultural factor, 60
        relative advantage and disadvantages, 59
        soil type, 56–57
        tradition/previous experience with irrigation, 60
        water availability, 57
        water quality, 57–58
    selection procedure, 63
Self-mulching, 155
Self-priming pumps, 448, 457t, 460
Sensitive crops, 277, 286, 292, 298, 313
Sensitivity analysis, 78, 380, 387, 390
Sensitivity index, determination of, 410
Serpa solar power plant, Portugal, 490, 491f
SGMP computer model, 322
Side roll sprinkler system, 48–49, 48f
Side slope, 10, 12, 15–16, 18–20, 38, 69
SIMPACT model, 322

Simple linear regression analysis, 417
Simpson's rule, 147
Simulation, 386–387
    analytical solution, 387
    boundary condition, 386
    confidence, 388
    constraint, 386
    control variable, 386
    driving variable, 386
    dynamic system, 387
    empirical result/solution, 387
    initial condition, 386
    modeling, 386
    model parameter, 386
    objective function, 386–387
    rate constant, 386
    rate variable, 386
    sensitivity analysis, 387
    state variable, 386
    uncertainty, 388
    variable, 386
    verification/validation, 388
Single stage pump, 440, 451
SIRMOD model, 77–78, 84, 97, 129
SIWATRE, 321
SIWM model, 322
Small wind systems, 496
Soaking, see Priming
Social conflicts of water use, 179
Socioeconomic indicators
    description, 126
    performance, 119
Sodic (alkali) soils, 272, 279t
Sodic and saline-sodic soils, management of
    management of saline-sodic soil, 319
    management of sodic soil, 317–319
        calcareous soils, 317
        cation exchange reaction, 317–319
        chemical amendments, 317
        use of gypsum, 317
Sodicity, 285
    on soil and plant growth, 294–295
Sodic soil
    characteristics, 281
    management of, 317–319
        cation exchange reaction, 317–319
Sodium adsorption ratio (SAR), 284
    estimation of, 290
Soft energy technologies, 515
Soil-crop-weather management
    changing crops, 157
    crop sowing based on weather forecast, 157
    seedling age manipulation, 157

Soil-crop-weather management (*cont.*)
  seed priming/soaking, 157
  using anti-transpirants, 157
Soil map, 428
Soil moisture storage capacity in crop root zone, increasing
  adding organic matter, 158
  tillage and subsoiling, 158
Soil salinity and sodicity, development of
  causes of salinity development, 273–277
    climatic condition, 275
    landscape feature, 275
    man-made activity, 275–276
    salinity from irrigation water, 273–274
    salinity from urban area, 276
    salts in soil, 273
    seepage salting/salty groundwater discharge, 276–277
    shallow saline groundwater, 274
  factors affecting salinity, 277–278
    climate, 278
    depth to water table, 277
    method of irrigation, 277
    soil factor, 277–278
    soil moisture, 277
  mechanism of salinity hazard, 278
  salt balance at farm level, 278–279
Solar energy, 479–491
  assessment, 484–485
    technical potential, 484–485
    theoretical potential, 484
  other uses, 489
  solar/photovoltaic cells, 485
  solar photovoltaic (SPV) pump, 485–489
    cost of, 488
    design for, 486–488
    examples, 488–489
    photovoltaic system types, 485–486
    solar photovoltaic systems to generate electricity globally, 490–491
Solar Energy Generating Systems (SEGS), 490, 515
Solar or photovoltaic cells, 485
Solarpark Calveron, 490
Solar photovoltaic pump, 483, 485–489
Solar power, *see* Solar energy
Solar power plant, 490, 491f
Solar pumps, 170, 510
Solar system, *see* Solar energy
Solar Systems, Australia, 490
Solid set sprinkler, 47–48, 48f, 133
Soluble salts, 272–273, 275, 279–283, 288–289, 293, 308, 319

Solute leaching control, 265–266
  cultural management, 266
    cover crop, 266
    zero tillage/minimum tillage, 266
  irrigation management, 265
  nitrogen management, 265–266
Solute load, 261–264
Solute transport parameters, measurement of, 258–260
  analysis of breakthrough curve, 260
  breakthrough curve, 259, 259f
  breakthrough experiment, 259–260
  different parameters, 258–259
Solute transport processes, 250–257
  basic solute transport processes, 251–253
    convective solute transport, 251
    diffusive solute transport, 252
    dispersive solute transport, 253
  convection-dispersion equation, 254
    assumptions and drawbacks, 254
    governing equation through homogeneous media, 254–256
    one-dimensional transport, 255–256
    two-dimensional equation, 254–255
  one-dimensional solute transport with nitrification chain, 256–257
  transport of solute through soil, 250–251
  water flow and solute transport in heterogeneous media, 257
    dual porosity model, 257
South East Anatolia Project, 183
Spacing of furrow, 95, 130
Spatial data, 428–429
Spatial modeling, 429–430
Specific ion effect, 294, 294t
Specific speed, 443, 446
Specific yield, 148, 308
Sprinkler irrigation, 38, 46–52, 46f
  capabilities and limitations, 50–51
    advantages and disadvantages, 51
    crops, 51
    efficiencies, 51
    field shape and topography, 51
    financial involvement and labor requirement, 51
    soil type, 50–51
    water quantity and quality, 51
  concept and features, 46–47
  considerations, 52
  evaluation, 131–133
  types, 47–50
    center pivot, 49–50
    LEPA systems, 50

portable, 47
side roll system, 48–49
solid set, 47–48
Sprinkler system designing, 98–106
    aspects, 98–99
    considerations, 101
    examples, 104–106
    principles, 101
    spacing, 47, 100–102, 104, 106, 132
    steps and procedures, 102–103
    theoretical aspects, 99–101
        factors consideration, 99–100
        water distribution pattern, 99, 99f
Stability issue, 388
Stakeholders, 141, 164, 171–172, 178, 187, 231–233, 236, 382, 424
Stand-alone hybrid system, 486
Stand-alone photovoltaic system, 486
Stand-alone wind turbines, 496
State variable, 386, 388–389
Static and dynamic head, 24
Static model, 384
Statistical indicators for model performance evaluation, *see* Model performance evaluation, statistical indicators for
Statistical models, 384
Statistical uniformity coefficient, 134
Steady state drainage theory
    horizontal flow assumption, 347
    mass continuity equation, 345
    radial flow assumption, 347
Stewart model, 409–410
    determination of response factor of, 409
Stochastic model, 385
Stochastic time series model, 385
Storage efficiency, 71, 77, 79, 114, 119, 123, 130, 135t
Stunting, 295
Submersible pump, 439, 442, 451–454
    wire, 453
Submersible turbine pump, 440, 449–450, 450f
    applications, 452
    features, 453
    horizontal in-line booster pumps, 452
    open well submersible pump, 453
    requirements for, 453
    single phase open well submersible pumps, 454
    suitability of, 452–453
    vertical booster pumps, 452
    vertical deep-well pumps, 452
    wide ranges of, 453
Subsurface drainage, 307

buried pipe drain, 332, 333f
deep open drain, 331, 332f
evaluation system, 372–373
length of pipe drain, 332
Subsurface drainage design
    controlled drainage system, 361–362
        management of control drainage system, 361
        principle of method, 361
    data requirement for, 357
    examples, 362–365
    factors affecting spacing and depth of, 356–357
        impact of soil texture on drain depth, 356
        soil salinity, 356
    interceptor drain, 362
    layout of, 357, 358f
    principles/steps/considerations in, 358–360
Subsurface irrigation, 37, 54–55, 310
Suction head (SH), 103, 106, 443–446, 462, 473, 487, 510
Suction lift, 443, 446, 449, 452, 456, 460, 463
Suction mode pump, 165, 438, 442
Suitable site for windmill, 495
Sump pumps, 439
Superabsorbents, 169
Supply management
    approaches of, 161–162
    description of different approaches, 162–169
    implementation of, 169–170
    intermediate, 169–170
    modern, 170
Supply source, development of new, 161
    groundwater, 162
    safe yield, 162–163
    surface water, 162
    sustainable yield, 163–164
Surface drainage, 307, 330, 331f
    design
        considerations/layout, 349
        examples, 350
        hydraulic design, 349–350
        runoff, estimation of, 349
    land forming, 330
    open field drains, 331
Surface irrigation, 37–40, 42–43, 45, 51, 54–60, 71, 76–78, 81, 84, 92–94, 107, 113t, 129, 134, 156, 186, 199–201, 204–205, 207, 310, 312
    system designing, 66–68
        design principle, 66–67

Surface irrigation (*cont.*)
    hydraulics, 67–68
        phases of water-front, 68f
        variables, 67, 67f
Surface-mounted pumps, 439
Surface runoff, 11, 69, 96, 100, 129, 147, 156, 186, 199, 212–213, 220, 338, 339f, 349, 360, 362, 368, 399
Surface water schemes, 169
Surge flow/irrigation, 90, 96
Sustainability
    achieving, 175–178
    concept of
        according to ASCE, 174
        according to Brundtland Commission, 173–174
    scales of, 175
Sustainable watershed management, 235–236
Sustainable yield
    definition of, 163–164
    fresh-water aquifer underlying saline groundwater, 165
    groundwater overabstraction/groundwater mining adverse effect, 165
SWAM, 322
SWASALT, 323
SWAT, 227
Swept area, 498
SWMS-3D, 321
Symptoms of salinity/sodicity/high soil pH, 287t
Synthetic envelopes, 366
System capacity, 99, 103

T

Tailwater recovery ratio, 69
TARGET model, 320
Thin-film solar modules, 490
Thumb rule for furrow design, 94–95
    furrow length, 94
    furrow shape, 95
    slope, 94
    spacing of furrow, 95
    stream size/flow rate, 95
Tidal power, 503
Tidal stream power, 503
Tide mill, 505
Tillage, 155, 158, 207, 214, 220, 223, 226, 237, 245–246, 251, 266, 312, 403
Time of advance, 68, 68f, 70, 80
Time ratio, 88–89, 92
Time series model, 385
    deterministic time series model, 385
    stochastic time series model, 385
Tolerance to crop, factors influencing, 295–297
    environmental factors, 297
        kharif crop (rainy season), 297
        rabi crop (dry season), 297
    growth stage, 296
    rootstocks and salinity tolerance, 297
    soil moisture, 297
    soil type, 297
    varietal differences in salt tolerance, 296
Top width (T), 2, 5, 10, 17, 19–21
Torrents, 196
Total dissolved salts (TDS), 279, 283, 289
Total dynamic head, 445
Total soluble salts (TSS), 279, 282–283
Total topographic station, 129
Traditional energy sources, 480, 482
Trans-boundary pollution, 182
Trapezoidal channel, 10f, 15–17, 16f
Traveling gun type sprinkler, 49, 49f
Trickle irrigation, 54–55, 113, 116, 123, 134
Trickle or localized irrigation, 37, 55
Turbine pumps, 440, 449–450
    deep-well turbine pump, 440, 450–451, 450f
        multistage, 440
        single stage, 440
    features of, 450–451
    submersible turbine pump, 440, 449–450, 450f
        applications, 453
        features, 453
        horizontal in-line booster pumps, 452
        open well submersible pump, 453
        requirements for, 453
        single phase open well submersible pumps, 454
        suitability of, 452–453
        vertical booster pumps, 452
        vertical deep-well pumps, 452
        wide ranges of, 453
Turbulent flow, 25–28
Two-dimensional model, rectangular basin, 83

U

UCC, 115–116
Uncertainty, 4, 195, 232, 387–388, 412
UNEP, 516
UNESCAP, 485
UNESCO, 181, 272
Uniformity co-efficient, 115, 367
Unit of EC, 285
    conductivity to mg/l, 285

Subject Index 543

conductivity to mmol/l, 285
conductivity to osmotic pressure
  in bars, 285
mmol/l (chemical analysis) to mg/l, 285
Unit plot, 222
Universal Soil Loss Equation (USLE),
  222–223
  modified forms (MUSLE), 223
UNSATCHEM, 321
UPFLOW model, 397
Users and stakeholders, 178
Utility-based wind farms, 497

V

Validation, 388, 390–391
Vapor pressure (VP), 394, 443–444, 473
Variable, definition, 386
Variable speed pump, 462
Velocity head (VH), 23–25, 104, 120, 445,
  455, 459, 473, 487
Velocity limitations, 11
Velocity of flow in porous media, 341
Velocity profile, 26, 26f
Verification, 388–389, 391
Vertical-axis wind machines, 495
Vertical booster pumps, 452
Vertical deep-well pumps, 452
Vertical drainage, 307
Vertical waterwheel, 505
Virtual water, 165, 169, 180, 187
Void ratio, 335, 340, 401
Volatilization, 246–247
Volume balance, 77, 84, 95–97, 131
Volumetric constraints, 162, 168
Volute pumps, 440, 456, 457t
  multistage (multiple impellers) pumps, 456
  radial-flow centrifugal pump, 456
  single or double-suction pumps, 456
  single-stage (single impeller) pump, 456
  uses, 456

W

Waldpolenz Solar Park, 490
Wastewater pumps, 439
Water and environment, 9, 141, 250
Water application efficiency, 59, 77, 112–113,
  113t, 119, 305
Water application methods, 35–63
  classification, 36–38
    system–A, 36
    system–B, 36–37
    system–C, 37–38
  comparison of irrigation systems, 61t–62t
  methods of irrigation, 38–55

basin irrigation, 40–43
border irrigation, 38–40
drip irrigation, 52–54
furrow irrigation, 43–45
other forms, 54–55
sprinkler irrigation, 46–52
selection of irrigation method, 56–63
  factors affecting, 56–62
  selection procedure, 63
Water audits, 151, 158
Water availability and shortage, 119
Water balance equation, 163–164, 262, 339
Water balance in drained soil, 338–340
  artificial/natural drained soil, 339f
  definition, 338
Water column/tank pressure distribution, 24
Water conservation measures
  reducing evaporation
    from crop field, 155
    mulching, 155
    from water surfaces, 154–155
Water conservation plans, 158
Water conveyance efficiency, 118, 135
Water conveyance loss and system designing
  conveyance loss in command area,
    estimation of, 7
  designing open irrigation channel
    considerations, 11–12
    definition sketch of, 10–11, 10f
    examples, 18–21
    flow velocity calculation, 12–13
    hydraulic design, 14–17
    irrigation channel/open channel
      flow, 10
  designing pipe for irrigation water flow
    fundamental theories, 21–23
    head loss/head loss calculation, 26–31
    hydraulic and energy grade line, 25, 25f
    static/dynamic head, water pressure,
      23–24
    types of flow, Reynolds number, 25–26
    velocity profile, 26, 26f
  designing pipe size, 31–32
  factors affecting seepage, 2–3
  lining, reducing seepage loss
    benefits, 8–10
    decision on, 9
    materials used in, 9–10
  seepage, definition of, 2
  seepage, expression of, 3
  seepage measurement, 4
    inflow–outflow method, 5–6
    ponding method, 4–5

Water conveyance (*cont.*)
    seepage meter method, 6–7
    seepage reduction methods
        biological, 8
        chemical, 8
        physical, 8
Water delivery performance (WDP), 119, 121–122, 135t
Water energy, 502–506
    forms of, 503
    potentials of, 505–506
    tide mill, 505
    watermill, 504–505
    wave energy, 503–504
Waterfront, 67–68, 67f–68f, 81, 82f, 88–89
Water horsepower (WHP), 120, 445, 464
Water-lifting devices, pumps, 433–477
    cavitation
        in axial-flow pumps, 463
        in radial flow and mixed flow pumps, 463
    centrifugal pumps
        affinity laws, 446–447
        cavitation, 449
        features and principles, 442–443
        priming of, 448–449
        pump efficiency, 445
        specific speed, 446
        terminologies, 443–445
    classification of pumps, 438–442
        based on means by which energy is added, 441–442
        based on mode of intake of fluid to pumps, 438–439
        based on position of prime mover, 439
        based on principle by which energy is added to fluid, 440–431
        based on type of water use (or field of use), 439
    classification of water-lifting devices
        animal-powered devices, 436
        human-powered devices, 435–436
        kinetic energy powered device, 436–437
        mechanically powered water-lifting devices, 437
    definition of pump, 437
    description of different types of centrifugal pumps
        advantage and disadvantage of, 457, 457t
        mono-block pumps, 454
        problems/causes/remedial measures, 457
        submersible pump, 451–454
        turbines pump, 449–450
        volute pumps, 456
    examples, 473–476
    hydraulics in pumping system
        dynamic water depth, 468–469
        elevation difference, 469
        pressure and head, 468
        pressure *vs.* flow rate, 468
    power requirement in pumping, 464
    practical suction lift of suction-mode pump, factors affecting, 442
    principles in water pumping, 438
    pump control, 467–468
    pumping purpose, 437–438
    pump installation, 465–468
    pump operation, 466–467
    pump performance
        factors affecting, 469–470
    pumps connected in series and parallel, 469
    pump selection
        procedure for, 470–472
    types of pumps
        air-lift pumps, 458–459
        booster pumps, 462
        displacement pumps, 462
        hydraulic ram pumps, 462
        jet pumps, 459–461
        reciprocating pump/bucket pump, 461–462
        variable speed pump, 462
Water loss minimization, 143, 153, 155–156
Watermill
    advantages, 506
    types, 505
    working principle, 504–505
Water movement theory
    examples, 350–351
    hydraulic conductivity of layered soil, 342–345, 344f
        resultant conductivity of horizontal/vertical direction, 345
        resultant horizontal hydraulic conductivity, 343–345, 343f
        resultant vertical hydraulic conductivity, 344
    Laplace's equation for groundwater flow, 345
    terminologies, 341–342

drainage intensity/coefficient/ requirement, 341
equipotential line, 342
head, 341
homogeneous and isotropic media, 342
water table, 341
theory of groundwater flow toward drain, 346–347
  examples, 347–349
  non-steady state drainage situation, 347
  steady state problems, 346–347
velocity of flow in porous media, 341
water-table position during flow into drain, 346
Water ordinance (by law), 152
Water pollution, 180, 196, 229, 242–243, 247, 501
Water pressure, 23–24, 128, 153, 341, 438, 462–463, 468
Water productivity (WP), 118–119, 124–125, 136t, 154, 156–158, 407
  estimation, 430
Water pumps, 434, 439
Water pumping windmill, 496
Water requirement efficiency, 114, 119
Water resources management
  challenges in, 188–189
  concept/perspective/objective of, 140–144
  conflicts in, 178–185
  demand and supply estimation, 144–150
  impact of climate change on, 185–188
  strategies for, 150–173
  sustainability issues, 173–178
Water-saving irrigation scheduling
  adopting efficient irrigation method, 154
  alternate furrow irrigation, 153
  alternate wetting and drying of rice fields, 153
  deficit irrigation, 154
  soil drying during crop ripening, 153
Water scarcity, 173, 178, 196, 408–409, 429
Watershed address, *see* Watershed's Hydrologic Unit Code (HUC)
Watershed, definition, 194
Watershed functions, 198, 227–232
Watershed management, 227–236, 430
  community-based approach, 231–234
  components, 228
  land use planning and practices, 230, 234–235
  pond management, 231
  problem identification, 227
  structural management, 230
  sustainable watershed management, 235–236
  tools for watershed protection, 230
  watershed planning and management, 229–230
Watershed planning, 229, 235
Watershed protection, 194–195, 230, 236
Watershed restoration, 230, 236–238
  and wetland management, 236–238
  drinking water systems using surface water, 236
Water supply–requirement ratio, 119, 121, 135t
Water surface elevations, 12, 82
Water tank, static pressure, 24f
Water use efficiency (WUE), 39, 125, 154, 159, 179, 407
Water-wise cultivation method, 156
WATSUIT, 321
Wave energy, 503–504, 512
Weather-based (yield) prediction model
  methods of formulation of, 413–414
    basic assumptions in, 414
    choice of weather variables, 413–414
    crop-production statistics, 413
    regression approach, 414
    removing trend component, 413
    typical form of model, 414
    zoning of crop area, 413
  need of, 411–412
  sample example of formulating, 415–419
    formulation, 416–418
    multiple regression analysis, 418
    trend (slope) of regression line, 416
    weather (single or composite) elements, yield on, 417–418
    yield data, 416–417
    yield on weather elements, 417
Weed control, 155
Well characteristics, 470, 472
Well pumps, 439, 452–454, 459
Wetlands, 149, 154, 163–164, 172, 182, 194, 197–198, 235–238, 322
Wetland management
  in agricultural watersheds, 236–237, 237f
  biological buffers, 237
  crop/tillage/nutrient management, 237–238
  integration of research information, 238
Wet-seeded/direct-seeded technique, *see* Water-wise cultivation method
Wetted perimeter, 2–3, 11, 13–14, 18, 85, 91–94
Wetting/ponding/storage phase, 68

WHO, 181
Wild life, hazards on, 249
Wind/air-flow, 491–492, 492f
Wind Amplified Rotor Platform (WARP), 495
Wind classes, 496
Wind energy, 170
   advantages, 493–494
   application of, 496–497
      mode of use, 496
         production of electricity, 496–497
         water pumping windmill, 496
   causes of wind flow, 492–493
   energy from wind, 493
   and environment, 501
   examples, 501–502
   intermittency problem with, 500
   potential assessment, 494–495
   renewable and environmentally friendly, 491
   suitable site for windmill, 495
   types of wind machines, 495
      horizontal-axis, 495
      vertical-axis, 495
   wind power calculation, 498–499
   wind power plants/wind farms, 498
   working principle of wind machines, 497, 497f
Wind machines, 497f
   horizontal-axis, 495
   types of, 495
   vertical-axis, 495
   working principle of, 497
Windmill, 481, 483, 492, 494–498, 499f, 500–501, 510
Wind power characterization, 496
Wind speed, 99, 131–132, 393, 396, 412–413, 493–496, 500–502
Wind turbines, 481, 493–494, 496–503, 499f, 513, 516
World Bank, 145, 183, 490
World Commission on Environment and Development, 163
World Energy Council, 503

Y

Yield rate, 119, 125, 413–416, 417t–418t
Yield reduction due to salinity
   piece-wise linear model, 299
   relative yield *vs.* EC, 300f
Yield response factor, 409

Z

Zero-inertia model, 83–84
Zero order rate constant, 255, 258

Printed by Printforce, the Netherlands